Limit analysis of structures at thermal cycling

Monographs and textbooks on mechanics of solids and fluids

editor in chief: G.Æ. Oravas

Mechanics of plastic solids

editor: J. Schroeder

1. Foundations of plasticity
   A. Sawczuk (ed.)

2. Problems of plasticity
   A. Sawczuk (ed.)

3. Introduction to the mechanics of plastic forming of metals
   W. Szczepiński

4. Limit analysis of structures at thermal cycling
   D.A. Gokhfeld and O.F. Cherniavsky

Limit analysis of structures at thermal cycling

D.A. Gokhfeld and O.F. Cherniavsky
Polytechnical Institute, Cheliabinsk, USSR

SIJTHOFF & NOORDHOFF   1980
Alphen aan den Rijn – The Netherlands
Rockville, Maryland – USA

Copyright © 1980 Sijthoff & Noordhoff International Publishers bv
Alphen aan den Rijn, The Netherlands

All rights reserved. No part of this publication may be reproduced, stored in a retrieval system, or transmitted, in any form or by any means, electronic, mechanical, photocopying, recording or otherwise, without the prior permission of the copyright owner.

ISBN 90 286 0455 3

Library of Congress Catalog Card Number: 80-83265

Original edition published under the title
"Nesushchaja sposobnost' konstrukciji v uslovijakh teplosmen" (D.A. Gokhfeld, Moscow, Mashinostrojenije, 1970) rearranged and expanded by the authors with utilization of the materials published in the second book
"Nesushchaja sposobnost' konstrukciji pri povtornykh nagruzhenijakh" (D.A. Gokhfeld, O.F. Cherniavsky, Moscow, Mashinostrojenije, 1979).
Translated from Russian by M. Kwieciński; edited by S. Świerczkowski.

Printed in the Netherlands

Authors' Preface

Over the last few decades there has been an increase of interest in the problem of strength of structures subjected to variable repeated actions of unsteady thermal fields coupled with mechanical loads. The methods of limit analysis of elastic-plastic systems, employing simple models of continuum, enable one to find a most clear picture of the possible limit states and thus establish a safe domain for the structure considered.

In this book we give a systematic presentation of the theorems and methods of the shakedown theory in their up-to-date interpretation. We apply these to solve a considerable number of problems concerning incremental collapse and alternating plastic flow of discs, plates, shells, pressure vessels and other structures. Special attention is paid to the examination of mechanisms of progressive geometry changes under thermal cycling. Some isothermal contact problems of practical interest are also considered.

The last chapter is devoted to the problem of direct determination (without studying the deformation history) of the steady stress cycle parameters in the presence of repeated inelastic deformations.

The authors wish to express their gratitude to Professor A. Sawczuk for his valuable comments to the manuscript and support in its publishing.

Cheliabinsk, USSR　　　　　　　　　　　　　　　　*D.A. Gokhfeld*
October, 1978　　　　　　　　　　　　　　　　　　*O.F. Cherniavsky*

Contents

Authors' Preface V

Preface by A. Sawczuk XIII

Foreword XVII

Introduction XIX
    References XXVI

1. Analysis of simple elastic-plastic systems under variable repeated loading and temperature 1
    1.1. One-parameter systems 3
    1.2. One-parameter systems. Some additional remarks . . . 13
    1.3. Two-parameter systems 20
    1.4. Influence of strain-hardening on the cycle stabilization and shakedown conditions at single and repeated deformations 30
    1.5. Peculiarities of structural behaviour under thermal actions 36
    References 40

2. Fundamental theorems of the shakedown theory 43
    2.1. Prerequisites from the mechanics of elastic-perfectly plastic solids 43
    2.2. Statical shakedown theorem. Generalization to thermal effects. Further development of the statical theorem . . 50
    2.3. Kinematical shakedown theorem 56
    2.4. Introduction of a fictitious yield surface and restatement of the statical theorem. Formulation of the shakedown problem connected with a non-trivial problem of limit analysis 62

Contents

2.5. Construction and properties of the fictitious yield surface 65
2.6. Restatement of the kinematical theorem. Conditions for limit states following from both restated shakedown theorems 71
2.7. On the uniqueness of stress state at the limiting cycle 79
2.8. On the boundedness of inelastic deformation preceding shakedown 81
References 87

3. Generalized variables in the shakedown analysis 91

3.1. General method of constructing an interaction surface 93
3.2. Fictitious interaction surfaces for a rod under cycling normal stress 96
3.3. Fictitious interaction surfaces for a stretched rod subjected to cycling torque 100
3.4. Approximate method of introducing generalized variables in shakedown problems 107
3.5. Generalized stresses in circular plates under tension and bending 114
3.6. Formulation of the statical theorem in terms of generalized stresses 127
References 129

4. Statical methods of solving shakedown problems 131

4.1. The Pontragin maximum principle and its application to the problems of limit analysis 134
4.2. Application of the mathematical programming methods to problems of incremental collapse 145
4.3. Simple examples of solving the shakedown problems via the statical approach by using the simplex method 150
4.4. Discretization of the continuum in the shakedown problems. Finite element method 164
4.5. Approximate statical methods 166
4.6. Examples illustrating the use of approximate statical methods in shakedown problems 169
References 179

Contents

5. **Kinematical methods of solving the shakedown problems** . . . 183

   5.1. Application of the mathematical programming methods to shakedown problems in the kinematical formulation . . 185
   5.2. Solution to the problem of incremental collapse of a plane disc by means of the simplex method 192
   5.3. Approximate kinematical method of determining parameters of the limiting cycle. Simple examples . . . 202
   5.4. Incremental collapse of a thick-walled tube subjected to thermal cycling and internal pressure 210
   5.5. Overload method. Conditions for incremental collapse of a spherical vessel under variable repeated temperature 213
   5.6. Full shakedown analysis of a beam under simultaneous steady tension and variable repeated bending 220
   References 225

6. **Carrying capacity of a turbine disc under single and repeated loading** 227

   6.1. Limit analysis of rotating disc. Global and partial collapse mechanisms 229
   6.2. Some additional remarks on the evaluation of carrying capacity of discs 235
   6.3. Shakedown of plane disc at cyclic variation of angular velocity and temperature 237
   6.4. Influence of creep on the shakedown conditions . . 245
   6.5. Shakedown analysis of rotating, non-uniformly heated discs of arbitrary profiles 249
   6.6. Peculiarities of incremental collapse analysis of rotor discs of radial-flow turbine 261
   6.7. Using the shakedown diagram to evaluate the load-carrying capacity of structures under variable repeated loading 264
   6.8. Analysis of some test results and adoption of shakedown safety factors 268
   6.9. Tests and analysis of the rotor disc of a radial-axial-flow turbine 274
   References 281

7. **Shakedown of plates** 283

   7.1. Alternating mechanical loading of a circular plate . . . 284

Contents

7.2. Shakedown of a simply supported circular plate at thermal cycling ... 288
7.3. Clamped circular plate ... 292
7.4. Circular plate with rigid boss ... 307
7.5. Rectangular plates and plates of arbitrary shapes. Upper bounds on the limiting cycle parameters ... 313
7.6. Shakedown of perforated plates ... 323
References ... 336

8. Shakedown of shells at mechanical and thermal cycling ... 339

8.1. Cylindrical shell under ring load and thermal cycling ... 340
8.2. Limit analysis of cylindrical elastic-plastic shells by the Pontragin maximum principle ... 355
8.3. Shakedown of cylindrical shells under moving mechanical load ... 366
8.4. Incremental collapse of a conical shell ... 372
8.5. Shakedown of spherical shells ... 383
8.6. Combustion chamber of liquid fuel rocket engine ... 391
8.7. Fast nuclear reactor fuel element ... 395
References ... 404

9. Strain accumulation at thermal cycling under negligible mechanical loading ... 407

9.1. Strain accumulation conditions based on the kinematical theorem of shakedown. Influence of the type of temperature field ... 408
9.2. Repeated action of moving heat source on a bar system ... 412
9.3. Ratchetting of a thin cylindrical shell under repeated action of quasi-stationary temperature field ... 417
9.4. Ratchetting of a crystallizer of a semi-continuous tube casting machine ... 425
9.5. Ratchetting of a conical shell-type structure (slag container) ... 434
9.6. Experimental investigation of deformations under variable repeated action of quasi-stationary (in the moving coordinate system) temperature fields ... 447
9.7. Incremental buckling at thermal fluctuations. Control and interpretation of experiments, the present-day concepts ... 456

Contents

9.8. Incremental buckling at thermal cycling as demonstrated on a bar system 466
9.9. Conditions for progressive warping in the light of shakedown theory 472
9.10. Examples of progressive warping analysis 475
References 484

10. Some contact problems in the shakedown theory 487

10.1. Contact problems in the shakedown theory. Description of the problem. Statical method of analysis 487
10.2. Contact problems in the shakedown theory. Kinematical method of analysis 495
10.3. Strain accumulation in a semi-plane on which a rigid disc is rolling 501
References 504

11. Inelastic shakedown in steady stress cycles 505

11.1. Existence and uniqueness of a steady stress cycle . . 507
11.2. Stress and strain rate analysis in steady cycles. Different formulations of the problem 511
11.3. Particular cases of the problem of steady stress cycle calculation. The Melan theorem. Strain accumulation conditions at advanced alternating plastic flow . . . 514
11.4. Stress and strain rate analysis in steady cycles. Another variational formulation and its consequences 521
11.5. Evaluation of steady cycle characteristics under quasi-stationary external actions 528
References 531

Index 533

Preface

Modern society imposes on engineers the need for more efficient and more pertinent methods of designing reliable structures for longer service. Restricted resources of materials, care about the environment as well as the growing social conscience and responsibility are at the origin of more stringent design requirements.

Engineering mechanics fulfils an important role in society. Its obligations and responsibilities are manifold. In the first place it has to estimate and to evaluate in a mathematically objective manner the mechanical properties of real materials under various physical agencies and environments. The next obligation is to synthetise the material behaviour and the disclosed couplings between mechanical and other physical properties in appropriate mathematical models. To be useful in engineering a model has to be pertinent and straightforward. Once a model is established and justified, suitable mathematical methods of mechanical analysis have to be developed accounting for cyclic loading, elevated temperatures, corrosive environments, combined actions of mechanical and physical agencies, and the like. Next follow the methods of solving boundary and initial value problems, appropriate methods for design for displacements, for the estimating of the load carrying capacity, failure, creep, reliability etc. Eventually the accumulated knowledge and experience lead to the elaborating of design rules and codes which nowadays must have world-wide acceptability.

More recently the inelastic analysis began to play an important role in engineering mechanics due to the requirements of modern technologies like nuclear power plants, chemical industry, metal forming, energy conversion and future fusion technology, coal gasification or solar and geothermal installations. Because of modern and future technologies, more stringent demands are imposed on the engineering mechanics education. For instance, there is a growing need and necessity for teaching the actual problems, results and methods of various branches of inelastic and non-linear engineering mechanics. The methods

Preface

of plasticity, for example, penetrate to design procedures at a growing pace.

Among the methods of plasticity employed in structural mechanics for the purpose of designing structures and structural components, the theory of limit analysis became already an engineering tool. The theory of limit analysis employs the rigid-plastic model of material and allows to evaluate the load intensity which the structure is able to carry under agencies monotonically increasing in proportion to a single parameter. Several monographs exist on this matter. Other branches of plasticity, including elastic-plastic and hardening behaviour under a single parameter loading, are also fairly well developed.

Until now no comprehensive treatment of elastic-plastic structures subjected to cyclic or repeated loading has been in existence, although the principles and theorems relevant to this type of material behaviour were known. This monograph presents the principles, methods and solutions concerning the behaviour of elastic-plastic structures under mechanical and thermal cycling. Its contents and the method of presentation make it undoubtedly an important contribution to the literature presenting the plastic analysis to the engineering community; at the same time the monograph has a significant research value.

The book follows a natural, thus logical, way of bringing the considered branch of inelastic analysis to the attention of engineers and eventually to utilization. It presents the principles, the fundamental theorems, the methods and available solutions currently furnished by the shakedown theory. The shakedown theory deals with the analysis of elastic-plastic structures subjected to cyclic, repeated or multiparameter loads. Its basic theorems are related to the fundamental theorems of limit analysis, as appropriate generalisations to time-dependent external agencies. This fact allowed to make references to limit analysis in presenting the methods of elastic-plastic analysis and design at cyclic actions. Such an approach permitted the application of the already available experience and knowledge in arriving at new solutions concerning cyclic and repeated agencies imposed on elastic-plastic, temperature sensitive structures.

The monograph contains the accumulated knowledge on the behaviour of elastic-plastic structures subjected to mechanical and thermal cycling, and it presents the results obtained by Professor Gokhfeld and his associates during many years of specialized research. It is written in a distinctive way starting with the mechanical concepts to be dealt with, unfolding with the basic theorems, then going on to

Preface

methods and eventually to solutions of problems. It is worthwhile to point out that the mechanical and analytical aspects of the problems are stressed. This allows the reader to grasp the principles before starting complicated computations. The text can thus serve for both mastering the shakedown theory and employing its methods in engineering.

The basic equations of structural mechanics are usually written in terms of stress resultants and the related displacement variables. The principles of shakedown are thus given in appropriate generalized variables, and an extension of this type is not trivial. The methods of shakedown are presented as consequences of both the statical and kinematical theorems. After the principles are developed and commented upon, the solutions to specific problems regarding structures subjected to mechanical and thermal cycling are given. Examples concerning turbine discs, plates, shells, and contact problems are worked out. It is characteristic for the book that attention is given to methods, including application to shakedown design of the principles of mathematical programming and of control theory.

Engineers as well as scientists will find in this monograph much useful information on what is presently known and an inspiration to use the appropriate approaches in their particular tasks concerning reliable design of elastic-plastic structures and components subjected to thermal and mechanical cycling.

Warsaw, June 1979 *Antoni Sawczuk*
of the Polish Academy of Sciences

Foreword

By writing the present book the authors have undertaken to expand the monograph [0.1] written by the senior author and published in Russian in 1970. Since that time more and more interest has been focused on the problems of strength of structures working in unsteady temperature fields; the present-day technology imposes more stringent requirements on the thermal and loading parameters of structures used in many applications. Also since that time new developments occurred in the experimental investigations into the mechanical behaviour of metal alloys under various programs of isothermal as well as non-isothermal loading. This, together with the progress in the numerical methods and computer-aided procedures, has made it possible to substantially further the studies of stress and deformation states in structures by using various models of deformable bodies.

A considerable progress has also been made in the limit analysis of elastic-plastic structures and, in particular, in the shakedown theory dealing with the unsafe states generated by variable repeated loading. The following two features of the theory have always been and are of appeal to the engineer: A relative simplicity of calculations and a clarity of the qualitative picture arrived at, due to proper emphasis placed on the basic properties of deformation processes and to the justified neglect (at least, at the first stage) of some second-order effects.

The present new edition of this book has been written with the objective of reporting the current state of the shakedown theory. Thus there have been included some recent results obtained after the monograph [0.1] had appeared. Fundamental theorems and methods of the shakedown theory have been presented in a consistent and complete manner. The number of examples has been increased in order to explain more fully the computational procedures and to illustrate more adequately the applications of the presented theory in various branches of technology.

Thus it is natural that in spite of our attempts at condensing the

Foreword

presentation, we had to substantially rearrange the text and expand it slightly. It must be remembered that new publications on the subject number hundreds, thus the references are not claimed to be complete. Only those items have been included which were, in the authors' opinion, most important and characteristic for the employed approach.

The shakedown theory of elastic-plastic structures is here presented in its greatest generality, so as to be applicable to any continuous body subjected to both thermal and mechanical actions. This applicability aspect of the theory is insisted upon since, from the authors' point of view, it is mainly the structures under thermal stresses that are most effectively dealt with within the framework of the established theory. Accordingly we give special consideration to some peculiarities in the behaviour of materials under elevated temperatures. However, all the essential results remain valid in the case of purely mechanical, variable repeated loads, provided sufficiently general computational methods are employed. Particularly, Chap. X is devoted to isothermal shakedown problems due to repeated rolling contact.

This is also the case with the methodical, explanatory examples given in the book. Accordingly, specific methods of analysing some complex bar structures, encountered mostly in civil engineering applications, are not to be found here. They are dealt with in a number of known monographs and in numerous papers, e.g. [0.2–0.5].

We would like to conclude this foreword by quoting J.I. Frenkel, a Soviet physicist, who said in 1946 that "a theoretical physicist much resembles a caricaturist. His aim is not to provide a photographic description, but one which is simple enough to exhibit clearly the most relevant and characteristic features. The photographic accuracy can (and in fact must) be required only in the case of a theoretical description of a simple system. A good theory of complex systems should only be a good "cartoon" showing the most typical features of the properties and ignoring the irrelevant details..."

Introduction

It is a characteristic feature of investigations of the strength analysis of structures subjected to thermal flux that the scientist has necessarily to face a complex of interconnected partial problems related to different fields of knowledge such as thermophysics, continuum mechanics and material science. A temperature field, apart from generating certain stress and strain states in a structure, exerts an influence on both the thermophysical and mechanical properties of the material. Under more pronounced thermal flux in structural metal alloys time-dependent effects come into play which at room temperature and moderate stress would not have been markedly exhibited.

At different stages of the development of structural mechanics various methods were found to evaluate the influence of thermal stresses on the serviceability of structures. In the existing literature there can be encountered different points of view on the subject, ranging from the insistence on the superposition of mechanical and thermal stresses, assuming perfectly elastic response, to proposals to completely neglect the thermal contribution to the stress. As can be readily seen, those approaches correspond to the two basic concepts employed in the strength analysis: The concept of local failure and the concept of limit equilibrium. Those contrary points of view reflect the actual situation since the influence of thermal stresses on the strength of structure can be different depending on particular properties of the material and the nature of thermal action [0.6–0.11 and others].

Experimental evidence shows that a single action of a non-uniform thermal field can cause a failure only in brittle materials such as glass, ceramics, cast iron. At the same time, it is worth remembering that a rapid application of stress accompanying a thermal shock will usually cause a transition, under ordinary circumstances, of the abruptly stressed material from the plastic to the brittle state. In such a situation the failure is preceded by three-dimensional tension. However, excluding singular cases, a single change in temperature is found to be safe, as far

Introduction

as materials with fairly ductile properties are concerned, even if the fictitious thermoelastic stresses considerably exceed the yield stress of the material. In general, the influence of thermal stress states on the strength is completely analogous to the influence of other self-equilibrated systems of stress states such as residual stresses. It is worth noting that in the considered situations there can take place some exceedingly large displacements of the structure.

The picture changes markedly when the character of the temperature field is cyclic. In this case the amplitude of resultant stresses plays a decisive role. As a rule, whenever the number of thermal cycles per lifetime of a real structure remains within the order of thousands, fatigue does not take place, provided that maximum stresses do not exceed the yield point and, in consequence, no plastic deformations in the macro sense occur. Fatigue is not found to occur before the thermal cycling is accompanied by repeated plastic deformation. This type of low cycle fatigue is usually referred to as thermal fatigue; an extensive literature is devoted to this subject alone. Of great relevance here is the relation between failure under given conditions and the properties of a cyclic, non-isothermal deformation process such as cyclic work hardening and work softening, cycling creep and relaxation, interaction of subsequent stages of cyclic and long-term loading under elevated temperature etc. provided those processes exert some influence on the parameters describing the damage accumulation.

The other effect that may take place under constant or variable mechanical loading of even low intensity is that of the accumulation of strains of the same sign, adding up with each cycle of the temperature field. This situation is termed an incremental collapse, in contrast with the instantaneous collapse of a structure made of an ideally plastic material. In the thin-walled structures the expected result of thermal cycling is clearly the more essential change of initial geometry.

An accumulation of strains acting in the same sense may clearly endanger the serviceability of a structure and result in its collapse. Sometimes a change in the initial geometry can bring about changes in the conditions of gas and liquid flow and heat exchange which substantially accelerate the deformation process. Failure due to the complete exhaustion of plastic properties of the material cannot be excluded.

The upsurge of interest in the progressive strain accumulation under thermal cycling, observed over the last decades is connected with the development of those branches of technology in which high temperatures have to be applied in non-stationary situations. This primarily

Introduction

refers to the nuclear reactors technology where stringent requirements are imposed on the stability of geometry of structural elements in spite of unavoidable temperature fluctuations which accompany the working regime. The corresponding research, devoted to the so-called thermal ratchetting, has received considerable attention in a number of papers [0.12–0.14 and others]. It is worth remembering that already four international conferences were devoted to advances in structural mechanics as applied to nuclear reactors technology (West Berlin 1971 and 1973, London 1975, San Francisco 1977).

Investigations of incremental deformation processes are also of importance in other branches of technology such as turbine, aircraft and spacecraft industries. Progressive deformation of installations is frequently encountered in metallurgical and chemical industries where high temperature technological processes are involved employing multiple thermal actions of great intensities (cinder tanks, crystallizers, metal moulds, furnace feeders and so on). In some of those technological processes it is vitally important to ensure that the shapes and dimensions of elements remain unchanged.

It is the engineer's task, when designing a structure intended to sustain thermal stresses, to safeguard against the occurrence of undesirable effects such as damage accumulation and strain accumulation leading eventually to impaired serviceability and a shorter life of the structure.

In situ investigations are rarely performed due to their prohibitive cost. On the other hand, smaller than life model investigations are rather impractical since it seems to be impossible to formulate the rules of thermo-mechanical similitude in a satisfying manner. This means that the importance of theoretical explorations can hardly be overestimated; but also without suitable analysis of the experimental investigations one cannot expect to obtain meaningful results.

As is well known, the most classical approach to the assessment of the strength of a structure consists in a detailed study of its stress, strain and displacement fields. Under the action of variable repeated mechanical and thermal loading the consecutive stages must be examined in which time-dependent stress and strain states develop. Those latter, in turn, are found to depend on the loading path which is usually far from being well defined, as well as the deformation state evolution caused by variable temperature and repeated loading, the interaction of creep with the short-term deformation processes and on other characteristic features of the load history.

Introduction

We conclude that a proper choice of a model of the considered medium is of primary importance. Models and theories, applicable in ordinary conditions, utterly fail to describe the complicated influences of time and loading history, especially in the problems of cyclic nonisothermal deformation. Accuracy of calculations based on these may turn out to be quite inadequate or doubtful. Here it is worth mentioning some developments achieved recently in the field of structural behaviour under working conditions, by means of the so-called structural models of the continuum [0.15]. Those models are based on the concepts of Masing [0.16], later developed by Afanasyev [0.17], Besseling [0.18] and others.

The necessity to work with more refined models and theories makes it even more difficult, despite the increased potentialities of present-day computational techniques, to perform the already cumbersome calculations in order to analyse the kinetics of deformation of the structure in question. The more so that it appears necessary to assess the influence of different parameters on the strength or to select an optimum structure among a number of possibilities. In following this avenue some additional difficulties arise resulting from the fact that, at the stage of designing, it is only the limits on the load parameters that usually are specified, without prior knowledge of the sequence of loads that are free to change within prescribed bounds.

The direct approach to the above problem, when used for the quantitative determination of local magnitudes or amplitudes of stresses, strains and displacements can take advantage of virtually unlimited possibilities to enhance the accuracy of results. However, the above mentioned difficulties make it meaningful to evaluate, even if only in an approximate manner and with the use of simple tools, the safety of structures against variable repeated isothermal as well as nonisothermal loading. With this aim in view, all the present-day knowledge of the subject should be used to select the most decisive factors entering the considered situation.

In the conditions of continuously increasing loading the limit equilibrium analysis is most widely used for the evaluation of ultimate load intensity an elastic-plastic structure is able to carry. The merits of this theory, as compared with the traditional approach via local stress intensities, have been emphasized by many authors [0.19–0.20 and others]. The theory is found to be applicable, to a certain degree, also to the structures made of work hardening materials.

When variable repeated loading is present, a suitable generalization

Introduction

of the limit analysis is introduced, known as the shakedown[1] theory. This branch of plasticity has rapidly developed over the last decade. Along with the assumptions made in the limit analysis, the shakedown theory in its classical formulation accepts that the stress–strain diagram remains unchanged under repeated loads of arbitrary type. On these premises the conditions are determined, e.g., the limits of variation of external load factors under which the variable repeated loading does not result in plastic deformations, except for those bounded ones which have occurred during the initial cycles and have generated a certain helpful state of self-equilibrated residual stresses. This state, superimposed on the working stress state, is found to ensure that the further behaviour of the structure is of a purely elastic type. On the other hand, a violation of the shakedown conditions means that further loading will be systematically accompanied by cyclic plastic deformations associated either with alternating strains that disappear after each cycle or with one-sided strains that accumulate as the number of cycles increases. The combined situation can evidently also occur.

Thus, the discussed theory based on a simple model makes it possible to describe the most important effects accompanying repeated loading and, in particular, thermal cycling. For structures subject to a relatively low number of loading cycles during their lifetime, the conditions for shakedown can be considered as a certain criterion together with a suitably selected safety factor. These carrying capacity conditions appear to be in fact approximate since they are obtained without taking into account connections between the magnitude and range of change of deformation and the number of cycles prior to failure. Actually, in the case of a low number of cycles this approach is conducive to estimating the lower bound on the limit load factors. But despite its inaccuracy, the method can be successfully applied in the design practice and the data thus derived can be used, in conjunction with experimental evidence, to provide information necessary for the standardization of safety factors to be assured in structures. It is worth noting that within this method only some limited information is required on the bounds of loading or other external agencies, in particular, virtually no knowledge on the sequence of loads is necessary to arrive at an acceptable solution.

Apart from its direct applications in the engineering calculations, the shakedown theory is recognized for its cognitive significance: Owing

---
[1] The term "shakedown" is due to Prager [0.21].

Introduction

to proper specification of the qualitative conditions for failure, the singular features of structural behaviour under repeated loading are uncovered in a clear way. In particular, this concerns the conditions for, and mechanisms associated with collapse, as related to the given parameters.

The initial stage of development of the shakedown theory dates back to the thirties and is mainly connected with bar structures as used in civil engineering. The statical shakedown theorem for a structure with single redundancy was formulated and proved by Bleich in 1932 [0.22]. This proof was extended by Melan in 1936 [0.23] to cover statically indeterminate trusses with arbitrary degree of redundancy. It was also Melan, who generalized this theorem to the case of three-dimensional continuum [0.24]. It is worth realizing that at that stage the shakedown theory was developed quite independently of the limit analysis theory. It is well known that in 1936 Gvozdev arrived at his fundamental results in the limit analysis of elastic-ideally plastic structures subjected to single loading [0.25].

It was Koiter, who first recognized the fact that the theorems on plastic collapse should be understood as limiting cases of the shakedown theorems corresponding to the coinciding of the upper and lower bounds for each of the contributing external actions. Based on this analogy, Koiter put forward the other (kinematic) shakedown theorem [0.26] and thus stated and proved the plastic analysis theorems, i.e. the limit analysis and the shakedown ones in the form used nowadays [0.27].

Application of the bounding theorems makes it possible, by solving an extremum problem, to assess the possibility, or impossibility for a structure to shake down in the presence of variable repeated loading. The first solutions to the shakedown problem in continuous media were arrived at in [0.28, 0.29]. In both papers the shakedown was limited by the alternating (i.e., of changing sign) yielding; the corresponding magnitudes of shakedown load factors were compared with those of the limit analysis. As to the incremental collapse, the known examples were confined to certain bar systems only and the approach to such systems was mainly intuitive.[1]

The investigations of the shakedown of structures under variable temperature fields were initiated by Prager in 1956, who suitably generalized the Melan statical theorem and further considered some

---

[1] Such a situation was first shown by Horne [0.30] to hold true in the case of repeated loading.

Introduction

examples [0.31]. It was shown that the principle of independence of the load-carrying capacity of elastic-perfectly plastic structures from the self-balanced residual stresses was no longer valid when the cyclic thermal stresses entered the picture.

At roughly the same time a similar development of the shakedown theory was achieved in the Soviet Union. In particular, it was Rosenblum [0.32], who independently indicated that the statical shakedown theorem may be used to cover the case of thermal actions. In 1965 he also generalized the kinematical theorem [0.33].[2]

The application of the original (classical) formulations of the fundamental theorems has proven to be rather complicated due to the fact that in the solutions of the corresponding extremum problems the variables, such as stresses and displacements (velocities), depend not only on the coordinates of the point considered, but also on current time. This circumstance alone slowed down the application of the shakedown theory, especially of the kinematical theorem. The first attempts to employ the latter were rather disappointing [0.27]. As soon as the kinematical theorem was restated [0.1, 0.36], the difficulties were overcome and new avenues were opened up to solve a broad class of shakedown problems in plates and shells subjected to both isothermal and non-isothermal loading. Further investigations led to a similar restatement of the statical theorem and to the introduction of the concept of a fictitious yield surface. This idea made it possible to reduce the shakedown problem to the limit analysis problem for a body possessing a fictitious nonhomogeneity of its plastic properties [0.37, 0.38]. Thus a new light was thrown on the issue of similarities between the theories of limit analysis and of shakedown. In particular, conditions for the occurrence of incremental collapse become known and the distinction of incremental collapse from the instantaneous collapse was made clear. On the other hand, the determination of alternating plasticity conditions does not require the solution of the extremum problem in each separate case. It appeared sufficient to consider the changes in the stress state over a cycle and at a generic point of the body.

The new interpretation of the fundamental theorems made possible the introduction of the generalized variables to the shakedown problems of plates and shells. This introduction has appeared to be, in the case of arbitrary loading, by no means trivial [0.38, 0.39]. The present book is primarily devoted to the exposition of the fundamental shakedown

---

[2] The same result was obtained by Maier [0.34] and then by Donato [0.35].

Introduction

theorems together with their up-to-date interpretation, the resulting simple as well as more complex methods and the shakedown analysis of typical structural elements acted upon by repeated temperature fields and mechanical loading.

Any further advancement of the shakedown theory is felt to be conditioned, on the one hand, by the corresponding development of computational procedures with the use of suitable mathematical tools and, on the other hand, by refinements of the theory itself in order to better reflect the actual properties of real materials, the changes in geometry and so on. It must be emphasized that, concerning the latter problem, the rejection of some assumptions accepted in the classical theory necessarily results in a different formulation of the shakedown situation and its interpretation at the cost of a more or less grievous loss in the simplicity and clarity of the general theory.

For some structures with prescribed lifetime of rather short duration for which the shakedown requirements can reasonably be relaxed, the determination of cyclic deformation parameters becomes of growing interest. The ranges and increments of inelastic deformations are then sought corresponding to steady stress cycles. The existence of the latter was pointed out by Frederic and Armstrong [0.40]. Now the problem consists in determining the ranges associated with a system of limiting relationships without tracing the whole loading history.

The above listed problems are also treated in the present monograph which, in general, is based on the specifically oriented investigations of its authors. In some cases we were forced for the sake of continuity of exposition, not to discuss, or even mention, numerous works devoted to various aspects of the shakedown theory. The reader is here referred to recently published survey papers [0.41–0.43].

References

[0.1] Gokhfeld, D.A., *Carrying capacity of structures under variable thermal conditions*, in Russian. Mashinostroenie, Moscow, 1970.

[0.2] Rzhanitzin, A.R., *Structural analysis taking account of plastic properties of materials*, in Russian. Gosstroiizdat, Moscow, 1954.

[0.3] Neal, B.G., *The plastic methods of structural analysis*. Chapman and Hall, London, 1956.

[0.4] Hodge, P.G., *Plastic analysis of structures*. McGraw-Hill, New York, 1959.

[0.5] Davies, J.M., Variable repeated loading and the plastic design of structures. *Struct. Eng.*, 1970, 48, Nr 5, 181–194.

References

[0.6] Gatewood, B.E., *Thermal stress with applications to airplanes, missiles, turbines and nuclear reactors.* McGraw-Hill, New York, 1957.

[0.7] Davidenkov, N.N., Likhachev, V.A., *Irreversible deformation of metals under thermal cycling,* in Russian. Mashgiz, 1962.

[0.8] Shorr, B.F., Dulnev, R.A., *Investigations of thermal stresses and creep under variable temperature (survey),* in Russian. Zawodskaya Laboratorya, Nr 3, 1964.

[0.9] Manson, S.S., *Thermal stress and low-cycle fatigue.* McGraw-Hill, New York, 1966.

[0.10] ASME boiler and pressure vessels Code, Section VIII, Div. I, 1971.

[0.11] *Thermal strength of machine parts,* in Russian. Ed. I.A. Birger and B.F. Shorr, Mashinostroenie, Moscow, 1975.

[0.12] Miller, D.R., Thermal stress ratchet mechanism in pressure vessels. *Trans. ASME,* ser. D, 81, Nr 2, 1959.

[0.13] Burgreen, D., Review of thermal ratchetting in fatigue at elevated temperatures, ASTM STP 520, American Society for Testing and Materials, 1973, 535–551.

[0.14] Bree, J., Elastic-plastic behavior of thin tubes subjected to internal pressure and intermittent high-heat fluxes with application to fast nuclear reactor fuel elements. *J. Strain Anal.* 2, Nr 3, 1967.

[0.15] Gokhfeld, D.A., Sadakov, O.S., A mathematical model of medium for analysing the inelastic deforming of structures subjected to repeated actions of load and temperature. *Trans. of the 3rd Intern. Conf. on Struct. Mech. in React. Techn.* 5, Part L. London, 1–5 Sept. 1975.

[0.16] Masing, G., Wissenschaftliche Veröffentlichungen aus dem Siemens-Konzern, S, 135, 1926.

[0.17] Afanasyev, N.N., Statistical theory of fatigue in metals, in Russian. *Izd. AN UkSSR,* Kiev, 1953.

[0.18] Besseling, J.F., Theory of elastic, plastic and creep deformations of an initially isotropic material showing anisotropic strain-hardening, creep recovery and secondary creep. *J. Appl. Mech.* 25, 1958, 529–536.

[0.19] Drucker, D.C., Plastic design methods – advantages and limitations. *Trans. Soc. Nav. Arch., and Marine Engrs.,* 65, 1957, 172–196.

[0.20] Rabotnov, Yu.N., Solid body mechanics and its development, in Russian. *Izvestia AN SSSR, Mekhanika i Mashinostroenie,* Nr 2, 1962.

[0.21] Prager, W., Problem types in the theory of perfectly plastic materials, *J. Aero Sci.,* 15, 1948.

[0.22] Bleich, H., Über die Bemessung statisch unbestimmter Stahltragwerke unter Berücksichtigung des elastischplastischen Verhaltens des Baustoffes. *Bauingenieur,* 19/20, 1932, 261.

[0.23] Melan, E., Theorie statisch unbestimmter Systeme aus idealplastischen Baustoff. *Sitz. Ber. Ac. Wiss.* IIa, Wien, 1936, 145–195.

[0.24] Melan, E., Der Spannungszustand eines Hencky-Misses'schen Kontinuums bei veränderlicher Belastung. *Sitz. Ber. Ak. Wiss.* Wien, IIa, 1938, 147, 73.

[0.25] Gvozdev, A.A., Determination of the value of collapse load for statically indeterminate systems undergoing plastic deformation, in Russian. *Proceedings of the Conference on Plastic Deformation. Izd. AN SSSR,* 1938.

[0.26] Koiter, W.A., A new general theorem on shakedown of elastic-plastic structures. *Proc. Kon. Ned. Ak. Wet.,* B. 59, 24, 1956.

[0.27] Koiter, W.T., General theorems for elastic-plastic solids, *Progress in solid mechanics,* I.I.N. Sneddon and R. Hill, Editors, North-Holland, Amsterdam, 1960.

Introduction

[0.28] Symonds, P.S., Shakedown in continuous media. *J. Appl. Mech.*, *1*, 1951.
[0.29] Hodge, Ph.G., Shakedown of elastic-plastic structures. In *Residual stresses in metal construction*, edited by W.R. Osgood. Reinhold, New York, 1954.
[0.30] Horne, M., Fundamental proportions in plastic theory of structures. *J. Inst. Civ. Engrs.*, *34*, 1950, 174–177.
[0.31] Prager, W., Plastic design and thermal stresses. *British Welding Journ.*, 3, Nr 8, 1956, 355–359.
[0.32] Rosenblum, V.I., On shakedown analysis of nonuniformly heated elastic-plastic bodies, in Russian. *Izvestia AN SSSR*, OTN, Nr 7, 1957.
[0.33] Rosemblum, V.I., On shakedown analysis of nonuniformly heated elastic-plastic bodies, in Russian. *Zh. Prikl. Mekh. Tekh. Fiz.* (PMTF), Nr 5, 1965.
[0.34] Maier, G., Shakedown theory in perfect elastoplasticity with associated and nonassociated flow-laws: a finite element, linear programming approach. *Meccanica*, 4, Nr 3, 1969, 250–260.
[0.35] Donato, O. de, Second shakedown theorem allowing for cycles of both loads and temperature. *Rend. Ist. Lombardo Accad. sci. e lett.*, *A104*, Nr 1, 1970, 265–277.
[0.36] Gokhfeld, D.A., Some problems of the shakedown theory for plates and shells, in Russian. *Proceedings of VI All-Union Conference on theory of shells and plates*, Baku, 1966, Nauka 1966.
[0.37] Gokhfeld, D.A., Cherniavsky, O.F., On conditions of incremental collapse in axisymmetric shakedown problems with the use of linear programming, in Russian. *Izvestia AN SSSR*, MTT, Nr 3, 1970.
[0.38] Gokhfeld, D.A., Cherniavsky, O.F., Methods of solving problems in the shakedown theory of continua, *Proc. Int. Symp. on Found. of Plasticity, Warsaw, 1972.* Noordhoff, Leiden, 1973.
[0.39] Gokhfeld, D.A., Czerniavsky, O.F., Generalized variables in the shakedown problems of plates and shells, in Russian. *Proceedings of VII All-Union Conference on theory of shells and plates, Dnepropetrovsk, 1969*, Nauka, 1970.
[0.40] Frederic, C.O., Armstrong, P.J. Convergent internal stresses and steady cyclic states of stress. *J. Strain Anal.*, Nr 2, 1966.
[0.41] Perzyna, P., Sawczuk, A., Problems of thermoplasticity, *Nuclear Eng. and Design*, 24, 1973, 1–55.
[0.42] Sawczuk, A., Shakedown analysis of elastic-plastic structures. *Nuclear Eng. and Design*, 28, I, 1974, 121–136.
[0.43] Gokhfeld, D.A., Cherniavsky, O.F., The shakedown theory, Its status, applicability and treads of development, in Russian. In *Problems of strength in machines*, (Zb. Nauch. Trud., Nr 151), Polytechnical Institute, Cheliabinsk, 1974.

1

Analysis of simple elastic-plastic systems under variable repeated loading and temperature

Our aim is to describe, as precisely as possible, the effects of variable repeated loading on the elastic-plastic structures. In order to do so, and to introduce certain necessary notions, let us make a suitable analysis for a simple bar system. A similar approach was applied by many authors in the considerations of the mechanical behaviour under various circumstances [1.1–1.3 and others]. This attitude was found to facilitate the desirable generalizations to more complex systems and continua.

From now on, if not stated otherwise, the analysis of the structural behaviour will be based on the assumption that we are dealing with an initially isotropic, elastic-perfectly plastic material whose stress–strain diagram (Fig. 1.1) does not change shape on repeated loading but can only undergo translation along the ϵ-axis. It is also for the time being assumed that the thermophysical and mechanical characteristics of the material do not depend on the temperature, creep is absent, and loss of stability of compressed bars is excluded. Loading is assumed to be quasi-statical and therefore dynamic effects are not taken into account.

One of the more important properties of statically indeterminate systems is their capability of redistributing stresses generated by external loading, by superimposing certain self-equilibrated stress fields that are usually termed the residual, initial, internal or self-stresses, depending on the formulation of the considered problem and the adopted terminology. Those stress states are caused, in particular, by inelastic deformation when the resulting irreversible strain state appears to be incompatible, i.e. for the plastic strains the continuity equations are not satisfied.

If the inelastic deformations are not known, the distribution of self-stresses in redundant systems can be determined only to within an

1 Analysis of simple elastic-plastic systems

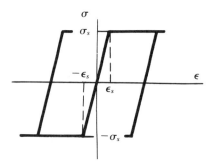

Fig. 1.1. Stress–strain diagram of elastic-perfectly plastic material.

accuracy of certain parameters, the number of which is equal to the degree of redundancy of the considered system. The most simple is a one-parameter system with single redundancy. However, it must be realized that a continuum represents a system for which the number of parameters necessary to determine the distribution of internal stresses is infinite. Thus it is felt essential to investigate those features of structural behaviour that are peculiar to an increasing number of relevant parameters.

Apart from the parameters that result from the properties of the structure itself, there must also be considered a number of loading parameters. As readily seen, the one-parameter proportional loading constitutes the simplest case. The problems considered in the present book deal mainly with the multi-parameter, variable repeated loading. However, for the purpose of introductory analysis of simple systems the number of loading parameters may be kept conveniently low.

The analytical study of the behaviour of a one- or two-parameter system under variable repeated loading and thermal actions represents a cumbersome, if not very complex problem.

Much clearer insight can here be gained by applying the graphical interpretation of the obtained results. Numerous methods of geometrical representation of the stress and deformation states in statically indeterminate systems were devised by Rzhanitzin, Prager and Hodge [1.2–1.5]. In this chapter one of those representations is generalized to cover the case of thermal actions and its further development is shown to apply to the two-parameter bar system.

The known analogy between the behaviour of an actual material and its model, consisting of elements with suitably idealized mechanical properties, makes it possible to illustrate qualitatively and sometimes also quantitatively the effects of such factors as strain-hardening under monotonic and cyclic loading. This problem is of great practical importance since a structural analyst can feel dubious as to the applicability of idealized models when dealing with real materials.

At substantially elevated temperatures an evolution of physical and mechanical properties of the material begins to enter the picture together with the phenomenon of creep. Also in those most complicated situations one derives by analogy with the study of somewhat simpler systems a description, at least a qualitative one, of the influence of singular characteristics of the material on the performance of structures under cyclic loading.

All the results arrived at in this chapter reflect the more general relationships developed later and following from the fundamental theorem of the shakedown theory.

1.1. One-parameter systems

Let a symmetric[1] system shown in Fig. 1.2 be subjected to cyclic loading P and temperature. Axial forces in the elements satisfy the equilibrium equation

$$N_1 + N_2 = P \tag{1.1}$$

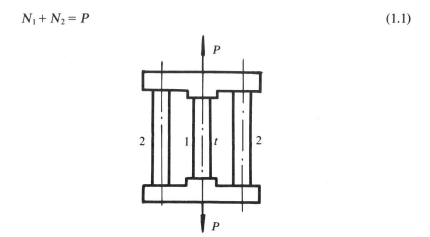

Fig. 1.2. One-parameter system.

---

[1] Due to its symmetry the system is considered to consist of two elements, one of which is split up into two identical parts.

1 Analysis of simple elastic-plastic systems

together with the compatibility equation in terms of elongations

$$\Delta l_1 = \Delta l_2. \tag{1.2}$$

The elongations are assumed to consist of three parts, elastic, plastic and thermal,

$$\Delta l_i = \Delta l_{ei} + \Delta l_{pi} + \Delta l_{ti}, \quad (i = 1, 2). \tag{1.3}$$

The elastic and the thermal parts are determined by the instantaneous magnitudes of load and temperature

$$\Delta l_{ei} = \frac{N_i}{c_i}, \quad \Delta l_{ti} = (\alpha t l)_i,$$

where $c_i = (EF/l)_i$ is the longitudinal stiffness of the element, F, l are the area of cross-section and length, E, α are Young's modulus and the coefficient of linear thermal expansion and t is the temperature increment measured with respect to that initial temperature at which no thermal stresses are present.

In order to determine the plastic component of the elongation the whole history of the loading process should be known.

Given the program of cyclic loading, a computation of internal forces, stresses and deformations from equations (1.1)–(1.3) is a fairly elementary procedure. However it is the graphical representation of results which, as already emphasized, provides a useful means to depict the situation.

Let us adopt new variables in such a way that both the forces and the elongations have the same dimensions,

$$s_i = \frac{N_i}{\sqrt{c_i}} = \Delta l_{ei}\sqrt{c_i}, \quad y_i = \frac{Y_i}{\sqrt{c_i}}, \quad \theta_i = \Delta l_{ti}\sqrt{c_i}, \quad \lambda_i = \Delta l_{pi}\sqrt{c_i}. \tag{1.4}$$

Here $Y_i = \sigma_{si}F_i$ is the yield load and σ_{si} is the yield point of the material. The equilibrium and the compatibility equations take the form

$$s_1\sqrt{c_1} + s_2\sqrt{c_2} = P, \tag{1.5}$$

$$\frac{s_1 + \theta_1 + \lambda_1}{\sqrt{c_1}} = \frac{s_2 + \theta_2 + \lambda_2}{\sqrt{c_2}}. \tag{1.6}$$

1.1. One-parameter systems

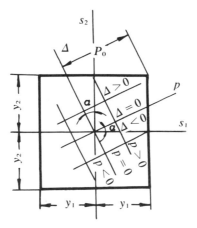

Fig. 1.3. Internal forces diagram for a one-parameter system.

In the s_1, s_2-system of coordinates the above equations represent two orthogonal families of straight lines, Fig. 1.3. The position of equilibrium line for the given system is determined by the load magnitude P, as in equation (1.5), the change in which results in a parallel translation of this line over the distance

$$p = \frac{P}{\sqrt{c_0}}, \quad (c_0 = c_1 + c_2). \tag{1.7}$$

The position of the compatibility line (1.6) depends on the values of both thermal and plastic deformations of the elements. When $\theta_1 = \theta_2 = \lambda_1 = \lambda_2 = 0$, that line crosses the origin of coordinates; changes in the above variables result in parallel translation of compatibility lines over the distance

$$\Delta = (\theta_1 + \lambda_1)\sqrt{\frac{c_2}{c_0}} - (\theta_2 + \lambda_2)\sqrt{\frac{c_1}{c_0}}. \tag{1.8}$$

The orthogonal net is inclined at the angle α satisfying

$$\operatorname{tg} \alpha = \sqrt{\frac{c_2}{c_1}}. \tag{1.9}$$

The given values of the load P together with the thermal and the

5

1 Analysis of simple elastic-plastic systems

plastic deformations of elements correspond to a certain point on the s_1, s_2-plane which will be termed the actions point. Its coordinates, determined by the intersection of appropriate equilibrium and compatibility lines, are

$$s_1 = \frac{1}{c_0}[P\sqrt{c_1} - c_2(\theta_1 + \lambda_1) + \sqrt{c_1 c_2}(\theta_2 + \lambda_2)],$$
$$s_2 = \frac{1}{c_0}[P\sqrt{c_2} - \sqrt{c_1 c_2}(\theta_1 + \lambda_1) - c_1(\theta_2 + \lambda_2)].$$
(1.10)

Those coordinates can be found once the initial state and the loading history of the system are known. It can be readily seen that under the conditions of ideal plasticity the actions point cannot lie outside the region bounded by the values of yield loads (Fig. 1.3), irrespective of the changes in external load and temperature.

The graphical construction is shown in Fig. 1.4. Let the point O correspond to an initial position of the actions point and the point A' to a fictitious position which would have been relevant had the deformations been elastic over the whole path of loading and temperature. The actual position of the actions point A can now be found by drawing from A' a straight line parallel to the axis Δ and intersecting it with the boundary of the elastic region. The segment AA' shows the plastic deformation that has taken place in the process. If, as shown in the example, it was the line $s_2 = y_2$ which was intersected, the plastic elongation occurred in the element 2. The magnitude of plastic deformation as well as the final coordinates of the actions point can easily be found by means of the relationships (1.10).

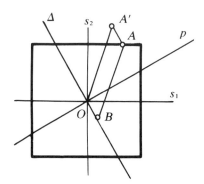

Fig. 1.4. Determination of the actions point.

1.1. One-parameter systems

The internal forces diagram makes it possible to investigate the various situations that can occur under variable repeated loading of the considered system. First, let us pay some attention to the capability of the system to stabilize its state under cyclic loading described by the limiting values and the program of its changes. The steady process is found to settle due to the formation, at the initial stage of loading, of a certain state of self-stresses caused by plastic yielding in any element of the system. This state does not vary as the process goes on. To throw more light on the situation an illustrative example will now be discussed.

Let the internal forces diagram, corresponding to specific values of bar stiffnesses and their yield forces, have the form shown in Fig. 1.5. Let us first confine ourselves to the case in which the load P and the thermal deformations of the elements θ_1 and θ_2 and also the corresponding temperatures remain all proportional to a common factor. Thus the first loading, starting from the initial, stress-free state corresponds to a certain radius OB whose direction depends on the relation between the intensities of loading and thermal actions, thus on the loading profile. The point B represents a fictitious state of perfect elasticity. Its actual position B', due to the plastic tension in the element 2 (since the side $s_2 = y_2$ has been touched), is given by the segment BB' parallel to the axis Δ. Further change in the loading parameters causes the actions point to move in the $B'D$-direction so that the fictitious state on complete unloading corresponds to the point D ($B'D = OB$), whereas for perfect

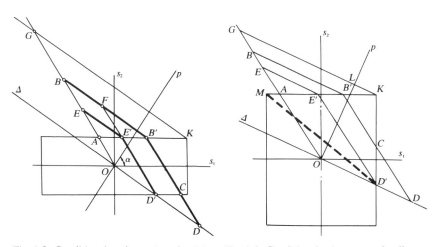

Fig. 1.5. Condition for alternating plasticity. Fig. 1.6. Condition for incremental collapse.

7

1 Analysis of simple elastic-plastic systems

plasticity the actual one to the point D'. Over the next cycles of loading and unloading, in accordance with the given parameters, the actions point will follow the path $D'E'B'CD'$ (fictitious states F and D) and the element 2 will suffer, at each cycle, plastic tension on loading and plastic compression on unloading. The whole deformation per cycle clearly vanishes, excluding the first cycle at which the initial (i.e. thermal) actions have been generated, as shown by means of the coordinates s_1, s_2 of point D'. Such a situation is usually termed alternating plasticity. This state can appear unsafe if, after a certain number of cycles, depending on the amplitude of plastic deformation, failure takes place due to the low-cycle (plastic) fatigue.

It can be readily shown that the stabilized system begins to behave in a purely elastic manner, provided the peak values of cycle parameters do not exceed those corresponding to the segment $OE = D'E'$. The plastic deformation is confined merely to the first loading in the initial half-cycle and the deformations per subsequent cycles of the same intensity turn out to be elastic. When this happens, the system is said to shake down to the given external agencies and this state of elastic adaptation is called the shakedown. It is found that the structures which are subjected to not more than about 10^4 cycles per their lifetime are rather safe so far as their strength is concerned. Usually, both the plastic deformation per the initial half-cycle (corresponding to the segment EE') and the final displacements are acceptable.

It thus follows that the state of shakedown is distinguishable from that of alternating plasticity simply by the length of the segment $D'E'$ on the internal actions diagram; the system will shake down if the loading parameters on the half-cycle OE do not exceed those corresponding to the segment $D'E'$. If the loading parameters vary within the length OA, the system will behave purely elastically right from the start of the loading program. When $OB > D'E'$, the cyclic loading results in alternating plasticity of the element 2 and the peak value of the internal action in this element is equal to twice its yield force ($D'E' = 2OA$).

It is worth noticing that the loading profile OG (or in fact OAK) represents a transition into the state of limit equilibrium and thus appears to be the most extensive loading program the elastic-plastic system can support. Depending on the parameters of the considered system and the loading profile, the length $OE = D'E'$ can be greater or smaller than OG.

On changing either the parameters of the structure (resulting in suitable changes in the dimensions of the rectangle of internal actions

1.1. One-parameter systems

and in the directions of the axes p, Δ) or the relation between the mechanical and thermal loads (resulting in suitable changes in the direction of the radius OB), we can determine in like manner the conditions for alternating plasticity which will occur in the element 1.

However, other steady states can be shown to exist. A different possible situation is depicted in Fig. 1.6. The initial (fictitious) loading profile OB shifts the actions point to the position B', as a result of yielding in the element 2. During unloading $B'D$ the actions point shifts from the fictitious point D to occupy the position D' clearly associated with the plastic deformation of the element 1, having the same sign as that in the element 2. Each subsequent loading cycle $D'E'B'CD'$ will now result in a steady state strain accumulation: Element 2 yields on every loading, element 1 on every unloading branch of the loading profile. The internal actions in the system at the end of a cycle (point D') are the same as at its beginning, i.e. the elastic deformations of elements per cycle do not change. Since the same applies to the thermal deformation, it follows that plastic deformation increments per cycle should satisfy the compatibility condition. Indeed, the calculations based on (1.10) readily confirm this statement: The plastic elongations of the elements 1 and 2 per cycle are equal to each other and therefore the condition (1.2) holds good.

The described situation is commonly termed the incremental collapse in contrast to instantaneous collapse which takes place in an elastic-plastic structure upon reaching the limit state.

Both in the former and in the latter case the term collapse is understood as the possibility of unlimited increase in deformation. The difference between the two situations consists in that the incremental collapse does not occur instantaneously but it develops as the number of loading cycles increases. Moreover, in both cases it is only the inception of the collapse that is in fact investigated since we have not allowed for geometry changes nor strain-hardening of the material, which are known to accompany the advanced deformation processes.

The most characteristic feature of the incremental collapse is the nonisochronism (nonsimultaneity) of the plastic yielding at different zones in the structure. The instantaneous collapse under given program of proportional loading would take place for a somewhat larger value of the load factor (corresponding to the fictitious actions point G and the actual point K).

However it is not always the case that the instantaneous collapse is preceded by the incremental one. Certain necessary conditions must be

1 Analysis of simple elastic-plastic systems

fulfilled ensuring that the nonisochronism does in fact take place, i.e. that the yielding does not commence at different points at the same time. In particular, the incremental collapse is found to be absent if on increase of the load factor the internal actions computed elastically are of the same sign in both elements of the system (the corresponding radius would be placed in the first and the third quadrant of the s_1, s_2-coordinate plane). If this is the case, the yield loads of both bars are reached during the same loading process and the only collapse mechanism that can form is that of an instantaneous type.

If the maximum value of the load factor is not greater than that corresponding to the segment $OE = D'E'$, the subsequent (steady) cycles are not accompanied by yielding. For values greater than OA this situation is due to the advantageous distribution of initial actions (translation of the initial position of the actions point from point O towards D') at the cost of some plastic deformation over the first half-cycle. The length of the segment $D'E'$, parallel to the load radius OG, is a measure to distinguish between the shakedown and the incremental collapse. The point D' represents the distribution of the initial actions. They are clearly seen to be self-equilibrated since the point D' lies on the axis $p = 0$. On change of the load profile, i.e. on rotation of the radius OAB, the length of the limiting segment changes from $D'M$ (coinciding with the alternating plasticity in the element 1) to $D'K$ (coinciding with the condition for the limit equilibrium). Further clockwise rotation of the load radius leads to shortening of the limiting segment down to OL, whereas the magnitude of the external load $p = p_0$, corresponding to the condition of instantaneous collapse, remains constant, being independent of the self-equilibrated, thermal internal actions.

It follows from the above considerations that the following three limit states can take place under repeated external actions:
1) alternating plasticity (in any of the components),
2) incremental collapse,
3) instantaneous collapse.

The segments 1, 2, 3 in Fig. 1.7a determine the parameters of respective limiting cycles which occur after the stabilization which was already shown in Fig. 1.5 and Fig. 1.6, under one-parameter variation of the load P and of the temperature of the elements.[1] The above simple

---

[1] It will be assumed that only one of the bars is subjected to thermal cycling, for instance the first one. It should be remembered that an additional heating which results in the same thermal elongations of both bars, $\theta_1/\sqrt{c_1} = \theta_2/\sqrt{c_2}$, leads to no change in the internal actions of the system.

1.1. One-parameter systems

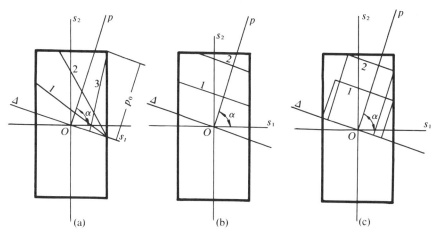

Fig. 1.7. Limiting cycles of various loading programs.

relationships make it possible to construct easily a shakedown diagram, i.e. to determine the relations between the limiting values of loading and temperature (Fig. 1.8, line a). Let us introduce the following notation: Let θ_* and P_* be the interconnected limiting values of varying thermal deformation and loading ($0 \leq \theta_1 \leq \theta_*$, $0 \leq P \leq P_*$), let P_0 be the limit load,

$$P_0 = Y_1 + Y_2 = y_1\sqrt{c_1} + y_2\sqrt{c_2}, \tag{1.11}$$

and let θ_1^0 be the limiting value of the amplitude of the thermal defor-

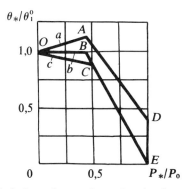

Fig. 1.8. Shakedown diagram for various loading programs.

11

1 Analysis of simple elastic-plastic systems

mation at which the thermal actions alone (in the absence of mechanical load) do not cause alternating plasticity (with the proportions of parameters adopted in Fig. 1.7; this condition has to do with the internal actions in the element 1). The value of θ_1^0 can be found from the fact that the thermal action in the element 1 is equal to

$$s_{1t} = \theta_1 \frac{c_2}{c_0}, \quad (N_{1t} = N_{2t} = N_t), \qquad (1.12)$$

hence it follows that

$$\theta_1^0 = 2y_1 \frac{c_0}{c_2}. \qquad (1.13)$$

The three segments of the piecewise linear diagram $OADE$ of Fig. 1.8, the third segment being clearly vertical, correspond to the three unsafe states defined above.

Other external actions programs can be studied in a similar way. Fig. 1.7b illustrates the possible limit states understood as the limits of elastic behaviour, of a cycle under the conditions of thermal changes accompanied by constant mechanical load ($0 \leqslant \theta_1 \leqslant \theta_*$, $P = P_* = $ const.). The situation is seen to depend on the level of the load. For relatively small values of the load the alternating plasticity appears critical, whereas for greater ones the strain accumulation is the corresponding limiting state. The process of stabilization that terminates after the first cycle can be considered in a fashion similar to that of the preceding case; also similarly one can determine the deformation prior to the shakedown. It is readily noticed that in the case of alternating plasticity, the range of thermal deformation, given by the length of the segment 1, does not depend on the magnitude of the constant load.

The corresponding shakedown diagram, based on the relations (1.8)–(1.10), is shown in Fig. 1.8 as line b and consists of two straight segments.

Fig. 1.7c shows the most general and common case of the shakedown problem in which it is necessary to assess, independently of the loading program, the peak values of external actions that correspond to the shakedown. The limiting cycle is represented now as a certain domain. In particular, for two-parameter loading, it constitutes a rectangle whose sides determine the admissible changes in the mechanical load (along the axis p) and in the thermal deformation (along the axis Δ).

1.2. One-parameter systems. Some additional remarks

It remains to point out that the calculations for an arbitrary loading program are in fact equivalent to the calculations of the strength-wise most disadvantageous sequence of actions. In our case such a program will develop for instance as follows: Application of load, heating of the first element, unloading, cooling, and so on. The presented approach is of value when the relations between the loads are not of a regular character. The irregularity can be either actual or caused by insufficient knowledge of the load.

The shakedown diagram for an arbitrary loading program is also shown in Fig. 1.8 (the line OCE). The shakedown region is the most confined one. Coincidence of limiting segments corresponding to the state of incremental collapse shows that thermal cycling under constant mechanical loading appears, for the system considered, to be the most disadvantageous loading program indeed. This circumstance, as well as the differences in the conditions for alternating plasticity under corresponding loading programs, could already be clearly seen on comparing Fig. 1.7b and 1.7c.

1.2. One-parameter systems. Some additional remarks

The above analysis of a simple system has shown that repeated loading usually brings about one of the two different unsafe states: Alternating plasticity or incremental collapse. The following simple condition causes the first state to occur: Under repeated loading, the change in the fictitious stresses, calculated in a purely elastic fashion, in any of the elements exceeds, in a loading cycle, more than twice the yield point of the material. The picture of the other state is more complex: It is necessary that the plastic yielding, developed during the loading cycle, spreads further to affect all the elements. (In the next chapters some more complex structures will be dealt with in which the incremental collapse will be a partial one, i.e. relating to only a finite part of the structure). An important feature of the incremental collapse, in contrast to the instantaneous collapse, is that the inelastic deformations of different elements take place non-simultaneously, i.e. are not isochronous.

It should be borne in mind that not only the causes but also the effects of alternating plasticity are substantially different from the causes and effects of incremental collapse.

In the first case the local failure may occur as a result of low-cycle

1 Analysis of simple elastic-plastic systems

fatigue. In the second case the effect is of a rather global nature and the actual collapse after a number of loading cycles can take place either due to the occurrence of unacceptable displacements endangering the serviceability of the structure or as a result of the exhaustion of the plastic properties of the material in any of the structural elements (quasi-statical failure). The durability in the case of the fatigue failure depends on the range of plastic deformation during the cycle and the relationships put forward by Coffin and Manson are here useful. In the case of quasi-statical failure, the durability depends on the magnitude of deformation accumulated per cycle.

Thus it is of interest to study the characteristic features of the incremental collapse in more detail even by means of an example of the simple, one-parameter system, in order to distinguish better the effects associated with the corresponding factors.

Let us return to the limiting cycle with the following program: Thermal changes in the element 1 (heating and cooling) under constant load, Fig. 1.7b. As indicated before, this program leads most likely to the incremental collapse.

From the geometrical relationships it follows that the limiting range of thermal deformation Δ_{max} and the force P obey the formula (Fig. 1.7b)

$$\Delta^* = \frac{p_0 - p}{\sin \alpha \cos \alpha}. \tag{1.14}$$

On account of (1.8), (1.9), we obtain

$$\theta_* = (p_0 - p)\frac{c_0}{c_2}\sqrt{\frac{c_0}{c_1}}. \tag{1.15}$$

Remembering that the thermal action is given by (1.12) and using the parameters introduced in (1.4), (1.7), we arrive to the limiting condition (1.15) in the form

$$P = P_0 - N_t^*, \tag{1.16}$$

where N_t^* is the maximum value of the thermal action in the system.

The condition obtained can be interpreted as responsible for a decrease in the limit load due to thermal changes, the limit load being understood as giving rise to an unrestricted increase in deformation,

1.2. One-parameter systems. Some additional remarks

irrespective of whether the deformation is instantaneous or develops as the number of loading cycles increases. The drop in the limit load is visualized in Fig. 1.9. Line 1 corresponds to the condition of alternating plasticity ($N_t^* = 2Y$) for various ratios of the yield forces $Y_1 : Y_2$. Line 2 represents the condition for incremental collapse (1.16). A simple analysis shows[1] that the coincidence of both types of cyclic plastic deformation in the considered system is possible only in the case which is represented by the abscissa of the point at which lines 1 and 2 intersect (for given ratio $Y_1 : Y_2$). The alternating and mounting components of the deformation process cannot be here separated.

Thus the diagram obtained consists of three domains: The shakedown domain A, the alternating plasticity domain B and the incremental collapse domain C. The piecewise linear diagram 2, 3 can be looked upon as a diagram showing the change in the limit load (in the sense indicated before). As the ratio $Y_1 : Y_2$ tends to unity, the alternating plasticity domain shrinks. On reaching unity, the line 1 degenerates to a point, as seen in Fig. 1.9. Then the limit load becomes zero as a result of thermal effects. This means that the thermal cycling in the case of $N_t^*/P_0 > 1$ can lead to an unlimited increase in deformation under arbitrarily small values of the

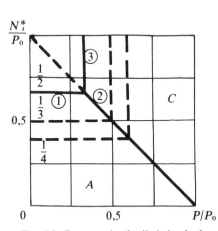

Fig. 1.9. Decrease in the limit load of a system subjected to thermal changes.

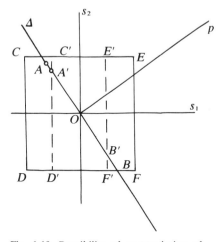

Fig. 1.10. Possibility of accumulation of thermal deformation in the absence of external load.

---

[1] This will be better seen later on, Fig. 1.11.

15

1 Analysis of simple elastic-plastic systems

load P. However, in view of the inseparability mentioned earlier, P must be kept finite.

There are some cases encountered in practice and in laboratory tests in which the strain accumulation under thermal cycling is clearly independent of the mechanical loading. In connection with this, there arises the problem of proper selection of relevant factors. Also the question should be answered whether the considered one-parameter system is capable of showing this effect.

In the absence of mechanical agency the loading and the unloading lines both coincide with the axis Δ, Fig. 1.10. It can be readily seen that this axis can intersect the diverse domain boundaries only when there exists a plastic anisotropy of at least one member of the system. The reason for anisotropy may be, in particular, a temperature-dependence of the yield point. In such a case the heating of the element 1 results in a translation of the actions point in the direction OA simultaneously with a translation of the line CD to the right, due to the decrease in the yield stress transmitted by this element. The segment AA' corresponds to the plastic compression, provided a steady state is considered beginning with the second cycle. Since on cooling the yield point is restored, the actions point reaches the line DF and the plastic compression of the element 2 will commence. Thus the thermal cycling is found to cause shortening of both elements accompanied by the two platens coming closer, Fig. 1.2.

In the general case of anisotropy the condition for the progressive accumulation of plastic deformation has the form

$$\frac{\sigma_{s1}^c - \sigma_{s2}^p}{\sigma_{s1}^p - \sigma_{s2}^c} < 0. \tag{1.17}$$

Plastic compression as well as plastic tension is now possible, depending on the signs of the numerator and the denominator of the above ratio.

Suitable analysis shows that, in particular, it is vital to allow for temperature dependence of the yield point in the non-isothermal shakedown problems since the influence of this dependence can turn out to be not only of quantitative, but also of qualitative character.

Bearing in mind (1.16), we conclude that this expression reveals some additional possibilities of direct determination of the conditions for incremental collapse, at least in simple systems with one-dimensional stress states.

On account of the equilibrium equation (1.1), the internal forces in

1.2. One-parameter systems. Some additional remarks

the members of the considered structure are in equilibrium with the external load P at each instant of time. The forces due to thermal actions appear to be self-equilibrating (equal in each element but acting in the opposite directions). When added to the tensile forces caused by the mechanical load, they clearly overload one of the elements while unloading the other. The expression (1.16) confirms that, when the unloading is disregarded, the new internal forces are in equilibrium with the load P_0. In other words, the limit equilibrium state will be reached.

The situation is illustrated in Fig. 1.11 in which there are shown two states corresponding to the termination of heating and of cooling in the limiting, steady cycle. The transition from one state to the other consists in imposing or releasing the thermal actions calculated on the assumption of perfect elasticity (Fig. 1.11). From (1.16) it is clearly seen that by following the procedure (constructing the stress or the internal actions diagrams that correspond to the relevant stages of the limiting cycle) one can similarly arrive at the solutions to somewhat more complex problems.[1]

Let us note that under the conditions of limiting cycle the structure behaves purely elastically; the strain accumulation will commence with an arbitrarily small (but finite) increase in the mechanical loading or in the intensity of thermal changes.

The construction of a picture similar to that of Fig. 1.11, corresponding to the proportional loading and temperature changes, is left to the reader.

The incremental collapse is well known to arise also in the case of purely mechanical, variable repeated loading [1.2, 1.9, 1.10 and others].

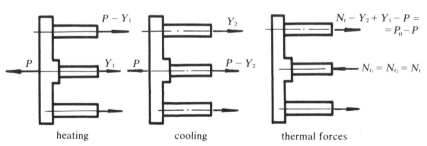

Fig. 1.11. Illustration of the incremental collapse mechanism by means of limit state diagrams.

---

[1] A development of the procedure for determination of the incremental collapse state will be described in Section 5.5.

17

1 Analysis of simple elastic-plastic systems

However, the described effect cannot occur in the system considered: the accumulating deformation should have the same sign in both elements and take place simultaneously with the acting load attaining its peak value. However, this situation does not mean that the realization of incremental collapse is impossible under one-parameter cycling load. Let us show, by employing the above presented method of analysis, one of the examples which illustrate the situation.

The characteristic feature of the one-parameter system shown in Fig. 1.12 is the discrepancy of the collapse mechanism (both instantaneous and incremental) with the distribution of the 'elastic' internal forces generated by the given load $P(\tau)$. The reason is that, in particular members, the signs of elastic deformations do not agree with the signs of plastic deformations accompanying the collapse. Thus it is possible to satisfy the necessary condition for the accumulation of deformation under repeated loading. We have here the nonisochronism of the internal forces attaining their respective yield values at different instants of time. The equilibrium conditions of the system shown in Fig. 1.12,

$$N_1 + N_2 + N_3 = P,$$
$$4N_1 + N_2 = 2P$$
(1.18)

together with the compatibility of elongations given by

$$\Delta l_1 + 3\Delta l_3 = 4\Delta l_2 \tag{1.19}$$

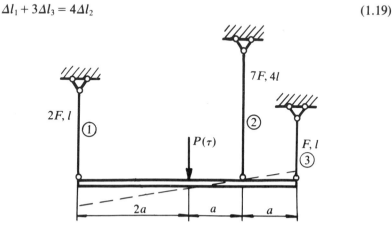

Fig. 1.12. *Probability of incremental collapse under repeated one-parameter mechanical loading.*

1.2. One-parameter systems. Some additional remarks

provide the formulae for the elastic tensile stresses in all the elements of the structure:

$$\sigma_1^{(e)} = \frac{53}{261}\frac{P}{F}, \quad \sigma_2^{(e)} = \frac{14}{261}\frac{P}{F}, \quad \sigma_3^{(e)} = \frac{57}{261}\frac{P}{F}. \tag{1.20}$$

Assuming all the elements to be made of the same material with equal yield points in tension and compression, the formulae (1.20) directly yield the limit value of the external load associated with alternating plasticity (maximum stress in bar 3)

$$P^* = P^0 = \frac{522}{57}F\sigma_s \approx 9{,}2F\sigma_s.$$

In order to determine the limit equilibrium load it is necessary to investigate three possible collapse mechanisms (a yielding of two out of three bars is clearly sufficient to start the plastic motion). The actual mechanism, as it is well known [1.2, 1.10], is that corresponding to the least value of the associated limit load. This value is found to be

$$P^* = P^0 = 7F\sigma_s$$

and it is associated with the mechanism in which the bar 2 remains rigid (as shown by broken line in Fig. 1.12).

Assuming that a similar mechanism is formed under repeated loading, the free-body diagram can now be plotted analogous to that of Fig. 1.11. In Fig. 1.13 are shown the states at the end of loading (in Fig. 1.13a bar 1 yields in tension) and at the end of unloading (in Fig. 1.13b bar 3 yields in compression). Those two states differ in the elastic actions generated by the load $0 \leq P(\tau) \leq P^*$ (Fig. 1.13c). Using the equilibrium conditions for each of the two states together with the formulae (1.20) and inspecting Fig. 1.13c it is easy to arrive at the condition stating that the incremental collapse will occur when $P^* > 5{,}75F\sigma_s$.

A suitable check confirms that the stresses in bar 2 stay elastic at any stage of the considered cycle. Thus the assumed mechanism appears to remain correct under repeated loading. The reduction of the carrying capacity of the system due to repeated loading is about 18 per cent.

It must be pointed out that, in general, the strain accumulation under repeated external actions is more typical in the case of thermal than under mechanical loading. In the presence of purely

1 Analysis of simple elastic-plastic systems

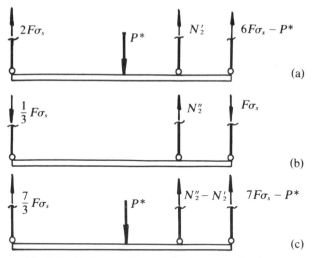

Fig. 1.13. Limit state diagrams for incremental collapse.

mechanical, also multi-parameter, loading the difference between the instantaneous plastic collapse load and the incremental collapse load is found to be insignificant (cf. Fig. 1.9 which illustrates the difference under thermal actions). Concerning this fact an opinion was formed [1.9] that the limit equilibrium calculations and the proper choice of the safety factor can be done without necessarily checking the conditions for the incremental collapse. However, some examples were found later proving that this point of view is incorrect, in particular, in the case of moving loads.

1.3. Two-parameter systems

A continuous body constitutes a system in which the admissible stress distributions cannot be determined by means of functions of a finite number of parameters. Thus it is important to realize what kind of changes will accompany the increase of the number of parameters, especially from one to two, as concerns the behaviour under variable repeated loading, the conditions for transition to limit states and the process of cycle stabilization.

In the symmetric system shown in Fig. 1.14 there can be distinguished two independent states of self-stresses; indeed only one

1.3. Two-parameter systems

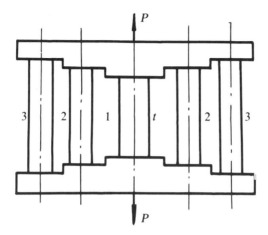

Fig. 1.14. Two-parameter system.

equilibrium equation is here non-trivial while three unknown internal forces are involved. Let us introduce the geometrical relationships based on the considerations of the previous section in which the one-parameter system was dealt with. In the adopted notation (1.14) the equilibrium and the compatibility equations assume now the form

$$s_1\sqrt{c_1} + s_2\sqrt{c_2} + s_3\sqrt{c_3} = P, \tag{1.21}$$

$$\frac{s_1 + \theta_1 + \lambda_1}{\sqrt{c_1}} = \frac{s_2 + \theta_2 + \lambda_2}{\sqrt{c_2}} = \frac{s_3 + \theta_3 + \lambda_3}{\sqrt{c_3}}. \tag{1.22}$$

In the s_1, s_2, s_3-space the first equation represents a set of parallel planes termed equilibrium planes, whereas the second one corresponds to a set of straight lines which are called compatibility lines and are perpendicular to the equilibrium planes. The forces in the elements of the considered structure are given by the points of intersections of the equilibrium planes with the compatibility lines giving the actions points which correspond to the prescribed values of mechanical load as well as the thermal and the plastic deformations of the elements. In particular, we have

$$s_1 = \frac{1}{c_0}[P\sqrt{c_1} - (c_0 - c_1)(\theta_1 + \lambda_1) + \sqrt{c_1 c_2}(\theta_2 + \lambda_2) + \sqrt{c_1 c_3}(\theta_3 + \lambda_3)], \tag{1.23}$$

where $c_0 = c_1 + c_2 + c_3$.

1 Analysis of simple elastic-plastic systems

The formulae for s_2 and s_3 can be readily obtained from (1.23) by cyclic permutation of the indices 1, 2, 3.

In the absence of strain-hardening of the material the possible positions of the actions point lie inside the parallelepiped shown in Fig. 1.15 whose sides are determined by the yield forces y_i. Those forces are assumed to be equal both in tension and compression but this assumption can be relaxed if necessary.

The spatial picture of the situation obtained above can be reduced to two dimensions by placing all the equilibrium planes (two of them can be seen in Fig. 1.15) in the plane of Fig. 1.16. The equilibrium planes are brought into contact by sliding them along one of the coordinate axes, for instance, the vertical axis s_3. The plane boundaries are then determined by the limiting conditions $s_1 = \pm y_1$, $s_2 = \pm y_2$, Fig. 1.16.

Let us introduce a skew coordinate system created by the intersections of the equilibrium planes with the coordinate planes $s_1 = 0$, $s_2 = 0$. The transformation formulae relating the new to the old coordinates are found to be

$$s_1 = x_1 \cos \alpha_1, \quad s_2 = x_2 \cos \alpha_2,$$
$$s_3 = \frac{P}{\sqrt{c_3}} - x_1 \sin \alpha_1 - x_2 \sin \alpha_2, \tag{1.24}$$

while

$$\operatorname{tg} \alpha_1 = \sqrt{\frac{c_1}{c_3}}, \quad \operatorname{tg} \alpha_2 = \sqrt{\frac{c_2}{c_3}}. \tag{1.25}$$

The angle between the axes x_1 and x_2 is given by

$$\cos \gamma = \sin \alpha_1 \sin \alpha_2 = \sqrt{\frac{c_1 c_2}{(c_0 - c_1)(c_0 - c_2)}}. \tag{1.26}$$

Let the initial position of the actions point coincide with the origin of the coordinates, Fig. 1.16. As the external load increases, the actions point moves along a certain axis p whose direction can be determined by means of the magnitudes adopted above. We obtain that

$$\operatorname{tg} \beta_1 = \sqrt{\frac{c_2 c_3}{c_1 c_0}}, \quad \operatorname{tg} \beta_2 = \sqrt{\frac{c_1 c_3}{c_2 c_0}}, \quad (\beta_1 + \beta_2 = \gamma). \tag{1.27}$$

1.3. Two-parameter systems

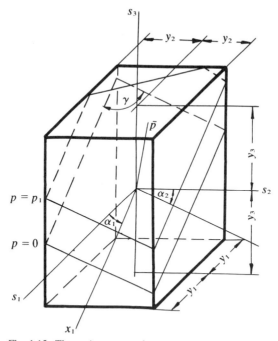

Fig. 1.15. The actions space for a two-parameter system.

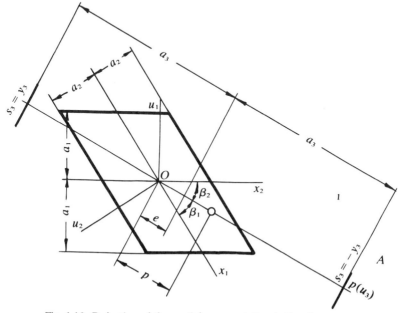

Fig. 1.16. Reduction of the spatial representation to the plane one.

1 Analysis of simple elastic-plastic systems

The axis p represents a projection of the compatibility line $\bar{p}$ on the equilibrium plane in the direction of the axis s_3. It is readily seen that the axis p remains perpendicular to the traces of the equilibrium plane on the limiting planes $s_3 = \pm y_3$ and therefore it is the line of maximum slope of the equilibrium plane in the s_1, s_2, s_3-space, where s_3 is kept vertical (Fig. 1.15).

As the load increases the actions point is displaced by

$$p = P\sqrt{\frac{c_0 - c_3}{c_0 c_3}} \tag{1.28}$$

in the direction of the axis p, irrespective of the values of the prescribed parameters.

The dimensions of the admissible domain of the actions which take place in the elements of the system are determined in the x_1, x_2-plane by the distances between suitable limiting lines,

$$a_i = y_i \sqrt{\frac{c_0}{c_0 - c_i}}, \tag{1.29}$$

and by the relative positions of the lines $s_3 = \pm y_3$ with respect to the origin of coordinates, dependent on the value of external load and given by the segment

$$e = P\sqrt{\frac{c_0}{c_3(c_0 - c_3)}}. \tag{1.30}$$

A simple argument shows that a change in the sum of the thermal and plastic deformation in any of the elements results in a shift of the actions point in the direction perpendicular to the limiting lines corresponding to that element. Thus the axes u_1, u_2, u_3 can be conveniently introduced, the last one coinciding with the axis p. The motion of the actions point as a result of the above mentioned sum of deformations is now given by the formulae

$$u_i = (\theta + \lambda)_i \sqrt{\frac{c_0 - c_i}{c_0}}, \quad (i = 1, 2, 3). \tag{1.31}$$

The specific position of the axis u_3, following from the above relationships, is explained by the fact that the diagram shown in Fig. 1.16

1.3. Two-parameter systems

was obtained by projecting all the relevant segments on the equilibrium plane in the direction of that very axis.

The diagram so constructed makes it possible to study the behaviour of the two-parameter system under repeated loading and temperature. Various loading programs can be analysed; for the sake of simplicity let us now confine ourselves to the case in which the loading is kept constant whereas the temperature of the bar 1 changes in a cyclic manner. The ensuing possibilities are shown in Fig. 1.17.

Let the yield force of the bar 3 be assumed the largest one and let the admissible domain of the actions point be bounded by the lines of intersection of the internal forces plane with the limiting planes $s_1 = \pm y_1$,

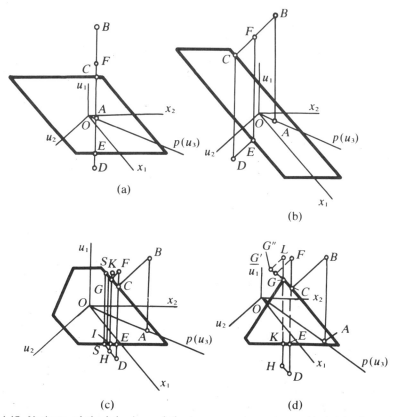

Fig. 1.17. Variants of the behaviour of the two-parameter system subjected to thermal cycling.

1 Analysis of simple elastic-plastic systems

$s_2 = \pm y_2$. Depending on the set of parameters that describe the considered system, the thermal changes can cause the alternating deformation of either the first (Fig. 1.17a) or the second (Fig. 1.17b) bar. The notation is as follows: $AB = CD = EF$ stands for thermal deformation, BC denotes plastic deformation of the relevant element in the initial semi-cycle (i.e. on the first loading), $DE = FC$ shows plastic deformation over the subsequent semi-cycles.

As can be seen in Fig. 1.17, the presence of the third, elastically stressed bar is of no consequence as far as the stabilization is concerned, which under the given conditions takes place as soon as the first cycle is over (this is also the case in the one-parameter system).

Consider the situation depicted in Fig. 1.17b, the amount of plastic deformation per semi-cycle in bar 2 excluding the initial one is, on account of the relationships (1.26), (1.29), (1.31), constant and equal to

$$\lambda_2 = \frac{c_0}{c_0 - c_2}\left(\theta_1 \frac{\sqrt{c_1 c_2}}{c_0} - 2y_2\right). \tag{1.32}$$

Assuming $\lambda_2 = 0$, we obtain the thermal elongation which corresponds to the maximum change in temperature enabling the structure to shake down

$$\theta_1^0 = 2y_2 \frac{c_0}{\sqrt{c_1 c_2}}. \tag{1.33}$$

Under larger values of the force P the thermal cycling can bring about strain accumulation, growing with each cycle (Fig. 1.17c, d). As clearly seen in Fig. 1.17c, the character of stabilization can here be essentially different, namely continuing for a finite number of cycles, or asymptotic. Attention must be drawn to the fact that the tensile internal force in bar 3 increases in the process of stabilization and the actions point is gradually translated upwards while staying in the inclined plane (see Fig. 1.15). If this bar remains elastic then (taking into account a rapid drop in the deformation increments) the structure is supposed to shake down after a few cycles (Fig. 1.17c); if bar 3 begins to yield after a number of loadings then every cycle, starting from the onset of its yielding, causes a constant strain increment (Fig. 1.17d).

When the system considered shakes down the strain increments are slowing down, starting with the first semi-cycle, according to the rule of geometrical progression (Fig. 1.17c). The final deformation for bar 1 is

1.3. Two-parameter systems

found to be the finite quantity

$$\lim_{1,3,5} \sum_{1,3,5}^{\infty} \lambda_{1k} = \frac{\lambda_{11}}{\sin^2 \gamma} = \lambda_{11} \frac{(c_0 - c_1)(c_0 - c_2)}{c_0 c_3}, \tag{1.34}$$

where λ_{11} is its deformation per first semi-cycle and $k = 1, 3, 5, \ldots$ is the number of semi-cycles; initial (zero), then first, second etc.

Some conclusions can now be drawn:

1. Alternating plasticity takes place, irrespective of the number of the self-stress parameters of the system considered, when the change in the fictitious 'elastic' stresses exceeds twice the yield point in any of the structural members, that is

$$\max \Delta\sigma > 2\sigma_s. \tag{1.35}$$

If the yield point is found to be temperature-dependent, the above inequality takes the form

$$\max \Delta\sigma > \sigma_{st1} + \sigma_{st2}, \tag{1.36}$$

where σ_{st1}, σ_{st2} are the values of the yield point corresponding to the actual temperatures at the beginning and at the end of the semi-cycle.

The parameters of the limiting cycle can be determined by introducing the equality signs in (1.35) and (1.36).

2. Incremental collapse under variable repeated loading can occur when the number of yielding members involved per cycle is such that if they yield simultaneously, the system would transform into a mechanism. In other words, the limit equilibrium state would have been reached corresponding to the 'instantaneous' collapse mechanism. This circumstance can be employed to determine the parameters of the limiting (steady) cycle. Non-isochronism of yielding in different members of the structure appears to be the main necessary condition for the incremental collapse to take place.

3. The steady stress cycle can settle either after the first cycle (alternating plasticity, including the limiting cycle of shakedown) or after a certain finite number of cycles or asymptotically (incremental collapse, including the suitable limiting cycle). The preceding analysis has shown that the process of stabilization is a rapid one and the first cycle usually suffices to terminate it. However, this problem will be studied in more detail in Chapter 11 in which the kinetics of cycling deformation will receive attention.

1 Analysis of simple elastic-plastic systems

4. The distribution of the internal forces during local alternating plasticity and in the corresponding limiting cycle fails to be unique, except for the member subjected to cyclic deformation. Under these circumstances the stabilization consists only in the change in the characteristics of the cycle: After the plastic deformation in the initial semi-cycle the loading cycle for the inelastically deformed member appears to be symmetric. As to the elastically deformed elements, the distribution of residual internal forces is not strongly determinate; it can vary within known limits depending on the deformation history. This is readily seen in Fig. 1.17a, b: The steady cycle can be 'displaced', without change of its parameters, along the limiting lines that correspond to the element subjected to alternating deformations.

In the presence of incremental collapse following the stabilization of the stress cycle (including the limiting cycle) the distribution of internal forces proves to be strongly determinate (Fig. 1.17c, d). It will be shown later that the insistence on the uniqueness of the internal forces distribution is appropriate only for those elements that undergo yielding within the cycle (Section 2.7).

In the preceding discussion we confined ourselves to the two-parameter system subjected to the one-parameter proportional loading pattern and a one-parameter temperature field. It is of interest to know what qualitative changes would entail the more complex, multi-parameter loading and temperature fields. It is evident that the above formulated conditions for alternating deformation (1.35), (1.36) remain valid. The necessary condition for incremental collapse, formulated in point 2, remains also valid. The probability that this situation will occur under multi-parameter external actions is expected to increase substantially.

Let us illustrate this fact on the example of the strain accumulation as a result of temperature changes in the absence of external loads. To this end, let us consider repeated actions of some typical non-stationary temperature field. All the three component members, as shown in Fig. 1.14, are assumed to have the same geometric and mechanical characteristics. Under vanishing external load the relevant diagram in the internal forces plane, as shown in Fig. 1.16, has the form depicted in Fig. 1.18. The limiting lines corresponding to the element are distinguished by numbers, the plus and minus superscripts denoting plastic tension and plastic compression, respectively.

The considered cycle is assumed to consist of heating followed by cooling in each of the members. In other words, the system is acted

1.3. Two-parameter systems

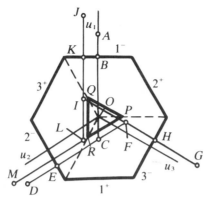

Fig. 1.18. Possibility of deformation accumulation under thermal changes unaccompanied by any mechanical load (repeated action of temperature field is caused by moving heat source).

upon by the temperature field of the type of the moving 'thermal wave'. If, under heating, the first member, subjected to thermal deformation OA reaches the yield point in compression (line 1^-, plastic deformation shown by the segment AB), then after cooling resulting in the thermal deformation $BC = OA$ there will be generated the self-equilibrating forces corresponding to the coordinates of the point C. It appears now that at the subsequent stage of the cycle at heating of the second member this new state is less favourable than the previous state; as seen in Fig. 1.18 the plastic deformation will now commence under smaller temperature difference since point C is closer to the limiting line 2^- than point O is to line 1^-. Thus, in the presence of the same thermal deformation $CD = OA$, the heating of the second member leads to the larger plastic deformation $DE > AB$. The situation after cooling of the second member is shown by point F and it can be seen that at the third stage, i.e. at heating and cooling of the last element, the plastic deformation, the segment GH, is still larger under the constant thermal deformation $FG = AO = CD$.

Starting from the second cycle, the situation is rapidly approaching the steady state in which the plastic deformations of the members are equal at all stages of the stabilized process. The asymptotic conditions correspond to the triangle QPR in Fig. 1.18, the sides of which are perpendicular to the appropriate limiting lines. The distances of the vertices Q, P, R from the lines 3^-, 2^-, 1^-, respectively, are identical and equal to the thermal deformation OA.

1 Analysis of simple elastic-plastic systems

A simple argument leads to the conclusion that, for the given system, in the steady situation the cycle consisting of three stages causes the contraction of each element to increase by

$$|\Delta\epsilon| = \frac{3}{E}(\sigma_* - \sigma_s) \qquad (1.37)$$

as the temperature varies within the range $0 \leq t \leq t_*$, where $\sigma_* = (2\alpha E t_*)/(3F)$ stands for the largest thermal stress under the assumption of perfect elasticity.

Thus a very interesting property, important in many applications, of the moving temperature fields (in particular, of the 'thermal wave') is uncovered. Under repeated actions of such fields the deformations can develop and accumulate with each cycle even if the external mechanical load is absent. In the example considered, as seen in Fig. 1.18 and as follows from (1.37), it is only necessary that the maximum thermoelastic stresses are larger than once the yield point, instead of being twice as large, as it was the case in the alternating plasticity situation. It is worth stressing that the dependence of the yield point on temperature which was of primary importance for thermal ratchetting conditions in the case of a one-parameter temperature field as studied in Section 1.2, is here unnecessary and leads here only to some qualitative corrections.

In Chapter 9 we shall give a more general discussion of the problem of strain accumulation under thermal cycling alone, i.e. when the mechanical loading is negligible.

1.4. Influence of strain-hardening on the cycle stabilization and shakedown conditions at single and repeated deformations

The shakedown theory, as a branch of the general theory of ideal elastic-plastic bodies, is based on an idealized stress–strain relationship disregarding the strain-hardening of the material. Such an idealization is most suitable for ordinary low-carbon steel with a distinct plastic plateau. The structural members made of such grades of steel are expected to behave under variable repeated loading according to the predictions of the shakedown theory, both qualitatively and quantitatively.

However, most alloys have stress–strain diagrams without plastic plateaus and the strain increments are possible only if accompanied by suitable stress increments. On the other hand, the strain-dependent

1.4. Influence of strain hardening

tangent modulus that characterizes the strain-hardening of the material is found to be 1 to 2 orders of magnitude smaller than the Young modulus.

In the shakedown theory the stress–strain diagram is assumed to remain unchanged after any type of repeated loading. Experimental evidence shows that this is, in general, not the case. Under cyclic deformation beyond the elastic limit the stress–strain relationship can, to a lesser or greater degree, change with the number of passing cycles.

The evolution of the stress–strain diagram has received much attention leading to the recognition of the following three types of materials: The cyclic strain-hardening, the cyclic strain-softening and the stable ones. To be more precise, the three states are being distinguished rather than materials since it is now a well established fact that one and the same material, depending on the previous thermal and mechanical treatment as well as the deformation conditions (temperature, strain rate), can become strain-hardening, strain-softening or stable.

An unavoidable question occurs now to the shakedown-conscious engineer: To what degree will the description of structural behaviour be affected by the neglect of the strain-hardening under single and repeated deformation, and to what extent does strain-hardening influence the magnitudes of load factors characterizing the unsafe situations. For this purpose the results can be used obtained in the preceding section in which the behaviour of two-parameter systems was analysed. It must be borne in mind that the bar systems consisting of idealized members are capable of modelling the behaviour of strain-hardening materials [1,1, 1.11]. In particular, the set of two ideally elastic-plastic elements 2, 3 in the considered two-parameter system (Fig. 1.14) behaves in a manner characteristic for the material with linear strain-hardening.

A thorough study of the behaviour of two-parameter systems shows that the strain-hardening properties can lead to different effects depending on which unsafe situation, out of two basic ones, is apt to occur under repeated loading. It follows that the account of strain-hardening under monotonically increasing load has no influence on the formulation of the condition for alternating plasticity, provided the length of the elastic deformation range under unloading and reversed loading remains unchanged. Suitable tests have indicated that under moderate deformations ($\epsilon/\epsilon_s \leq 10$ where $\epsilon_s = \sigma_s/E$) and isothermal loading the cyclic yield stress for the majority of structural metallic alloys can be taken as roughly twice the elastic limit corresponding to the initial semi-cycle (first loading). This is just what is termed the ideal Bauschinger effect and it appears to be in agreement with the Masing principle based on his concept of a simple model of polycrystalline material [1.11].

From this it follows that in the case of the strain-hardening material with stable diagram of cyclic deformation the condition for alternating plasticity remains the same as in the case of the ideal elastic-plastic body. It is worth mentioning that when the ratio of elastic limits under cyclic and single loading is different from two, a suitable correction must be made in the condition, provided the necessary experimental data are gathered.

The limiting strain amplitude (for alternating plasticity condition) has to be taken as the half of the cyclic yield stress determined for the corresponding material related to a given width of a hysteresis loop. The latter is related to the required durability of a structure and depends on the Coffin–Manson law [1.6, 1.26].

The plastic strain amplitude during the cycle depends on the tangent modulus. As the latter increases, the strain amplitude decreases. It systematically shrinks as the cyclic (isotropic) strain-hardening develops. Theoretically, the possibility of shakedown is admitted as a certain asymptotic property of the situation under arbitrary stress amplitude corresponding to the ideal elasticity, unless this is prevented by earlier failure due to plastic fatigue. This possibility can also be limited by the parameters of the stabilized diagram, in the case when strain-hardening decreases rapidly and practically ceases to exist after a certain number of initial cycles.

Thus the structures made of strain-hardening material possess some additional capabilities of shaking down. The assessments of the allowable values of load factors which are usually made, disregarding the 'physical' shakedown and taking into account merely the cyclic strain-hardening, i.e. the diagram of initial deformability, are found to be too conservative. On the other hand, when allowance for cyclic strain-hardening in the shakedown problems is made, then it becomes also necessary to investigate the durability (damage accumulation as a result of the actual plastic strain amplitudes and their variations with the number of cycles) because the notions of shakedown and strength cease to be adequate for such an approach.

The considered cases can be analysed both qualitatively and (given the necessary test data) numerically with the help of the construction already used in Fig. 1.17a, b; the boundaries of the elastic deformation domain for each element subjected to alternating plasticity should now be assumed movable.

1.4. Influence of strain hardening

Deformation accompanying incremental collapse does not change sign, while it increases with each cycle. Strain-hardening of the material, resulting in the cycle-by-cycle growth of the elastic limit, exerts here a marked influence on the process. A gradual decreasing of plastic deformation caused by strain-hardening is illustrated in Fig. 1.17c in which strain-hardening is of a structural character, being connected with the redistribution of internal forces in the system. The effect of physical strain-hardening would be wholly analogous. Hypothetically assumed uncontained strain-hardening would lead to a situation in which the incremental collapse does not develop, when understood as a process of continuing plastic deformation. However, the accumulation of displacements would then turn out unacceptable from the point of view of the serviceability of the structure in question, and strains might exceed those sufficient to cause failure under monotonical deformation.

Thus the structural strain-hardening appears to enhance the shakedown capabilities of the system similarly as was the case with the material strain-hardening. However, the applicability of this fact is rather limited for the reasons shown above: The shakedown calculations should clearly be accompanied by the analysis of strain and displacement states that develop in the shakedown process and checked against the respective allowable magnitudes. This last problem is outside the framework of the limit analysis of structures.

To illustrate the above exposition let us show an example of a structure shaking down in the presence of strain-hardening. We shall assume that the elements of the one-parameter structure shown in Fig. 1.2 are made of a material possessing linear strain-hardening properties. Thus, on account of Fig. 1.19, the elastic limit depends on the current

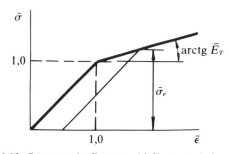

Fig. 1.19. Stress–strain diagram with linear strain-hardening.

1 Analysis of simple elastic-plastic systems

plastic deformation according to the formula

$$\bar{\sigma}_e = 1 + \bar{\epsilon}_p \frac{\bar{E}_T}{1 - \bar{E}_T}, \tag{1.38}$$

where $\bar{\sigma}_e = \sigma_e/\sigma_s$, $\bar{\epsilon}_p = \epsilon_p/\epsilon_s$.

The limiting condition for the incremental collapse of the considered system has, according to (1.16), the form

$$P + N_t^* = \sigma_{e1}F_1 + \sigma_{e2}F_2, \tag{1.39}$$

where N_t^* is the maximum thermal force in the system.

With the help of (1.38) we can express the elastic limits in the last equation in terms of the plastic deformations ϵ_{p1}, ϵ_{p2} accumulated in the respective elements. In order to solve the problem thus stated, we have to establish the relationship between the latter magnitudes. This is given by the following equations:

$$N_{1p} + N^0 = \sigma_{e1}F_1, \tag{1.40}$$

$$N^0 = \frac{1}{\delta_1 + \delta_2}(\epsilon_{p2}l_2 - \epsilon_{p1}l_1), \tag{1.41}$$

where N_{1p} is the 'elastic' action in element 1 under load P, N^0 is the residual (final) internal force in the system and $\delta_i = 1/c_i$ are the elastic compliances of the elements.

Equation (1.40) describes the fact that in the limiting cycle the stress in element 1 over cooling is equal to the elastic limit. In element 2 it is at the end of heating that the stress reaches its elastic limit. Equation (1.41) supplies a relationship between the accumulated plastic deformations and the residual action in the system after a number of cycles preceding the shakedown.

The solution of the system of equations (1.38)–(1.41) is

$$\epsilon_{p1}l_1 = \frac{\delta_1(1 - \bar{E}_{T1})}{(1 + \alpha)\bar{E}_{T1}}[P + (N_t - Y_2)(1 - \bar{E}_{T2}) + Y_1(1 + \alpha\bar{E}_{T1}),$$

$$\epsilon_{p2}l_2 = \frac{\delta_2(1 - \bar{E}_{T2})}{(1 + \alpha)\bar{E}_{T2}}[\alpha P + (N_t - Y_2)(\alpha + \bar{E}_{T2}) - \alpha Y_2(1 - \bar{E}_{T1}), \tag{1.42}$$

1.4. Influence of strain hardening

where

$$\alpha = \frac{\delta_1 \bar{E}_{T2}}{\delta_2 \bar{E}_{T1}}, \quad Y_1 = \sigma_{e1} F_1, \quad Y_2 = \sigma_{e2} F_2.$$

Next, one can compute the residual stresses together with the actual strains (including elastic ones) accompanying shakedown.

As it was pointed out before, the incremental collapse analysis accounting for strain-hardening can be employed to assess strength in two ways: Either comparing the plastic deformations prior to shakedown with the strength reserves of the material or comparing the accumulated displacements with their acceptable values following from the serviceability requirements. This is the manner in which the problem of life expectancy of fuel elements in nuclear reactors should be approached ([1.12], see also Chapter 8).

From the above considerations it follows that the peculiarities of the stress-strain diagram of the material exert a definite quantitative influence on the ranges of the structural behaviour under repeated loading (alternating plasticity, incremental collapse, shakedown).

The applications of idealized models of the material are fully justified though the obtained results constitute of necessity only approximate evaluations. This is the case with the lower bound solutions to the actual load-carrying capacity if strain-hardening is disregarded. On the other hand, the two unsafe states, namely alternating plasticity and incremental collapse, are distinguishable not only by the conditions causing them to occur and the possible consequences but also by the effects of various factors neglected in the basic assumptions. Thus the strength of structures should be ensured by means of different safety margins corresponding to either state. Hence the notion of a unique shakedown safety factor can no longer be insisted upon, at least not in applications to engineering.

The first attempt to account for material strain-hardening, and, in consequence, to abandon the concept of perfectly plastic body is due to Neal [1.13]. Shakedown of structures made of cyclic strain-hardening materials was investigated by Moskvitin [1.14]. Among other papers devoted to investigations of strain-hardening influences on the shakedown conditions [1.15] is especially worth mentioning.

1 Analysis of simple elastic-plastic systems

1.5. Peculiarities of structural behaviour under thermal actions

As we have already emphasized, the strain accumulation under cycling thermal stresses appears to be a characteristic effect, and so is the alternating plasticity leading to thermal fatigue. Variation of the physical-mechanical properties of materials during their heating are found to influence substantially the conditions for the occurrence of unsafe states as well as the intensity of cyclic plastic deformation. If the actual temperatures are high enough ($T_{max} > (0.2-0.3)T_m$, where T is the absolute temperature in ° Kelvin and T_m is the melting point) the changes in the stress–strain diagram begin to play an important role, especially in the variation of the yield point as well as in other plastic characteristics. The elastic characteristics are found to be less sensitive and still higher temperatures are necessary to change them appreciably. As the temperature rises, the loading rate becomes more and more important and so does the duration of loading. Resistance of the material against deformation and failure must both be considered as time-dependent phenomena.

The temperature-dependence of the yield point has already been mentioned; an account of it in the shakedown problems is of obvious importance and no difficulties arise here, in spite of the allowance for the variation of elastic moduli. The latter problem requires the determination of stresses generated by the external actions and supported by a body whose elastic parameters are both time and point-dependent. The residual stress state does also depend on temperature, even in the presence of constant plastic strains.

The influence of temperature-dependence of elastic moduli on the occurrence of unsafe states under repeated loading was investigated by König [1.16, 1.17] in connection with the fundamental shakedown theorems. This problem will be touched upon again in Chapter 9.

In comparison, creep can influence the process to a greater extent. The most thoroughly investigated is the problem of creep strain accumulation under constant load and moderate stress. A vast literature is devoted to this subject alone, especially the excellent monograph [1.18]. Creep under non-stationary conditions of both loading and heating is not so well understood, including the interaction between creep strain and short-term plastic strain in structures subjected to cyclic loading.

Effects of creep on the shakedown conditions under repeated loading will first be studied by way of a simple system, employing its

1.5. Peculiarities of structural behaviour under thermal actions

graphical representation as shown in Sec. 1.1. Several possibilities will be shown to exist.

The segments AB, CD, EF shown in Fig. 1.20 correspond to three steady cycles of the type $p = $ const., $t = $ var., e.g. periodical heating and cooling of one element, for instance the first one, under constant values of the force P. The cycles would have been the limiting ones had the time-dependent strain state been absent. The points B, D, F correspond to the termination of heating. If the cycle includes hold-time under peak temperature, the creep strain will develop in the element 1. As a result of the cycle ABA, the compressive stress in the heated element is becoming smaller and the actions point is accordingly shifted along BA in the south-east direction by a segment the length of which depends on the duration of the hold-time. In consequence, during cooling the actions point can remain within the admissible domain only at the cost of a reverse short-term plastic strain in the same element (segment AA').

Thus, in the considered cycle ABA, including annealing, the relaxation of a favourable distribution of self-equilibrating internal actions, which would have taken place in the system in the absence of time dependent strain, results in the long-term and the short-term inelastic strains of different signs. These are generated consecutively in one of the elements. Alternating plasticity of such a type is definitely unsafe since low-cycle failure can occur due to thermal fatigue. Interaction of creep and plasticity processes can induce, in some conditions, damage accumulation and failure.

In the limiting cycle CD the 'sliding' of the point D during

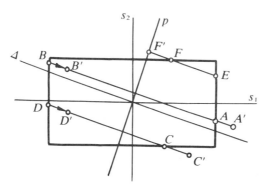

Fig. 1.20. Some possible situations related to the creep effects.

1 Analysis of simple elastic-plastic systems

hold-time will be similar. However, the decrease in the compressive action in the heated element will be here accompanied by an increase in the also compressive action in the element 2. Obviously at the end of cooling the respective boundary of the domain will be reached and element 2 will receive its share of plastic strain shown by the segment $C'C$. The effects of consecutive cycles prove to be the same, provided neither the gradual stabilization of creep rate in element 1 nor the strain-hardening of element 2 are taken into account.

The situation considered above shows that there exists a possibility of occurrence of 'mixed' incremental collapse mechanisms: The compatible strain state generated over every cycle is in certain elements the short-term state (corresponding to the yield stress) while in other elements it is the creep state due to long-term loading. Unlike in the case considered before, in which plastic and creep processes interacted only locally as a result of definite properties of the material, an interaction between plasticity and creep is here encountered on the structural level. Under specific circumstances both the cyclic alternating strains and the accumulating incremental strains can occur simultaneously.

In the last of the considered cycles EFE a peculiar limiting state can be distinguished under hold-time at the end of the heating (point F): The creep of element 1 should cause here a gradual decrease in its tensile action accompanied by the translation of point F towards point F'. However, such a movement is clearly impossible; such a strain will be 'compensated' without delay by plastic yielding in element 2 and therefore the actions point F will continue to occupy its original position. Strains will accumulate with time (as they would in the case of creep spreading), at suitable temperature of heating. The accumulation takes place on both elements and, if unloading (cooling) is not accompanied by plastic strain, the cyclic character of loading is of no significance to the process.

Thus it is possible that creep under repeated loading (cycles AB and CD) will prevent a structure to shake down. This can be recognized, on one hand, by the relaxation of the favourable distribution of self-equilibrating stresses and, on the other hand, by the appearance of cycles of inelastic strains consisting both of rheonomic, i.e. time-dependent and scleronomic, i.e. time-independent components. Those two phenomena are interconnected during the process.

However, each structural alloy, suitably used in specified conditions, possesses a definite stress range within which creep is practically insignificant under given temperature and duration of loading, i.e.

1.5. Peculiarities of structural behaviour under thermal actions

strains are small enough to be ignored so far as a considered structure is concerned. Under such circumstances the influence of creep under repeated semi-cycles can be reduced to an appropriate restriction of the shakedown domain. The most simple approximative approach to the determining of this creep-dependent domain consists in replacing the yield stress by the creep stress obtained by means of the isochronous creep curves approximated according to the ideal elastic-plastic diagram of the body, thus on allowing for inelastic strains. The preceding analysis of the incremental collapse conditions is entirely analogous to that made in order to determine the yield stress as a temperature-dependent magnitude. A suitable procedure will be shown in Chapter 6 by applying the fundamental shakedown theorems to the problem of a rotating flat disc and in Chapter 7 to the problem of the fuel element shell.

The range of purely elastic behaviour of structures exposed to elevated temperatures can frequently appear to be extremely narrow (the narrower, the higher the working temperatures). Therefore the shakedown condition adopted as a strength criterion might lead to unrealistic designs, at least in the case of ordinary commercial materials. Thus the safety margin for the performance of a structure outside the shakedown domain should be frequently taken reasonably low. There is emerging now a separate problem of determining the parameters of service life, i.e. ranges and increments of inelastic deformations. Some discoveries of the shakedown theory are found to remain valid here; their further development will be shown in Chapter 11 where the steady stress cycles will be discussed.

A somewhat more adequate analysis of the structural behaviour under transient and stationary regimes should be based on a model of continuum in which both the short-term and the long-term deformation processes can be incorporated together with other effects related to the loading and the temperature histories [1.6, 1.7, 1.19–1.23 and others].

It has been also both experimentally and theoretically established that strain accumulation can take place under cyclic loading [1.14, 1.26–1.28]. A great deal of data on the strain accumulation in some simple specimens subject to thermal cycling is presented in the monograph [1.29], in which there are also put forward some hypotheses to explain the internal mechanisms of the investigated effects.

Let us finally mention that an increasing attention paid to the deformation processes generated by variable repeated loading and thermal cycling has stimulated a number of experimental investigations to be initiated not only on specimens, as discussed before, but also on

1 Analysis of simple elastic-plastic systems

some simpler structures. The papers [1.30–1.33] should be mentioned in this connection, where the reported results are, to a greater or lesser degree, related to the shakedown theory.

References

[1.1] Hencky, H., Theorie plastischen Verformungen und der hier durch im Material hervorgerufenen Nachspannungen. *ZAMM, Bd.* 4, H. 4, S. 1924, 323–334.
[1.2] Rzhanitzin, A.R., *Structural analysis taking account of plastic properties of materials*, in Russian, Gosstroiizdat, Moscow, 1954.
[1.3] Prager, W., *Probleme der Plastizitätstheorie*. Birknäuser, Basel und Stuttgart, 1955.
[1.4] Prager, W., Problem types in the theory of perfectly plastic materials. *J. Aero Sci.*, 15, 1948.
[1.5] Hodge, Pn.G., Jr., Shakedown of elasto-plastic structures, in *Residual stresses in metal construction*, edited by W.R. Osgood, Reinhold, New York, 1954.
[1.6] Manson, S.S., *Thermal stress and low-cycle fatigue*. McGraw-Hill, New York, 1966.
[1.7] Serensen, S.V., Shneiderovitch, R.M., Kogaev, V.P., *Carrying capacity and strength of machine parts*, in Russian, Mashinostroenje, 1975.
[1.8] Prager, W., Shakedown in elastic-plastic media subjected to cycles of load and temperature. *Symp. su la plasticita nella scienza delle costrutioni in onore de Arturo Danusso, Varenna, Set., 1956*, Bologna, 1957, 239–244.
[1.9] Horne, M., Fundamental proportions in plastic theory of structures. *J. Inst. Civ. Engrs.* 34, 1950, 174–177.
[1.10] Neal, B.G., *The plastic methods of structural analysis*. Chapman and Hall, London, 1956.
[1.11] Masing, G., Wissenschaftliche Veröffenlichungen aus dem Siemens-Konzern, 5, 135, 1926.
[1.12] Bree, J., Elastic-plastic behavior of thin tubes subjected to internal pressure and intermittent high-heat fluxes with application to fast-nuclear reactor fuel elements. *J. Strain. Anal.*, 2, Nr 3, 1967.
[1.13] Neal, B.G., Plastic collapse and shakedown theorems for structures of strain-hardening material. *J. Aero Sci.*, 17, 1950, 297.
[1.14] Moskvitin, V.V., *Plasticity under variable loads*, in Russian, Moscow State University Press, Moscow, 1965.
[1.15] Maier, G., A shakedown matrix theory allowing for work-hardening and second-order geometric effects. *Proc. Int. Symp. on Foundations of Plasticity. Warsaw, 1972.* Noordhoff, Leiden, 1973.
[1.16] König, J.A., A shakedown theorem for temperature dependent elastic moduli. *Bull. Acad. Polon. Sci. Ser. Sci. Techn.*, 17, Nr 3, 1969.
[1.17] König, J.A., Shakedown design of structures with temperature-dependent elastic moduli, in Polish, *Rozpr. Inż.*, 20, 1972, 423–434.
[1.18] Rabotnov, Yu.N., *Creep problems in structural members*, in Russian, Nauka, 1966. (Also in English, North Holland, Amsterdam, 1969).
[1.19] Gokhfeld, D.A., Ivanov, I.A., Rebyakov, Yu.N., On interaction of plastic and viscous deformations under cyclic loading, in *Thermal strength of materials and structural members*, in Russian, Nr 5, Naukova Dumka, Kiev, 1969.

References

[1.20] Gokhfeld, D.A., Kononov, K.M., Sadakov, O.S., Behaviour features of metals under cyclic deformation at high temperatures and the possibilities of their mathematical interpretation. *Proc. 4th Conf. on Dimensioning, Budapest, 1971.* Akademiai Kiado, Budapest, 1973.

[1.21] Campbell, R.D., Creep fatigue interaction for 304 stainless steel subjected to strain controlled cycling with hold times at peak strain. *Trans. ASME*, sec. B, Nr 4, 1971.

[1.22] Krempl, E., The interaction of rate and history-dependent effects and its significance for slow cyclic inelastic analysis at elevated temperature. *Nucl. Eng. and Design*, 29, Nr 1, 1974, 125–134.

[1.23] Gokhfeld, D.A., Sadakov, O.S., A mathematical model of medium for analysing the inelastic deforming of structures subjected to repeated actions of load and temperature. *Trans. of the 3rd SMIRT Conf.*, London, 1975.

[1.24] Arutyunyan, R.A., Vakulenko, A.A., On multiple loading of elastic-plastic bodies, *Izv. AN SSSR, Mekhanika*, Nr 4, 1965.

[1.25] Martinenko, M.E., On one-sign plastic strain accumulation under nonsymmetric mildly cyclic loading, *Materials of the All-Union Symposium on low-cycle fatigue at elevated temperature*, Nr 4, Cheliabinsk, 1974.

[1.26] Coffin, L.F., Jr., The stability of metals under cyclic plastic strain. *Trans. ASME*, ser. D, Vol. 82, 1960.

[1.27] Schwiebert, P.D., Moyar, G.J., An application of linear hardening plasticity theory to cycle and path dependent deformation. Univ. of Illinois T.A.M., Report Nr 212, Jan, 1962.

[1.28] Gokhfeld, D.A., Ivanov, I.A., Sadakov, O.S., Description of complex loading effects as based on the structural model of continuum, in *Problems of deformable body mechanics*, in Russian, AN SSSR, Nauka, Moscow, 1975.

[1.29] Davidenkov, N.N., Likhachev, V.A., Irreversible deformation of metals under thermal cycling, in Russian, Mashgiz, Moscow–Leningrad, 1962.

[1.30] Gatewood, B.E., Grothouse, A.P., v. Hausen, W.W., Experimental data on strain accumulation under equivalent thermal cycling. *J. Aero Sci.*, 28, Nr 6, 1961.

[1.31] Gokhfeld, D.A., Some test results concerning shakedown under thermal actions, in *Thermal stresses in turbines*, Nr 2, AN SSSR, Kiev, 1962.

[1.32] Gokhfeld, D.A., Kononov, K.M., Study on strain increase at cyclic fluctuations of temperature, in *Problems of high-thermal strength in machinery*, AN SSSR, Kiev, 1963.

[1.33] Stentz, R.N., A study of thermal ratchetting using closed-loop, servo-controlled test machines. *Cyclic stress–strain behaviour: analysis, experimentation and failure prediction. Symposium ASTM, Bal Harbour, Fla., Dec. 1971.* ASMT, May, 1973.

2

Fundamental theorems of the shakedown theory

The following exposition of the shakedown theory is based on the fundamental theorems formulated for the case of a three-dimensional elastic-perfectly plastic body. The theorems enable us to determine the conditions for shakedown together with the conditions for the occurrence of limit states such as alternating plasticity or incremental collapse, without the need to carry out the calculations of the development of deformation in the process of stabilization of a stress cycle.

Before the theorems are considered and the ensuing computational procedures are described it is worthwhile to discuss briefly some useful notions of ideal plasticity and to fix the notation. To get a deeper insight into the theory of ideal elastic-plastic continuum the reader is referred to well-known monographs [2.1–2.4].

2.1. Prerequisites from the mechanics of elastic-perfectly plastic solids

The stress state at a generic point of a body is conveniently described by a tensor whose components are stresses acting in three mutually orthogonal planes. In continuum mechanics and, in particular, in the theory of plasticity it is commonly agreed upon to refer the tensors to a Cartesian coordinate system. Latin indices will consistently be employed with the three numbers 1, 2, 3 related to the coordinate axes x_1, x_2, x_3. Thus σ_{ij} denotes a stress tensor component and $i = j$ indicates a normal stress whereas $i \neq j$ means a tangential (shearing) stress. The condition of equilibrium requires that $\sigma_{ij} = \sigma_{ji}$ which means that the stress tensor is symmetric.

To shorten the introduced notation we shall omit the summation sign (the summation convention as devised by Einstein). This simply

2 Fundamental theorems of the shakedown theory

means that an index appearing twice in any term implies summation over 1, 2, 3.

For instance, the well-known stress boundary conditions

$$p_x = \sigma_x l + \tau_{yx} m + \tau_{zx} n$$

(one out of the three equations is only shown) can be written down in an abbreviated form as

$$p_i = \sigma_{ij} n_j,$$

where n_j are the directional cosines of a normal and p_i are the components of surface tractions.

The Latin index j repeated in the product is usually called the dummy index (or the summation index). It should not appear in any term more than twice. The index i which changes from equation to equation is termed the free index.

It will be convenient to introduce a unit tensor δ_{ij} called the Kronecker delta which is defined by

$$\delta_{ij} = \begin{cases} 1 & \text{if } i = j, \\ 0 & \text{if } i \neq j. \end{cases}$$

In this notation the deviatoric components of a stress tensor can be written down in the form

$$s_{ij} = \sigma_{ij} - \sigma \delta_{ij},$$

where $\sigma = \sigma_{kk}/3$ is the mean normal stress.

The time derivative is shortly denoted by a dot over the relevant symbol, hence $\dot{\sigma}_{ij}$ stands for the stress rate at small strains and rotations. The space derivative is denoted by a comma preceding an index, for instance $f_{,i} = \partial f/\partial x_i$. In particular, the three partial differential equations of equilibrium, the first of which has a long-hand form

$$\frac{\partial \sigma_x}{\partial x} + \frac{\partial \tau_{yx}}{\partial y} + \frac{\partial \tau_{zx}}{\partial z} + X = 0,$$

can be now written as

$$\sigma_{ij,j} + X_i - 0,$$

2.1. Prerequisites from elastic-perfectly plastic solids

where X_i are the projections of the body forces. Differentiation with respect to an index which appears twice in a term automatically implies summation of derivatives.

The strain tensor components have at small displacements the form

$$\epsilon_{ij} = \frac{1}{2}(u_{i,j} + u_{j,i}), \tag{2.1}$$

where u_i are the displacement components.

Similar relationships hold true for the displacement rates $\dot{u}_i$ and the strain rates $\dot{\epsilon}_{ij}$.

In the theory of elastic-plastic solids considered here the total strain tensor ϵ_{ij} is assumed to be composed of its elastic part ϵ'_{ij} and its plastic part ϵ''_{ij}.

The elastic part is related to the stress tensor by means of Hooke's law

$$\epsilon'_{ij} = A_{ijhk}\sigma_{hk}, \tag{2.2}$$

where A_{ijhk} is the fourth order tensor of elastic moduli, symmetric with respect to each of the indices i, j, h, k as well as with respect to their pairs, hence

$$A_{ijhk} = A_{jihk} = A_{ijkh} = A_{hkij}.$$

Plastic deformation can develop when the stress components satisfy a yield condition

$$\phi(\sigma_{ij}) = k^2, \tag{2.3}$$

where k is a reference yield stress.

To visualize the yield condition it is usual to employ its graphical representation. The yield condition clearly represents a certain surface in the nine-dimensional stress space. In this space the stress tensor can be treated as a vector whose origin is at the origin of the coordinate system and whose head is at the stress point with coordinates σ_{ij}.

This vectorial representation of the tensor, convenient as it is, has a limited significance since one and the same tensor referred to another coordinate system is represented by a different vector, although of the same length. This length, being the only invariant of the vector, cor-

2 Fundamental theorems of the shakedown theory

responds to the second invariant of the tensor. However, to facilitate some simpler tensorial operations encountered in the theory of plasticity, we can employ similarities with vectorial operations. In particular, this is true for the multiplication of a tensor by a scalar, addition of tensors, contraction of two tensors (corresponding to the scalar product of vectors in the nine-dimensional space).

The term 'yield surface' means that the concept of the yield stress under simple tension is suitably generalized to cover an arbitrary spatial stress situation. So far as the elastic-perfectly plastic material shown in Fig. 1.1 is concerned, the yield surface remains unaltered over the deformation process. The end of the stress vector can either lie inside the yield surface (elastic domain) and then the plastic strain rates are equal to zero or can touch the yield surface thus generating non-vanishing plastic strain rates. The situation in which the the stress vector would 'pierce' the yield surface to lie partly outside is impossible in the case of ideal plasticity.

The stress state corresponding to a stress point lying inside the yield surface is usually termed a safe stress state and denoted by $\sigma_{ij}^{(s)}$, whereas the stress states corresponding to positions of the stress point including those on the yield surface are termed admissible stress states $\sigma_{ij}^{(a)}$.

The equation of a yield surface may be written down in the form

$$f(\sigma_{ij}) = 0 \tag{2.4}$$

instead of (2.3). Hence one has

$$f(\sigma_{ij}^{(s)}) < 0, \quad f(\sigma_{ij}^{(a)}) \leq 0. \tag{2.5}$$

The distinction between the safe and the admissible stress states will turn out to be important for the proofs of the theorems.

The yield surface is always convex. In the three-dimensional Euclidean space an arbitrary plane intersection of a convex surface gives a convex curve, i.e. a curve that can be intersected by a straight line in two points only. The definition of a convex surface in the nine-dimensional space is analogous.

The above assumption of convexity is of primary importance in the theory of plasticity. It was introduced in a number of ways. The present-day approach is based on the Drucker postulate of stability, sometimes termed the quasi-thermodynamic postulate. It can be formulated as follows [2.2, 2.5].

2.1. Prerequisites from elastic-perfectly plastic solids

Let an elastic-plastic element under a certain initial stress state be subject to a slow loading and unloading process[1] by means of external actions other than those generating the initial state. The postulate asserts that the work of additional stresses on a closed path of stress is non-negative, or that no energy can be recovered from the initially stressed element. The postulate is given the form

$$[\sigma_{ij} - \sigma_{ij}^{(s)}]\dot{\epsilon}_{ij}'' > 0, \tag{2.6}$$

$$[\sigma_{ij} - \sigma_{ij}^{(a)}]\dot{\epsilon}_{ij}'' \geq 0, \tag{2.7}$$

where σ_{ij} is the yield surface stress state generating the plastic strain rates $\dot{\epsilon}_{ij}''$.

The convexity of the yield surface follows from the weak inequality (2.7). We must distinguish now between regular (smooth) and singular (with ridges or corners) yield surfaces. For regular yield surfaces (or regular portions of them) the above formulated postulate leads to the following statement: If the plastic strain rates are represented in the nine-dimensional stress space along the respective axes, then the plastic strain rate tensor (as a vector in the above space) is directed along the outer normal to the yield surface. The strain rate vector originates at the relevant point of the yield locus. If the point turns out to be singular, lying at the intersection of neighbouring smooth (regular) surfaces, the direction of the plastic strain rate vector is contained between the two normals corresponding to the two meeting surfaces.

This statement is commonly called the associated flow rule since it relates the plastic regime at a point in the body (ratios between the plastic strain rates) with the location of the stress point on the yield surface and with the equation of the surface (2.3) itself. The flow rule associated with the yield condition determines to within a certain common factor λ_α the plastic strain rates:

$$\epsilon_{ij}'' = \sum_\alpha \lambda_\alpha \frac{\partial f_\alpha}{\partial \sigma_{ij}}, \tag{2.8}$$

where $\lambda_\alpha = 0$ if $f_\alpha < 0$ (elastic state) or $f_\alpha = 0$ and $\dot{f}_\alpha = (\partial f_\alpha/\partial \sigma_{ij})\dot{\sigma}_{ij} < 0$ (unloading from the plastic state), and $\lambda_\alpha > 0$ if $f_\alpha = 0$ and $\dot{f}_\alpha = 0$ (plastic loading); f_α, $\alpha = 1, 2, \ldots$ being the yield functions describing a singular yield surface (in the regular case we have $\alpha = 1$).

---

[1] In other words, a quasi-statical isothermal loading process is here assumed.

2 Fundamental theorems of the shakedown theory

It is important to realize that no conclusions can be drawn as to the total plastic strains before the whole deformation history of the considered element is known.

The most widely used and experimentally corroborated yield conditions are the following two:

$$(\sigma_1 - \sigma_2)^2 + (\sigma_2 - \sigma_3)^2 + (\sigma_3 - \sigma_1)^2 = 2\sigma_s^2, \qquad (2.9)$$

$$\sigma_1 - \sigma_3 = \sigma_s, \qquad (2.10)$$

where $\sigma_1 \geq \sigma_2 \geq \sigma_3$ are the principal normal stresses and σ_s is the yield point. The first, called the Huber–Mises[1] yield condition is based on the assumption that yielding at a certain point commences as soon as the octahedral tangential stress attains a specified limiting value. In the shorthand notation we have

$$s_{ij}s_{ij} = 2k^2, \quad (k = \sigma_s/\sqrt{3}). \qquad (2.11)$$

In the second, the Tresca–Saint Venant yield condition, yielding is assumed to take place immediately after the maximum principal tangential stress attains a definite value. Assuming that $\sigma_1, \sigma_2, \sigma_3$ are the principal normal stresses with respect to the fixed principal axes x_1, x_2, x_3, we arrive at another form of the second yield condition (2.10)

$$\max\{|\sigma_1 - \sigma_2|, |\sigma_2 - \sigma_3|, |\sigma_3 - \sigma_1|\} = 2k, \quad (k = \sigma_s/2). \qquad (2.12)$$

Here the singular character of condition (2.10) can best be seen in contrast to the regularity of condition (2.9).

Experimental evidence shows that for numerous materials the Huber–Mises yield condition is in better agreement with the test data than the Tresca yield condition. However, the second yield condition seems to be more fitting when the yield stress corresponds to a distinct yield plateau characteristic for perfectly plastic materials. In general, the discrepancies between the predictions of the two yield criteria are rather small and do not exceed 16 per cent, and on suitable specification of the yield stress, are confined to within ±8 per cent of that stress. Thus the selection of a yield condition to be employed follows mainly from reasons of convenience, concerning its application to the solution of

---

[1] This yield condition was introduced by M.T. Huber in 1904 and then, independently, by Mises (translator's remark).

2.1. Prerequisites from elastic-perfectly plastic solids

problems of a given type. In the applied theory of ideal plasticity it is preferable to use the Tresca yield condition since working with piecewise linear functions is often (but not always!) simpler than with nonlinear ones.

During plastic deformation a dissipation of energy is found to take place in the form of heat. The energy dissipation rate per unit volume (the dissipation power) can be shown to be

$$F(\dot{\epsilon}_{ij}'') = \sigma_{ij}\dot{\epsilon}_{ij}'' > 0, \quad (\dot{\epsilon}_{ij}'' \neq 0), \tag{2.13}$$

where σ_{ij} is the yield surface stress state associated by the flow rule with the plastic strain rates $\dot{\epsilon}_{ij}''$.

By using the yield condition (2.10) together with the associated flow rule which specifies the relations between the plastic strain rate components, the plastic energy dissipation rate can be expressed as

$$F(\dot{\epsilon}_{ij}'') = \sigma_s |\dot{\epsilon}_{ij}''|_{\max}, \tag{2.14}$$

where $|\dot{\epsilon}_{ij}''|_{\max}$ is the absolute value of the maximum component of the state of plastic strain rate at a considered point of the body.

Since the stress state in a perfectly plastic medium is bounded by an appropriate yield condition, the external loads supported by a body in agreement with equilibrium requirements must also be bounded by certain limiting values. Application of the limit load brings about the so-called plastic (or instantaneous or short-term) collapse which manifests itself by unlimited increase in deformation under steady load. It should be borne in mind that the above description concerns an incipient stage of collapse rather than an advanced one since no geometry changes accompanying the advancement of plastic deformation are accounted for in the equilibrium conditions. Such a concept of plastic collapse, intrinsic as it is in the limit analysis, remains unaltered in the shakedown theory.

A distribution of deformation rates (or increments) obeying the compatibility equations (2.1) and the kinematic boundary conditions is termed a kinematically admissible distribution. When the distribution of plastic deformation rates (or increments) is studied at collapse, we are dealing with the so-called kinematically admissible collapse mechanism, conveniently abbreviated to collapse mechanism.

In order to formulate the fundamental theorems of the limit analysis and the shakedown theories, the principle of virtual work is extensively

2 Fundamental theorems of the shakedown theory

employed in the form

$$\int X_i u_i \, dv + \int p_i u_i \, ds = \int \sigma_{ij} \epsilon_{ij} \, dv. \qquad (2.15)$$

This equation is a generalization of the equilibrium conditions and it remains valid for an arbitrary system of external body forces X_i and surface tractions p_i supported by the stress state σ_{ij} and for an arbitrary displacement field u_i compatible with a kinematically admissible strain distribution ϵ_{ij}.

From now on, if not stated otherwise, the volume integrals will be taken over the whole volume of the body and the surface integrals over its entire boundary.

Equation (2.15) continues to hold good when the displacement field and the corresponding strain distribution are both replaced by the displacement rates $\dot{u}_i$ and the strain rates $\dot{\epsilon}_{ij}$, respectively.

As follows from (2.15), the work done by a self-equilibrating stress system on a kinematically admissible strain (or strain rate) field is nil.

2.2. Statical shakedown theorem. Generalization to thermal effects. Further development of the statical theorem

The statical approach to the analysis of shakedown conditions in the presence of repeated loading is based on the first, or statical, theorem formulated for a three-dimensional situation by Melan in 1938 [2.2, 2.6]. The theorem is composed of the following two statements.

1. The structure will shake down to the variable repeated loading, i.e. its behaviour after a number of initial loading cycles will become purely elastic, if there exists a time-independent distribution of residual stresses $\bar{\rho}_{ij}$ such that its superposition with 'elastic' stresses $\sigma_{ij}^{(e)}$ results in a safe stress state $\sigma_{ij}^{(s)}$ at any point of the structure,

$$\sigma_{ij}^{(e)} + \bar{\rho}_{ij} = \sigma_{ij}^{(s)}, \qquad (2.16)$$

under any combination of loads inside prescribed limits. An essential part of this statement is the assertion that, if shakedown is at all possible under a given repeated loading, then it will certainly take place. The shakedown will be caused by plastic deformations during the first cycles

2.2. Statical shakedown theorem

which are followed by a certain steady, independent of a further loading program, distribution of residual stresses.

2. On the other hand, shakedown never takes place unless a time-independent distribution of residual stresses can be found such that under all the possible load combinations the sum of the residual and 'elastic' stresses proves to be an admissible stress state (i.e. corresponding to a stress point that lies either inside, or on, the yield surface).

The second statement is obviously true whereas the first one requires a formal proof. The following argument is a typical one in the whole theory of shakedown.

Consider a positive, elastic strain energy A corresponding to the self-equilibrating stress state $\rho_{ij} - \bar{\rho}_{ij}$,

$$A = \int \frac{1}{2} A_{ijhk}[\rho_{ij} - \bar{\rho}_{ij}][\rho_{hk} - \bar{\rho}_{hk}] \, dv, \tag{2.17}$$

$$\rho_{ij} = \sigma_{ij} - \sigma_{ij}^{(e)}, \tag{2.18}$$

denoting by ρ_{ij} the current magnitudes of actual residual stresses, σ_{ij} the current stresses in a considered cycle reaching the yield surface at an instant at which the plastic strain rates at appropriate points of the body cease to be equal to zero, $\bar{\rho}_{ij}$ the time independent residual stresses satisfying the shakedown condition (2.16) at a given loading cycle, and by $A_{ijhk} = A_{jihk} = A_{ijkh} = A_{hkij}$ the tensor of elastic moduli.

The time derivative of the energy A is expressed by

$$\dot{A} = \int [\rho_{ij} - \bar{\rho}_{ij}] \dot{\epsilon}'_{ijr} \, dv, \tag{2.19}$$

where $\dot{\epsilon}'_{ijr}(\tau) = A_{ijhk}\dot{\rho}_{ij}$ are the elastic strain rates corresponding to the residual stress rates that result from the plastic deformation.

Since the distribution of the total residual strain rates, including both the elastic $\dot{\epsilon}'_{ijr}$ and the plastic $\dot{\epsilon}''_{ij}$ components,

$$\dot{\epsilon}_{ijr} = \dot{\epsilon}'_{ijr} + \dot{\epsilon}''_{ij}, \tag{2.20}$$

is kinematically admissible (i.e. satisfying the continuity requirements), we can employ the virtual work principle (2.15) in the form

$$\int [\rho_{ij} - \bar{\rho}_{ij}] \dot{\epsilon}_{ijr} \, dv = \int [\rho_{ij} - \bar{\rho}_{ij}][\dot{\epsilon}'_{ijr} + \dot{\epsilon}''_{ij}] \, dv = 0. \tag{2.21}$$

2 Fundamental theorems of the shakedown theory

Now, instead of (2.19), we have

$$\dot{A} = -\int [\rho_{ij} - \bar{\rho}_{ij}]\dot{\epsilon}''_{ij}\,dv. \tag{2.22}$$

Using (2.16) and (2.17), we can write the expression (2.22) in the final form

$$\dot{A} = -\int [\sigma_{ij} - \sigma^{(s)}_{ij}]\dot{\epsilon}''_{ij}\,dv. \tag{2.23}$$

According to condition (2.6) the integrand in (2.23) is positive only when σ_{ij} is the yield surface stress state. Hence it follows that the derivative (2.23) is negative at those instants of time at which, under the actual loading program, non-vanishing plastic strain rates are appearing at any point of the body whereas at other instants of time these rates vanish. Since the elastic strain energy cannot be negative, the plastic yielding under repeated loading cannot go on unlimitedly, i.e. the total deformation of the structure must be bounded. Termination of yielding will mean that the shakedown is about to take place. The first statement of the statical theorem is thus proved.

The above formulated statical theorem on shakedown of structures subjected to variable repeated loading can be used to determine the parameters of the limiting cycle which separates shakedown from unrestrained cyclic plastic deformation. For the parameters one may select either some local factors determining the range of variation of one or more load components (if the remaining components are given) or quantities can be found determining the mechanical properties of the material (the yield point), depending on the specific purpose of the calculations to be made. A detailed discussion of the corresponding statical formulation will be given later.

It is worth noticing that the origin of the actual stresses was totally irrelevant as the kinematically admissible distribution of total residual strains and their rates (2.20) was employed in the proof. These stresses could be generated either by external mechanical loading or by a temperature field or by a simultaneous action of both; actions of other types being also conceivable. Thus, the generalization of the theorem to cover thermal cycling, given by Prager [2.7] requires no separate proof.

In the proof of the theorem no assumptions were made about the regularity of the yield surface. This simply means that in the shakedown

2.2. Statical shakedown theorem

analysis piecewise continuous yield surfaces with singularities can be considered as well. In particular, the Tresca–Saint Venant yield surface (2.10) can be used.

It is also not necessary to postulate the constancy of the elastic and plastic moduli within the interior of the body; the theorem remains valid for nonhomogeneous materials. On the other hand, the time independence of the elastic moduli tensor A_{ijhk} is essential in the proof of the theorem presented above.

In connection with the applicability of Melan's theorem to the determination of the shakedown conditions at thermal cycling there arises the task of investigating the influence of temperature on the physical-mechanical properties of the material. This influence is found to be considerable within the range of working temperatures, especially at the yield point. To neglect it would lead in general to unsafe designs since the yield stress is known to drop as the temperature rises. Thus, from the point of view of practical applications, this problem can hardly be set aside.

It is a well-known fact that the theory of elastic-perfectly plastic continuum is merely an approximation to the behaviour of real materials. Thus it should not be considered surprising that the stress–strain diagrams of metals, including low-carbon steel, are found to lose, at elevated temperatures, their previously horizontal plastic platform. The validity of the concept of an elastic-perfectly plastic body with temperature dependent yield stress can be questioned because of the appearance of time dependent properties or owing to the possibility of the deformation process to become unstable, as pointed out in [2.8]. Some possibilities of accounting for the occurrence of creep under repeated loading have already been discussed in Chapter 1 where the influence of creep has been partially accounted for in suitably approximated isochronous stress–strain diagrams (in the case of incremental collapse) and corresponding cyclic stress–strain diagrams (in the case of alternating plasticity). As for the second reason for having objections against the concept of an ideal material, it must be emphasized that in the present-day literature the problem of non-isothermal plastic deformation is, as a rule, considered to be independent of the laws of thermodynamics (this approach is specifically recommended in [2.8, 2.9]).

In this approach to the shakedown problems the yield surface can be described by some parameters depending on the actual temperature at a point of the body. Moreover the yield surface has to be convex and the normality law of the plastic strain rate vector has to apply [2.9]. In particular, it follows that the second statement (on the conditions for

2 Fundamental theorems of the shakedown theory

impossibility of shakedown) is valid and it applies to the case of temperature dependent yield point.

Bearing in mind the proof of the statical theorem given above, we conclude that the assertion of the theorem holds true if the inequality (2.6) is satisfied at each instant of time and at the respective temperature. This is to say that the yield surface remains convex during its isotropic change caused by the temperature. As already mentioned, the tensor of elastic moduli should not be temperature-dependent since temperature is a function of time.

It is readily seen that the Melan theorem ceases to hold good as soon as the elastic moduli of the material vary during the cycle as a result of temperature fluctuations. The statical theorem was suitably restated by König [2.10, 2.11]. In accordance with this restatement a structure will shake down if there exists a distribution of time-independent plastic strains ϵ_{ij}'' such that the total strain state, including also the variable thermal strain components $\epsilon_{ij\tau}^{(T)}$ together with the elastic strain components, satisfies the compatibility equations (2.1) at each arbitrary instant of time,

$$\epsilon_{ij}'' + \epsilon_{ij\tau}^{(T)} + A_{ijhk}\sigma_{hk\tau} = \frac{1}{2}(u_{i,j} + u_{j,i}). \tag{2.24}$$

The stresses $\sigma_{ij\tau} = \sigma_{ij\tau}(\tau)$ should obey the conditions

$$f(\sigma_{ij\tau}) \leq 0, \tag{2.25}$$

$$\sigma_{ij\tau,j} + X_i(\tau) = 0, \quad \sigma_{ij\tau}n_j = p_i(\tau). \tag{2.25a}$$

Considerable difficulties are arising when one tries to apply to specific problems the theorem as stated in [2.24, 2.25]. Then a somewhat more convenient, alternative formulation can be proposed:

A structure shakes down to a prescribed external load if a time-independent self-equilibrating stress distribution $\bar{\rho}_{ij}$ can be found (at a certain instant of time assumed to be an initial one) such that, at any point of the body and at any instant of time, the following requirements are satisfied:

$$f(\sigma_{ij}^{(e)} + \bar{\rho}_{ij} + \rho_{ij\tau}) < 0, \tag{2.26}$$

$$A_{ijhk}(\bar{\rho}_{hk} + \rho_{hk\tau}) - \bar{A}_{ijhk}\bar{\rho}_{hk} = \frac{1}{2}(u_{i\rho,j} + u_{j\rho,i}), \tag{2.26a}$$

2.2. Statical shakedown theorem

$$\rho_{ij\tau,j} = 0 \quad \text{the interior } V, \tag{2.26b}$$

$$\rho_{ij\tau} n_j = 0 \quad \text{on } S_p. \tag{2.26c}$$

The tensor $\bar{A}_{ijhk}$ is fixed and corresponds to the temperature at the beginning of a cycle; the tensor A_{ijhl} is temperature-dependent; the stresses $\sigma_{ij}^{(e)}$ are calculated with the help of the latter; $\rho_{ij\tau}$ stand for changes in the residual stresses caused by changes in the elastic constants.

This formulation shows that the temperature-dependence of the elastic moduli influences not only the stresses $\sigma_{ij}^{(e)}$ but also results in changes in the self-equilibrating residual stresses in the presence of constant plastic strains.

The influence of cyclic (due to varying elastic modulus) components of the residual stresses on the behaviour of the structure is similar to the influence of variable thermo-elastic stresses.

On the basis of the numerical examples given in [2.11], we may conclude that an influence of variability of elastic moduli over a cycle on the shakedown conditions is relatively weak, at least within the range of working temperatures and for structures made of common materials.

In [2.12–2.14] there was investigated the possibility of extending the shakedown theory and, in particular, the statical theorem to cover the case of repeated dynamical actions. This was carried out under the assumption of an isothermal loading process, i.e. heat generation at multiple cyclic deformations was disregarded.

Ceradini formulated in [2.12] a sufficient condition for shakedown as follows: A structure subject to dynamical action will shake down if, among all the systems of initial conditions (including the self-equilibrating state as well as the distribution of initial displacements and their rates), there exists at least one system leading to an elastic behaviour of the structure at any instant of time during the dynamical process brought about by this system of initial conditions.

A proof of the above statement can be constructed in a way similar to that used by Melan and with the use of more general arguments. This covers a more general situation of the shakedown capacity of structures under repeated agencies than that following from the Melan theorem. However, as it was observed by Gavarini in [2.13], the possibility of shakedown under dynamic periodically variable loading is in fact determined by the forced vibrations, whose amplitude does not depend on the initial conditions of motion. Free vibrations will fade away at the

2 Fundamental theorems of the shakedown theory

first stage due to plastic deformation and/or viscous properties of the system. Thus the dynamic shakedown problem is reduced to an equivalent statical one. Indeed, such an approach is employed in [2.14].

In the majority of available papers the shakedown problem is dealt with in the deterministic formulation. Horne [2.15] was the first to discuss the probability of incremental collapse as related to the probability of repeated occurrence of a definite sequence of loads. In [2.16], based on the premises of similar results obtained by the authors with the use of the limit equilibrium conditions [2.17], some assessments are given of the probability of incremental collapse for bar structures with the yield stress treated as a random quantity.

2.3. Kinematical shakedown theorem

The second (kinematical) shakedown theorem was established in 1956 by Koiter [2.18], who used the analogy between the theorems of limit analysis and those of shakedown. The relevant similarities were not duly appreciated before Koiter's study. He assumed that, like the respective theorem of limit analysis, the second theorem was a powerful means to simplify the shakedown analysis and to facilitate its applications.

The kinematical theorem is based on the fundamental concept of an admissible cycle of plastic strain rates $\dot{\epsilon}''_{ij0}(\tau)$. In accordance with the definition of the admissible strain rate cycle, the plastic strain increment,

$$\Delta \epsilon''_{ij0} = \int_0^T \dot{\epsilon}''_{ij0} \, d\tau \tag{2.27}$$

represents a kinematically admissible strain distribution over a certain time interval. This is derived from

$$\Delta \epsilon''_{ij} = \frac{1}{2}(\Delta u_{i,j} + \Delta u_{j,i}), \tag{2.28}$$

i.e. by means of the equations of the type (2.1), and from the increments of residual displacements

$$\Delta u_{i0} = \int_0^T \dot{u}_{i0} \, d\tau \tag{2.29}$$

that satisfy the kinematic boundary conditions.

2.3. Kinematical shakedown theorem

Since the plastic deformation increments per cycle of period T are kinematically admissible, they generate no changes in the elastic strains and stresses. This means that the residual stresses at an instant of time $\tau = T$ assume again their initial values at $\tau = 0$. Thus the elastic strain increments per cycle are equal to

$$\Delta \epsilon'_{ij0} = \int_0^T \dot{\epsilon}'_{ij0} \, d\tau = 0. \tag{2.30}$$

Under given time-dependent external agencies there exists a certain steady stress cycle accompanied by a definite admissible cycle of the plastic strain rates.[1] Thus, due to the reversibility (per cycle) of the residual stresses it can be concluded that the corresponding energy per cycle period T suffers no increment at any point of the body, i.e.

$$\int_0^T \frac{1}{2} A_{ijhk} \dot{\rho}_{hk} \rho_{ij} \, d\tau = \int_0^T \frac{1}{2} \rho_{ij} \dot{\epsilon}'_{ij} \, d\tau = 0. \tag{2.31}$$

Consequently, one can write

$$\int dv \int_0^T \rho_{ij} \dot{\epsilon}'_{ijr} \, d\tau = 0, \tag{2.32}$$

where integration is over the whole interior.

Recalling that the residual strain rates (2.20), consisting of the elastic components $\dot{\epsilon}'_{ijr}$ and the plastic components $\dot{\epsilon}''_{ijr}$ constitute a kinematically admissible distribution over the interior of the body and that the residual stresses, on account of (2.18), can be expressed as a difference between the actual stresses σ_{ij} and the 'elastic' stresses $\sigma_{ij}^{(e)}$, the steady cycle identity (2.32) yields the following equation:

$$\int_0^T d\tau \int (\sigma_{ij} - \sigma_{ij}^{(e)}) \dot{\epsilon}''_{ij} \, dv = 0. \tag{2.33}$$

This is based on the fact that the work done by the self-equilibrating stresses on the kinematically admissible distribution of strain rates (over the interior of the body) does vanish according to the virtual work principle (2.15). Also the mutual independence of integration with respect to volume and with respect to time has been employed.

---

[1] The proof of the stabilization process under cyclic loading was given by Frederic and Armstrong [2.19]. To this problem we shall devote Chapter 11.

2 Fundamental theorems of the shakedown theory

The equality (2.33) for the steady cycle remains valid, in the conditions of shakedown, i.e. elastic stress stabilization, and cyclic inelastic deformation. It also remains valid during purely elastic deformations starting from the initial state when $\sigma_{ij} = \sigma_{ij}^{(e)}$. In general, the stresses σ_{ij} are related to the plastic strain rates $\dot{\epsilon}_{ij}''$ by means of the associated flow law (2.8) which for smooth portions of the yield surface has the form

$$\dot{\epsilon}_{ij}'' = \lambda \frac{\partial f}{\partial \sigma_{ij}}. \tag{2.34}$$

It should be remembered that the factor $\lambda \geq 0$ is positive when the actual stresses at a point of the body satisfy the yield condition and unloading does not take place.

The kinematical shakedown theorem will now be formulated in the form of two statements:

1. Shakedown never takes place, i.e. a body will eventually collapse as a result of cyclic plastic deformation, if, under the stresses $\sigma_{ij}^{(e)}$ determined by the assumption of perfect elasticity and generated by external actions that vary inside prescribed limits, there can be found an admissible cycle of plastic strain rates $\dot{\epsilon}_{ij0}''(\tau)$ such that the following inequality is satisfied:

$$\int_0^T d\tau \int (\sigma_{ij} - \sigma_{ij}^{(e)}) \dot{\epsilon}_{ij0}'' \, dv \leq 0. \tag{2.35}$$

It is here taken for granted that the yield surface stresses σ_{ij} are determined by the plastic strain rates in accordance with the associated flow rule (2.8).

The latter statement is often referred to as the non-shakedown theorem.

2. A body will shake down if, at stresses $\sigma_{ij}^{(e)}$ generated by external actions which vary within prescribed limits, during arbitrary admissible cycles of non-vanishing plastic strain rates $\epsilon_{ij0}''(\tau)$ the following inequality holds true:

$$\int_0^T d\tau \int (\sigma_{ij} - \sigma_{ij}^{(e)}) \dot{\epsilon}_{ij0}'' \, dv > 0. \tag{2.36}$$

Following the standard procedure [2.2], the first statement of the

2.3. Kinematical shakedown theorem

kinematical theorem will be proved by reduction to absurdity, i.e. by the disproving of the converse statement.

Thus let us assume the shakedown state to take place although there exists a certain admissible cycle of plastic strain rates $\dot{\epsilon}''_{ij0}(\tau)$ satisfying the inequality (2.35). Thus, according to the Melan theorem, there exists a time-independent distribution of residual stresses $\bar{\rho}_{ij}$ which, after adding up to the elastic stresses $\sigma^{(e)}_{ij}$ at any instant of time, represents at any point of the body the stresses

$$\sigma^{(e)}_{ij} + \bar{\rho}_{ij} = \sigma^{(s)}_{ij}. \tag{2.37}$$

Let us now rewrite the left-hand side of the inequality (2.35) with the help of the relationships (2.37) and (2.25a); we get

$$\int_0^T d\tau \int (\sigma_{ij} - \sigma^{(e)}_{ij}) \dot{\epsilon}''_{ij0} \, dv = \int_0^T d\tau \int (\sigma_{ij} - \sigma^{(e)}_{ij} - \bar{\rho}_{ij} + \bar{\rho}_{ij}) \dot{\epsilon}''_{ij0} \, dv =$$
$$= \int_0^T d\tau \int (\sigma_{ij} - \sigma^{(s)}_{ij}) \dot{\epsilon}''_{ij0} \, dv + \int \bar{\rho}_{ij} \Delta \epsilon''_{ij0} \, dv. \tag{2.38}$$

The second integral in the right-hand side of the expression (2.38) vanishes on account of the virtual work equation (2.15), as soon as the first integral becomes positive in accordance with Drucker's postulate (2.6). Thus the assumption on the existence of shakedown is found to lead to a contradiction with inequality (2.35).

The second statement is almost obviously true on account of condition (2.33), following from the existence of a steady cycle. Indeed, if the inequality (2.36) is to remain good for all the conceivable admissible cycles of non-vanishing plastic strain rates, then the stress cycle stabilization is possible only when $\dot{\epsilon}''_{ij} = 0$, i.e. when no further changes in the plastic deformation can take place. This is, by definition, the shakedown itself.

Drawing an analogy with respective theorems of limit analysis, Koiter [2.2, 2.18] formulated the kinematical shakedown theorem in purely kinematical terms. His formulation was in the form of two relationships between the functionals representing the work done by the body forces and surface tractions and the plastic energy dissipation per cycle,

$$\int_0^T \left\{ \int X_i \dot{u}_{i0} \, dv + \int_{S_p} p_i \dot{u}_{i0} \, dS \right\} d\tau \geq \int_0^T d\tau \int F(\dot{\epsilon}''_{ij0}) \, dv, \tag{2.39}$$

59

2 Fundamental theorems of the shakedown theory

$$\int_0^T \left\{ \int X_i \dot{u}_{i0} \, dv + \int_{S_p} p_i \dot{u}_{i0} \, dS \right\} d\tau < \int_0^T d\tau \int F(\dot{\epsilon}_{ij0}'') \, dv. \tag{2.40}$$

Using the virtual work principle (2.15) and the expression (2.13) for the dissipative function $F(\dot{\epsilon}_{ij0}'')$, it is easy to express the relationships (2.39), (2.40) in the form of (2.35), (2.36), respectively. Account has only to be taken of the fact that

$$\int \sigma_{ij}^{(e)} \dot{\epsilon}_{ij0} \, dv = \int \sigma_{ij}^{(e)} \dot{\epsilon}_{ij0}'' \, dv, \quad (\dot{\epsilon}_{ij0} = \dot{\epsilon}_{ij0}' + \dot{\epsilon}_{ij0}'') \tag{2.41}$$

since

$$\int \sigma_{ij}^{(e)} \dot{\epsilon}_{ij0}' \, dv = \int \dot{\rho}_{ij0} \epsilon_{ij}^{(e)} \, dv = 0, \tag{2.42}$$

where $\dot{\rho}_{ij0}$ is the self-equilibrating distribution of stress rates corresponding to the elastic components of the residual strain rates,

$$\dot{\rho}_{ij0} = A_{ijhk}^{-1} \dot{\epsilon}_{hk0}', \tag{2.43}$$

and $\epsilon_{ij}^{(e)}$ is the kinematically admissible strain distribution

$$\epsilon_{ij}^{(e)} = A_{ijhk} \sigma_{ij}^{(e)}. \tag{2.44}$$

It appears that the proof of the kinematical theorem presented in the form of the relationships (2.35), (2.36) is simpler than by using the relationships (2.39), (2.40). Moreover, the form (2.35), (2.36) is more general since the 'elastic' stresses $\sigma_{ij}^{(e)}$ involved can be generated by mechanical forces X_i, p_i as well as by the thermal or some other actions. Thus the adopted formulation entails no necessity of specially developing the kinematical theorem to cover the case of thermal actions, whereas Koiter's original statement [2.2, 2.18] had to be suitably generalized by Rosenblum and, later on, by Maier and Donato. To this end, an additional term [2.20, 2.21] was introduced into the left-hand sides of (2.39), (2.40), in the form

$$\int_0^T d\tau \int \dot{\rho}_{ij0} \delta_{ij} \alpha t \, dv, \tag{2.45}$$

2.3. Kinematical shakedown theorem

where αt is the thermal deformation and δ_{ij} is the Kronecker delta. Due to the identity

$$\int \dot{\rho}_{ij0}\delta_{ij}\alpha t \, \mathrm{d}v = \int \sigma_{ij}^{(e)}\dot{\epsilon}_{ij0}'' \, \mathrm{d}v, \qquad (2.46)$$

the relevant expressions reduce to those of (2.35), (2.36).

The identity (2.46) may be looked upon as the law of reciprocal work because there corresponds to an instantaneous plastic strain rate field $\dot{\epsilon}_{ij0}''$ a unique distribution of residual stresses. Similarly the thermal strains $\delta_{ij}\alpha t$ are related to a unique distribution of thermal stresses $\sigma_{ij}^{(e)}$.

In the new formulation of the kinematical theorem (2.35), (2.36) the elastic stresses are induced by the applied agencies; it is this fact that differentiates the shakedown theory from the limit analysis theory as resulting from variable loading. In the shakedown theory the preliminary analysis of elastic stresses generated solely by the time dependent actions to within an accuracy of suitably introduced multipliers proves to be necessary irrespective of which theorem, the statical or the kinematical, is used. Constant stresses can be isolated and, with the use of the principle of virtual work (2.15) in a reciprocal manner, they can be expressed in terms of corresponding loads X_i^0, p_i^0. Thus the non-shakedown condition assumes the form

$$\int_0^T \left\{ \int X_i^0 \dot{u}_i \, \mathrm{d}v + \int_{S_p} p_i^0 \dot{u}_{io} \, \mathrm{d}S \right\} \mathrm{d}\tau \geq \int_0^T \mathrm{d}\tau \int (\sigma_{ij} - \sigma_{ij\tau}^{(e)}) \dot{\epsilon}_{ij0}'' \, \mathrm{d}v. \qquad (2.47)$$

The equality sign clearly corresponds to the limiting cycle that separates the shakedown behaviour from plastic flow.

It follows from the presented derivation that the time independent self-equilibrating stress components as induced by a time independent temperature field, an initial plastic deformation or a change of volume due to structural transformations, have no influence whatsoever on the shakedown conditions.

The kinematical as well as the statical shakedown theorems can be generalized to allow for dynamic effects. A suitable formulation was proposed in [2.35]. The temperature-dependence of the elastic moduli can also be accounted for. Let us give without proof the statement of the first part (analogously as in the statical formulation (2.26)–(2.26c)):

Shakedown will not take place, i.e. under given external actions the body will collapse due to cyclic plastic deformation, if there can be

2 Fundamental theorems of the shakedown theory

found an admissible cycle of plastic strain rates $\dot{\epsilon}_{ij0}''$ and a number $m > 1$ such that the following inequalities hold:

$$\int_0^T d\tau \int (\sigma_{ij} - \sigma_{ij\tau})\dot{\epsilon}_{ij0}'' \, dv < 0, \qquad (2.48)$$

$$\phi(\sigma_{ij\tau}) \leq mk^2. \qquad (2.49)$$

The stresses $\sigma_{ij\tau}$ can be determined from the conditions

$$\epsilon_{ij}'' + \epsilon_{ij}^{(T)} + A_{ijhk}\sigma_{hk\tau} = \frac{1}{2}(u_{i,j} + u_{j,i}), \qquad (2.50)$$

$$\sigma_{ij\tau,j} + X_i = 0 \quad \text{in the interior } V, \qquad (2.51)$$

$$\sigma_{ij\tau}n_j = p_i \quad \text{on } S_p. \qquad (2.52)$$

Here k stands for the plastic constant in the yield condition of type (2.3). The constraints (2.49) should be satisfied at all points and at all instants of time.

It is worth noting that, neither in the formulation nor in the proof of the first (statical) and the second (kinematical) fundamental theorems, hardly any reference was made to the magnitudes of the plastic strains prior to shakedown. This circumstance can be of importance when the correctness of the shakedown and the safety notions is discussed as referred to structures. We shall return to this problem towards the end of this chapter.

2.4. Introduction of a fictitious yield surface and restatement of the statical theorem. Formulation of the shakedown problem connected with a non-trivial problem of limit analysis

It seems to be more difficult to apply the fundamental shakedown theorems than the related limit analysis theorems. The prime difficulty consists in having to solve a specific extremum problem, i.e. in an evaluation of the largest possible values of the parameters describing the loading for which an adaptation of a structure is possible, or the smallest possible values for which shakedown does not occur. This entails the necessity of a detailed analysis of both stress and strain states during the cycle of applied loading. Those were the circumstances responsible

2.4. Introduction of a fictitious yield surface

for a rather slow penetration of the shakedown theory into the range of practical problems. In particular, this was so for the kinematical theorem which could not have been applied directly in its original formulation [2.2].

It was later discovered that both the kinematical [2.22, 2.23] and the statical [2.24, 2.25] theorems can be reformulated so that the analysis of the stress state varying over a cycle could be separated from the solution of the extremum problem. The new approach made it possible to work out methods of solving sufficiently complex problems representing actual engineering situations. Particular methods were developed to deal with the expected type of cyclic plastic deformation such as incremental collapse and alternating plastic flow.

The restatement of the theorems is based on the already adopted idea of the limiting loading cycle forming a separation between the shakedown and the unrestrained cyclic plasticity.

Let us separate the applied actions as well as the stresses supporting them into constant and time-dependent components,

$$X_i = X_i^0 + X_{i\tau}, \quad p_i = p_i^0 + p_{i\tau}, \quad \sigma_{ij}^{(e)} = \sigma_{ij0}^{(e)} + \sigma_{ij\tau}^{(e)}. \tag{2.53}$$

In the shakedown situation we have

$$\sigma_{ij}^0 = \sigma_{ij0}^{(e)} + \bar{\rho}_{ij}, \quad \sigma_{ij} = \sigma_{ij}^0 + \sigma_{ij\tau}^{(e)}, \tag{2.54}$$

where $\bar{\rho}_{ij}$ are the time-independent self-equilibrating stresses leading to shakedown in accordance with the Melan theorem, $\sigma_{ij}^{(e)}$ are the stresses evaluated under the assumption of ideal elasticity, $\sigma_{ij\tau}^{(e)}$ are their varying components, and σ_{ij}^0 are the constant stresses supporting the time-independent load components.

A further restatement of the statical theorem claims that the necessary and sufficient basic shakedown condition (2.16) is satisfied if and only if at any instant of time and at any point the maximum value of the yield function (per cycle) is non-positive, i.e.

$$\max_\tau f(\sigma_{ij}^0 + \sigma_{ij\tau}^{(e)}) \leq 0. \tag{2.55}$$

The above replacement of the strong inequality in condition (2.16) by a weak one would have formally led to considerable difficulties in proving the Melan theorem. In practice, however, this is immaterial since the presence of the equality sign can be compensated for by a small adjustment of the yield stress. Moreover, in what follows, the above form of the condition will be more convenient.

The equality

$$\max_{\tau} f(\sigma^0_{ij*} + \sigma^{(e)}_{ij\tau}) = f(\sigma^0_{ij*} + \sigma^*_{ij\tau}) = 0, \tag{2.56}$$

whose left-hand side is free from any time-dependent magnitudes, can be considered as the parametric equation of a certain surface

$$\varphi(\sigma^0_{ij*}) = 0 \tag{2.57}$$

enclosing in the nine-dimensional stress space the region of admissible, time-independent stresses ($\sigma^0_{ij} \leq \sigma^0_{ij*}$) under given stresses induced by variable agencies. This surface will be in short referred to as the fictitious yield surface. Specific and stationary for a given cycle values of variable stresses $\sigma^*_{ij\tau}$ enfolding the domain enclosed by the surface (2.57) will be termed the 'determining' or the 'enveloping' variable stresses. The latter term is adopted from the structural mechanics of bar systems.

If the variable stresses are evaluated to within some multipliers, these will appear also in equation (2.57). We arrive thus at the following restatement of the statical theorem:

The body shakes down if there exists a field of time-independent stresses σ^0_{ij} which are in equilibrium with the constant loads X^0_i, p^0_i and belong to the region bounded by the fictitious yield surface, i.e. the region of those constant stresses which constitute an admissible stress state if added to the fluctuating 'elastic' stresses $\sigma^{(e)}_{ij\tau}$ at any instant of time.

The fictitious yield surfaces should be determined beforehand for each point of the body in accordance with the variable stress state at that point.

In the general case, both constant and variable external actions are prescribed to within some multipliers $m^0, m^{(\tau)}$. Under multi-parameter loading, these are vectors

$$m^0 = (m^0_1, m^0_2, \ldots, m^0_k, \ldots), \quad m^{(\tau)} = (m^{(\tau)}_1, m^{(\tau)}_2, \ldots, m^{(\tau)}_k, \ldots). \tag{2.58}$$

The extremum problem to determine the shakedown conditions takes

2.5. Construction and properties of the fictitious yield surface

now the form

$$\max_{\sigma_{ij}^0} m_i^0 = ? \quad \text{or} \quad \max_{\sigma_{ij}^0} m_l^{(\tau)} = ? \tag{2.59}$$

under the constraints

$$\sigma_{ij,j}^0 + m_k^0 X_{ik}^0 = 0, \quad \sigma_{ij}^0 n_j = m_k^0 p_{ik}^0, \tag{2.60}$$

and

$$\varphi(\sigma_{ij}^0) \leq 0, \tag{2.61}$$

where

$$\varphi(\sigma_{ij*}^0) = f(\sigma_{ij}^0 + m_k^{(\tau)} \sigma_{ij\tau k}^*) = 0. \tag{2.62}$$

Here $\sigma_{ij\tau k}$ is understood as the 'elastic' stress state induced by 'unit' variable actions and it is assumed that all the multipliers except one, say m_i^0 or $m_l^{(\tau)}$, are known.

The above restatement of the statical theorem reduces the shakedown problem to a certain non-trivial problem of limit analysis. If the sought for parameter (2.59) of the limiting cycle appears to be one of the parameters of the constant external actions m_i^0, the shakedown problem corresponds to that of evaluating from the fictitious yield surface (2.57) the collapse load of a fictitiously nonhomogeneous and anisotropic body. When the parameters of variable external actions are sought, the shakedown problem is analogous to a limit analysis problem for a nonhomogeneous and anisotropic body in which the yield stresses must necessarily be adjusted in such a way that the body be at limit equilibrium under the given constant load.

2.5. Construction and properties of the fictitious yield surface

Both the construction and the properties of the fictitious yield surfaces will be demonstrated on some simple examples. Let the directions of the principal stresses σ_x, σ_y in the plane stress be fixed during the cycle and let the stresses generated by the variable load be linearly related to a

2 Fundamental theorems of the shakedown theory

single parameter

$$\sigma_{x\tau}^{(e)} = p(\tau)\sigma_s, \quad \sigma_{y\tau}^{(e)} = 1{,}5p(\tau)\sigma_s, \tag{2.63}$$

where

$$0 \leqslant p(\tau) \leqslant p^* \tag{2.64}$$

and σ_s is the yield point.

From the Tresca–Saint Venant yield condition (2.12),

$$\max[|\sigma_x|, |\sigma_y|, |\sigma_y - \sigma_x|] - \sigma_s = 0, \tag{2.65}$$

we conclude that the equation of the fictitious yield surface, in view of (2.56), takes the form

$$\max[\max_{\tau}(|\sigma_{x*}^0 + \sigma_{x\tau}^{(e)}|, |\sigma_{y*}^0 + \sigma_{y\tau}^{(e)}|, |\sigma_{y*}^0 - \sigma_{x*}^0 + \sigma_{y\tau}^{(e)} - \sigma_{x\tau}^{(e)}|)] - \sigma_s = 0, \tag{2.66}$$

where $\sigma_{x*}^0, \sigma_{y*}^0$ are the stresses belonging to that surface.

On substituting the values of variable stresses (2.63) into equation (2.66) and using the condition (2.64), we can find the parameter $p(\tau)$ for which the left-hand side of equation (2.66) attains its maximum; thus the equation of the fictitious yield surface assumes the form

$$\max(\sigma_{x*}^0 + p^*\sigma_s, -\sigma_{x*}^0, \sigma_{y*}^0 + 1{,}5p^*\sigma_s, -\sigma_{y*}^0,$$
$$\sigma_{x*}^0 - \sigma_{y*}^0, \sigma_{y*}^0 - \sigma_{x*}^0 + 0{,}5p^*\sigma_s) - \sigma_s = 0. \tag{2.67}$$

This is illustrated in Fig. 2.1 in the centre of which there is shown a variable stress profile OK by a heavy solid line corresponding to the absence of constant stresses. The imposing of the non-vanishing constant stresses results in a parallel translation of the segment OK. Now, if the segment $O_1K_1 = OK$ slides by parallel translation around the actual yield surface $ABCDEFA$, the point O_1 plots the fictitious yield surface (2.67) which for $p^* = 0{,}4$ is shown in Fig. 2.1 as a hexagon $A_1B_1C_1D_1EFA_1$.

In the limiting cycle, after which the unrestrained cyclic plastic deformation commences, the sum of the constant and the variable stresses reaches the yield surface at isolated instants of time only; these correspond to the moments when the variable components attain their extremum values. Some of the possible situations are shown in Fig. 2.1.

2.5. Construction and properties of the fictitious yield surface

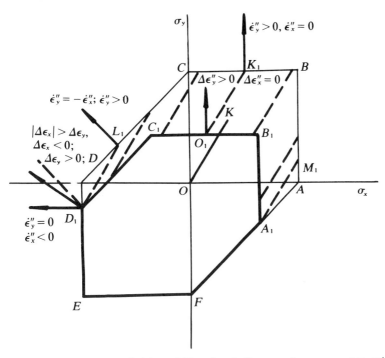

Fig. 2.1. Determination of the fictitious yield surface in the case of one-parameter variable loading.

If, for instance, the constant limiting cycle stresses are at a given point of the body such that the segment OK moves to take the position O_1K_1 then the yield surface is reached at a single value of the variable stresses (here at $p(\tau) = p^*$). In accordance with the associated flow rule (2.8) the plastic strain rate vector is directed along the normal to the side BC,

$$\dot{\epsilon}_y'' = \lambda \ (\lambda > 0), \quad \dot{\epsilon}_x'' = 0. \tag{2.68}$$

The vector of the plastic strain increment, defined by (2.27) will be referred to the fictitious yield surface. In the present case it remains parallel to the rate vector, whose direction stays unchanged during the cycle remaining normal to the segment B_1C_1 of the fictitious yield surface,

$$\Delta\epsilon_y'' > 0, \quad \Delta\epsilon_x'' = 0. \tag{2.69}$$

2 Fundamental theorems of the shakedown theory

When the constant stresses over the limiting cycle are represented by the corner D_1 (or A_1) of the fictitious yield surface, the total stress profile is shown by D_1L_1 (or A_1M_1). From this it follows that the yield surface for a given point of the body is reached at least twice per cycle, at two different values of variable stresses. The corresponding vectors of plastic strain rates and increments are shown in Fig. 2.1; it can be readily seen that, on account of the relationship (2.27), the plastic strain increment vector per cycle should be located between the normals to the smooth portions C_1D_1 and D_1E of the fictitious yield surface, meeting at the corner D_1.

Let us now consider a more general case when the stresses are generated by a variable load which is linearly related to two time-dependent, but mutually independent, parameters:

$$\frac{\sigma_{x\tau}^{(e)}}{\sigma_s} = 0{,}2p_1(\tau) - 0{,}4p_2(\tau), \quad \frac{\sigma_{y\tau}^{(e)}}{\sigma_s} = 0{,}5p_1(\tau) - 0{,}1p_2(\tau), \tag{2.70}$$

$$-p_1^* \leqslant p_1(\tau) \leqslant p_1^*, \quad 0 \leqslant p_2(\tau) \leqslant p_2^*. \tag{2.71}$$

Substituting (2.70) into equation (2.66) and employing conditions (2.71), we obtain the following equation for the fictitious yield surface:

$$\max[\sigma_{x*}^0 + 0{,}2p_1^*\sigma_s, -\sigma_{x*}^0 + (0{,}2p_1^* + 0{,}4p_2^*)\sigma_s, \sigma_{y*}^0 + 0{,}5p_1^*\sigma_s,$$
$$-\sigma_{y*}^0 + (0{,}5p_1^* + 0{,}1p_2^*)\sigma_s, \sigma_{x*}^0 - \sigma_{y*}^0 + 0{,}3p_1^*\sigma_s,$$
$$-\sigma_{x*}^0 + \sigma_{y*}^0 + 0{,}3(p_1^* + p_2^*)] - \sigma_s = 0. \tag{2.72}$$

The surfaces (2.72) are shown in Fig. 2.2, where in the centre a bounding hodograph KLL_1K_1 of varying elastic stresses (2.70) is depicted. The starting point of the hodograph plots the fictitious yield surface (2.72) by sliding without rotation the parallelogram KLL_1K_1 over the actual yield surface $ABCDEFA$. The results for different values of the variable loading parameters p_1^*, p_2^* are shown in Fig. 2.3.

For sufficiently small values of p_1^*, p_2^* the fictitious yield surface is a hexagon, like the original one; for larger values, the number of sides diminishes, e.g. the parallelogram $A_1B_1C_1D_1$ in Fig. 2.3. This occurs when the yield regimes, corresponding to any side of the actual yield surface, cannot be realized in the limiting cycle for any values of the constant stresses. The situation is similar to that we already visualized in Fig. 2.1, where the vector of total stresses could not pierce the yield surface corresponding to the segments D_1L_1 and A_1M_1.

2.5. Construction and properties of the fictitious yield surface

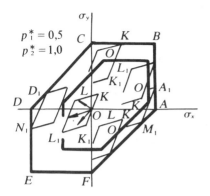

Fig. 2.2. The case of two-parameter variable loading.

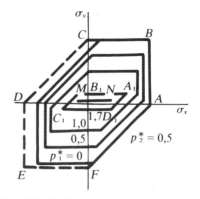

Fig. 2.3. Fictitious yield surface for various values of the variable loading parameters.

Finally, for sufficiently large values of the variable stresses the fictitious yield surface shrinks to a line (the straight line in Fig. 2.3) or even to a point. This means that no time-independent stresses can be found that, added up to the given (extremum) variable stresses, represent an admissible stress state at the given point of the body. In the case as shown in Fig. 2.4 the bounding hodograph of the limiting cycle variable stresses is touching two opposite sides of the yield surface, i.e. the plastic strain rates are becoming non-zero twice per cycle (at two different values of variable stresses) and are of opposite signs.

The graphical construction of the fictitious yield surface is similar when the material obeys the Huber–Mises yield condition (2.9), which in plane stress can be expressed by

$$\sigma_x^2 + \sigma_y^2 - \sigma_x\sigma_y - \sigma_s^2 = 0. \tag{2.73}$$

The results for cycles (2.63), (2.64) are plotted in Fig. 2.5. In order to obtain the analytical form of the fictitious yield surface, let us rewrite equation (2.56) so that it resembles (2.73):

$$\max_\tau [(\sigma_{x*}^0 + \sigma_{x\tau}^{(e)})^2 + (\sigma_{y*}^0 + \sigma_{y\tau}^{(e)})^2 - (\sigma_{x*}^0 + \sigma_{x\tau}^{(e)}) \times$$
$$\times (\sigma_{y*}^0 + \sigma_{y\tau}^{(e)}) - \sigma_s^2] = 0. \tag{2.74}$$

On substituting here the stresses (2.63) and taking into account the

2 Fundamental theorems of the shakedown theory

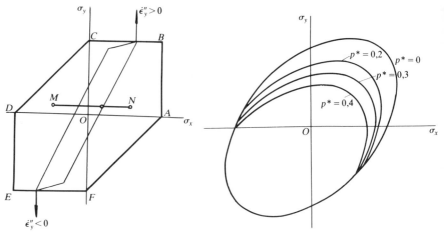

Fig. 2.4. Conditions for alternating plastic flow.

Fig. 2.5. Fictitious yield surface for the Mises criterion.

conditions (2.64), one obtains easily the equations of the two arcs AMC and ANC enclosing the fictitious yield surface:

$$(\sigma^0_{x*} + p^*\sigma_s)^2 + (\sigma^0_{y*} + 1{,}5p^*\sigma_s)^2 - (\sigma^0_{x*} + p^*\sigma_s) \times$$

$$\times (\sigma^0_{y*} + 1{,}5p^*\sigma_s) - \sigma_s^2 = 0,$$

$$(\sigma^0_{x*})^2 + (\sigma^0_{y*})^2 - \sigma^0_{x*}\sigma^0_{y*} - \sigma_s^2 = 0. \qquad (2.75)$$

We shall indicate now some general properties of the fictitious yield surface which can be noticed on the shown specific examples:

A. If the actual yield surface or the bounding hodograph of the variable elastic stresses is piece-wise linear, the fictitious yield surface will be generated by a parallel translation of particular sides of the actual yield surface toward the origin of coordinates. The fictitious yield surface remains convex if the actual yield surface is convex.

This property can be generalized to cover arbitrary shapes of the yield surfaces and of the bounding hodographs by considering a non-linear surface to be a limit of piece-wise linear ones.

B. If the actual yield surface or the bounding hodograph of the variable elastic stresses is piece-wise linear, the plastic strain increments $\Delta\epsilon''_{ij}$ per limiting cycle are related to the fictitious yield surface stresses

2.6. Restatement of the kinematical theorem

σ_{ij*}^0 by means of the equations

$$\Delta\epsilon_{ij}'' = \sum_\alpha \mu_\alpha \frac{\partial \varphi_\alpha(\sigma_{ij*}^0)}{\partial \sigma_{ij*}^0}, \quad \mu_\alpha \geq 0, \quad \alpha = 1, 2, \ldots \qquad (2.76)$$

which are analogous to the associated flow rule (2.8).

In a more general case of non-linear yield surfaces and non-linear bounding hodographs the associated flow rule leads to the following relationship between the strain increments and stresses at the limiting cycle:

$$\Delta\epsilon_{ij}'' = \sum_\alpha \mu_\alpha \frac{\partial f_\alpha(\sigma_{ij*}^0 + \sigma_{ij\tau}^*)}{\partial(\sigma_{ij*}^0 + \sigma_{ij\tau}^*)}. \qquad (2.77)$$

C. Irrespective of whether the actual yield surface is regular or singular, there will appear on the fictitious yield surface some ridges or corners whenever more than one plastic regime can take place (Fig. 2.5) as a result of a mere variation of variable stresses over an infinitesimal volume.

D. If the variable stresses increase, certain portions of the fictitious yield surface may vanish. This means that the related stress profiles cannot be realized within the limiting cycle. However the strain vector can assume the direction normal to a vanishing side of the polygon. It is due to the possibility of the realization of plastic regimes corresponding to the adjacent stress profiles.

2.6. Restatement of the kinematical theorem. Conditions for limit states following from both restated shakedown theorems

Let us employ the introduced earlier partition of stresses into those varying in time and the time-independent ones, as defined in (2.54). The latter can be singled out by condition (2.35) and expressed in terms of acting loads by using the principle of virtual work. The condition of inadaptation (2.35) takes ultimately the form

$$\int_0^T \left\{ \int X_i^0 \dot{u}_{i0} \, dv + \int_{S_p} p_i^0 \dot{u}_{i0} \, dS \right\} d\tau \geq \int_0^T d\tau \int (\sigma_{ij} - \sigma_{ij\tau}^{(e)}) \dot{\epsilon}_{ij0}'' \, dv. \qquad (2.78)$$

2 Fundamental theorems of the shakedown theory

Bearing in mind that integrations with respect to time and space are mutually independent, we can rearrange the left-hand side of the inequality with the help of (2.29), to become

$$\int_0^T d\tau \int X_i^0 \dot{u}_{i0} \, dv = \int X_i^0 \, dv \int_0^T \dot{u}_{i0} \, d\tau = \int X_i^0 \Delta u_{i0} \, dv,$$

$$\int_0^T d\tau \int_{S_p} p_i^0 \dot{u}_{i0} \, dS = \int_{S_p} p_i^0 \, dS \int_0^T \dot{u}_{i0} \, d\tau = \int_{S_p} p_i^0 \Delta u_{i0} \, dS. \qquad (2.79)$$

On the other hand for inadaptation it is necessary and sufficient that inequality (2.78) holds for certain values of stresses σ_{ij} and $\sigma_{ij\tau}^{(e)}$ varying within the prescribed limits which minimize with respect to time τ the integrand at the right-hand side of inequality (2.78) at any given admissible strain rate cycle $\dot{\epsilon}_{ij0}''$. Making use of the relationships (2.79) and of the above requirement, we can write the condition (2.78) for shakedown not to occur in the form

$$\int X_i^0 \Delta u_{i0} \, dv + \int_{S_p} p_i^0 \Delta u_{i0} \, dS \geq \int_0^T d\tau \int \min_\tau [(\sigma_{ij} - \sigma_{ij\tau}^{(e)}) \dot{\epsilon}_{ij0}''] \, dv. \qquad (2.80)$$

Transition to the limiting cycle means that the plastic strain rates $\dot{\epsilon}_{ij0}''$ can differ from zero only at those instants of time when the corresponding elastic stresses attain certain stationary values corresponding to the limits of the intervals inside which they vary. Otherwise, should the stress vector at a point of the body touch the yield surface, then it will be always possible to find a program of further loading under which the plastic flow commences in the neighbourhood of that point.

Inequality (2.80) describes also the specific case when stresses at each point of the body can reach the yield surface only at a single value of the variable applied stresses. Assume that for a given infinitesimal volume of the body, there exists a yield surface with a portion as shown in Fig. 2.6. Assuming a certain kinematically admissible collapse mechanism of the body (admissible cycle of plastic strain rates), we can determine the directions of the plastic strain rate vectors at all its points. According to the associated flow rule (2.8), the plastic strain rates at a given point of the body can become non-zero only when the head of the total stress vector $\sigma_{ij}^0 + \sigma_{ij\tau}^{(e)}$ reaches the point A of the yield surface whereas the location of A is determined by the vector $\dot{\epsilon}_{ij0}''$. All the remaining points of the yield surface are insignificant unless other collapse mechanisms are admitted.

2.6. Restatement of the kinematical theorem

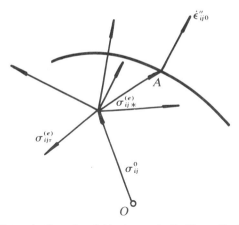

Fig. 2.6. Determination of variable stresses in the kinematical theorem.

In accordance with the adopted formulation of the problem, the vector of total stresses can reach point A only at that instant of time when the projection of the vector of additional 'elastic' stresses $\sigma_{ij\tau}^{(e)}$, induced by variable external actions, on the vector of plastic strain rate becomes maximum in the algebraic sense. The significance of the above condition is clear enough once a simple stress state is considered: For instance, plastic deformation at simple tension can develop only when additional, suitably large tensile (in contrast to compressive) stresses are imposed.

It follows that (see Fig. 2.6)

$$\sigma_{ij\tau}^{*}\dot{\epsilon}_{ij0}^{''} = \max_{\tau}(\sigma_{ij\tau}^{(e)}\dot{\epsilon}_{ij0}^{''}) \geq 0. \tag{2.81}$$

For the actual collapse mechanism the condition (2.81) can be concluded from the relationship (2.7), which in turn follows from the Drucker postulate.

The above considerations are similar to those used in solving limit analysis problems via the kinematical approach. In both cases the various admissible collapse mechanisms lead to the upper bound assessments of the limit loads. These are calculated from the respective kinematical theorems.

Since both the stresses $\sigma_{ij\tau}^{*}$ and the yield surface stresses depend only on the postulated flow regime at a point and not on the current

2 Fundamental theorems of the shakedown theory

time, we can use equality (2.27) in order to integrate with respect to time the right-hand side of inequality (2.80):

$$\int_0^T \min_\tau [(\sigma_{ij} - \sigma_{ij\tau}^{(e)})\dot{\epsilon}_{ij0}''] \, d\tau = \int_0^T (\sigma_{ij} - \sigma_{ij\tau}^*)\dot{\epsilon}_{ij0}'' \, d\tau =$$
$$= (\sigma_{ij} - \sigma_{ij\tau}^*) \int_0^T \dot{\epsilon}_{ij0}'' \, d\tau = (\sigma_{ij} - \sigma_{ij\tau}^*)\Delta\epsilon_{ij0}''. \tag{2.82}$$

Condition (2.81) remains also valid for the plastic strain increments in the limiting cycle.

Making use of the last result, we obtain

$$\int X_i^0 \Delta u_{i0} \, dv + \int_{S_p} p_i^0 \Delta u_{i0} \, dS \geq \int (\sigma_{ij} - \sigma_{ij\tau}^*)\Delta\epsilon_{ij0}'' \, dv. \tag{2.83}$$

When dealing with non-isothermic loading, the variation of the yield surface stresses over a cycle (see Sec. 2.2) should be properly accounted for. Instead of the condition (2.81), we have then

$$\min_\tau [(\sigma_{ij} - \sigma_{ij\tau}^{(e)})\dot{\epsilon}_{ij0}''] = (\sigma_{ij} - \sigma_{ij\tau}^{(e)})^* \dot{\epsilon}_{ij0}'', \tag{2.84}$$

which means that the knowledge of the variable stresses $\sigma_{ij\tau}^{(e)}$ is not enough in order to determine the minimum of the bracketed expression.

It must necessarily be borne in mind that, in the general case, at some points of the body more than one flow regime can take place during a single cycle, owing to fluctuations of variable stresses at those points. Thus the kinematical condition for shakedown not to occur can now be rewritten in the form

$$\int X_i^0 \Delta u_{i0} \, dv + \int_{S_p} p_i^0 \Delta u_{i0} \, dS \geq \int \min_{\tau,k} \sum_1^n [(\sigma_{ij} - \sigma_{ij\tau}^{(e)})\Delta\epsilon_{ij0}'']_k \, dv, \tag{2.85}$$

where $k = 1, 2, \ldots, n$ is the number of possible flow regimes at a point of the body, $\min_k$ is the least value of the sum of products shown in square brackets reached for a certain number k of plastic flow regimes and $(\Delta\epsilon_{ij0}'')_k$ are the plastic strain increments conditioned by (2.84) or, in particular, occurring at stresses which obey (2.82) for each possible flow regime; the sum of these increments corresponds to the kinematically admissible distribution of strain.

2.6. Restatement of the kinematical theorem

The expression in square brackets represents the work done by the stress vector $\sigma_{ij} - \sigma_{ij\tau}^{(e)}$ on the deformation $\Delta\epsilon_{ij0}''$ over the time interval in which its projection on the direction of associated strain rate vector is kept at minimum. The set of heads of the stress vectors $\sigma_{ij} - \sigma_{ij\tau}^{(e)}$, obeying the above stated property and corresponding to all the possible directions of the plastic strain rate vector per cycle, constitutes in the stress space a closed surface that is found to coincide with the fictitious yield surface (2.57) considered in Secs. 2.4 and 2.5.

Making use of the properties of the fictitious yield surface, we can shorten condition (2.85) so that it becomes

$$\int X_i^0 \Delta u_{i0} \, dv + \int_{S_p} p_i^0 \Delta u_{i0} \, dS \geq \int \sigma_{ij*}^0 \Delta\epsilon_{ij0}'' \, dv. \tag{2.86}$$

If the field of residual displacement increments appears to suffer discontinuities $\Delta' u_{i0}$ across some internal surfaces S_μ, the right-hand side of the inequality (2.86) should be supplemented by a term accounting for a suitable amount of work done by the fictitious yield surface stresses

$$\sum_\mu \int_{S_\mu} \sigma_{ij*}^0 n_j \Delta' u_{i0} \, dS_\mu. \tag{2.87}$$

Thus the most general form of the condition for shakedown not to occur is the following:

$$\int X_i^0 \Delta u_{i0} \, dv + \int_{S_p} p_i^0 \Delta u_{i0} \, dS \geq \int \sigma_{ij*}^0 \Delta\epsilon_{ij0}'' \, dv + \sum_\mu \int_{S_\mu} \sigma_{ij*}^0 n_j \Delta' u_{i0} \, dS_\mu. \tag{2.88}$$

The kinematical shakedown theorem can now be reformulated as follows: The body never shakes down if there exists any kinematically admissible field of plastic strain increments per cycle $\Delta\epsilon_{ij0}''$ satisfying (2.28), for which the work done over the whole body by the stresses belonging to the fictitious yield surface $\varphi(\sigma_{ij*}^0) = 0$, including surfaces of displacement discontinuities, is less than the work done by the time-independent components of the external loads on corresponding displacement increments Δu_{i0}. It is here assumed that the strain increment vector $\Delta\epsilon_{ij0}''$ is related to the fictitious yield surface stresses σ_{ij*}^0 by means of (2.76) or (2.77) in the way implied by the associated flow rule.

2 Fundamental theorems of the shakedown theory

The presented restatements of the fundamental statical (2.61) and kinematical (2.88) theorems permit one to split the solution to the problem into two consecutive stages. First, the fictitious yield surface should be constructed for every point of the body[1] and, secondly, the extremum problem for external actions parameters should be solved independently of the current time. The second stage of the computations differs from the ordinary limit analysis procedure only in that, instead of the actual yield surface, the fictitious yield surfaces must be employed, individually constructed for every point or line or region of the body in accordance with the applied time-dependent stresses. Those circumstances can be illustrated by means of the kinematical theorems by comparing condition (2.86) with the corresponding condition of the limit analysis theory [2.2],

$$\int X_i \dot{u}_i^0 \, dv + \int_{S_p} p_i \dot{u}_i^0 \, dS \geq \int \sigma_{ij} \dot{\epsilon}_{ij}^0 \, dv \quad \text{where } \dot{\epsilon}_{ij}^0 = \dot{\epsilon}_{ij}''. \tag{2.89}$$

For simplicity, the possibility of a discontinuous displacement field has been excluded in either case.

However, in contrast to the shown analogy, we wish to emphasize the following peculiarity of the shakedown problem for a fictitiously non-homogeneous body: Two essentially different types of limiting states are possible, one of which proves unacceptable for a 'homogeneous' body, i.e. when variable actions generating fictitious non-homogeneity are absent.

The ultimate state of the first type is reached when for at least one of the points of the body the region enclosed by the fictitious yield surface shrinks to a line or to a point. In this case there certainly cannot exist any residual stresses which, added up to the stresses induced by external actions, result in a safe stress state at that point, or points of the body.

It is worth noting that the residual stresses are here significant mainly because of their effect on the alteration of characteristics of the stress cycle in those parts of the body which happen to be in an unsafe state; their distribution at all the remaining points appears to be non-unique.[2] Suitable residual stresses, in accordance with the Melan

---

[1] Full construction of the fictitious yield surfaces is sometimes unnecessary, in particular, when using approximate kinematical methods (see Chapter 5).
[2] This was already demonstrated on simple systems in Chapter 1. More general considerations on the uniqueness of stresses in the limiting cycle will be given in Sec. 2.7.

2.6. Restatement of the kinematical theorem

theorem, will of course appear provided the stress boundary conditions are still satisfied.

The kinematical interpretation shows that the degeneration of the fictitious yield surface into a line or a point is associated with a plastic flow for which plastic strain increments per cycle vanish throughout the body ($\Delta \epsilon_{ij}'' = \Delta u_i = 0$), although the plastic strain rates at points of the body differ from zero during the cycle. The ultimate state of such a type is known under the term of alternating plastic flow.

In order to evaluate the limit load parameters under those conditions it suffices to analyse the fluctuations of the variable elastic stresses at each particular point of the body; direct use of the shakedown theorems appears to be unnecessary to solve specific problems. In general, it is proved (the form (2.85) of the kinematical theorem can here be employed) that alternating plastic flow must take place when the stress variation per cycle exceeds the double value of the yield stress at any point of the body.

Making use of the yield condition (2.3), we can express the limiting condition in the suitable form

$$2\sigma_{ij}\dot{\epsilon}_{ij0}'' - \max_{\tau}[\sigma_{ij\tau}^{(e)}\dot{\epsilon}_{ij0}''] + \min_{\tau}[\sigma_{ij\tau}^{(e)}\dot{\epsilon}_{ij0}''] = 0, \qquad (2.90)$$

or, under non-isothermal loading, as

$$\min_{\tau}[(\sigma_{ij} - \sigma_{ij\tau}^{(e)})\dot{\epsilon}_{ij0}''] - \max_{\tau}[(\sigma_{ij} - \sigma_{ij\tau}^{(e)})\dot{\epsilon}_{ij0}''] = 0, \qquad (2.91)$$

where $\dot{\epsilon}_{ij0}''$ indicates the direction of a normal to the yield surface.

The conditions (2.90) and (2.91) appear to be sufficient. Numerous descriptions of alternating plasticity made by different authors with the help of various analytical and numerical procedures are found to reduce to condition (2.90).

Under multiparameter loading the alternating plastic flow can also take place at lower absolute values of the stress amplitudes. The necessary conditions may easily be established in every particular problem, using the conditions of fictitious yield surface degeneration into a line or a point.

When the fictitious yield surface is nowhere degenerated, shakedown can take place only if, in the region of admissible constant stresses

2 Fundamental theorems of the shakedown theory

(i.e. inside the fictitious yield surface), there can be found a statically admissible stress distribution in equilibrium with constant load components, in accordance with (2.60).

The dangerous state neighbouring with shakedown arises by the accumulation of plastic deformations of the same sign at each cycle, i.e. this is the state of incremental collapse. This is the case when the shakedown problem reduces to the non-trivial problem of limit equilibrium (2.88) for a fictitiously non-homogeneous and anisotropic body.

From the above considerations it follows, in particular, that the determination of the incremental collapse conditions is the basic task in the shakedown theory, a task of prime importance to a number of technological applications. This fact alone is enough to justify the attempts at developing the theory with a proper use of the present-day mathematical tools.

The existence of a kinematically admissible collapse mechanism is, as stressed before, a common characteristic feature of both incremental collapse and limit analysis of elastic-plastic structures. However, while in the latter case the mechanism is formed in an 'instantaneous' manner, in the former the stresses at different points of the structure reach the yield surface at different instants of time during the cycle. We conclude that non-isochronism is the necessary condition for incremental collapse to take place.

It can be readily shown that, if the stress distribution $\sigma_{ij\tau}^*$ or $(\sigma_{ij} - \sigma_{ij\tau}^{(e)})^*$ takes place simultaneously, i.e. for the same load pattern at all the points of the body at which the plastic strain increments $\Delta\epsilon_{ij0}''$ accompanying the considered collapse mechanism differ from zero, then the inequality (2.86) fully coincides with the inequality (2.89) for the 'instantaneous' collapse.

It is interesting to note that the presented necessary condition for the incremental strain accumulation is sometimes also satisfied under proportional (one-parameter) loading; this is the case when the stress distribution in the elastic state is not in full agreement (sign-wise) with the actual plastic collapse mechanism of the structure in question. Suitable examples were given in Chapter 1; one of them (concerning a purely mechanical proportional loading) is shown in Fig. 1.13.

On the other hand, it follows from the formulated necessary condition that in some situations we can postulate a priori that incremental collapse will not take place. Then there is no need to make any attempts to obtain a solution within the framework of the shakedown theory. Regrettably enough, this fact was unknown to the author of the paper

[2.34] which is devoted to the study of a simply supported circular plate subjected to time-dependent uniform pressure.

We can treat the differences between the right-hand sides of (2.86) and (2.89), following Prager [2.7], as a decrease in the load-carrying capacity of the structure, resulting from variations of certain parts of external actions. This is best pronounced in the case of time-dependent temperature fields since the constant thermal stresses do not influence the limit equilibrium conditions whatsoever. The thermal variations are found to lead frequently to the strain accumulation with every cycle. In addition, this effect appears to take place not only under simultaneous actions of constant and repeated loadings but also when the former are absent provided some definite conditions are met. Somewhat simpler examples were given in Chapter 1; the problem will again be attacked in the following chapters, starting from more general premises.

2.7. On the uniqueness of stress state at the limiting cycle

The statical and kinematical theorems enable us, in the general case, to assess both from above and from below the limiting values of loading parameters at which shakedown takes place. In the following chapters there will be presented suitable methods which make it possible to narrow the bounds inside which the exact or complete, i.e. satisfying all the statical and kinematical conditions, answer should lie. In this context, the problem of uniqueness of stresses at the limiting cycle is essential.

Confining ourselves, for the time being, to the fixed program of cyclic loading, we shall show that the residual stress distribution in those regions of the body where the plastic strain rates during the limiting cycle are non-vanishing is unique, being independent of the preceding deformation history. From this fact it follows that the instantaneous distribution of actual, i.e. 'elastic' plus residual, stresses is also unique. The proof will be constructed by reduction to absurdity.

Let $\sigma_{ij}(\tau)$ represent the actual stresses in the limiting cycle, i.e. in the state preceding the occurrence of cyclic plastic deformation under the prescribed loading program; let $\dot{\epsilon}_{ij}''(\tau)$ be the corresponding, in accordance with the associated flow rule (2.8), admissible cycle of plastic strain rates; and let $\pi_{ij}(\tau)$ denote the probably existing, differing from $\sigma_{ij}(\tau)$, admissible distribution of stresses supporting the same program of cyclic loading but preceded by different history.

2 Fundamental theorems of the shakedown theory

On account of the Melan theorem, in the shakedown situation we can write

$$\sigma_{ij} = \sigma_{ij}^{(e)} + \bar{\rho}_{ij}, \quad \pi_{ij} = \sigma_{ij}^{(e)} + \bar{\mu}_{ij}, \tag{2.92}$$

where $\sigma_{ij}^{(e)}$ are the 'elastic' stresses, the same in both distributions under the given loading program, whilst $\bar{\rho}_{ij}$ and $\bar{\mu}_{ij}$ are time-independent residual stresses, assumed to differ from each other.

Using the virtual work principle (2.15) we get

$$\int_0^T d\tau \int (\sigma_{ij} - \pi_{ij}) \dot{\epsilon}_{ij}'' \, dv = \int_0^T d\tau \int (\bar{\rho}_{ij} - \bar{\mu}_{ij}) \dot{\epsilon}_{ij}'' \, dv =$$
$$= \int (\bar{\rho}_{ij} - \bar{\mu}_{ij}) \, dv \int_0^T \dot{\epsilon}_{ij}'' \, d\tau = \int (\bar{\rho}_{ij} - \bar{\mu}_{ij}) \Delta \epsilon_{ij}'' \, dv = 0 \tag{2.93}$$

since $(\bar{\rho}_{ij} - \bar{\mu}_{ij})$ is a self-equilibrating distribution of time-independent stresses.

Now, according to the Drucker postulate (2.7), we have

$$(\sigma_{ij} - \pi_{ij}) \dot{\epsilon}_{ij}'' \geq 0, \tag{2.94}$$

the inequality sign applying at all those points of the body where the stresses σ_{ij} reach the yield surface during the cycle and $\pi_{ij} \neq \sigma_{ij}$ at the corresponding instants of time. Thus, the conditions (2.93) and (2.94) are clearly contradicting each other, unless the distribution of stresses (total as well as self-equilibrating) is unique in those regions of the body in which the stresses over the cycle reach the yield surface. Uniqueness is here meant to within an accuracy of that component which does not affect the yielding.

The above considerations are also valid in the case of an alternating plastic flow satisfying $\Delta \epsilon_{ij}'' = 0$ everywhere within the body, however when the plastic strain rates at particular points have non-zero values during the cycle. The uniqueness of the self-equilibrating stresses (also to within an accuracy of those components which have nothing to do with the plastic flow) at appropriate points of the body is obvious. The question of which components are irrelevant, in a given situation, can be answered in the light of the adopted yield condition and the characteristics of the cycle. Some specific examples are shown in Fig. 2.7 in which the points O indicate the time-independent stress states. Segments OA ($O'A'$), OB ($O'B'$), OC correspond to the imposed variable

2.8. On the boundedness of inelastic deformation

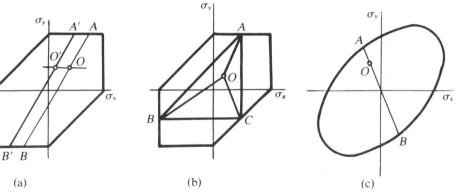

Fig. 2.7. Local uniqueness of stresses at a limiting cycle in the conditions of alternating plastic flow.

stresses. The checking of each of the shown examples is left to the reader.

Although the above proof is virtually valid for a prescribed loading program, the argument concerning the uniqueness of self-equilibrating stresses over the limiting cycle can be extended to an arbitrary loading program, i.e. for loading parameters arbitrarily varying inside prescribed limits, including the most unfavourable with respect to the actual collapse mechanism sequence of external load applications. This sequence can easily be found with the use of relationship (2.81) provided the 'elastic' stresses, supporting each of the actions, are known.

On the other hand, it must be remembered that it was just the limiting cycle which was considered; if shakedown takes place at a cycle whose parameters are smaller than the limiting ones, there are no grounds for stating that the stabilized stress state is unique.

The problem of stress state uniqueness in bar structures at the limiting cycle is investigated in [2.27].

2.8. On the boundedness of inelastic deformation preceding shakedown

Of importance in practical applications of the theory is the problem of strains and displacements that accumulate in the process of formation of a favourable distribution of the self-equilibrating stresses which ensure subsequent purely elastic behaviour of a structure. The problem was

2 Fundamental theorems of the shakedown theory

first approached by Koiter [2.2], who adopted the amount of plastic work done in the shakedown process as the estimate of an average level of strain accumulation. Following the quoted monograph, we shall show that, for the shakedown safety coefficient $m > 1$ (referred to the sum of 'elastic' stresses supporting all the applied actions), the over-all deformation of the structure remains finite.

The occurrence of the inequality $m > 1$ means that, after being multiplied by m, the 'elastic' should at least satisfy the necessary shakedown condition

$$m\sigma_{ij}^{(e)} + \bar{\rho}_{ij} = \sigma_{ij}^{(a)}, \tag{2.95}$$

where $\sigma_{ij}^{(a)}$ denotes the admissible stress state, as in the corresponding safe (working) cycle the equality

$$\sigma_{ij}^{(e)} + \bar{\rho}_{ij} = \sigma_{ij}^{(s)} \tag{2.96}$$

holds.

Here, unlike as in the limiting cycle (2.95), $\bar{\rho}_{ij}$ is not necessarily unique (see Sec. 2.7), thus the right-hand side of the equality (2.95) can be at some points of the body greater than, and at other points smaller than, $m\sigma_{ij}^{(s)}$. Only in the case of $\bar{\bar{\rho}}_{ij} = (1/m)\bar{\rho}_{ij}$ can we write $\sigma_{ij}^{(a)} = m\sigma_{ij}^{(s)}$ and rearrange the relationship (2.7), resulting from the Drucker postulate, to take the form

$$(\sigma_{ij} - \sigma_{ij}^{(s)})\dot{\epsilon}_{ij}'' \geq \frac{m-1}{m} \sigma_{ij}\dot{\epsilon}_{ij}''. \tag{2.97}$$

On combining the expressions (2.23) and (2.97), we get

$$\dot{W}_p = \int \sigma_{ij}\dot{\epsilon}_{ij}'' \, dv \leq \frac{m}{m-1} \int (\sigma_{ij} - \sigma_{ij}^{(s)})\dot{\epsilon}_{ij}'' \, dv = -\frac{m}{m-1} \dot{A}. \tag{2.98}$$

Integrating the inequality (2.98) with respect to time from $\tau = 0$ to $\tau = \tau_*$ (where τ_* is an instant of time at which the structure is recognized to shake down) and taking account of (2.17) we are lead to the condition

$$W_p \leq \frac{m}{m-1}[A(0) - A(\tau_*)] = \frac{m}{m-1}\left\{\int \frac{1}{2} A_{ijhk}\bar{\rho}_{ij}\bar{\rho}_{hk} \, dv - \int \frac{1}{2} A_{ijhk}[\rho_{ij}(\tau_*) - \bar{\rho}_{ij}][\rho_{hk}(\tau_*) - \bar{\rho}_{hk}] \, dv, \tag{2.99}$$

valid for loading imposed on the natural state of the structure.

2.8. On the boundedness of inelastic deformation

Since the second integral in the right-hand side of the inequality is non-negative, we can also write

$$W_p \leq \frac{m}{m-1} \int \frac{1}{2} A_{ijhk} \bar{\rho}_{ij} \bar{\rho}_{hk} \, dv. \tag{2.100}$$

It is worth noting that the neglected integral is equal to zero just for the limiting cycle, owing to the uniqueness of self-equilibrating stresses proved in Sec. 2.7.

From inequality (2.100) there clearly follows a sufficient condition for the boundedness of plastic work in the shakedown process; it is for this reason that the safety coefficient with respect to shakedown should be slightly greater than one.

It is proper to mention here that some attempts have been made to include the requirement of plastic work finiteness into the very formulation of the statical shakedown theorem [2.20]. This requirement is also employed in the statement of the kinematical theorem [2.18].

However, as mentioned before, the proof of boundedness of the over-all plastic work does not suggest a method to evaluate any magnitudes of local strains and displacements that can be generated in the process. Although the general proof of the finiteness of the latter has so far been lacking, the strain and displacements are unlikely to attain extensively large values in view of the boundedness of the self-equilibrating stresses in the structure and the compatibility of deformation (per cycle) in the actual collapse mechanism.

In engineering problems one would expect, though, to have a somewhat more specific assessment, especially of the displacements in the shakedown state.

The problem of assessing displacements at shakedown can be formulated in a number of ways. One of the simpler and application-oriented ways is to employ the inequalities of the type (2.99) and (2.100) together with the constraints on stresses and strain increments in the limiting state of shakedown (but by no means during the shakedown process!).

Remembering that the distribution of self-equilibrating stresses $\bar{\rho}_{ij}$, satisfying the condition (2.96) at a given value of the safety coefficient m, can turn out to be non-unique, we shall rewrite inequality (2.99) in the form

$$W_p = \int_0^{\tau_*} d\tau \int \sigma_{ij} \dot{\epsilon}_{ij}'' \, dv \leq \frac{m}{m-1} \min_{\bar{\rho}_{ij}} \left\{ \int \frac{1}{2} A_{ijhk} \bar{\rho}_{ij} \bar{\rho}_{hk} \, dv - \int \frac{1}{2} A_{ijhk} [\rho_{ij}(\tau_*) - \bar{\rho}_{ij}][\rho_{hk}(\tau_*) - \bar{\rho}_{hk}] \, dv \right\}. \tag{2.101}$$

2 Fundamental theorems of the shakedown theory

The actual residual stresses $\rho_{ij}(\tau_*)$ at the end of the shakedown process should obey the following requirements:

$$\rho_{ij,j}(\tau_*) = 0 \quad \text{in the interior } V, \tag{2.102}$$

$$\rho_{ij}(\tau_*)n_j = 0 \quad \text{on the surface } S_p,$$

$$f[\sigma_{ij}^{(e)} + \rho_{ij}(\tau_*)] \leq 0 \tag{2.103}$$

for all the values of $\sigma_{ij}^{(e)}$ which are possible during the cycle at any point of the body, and

$$\Delta\epsilon_{ij}'' + A_{ijhk}\rho_{hk}(\tau_*) = \frac{1}{2}(\Delta u_{i,j} + \Delta u_{j,i}), \tag{2.104}$$

and lastly the kinematical boundary conditions must be satisfied on the surface S_u.

The magnitudes

$$\Delta\epsilon_{ij}'' = \int_0^{\tau_*} \dot{\epsilon}_{ij}'' \, d\tau, \quad \Delta u_i = \int_0^{\tau_*} \dot{u}_i \, d\tau \tag{2.105}$$

denote here respectively the plastic strain and the displacement increments over the time of the shakedown process, i.e. from $\tau = 0$ to $\tau = \tau_*$.

During the shakedown process the actual stresses $\sigma_{ij} = \sigma_{ij}^{(e)} + \rho_{ij}(\tau)$ are related to the plastic strain rates $\dot{\epsilon}_{ij}''$ (the left-hand side of the inequality (2.101)) by means of the associated flow rule (2.8):

$$\dot{\epsilon}_{ij}'' = \lambda \frac{\partial f}{\partial \sigma_{ij}}, \quad \lambda \geq 0. \tag{2.106}$$

Imposing no other constraints on either the stresses or the plastic strain rates in the shakedown process, we shall now formulate the problem of upper bound to the final displacements as follows:

It is required to find a distribution of plastic strain rates $\dot{\epsilon}_{ij}''$, varying with time from $\tau = 0$ to $\tau = \tau_*$, which yields the absolute maximum of the displacement increment $\Delta u_i(C)$ at a given point C and in a given direction,

$$\max_{\dot{\epsilon}_{ij}''} \Delta u_i(C) = ?, \tag{2.107}$$

so that the conditions (2.101)–(2.106) are at the same time satisfied.

2.8. On the boundedness of inelastic deformation

It is obvious that the constraints (2.102)–(2.106) include the actual loading program but are not restricted to it. In other words, those constraints appear to be weaker than the constraints corresponding to the actual program alone. This is the reason why the solution to the problem (2.107) yields an upper bound to the actual displacements that can be accumulated over the shakedown process; this is also the reason why inequality (2.101) will unlikely become an equality as it was necessarily assumed in the general case.

Let us indicate that in the case of a piece-wise linear yield surface the strain rates can be replaced by the strain increments in the left-hand side of the inequality (2.101). If, in addition, the relationship (2.103) is rewritten to take the form (see Sec. 2.4)

$$\max_{\tau} f[\sigma_{ij}^{(e)} + \rho_{ij}(\tau_*)] = f[\sigma_{ij}^* + \rho_{ij}(\tau_*)] \leq 0,$$

it turns out that the time has disappeared from the description of the problem.

Other possible formulations of the problem to estimate the bounds for the limits of the displacements, which develop during the shakedown process, were investigated by Vitiello [2.29] and Maier [2.30], who used the finite element approach together with the piece-wise linear yield surface, as well as by König [2.31], who dealt with deflections in one-dimensional systems. A somewhat different, although stemming from the same approach, formulation of the problem was suggested by Capurso [2.32], who replaced the relationship (2.101) by a stronger inequality which prevents the obtained upper bound to coincide with the actual displacement under any conditions. In his description neither the actual stresses nor the strain rates are present.

It must be emphasized that the lack of a formal proof concerning the finiteness of the local values of final strains and displacements is by no means an obstacle to the solutions to the problem (2.107); if the above magnitudes are not bounded from above, then the corresponding upper assessment would be unbounded; such situations were never encountered when $m > 1$ and it is therefore highly probable that they do not take place at all.

One of the main drawbacks of the considered formulation is the impossibility of determining a specific loading program (consisting in load variations inside prescribed limits) under which the accumulated final displacements are at their maximum. Apart from this, when the inequalities of the type (2.99) and (2.100) are used, it is virtually impossible

2 Fundamental theorems of the shakedown theory

to seek a loading program at which the relevant displacement is at minimum. Eventually, the gap between the obtained upper bound and the actual maximum value of the final displacement is hard to evaluate within the presented approach.

In order to overcome the indicated disadvantages, a basically different formulation of the problem can be put forward in which the shakedown process itself is considered:

It is required to find a distribution of plastic strain rates and the corresponding residual stress rates which leads to a transition from the given initial state ($\epsilon_{ij}''(0) = 0$, $\rho_{ij}(0) = 0$) to the shakedown state ($\dot{\epsilon}_{ij}'' = 0$, $\rho_{ij} = \rho_{ij}(\tau_*)$) at which the final displacement $\Delta u_i(C)$ of a point C attains its maximum (minimum) value,

$$\max_{\dot{\epsilon}_{ij}''} \Delta u_i(C) = ?, \quad [\min_{\dot{\epsilon}_{ij}''} \Delta u_i(C) = ?], \tag{2.108}$$

under the following constraints:

$$\dot{\rho}_{ij,j} = 0, \quad \dot{\rho}_{ij} n_j = 0, \tag{2.109}$$

$$f\left[\sigma_{ij}^{(e)} + \int_0^\tau \dot{\rho}_{ij} \, d\zeta\right] \leq 0 \quad \text{for } 0 \leq \tau \leq \tau_*. \tag{2.110}$$

$$\dot{\epsilon}_{ij}'' + A_{ijhk}\dot{\rho}_{hk} = \frac{1}{2}(\dot{u}_{i,j} + \dot{u}_{j,i}), \tag{2.111}$$

$$\dot{\epsilon}_{ij}'' = \lambda \frac{\partial f}{\partial \sigma_{ij}}, \tag{2.112}$$

where $\lambda \geq 0$ for $f = 0$ and $\dot{f} = 0$, otherwise $\lambda = 0$.

The above statement differs from the usual formulation of the step-by-step procedure of investigating the elastic-plastic behaviour, mainly in that it does not specify the loading program, whilst prescribing only the limits of the external load variations. The program, fitting one of the extremum conditions (2.108), is being determined in the course of the solution. Moreover, the extremum character of the problem makes it possible to simplify the calculations in comparison with those used in the analysis of the kinetics of elasto-plastic deformation. The upper and lower bounds to the magnitudes sought are eventually obtained at the cost of modifications in the system of constraints.

It is clear that this approach enables us not only to find the extremum values of the final displacements in the shakedown state but also, if necessary, the extremum values of those displacements obtained in the course of the solution that vary during the plastic deformation process. For instance, one may look for the maximum total displacement of a point in the prescribed direction although this maximum is not necessarily expected to take place in the shakedown state:

$$\max_{\epsilon_{ij}''} \left[\int_0^\tau \dot{u}_i \, d\zeta + u^{(e)}(\tau) \right] = ?.$$

The system of constraints (2.109)–(2.112) remains here unaltered.

The subject matter of Ikrin's papers (for example, in his joint paper [2.33] with Filippov) can be recognized as one of the variants of the considered approach to the assessment of final displacements. So far as computing time is concerned, this variant can be placed somewhere in between the procedure (2.101)–(2.107) and the direct determination of the kinetics of the elastic-plastic deformation.

In the shown formulations the deformations accumulated prior to shakedown were assessed within the simple concept of ideal plasticity. However, bearing in mind the applications of the shakedown theory to the structural analysis, it is necessary to be aware of the errors due to the approximate character of the stress–strain relationships. The differences between the actual strains and displacements and those evaluated under the assumption of an elastic-perfectly plastic body can turn out to be substantial.

References

[2.1] Prager, W., Hodge, Ph.G., Jr., *Theory of perfectly plastic solids*. Prentice-Hall, New York, 1951.
[2.2] Koiter, W., General theorems for elastic-plastic solids. *Progress in solid mechanics*, VI, Edited by I.N. Sneddon and R. Hill. North-Holland, Amsterdam, 1960.
[2.3] Kachanov, L.M., *Theory of plasticity*, in Russian, Nauka, Moscow, 1969. English version, *Foundations of the theory of ideal plasticity*, North-Holland, Amsterdam, 1971.
[2.4] Ivlev, D.D., *Theory of perfect plasticity*, in Russian, Nauka, Moscow, 1966 (Also in English, North-Holland, Amsterdam, 1971).
[2.5] Drucker, D.C., A more fundamental approach to plastic stress–strain relations. *Proc. Ist. U.S. Nat. Congr. Appl. Mech.* 487, 1951.
[2.6] Melan, E., Zur Plastizität des räumlichen Kontinuums. *Ing. Arch.*, Nr 9, 116, 1938.
[2.7] Prager, W., Shakedown in elastic-plastic media subjected to cycles of load and temperature. *Symp. su la Plasticita nella Scienza delle Construzioni*. Bologna, 1957.

2 Fundamental theorems of the shakedown theory

[2.8] Drucker, D.C., Extension of the stability postulate with emphasis on temperature changes. *Plasticity*. Proc. 2nd Symp. on Naval Struct. Mech., 1960.
[2.9] Naghdi, P.M., Stress–strain relations in plasticity and thermoplasticity, *Plasticity*. Proc. 2nd Symp. on Naval Struct. Mech., 1960.
[2.10] König, J.A., A shakedown theorem for temperature dependent elastic moduli. *Bull. Acad. Polon. Sci. Techn.*, *17*, Nr 3, 1969.
[2.11] König, J.A., O projektowaniu konstrukcji z uwzględnieniem przystosowania w przypadku gdy stałe sprężyste zależą od temperatury. *Rozpr. Inż*, *20*, Nr 3, 1972, 423–434.
[2.12] Ceradini, G., Sull adattamento dei corpi elasto-plastici soggetti ad azioni dinamiche. *Giornale del genio civile*, Nr 4–5, 1969.
[2.13] Gavarini, C., Sul riento in fase elastica delle vibrazioni forzate elasto-plastiche. *Giornale del genio civile*, Nr 4–5, 1969.
[2.14] Ho, H.-Sh., Shakedown in elastic-plastic systems under dynamic loadings. *Paper ASME*, Nr 71-APMW, *27*, 1971.
[2.15] Horne, M.K., The effect of variable repeated loads on the plastic theory of structures. *Research Eng. Struct. Supp.* (Colstone Papers, 2), *141*, 1949.
[2.16] Augusti, G., Baratta, A., Plastic shakedown of structures with stochastic local strength. *Assoc. Int. Ponts et Charpentes*. *13*, S.L. 1973, 287–292.
[2.17] Augusti, G., Barratta A., Theory of probability and limit analysis of structures under multiparameter loading. *Proc. Int. Symp. on Found. of Plasticity. Warsaw, 1972*. Noordhoff, Leyden, 1973.
[2.18] Koiter, W.T., A new general theorem on shakedown of elastic-plastic structures. *Proc. Kon. Ned. Ak. Wet.*, B. 59, 24, 1956.
[2.19] Frederic, C.O., Armstrong, P.J., Convergent internal stress and steady cyclic states of stress. *J. Strain Anal.*, Nr 2, 1966.
[2.20] Rosenblum, V.I., On shakedown analysis of nonuniformly heated elastic-plastic bodies, in Russian, *Zhurn. prikl. mekh. tekh. fiz.* (PMTF), Nr 5, 1965.
[2.21] Donato, O., Second shakedown theorem allowing for cycles of both loads and temperature. *Rend. Ist. Lombardo. Accad. sci. e lett.*, A 104, Nr 1, 1970, 265–277.
[2.22] Gokhfeld, D.A., Some problems of the shakedown theory for plates and shells, in Russian. *VI All-Union Conf. on Plate and Shell Theory*, (Baku, 1966), Nauka, Moscow, 1966.
[2.23] Gokhfeld, D.A., On application of Koiter's theorem to the shakedown problems for nonuniformly heated elastic-plastic bodies, in Russian, *Prikl. Mekh.*, *3*, Nr 8, 1967.
[2.24] Gokhfeld, D.A., Cherniavsky, O.F., On the incremental collapse conditions in the axi-symmetrical shakedown problems dealt with by linear programming methods, in Russian, *Izv. AN SSSR, MTT*, Nr 3, 1970.
[2.25] Gokhfeld, D.A., Cherniavsky, O.F., Methods of solving problems in the shakedown theory of continua. *Proc. Int. Symp. on Found. of Plasticity. Warsaw, 1972*. Noordhoff, Leiden, 1973.
[2.26] Gvozdev, A.A., Determination of limit load intensities for redundant structures at plastic deformation, in Russian, *Proceedings of Conf. on Plastic Deformation*, AN SSSR, Moscow, 1938.
[2.27] Neal, B.G., *The plastic methods of structural analysis*. Chapman and Hall, New York, 1956.
[2.28] Konieczny, L., Shakedown theorems for beam structures. *Bull. Acad. Polon. Sci. Ser. Sci. Techn.*, *22*, Nr 5, 1974, 353–362.

References

[2.29] Vitiello, E., Upper bounds to plastic strains in shakedown of structures subjected to cyclic loads. *Meccanica*, 7(3), 1972, 205.
[2.30] Maier, G., A shakedown matrix theory allowing for work-hardening and second order geometric effects. *Proc. Int. Symp. Foundations of Plasticity, Warsaw, 1972*, Noordhoff, Leiden, 1973.
[2.31] König, J.A., Shakedown deflection. A finite element approach. *Prilozhna Mekh.* 3, Nr 2, 1972, 65–69.
[2.32] Capurso, M., A displacement bounding principle in shakedown of structures subjected to cyclic loads. *Int. Solids and Struct.* 10, Nr 1, 1974, 77–92.
[2.33] Ikrin, V.A., Filippov, V.V., Assessment of deflection at shakedown of plane frames, in Russian, *Stroi, t. mekh. rasch. Sooruzh*, Nr 3, 1974.
[2.34] Muspratt, M.A., Shakedown of steel plates. *Trans. ASME*, 38, Nr 4, 1971.
[2.35] Corradi, L., Maier, G., Dynamic inadaptation theorem for elastic-perfectly plastic continua. *J. Mech. Phys. Solids*, 22, 1975, 401–413.

3

Generalized variables in the shakedown analysis

It is useful to introduce generalized variables in order to determine the conditions for shakedown in various engineering structures such as shells and plates, obeying suitable yield surfaces, often called the interaction surfaces. The use of generalized variables in branches of continuum mechanics such as the strength analysis of materials, the theory of elasticity and the limit analysis enables us to reduce the number of independent variables by simply excluding from the governing equations one of the arguments, e.g. the coordinate of a point across the thickness of a shell or a plate, or across the depth of a beam. The generalized variables were first employed in the presence of alternating repeated loading in the shakedown problems, just by direct analogy with the limit analysis theory. It was not immediately recognized that this problem is in the case of general loading conditions far from being trivial.

The introduction of generalized strains (or strain rates) is necessarily based on certain assumptions regarding their distribution. These hypotheses are of a kinematical nature such as the assumption of preservation of plane sections (for beams) or of straight normals (for shells and plates). The generalized stresses (or internal forces) are chosen so that the work which they do on the generalized strains (or strain rates) equals the work of the corresponding stresses done on the local (physical) strains (or strain rates).

Under circumstances of repeated loading which causes local alternating plastic flow there are, as a rule, no grounds for any assumptions to be made as to the plastic strain distribution across the thickness (excluding the trivial case of uniform compression-tension). Moreover, in this situation there arises no necessity to apply generalized variables as the problem of determining the limiting cycle parameters can be properly solved in the physical stresses; the knowledge of their extre-

3 Generalized variables in the shakedown analysis

mum values (amplitudes) suffice to find the limiting condition. In particular, it is here advisable to employ criteria for degeneration of the fictitious yield surface into a line or a point (see Sec. 2.4) or to apply the sufficiency conditions (2.90), (2.91).

Thus the introduction of generalized variables makes sense only in situations in which the conditions for incremental collapse are decisive. The conventional determination of generalized stresses together with a suitable interaction surface as a function of 'instantaneous' values of physical stresses characteristic for the limit analysis remains meaningful under variable repeated loading only when some definite conditions are fulfilled which will be formulated below. The breakdown case was first shown by Koiter [3.1] in his comments to the shakedown theorem for bar structures, formulated by Neal [3.2] in terms of generalized variables. Koiter observed that the considered theorem holds no longer good when the location of the neutral axis of a cross-section of the beam does not remain fixed in the process of elastic-plastic deformation. This is particularly true for the bending of sections which are non-symmetric with respect to their neutral axes.

The example shown in Fig. 3.1 indicates that in this portion of the cross-section which is bounded by the initial (corresponding to elastic behaviour) and the final locations of the neutral axis, the stresses increase while the global unloading takes place, i.e. the external bending moment decreases. From this it follows that in the limiting cycle there exist stresses at points belonging to one and the same cross-section of the

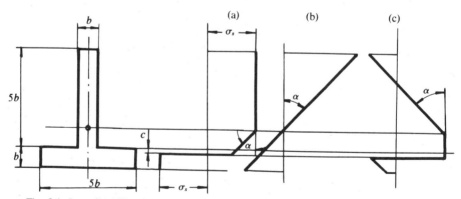

Fig. 3.1. Inapplicability of trivial approach to the determination of generalized variables under repeated loading.

3.1. General method of constructing an introduction surface

beam that reach the yield stress non-simultaneously. This situation cannot be appropriately described in terms of the bending moment alone.

It is worth noting that, similarly as it was the case with the bar system shown in Fig. 1.12, the present example illustrates the possibility of strain accumulation under cyclic one-parameter loading. Fig. 3.1a shows the stress distribution at the limiting cycle under $M = M_{max}$, Fig. 3.1c corresponds to $M = 0$, i.e. to the end of the loading cycle, and the distribution of 'elastic' stresses in equilibrium with the moment $M = M_{max}$ is seen in Fig. 3.1b. The determination of the limiting cycle parameters with the help of diagrams shown and the methods put forward in the first two chapters is left to the reader.

The non-isochronous attainment, in the limiting cycle, of the yield surface by the stresses at points belonging to the same normal to the middle surface is a characteristic feature of plates and shells subjected to variable repeated actions of mechanical and, in particular, thermal nature. It is only in those conditions in which the interaction between tension, bending and torsion is absent or negligible, that the conventional introduction of the generalized stresses can serve its purpose [3.3]–[3.5].

The general approach to the determination of the interaction surfaces in terms of the generalized stresses including the above discussed peculiarities of shakedown problems was presented in [3.6, 3.7].

3.1. General method of constructing an interaction surface

Let us briefly recall the method used for introducing the generalized variables in the limit analysis theory (see, for example, [3.10]).

The basic inequality of the kinematical limit state theorem has the form [3.9]

$$\int X_i \dot{u}_i^0 \, dv + \int_{S_p} p_i \dot{u}_i^0 \, dS \geq \int \sigma_{ij} \dot{\epsilon}_{ij}^0 \, dv, \tag{3.1}$$

where $\dot{\epsilon}_{ij}^0 = \dot{\epsilon}_{ij}''$ is the field of kinematically admissible plastic strain rates, $\dot{u}_i^0$ is the corresponding field of displacement rates and σ_{ij} are the stresses belonging to the yield surface.

On the basis of Kirchhoff's hypothesis, the distribution of plastic strain rates across the thickness z of the shell or plate can be written

3 Generalized variables in the shakedown analysis

as

$$\dot{\epsilon}''_{\alpha\beta} = z\dot{\kappa}_{\alpha\beta} + \dot{e}_{\alpha\beta} \quad (\alpha, \beta = 1, 2), \tag{3.2}$$

where $\dot{\kappa}_{\alpha\beta}, \dot{e}_{\alpha\beta}$ are the generalized strain rates.

Integrating the right-hand side of inequality (3.1) over the thickness and taking account of (3.2), we obtain

$$\int_{-h}^{h} \sigma_{ij}\dot{\epsilon}^0_{ij} \, \mathrm{d}z = \int_{-h}^{h} \sigma_{\alpha\beta}(z\dot{\kappa}_{\alpha\beta} + \dot{e}_{\alpha\beta}) \, \mathrm{d}z = M_{\alpha\beta}\dot{\kappa}_{\alpha\beta} + N_{\alpha\beta}\dot{e}_{\alpha\beta}, \tag{3.3}$$

where

$$M_{\alpha\beta} = \int_{-h}^{h} \sigma_{\alpha\beta}z \, \mathrm{d}z, \quad N_{\alpha\beta} = \int_{-h}^{h} \sigma_{\alpha\beta} \, \mathrm{d}z. \tag{3.4}$$

The distribution of the yield surface stresses $\sigma_{\alpha\beta}$ over the thickness of the shell for any possible relationships between $\dot{\kappa}_{\alpha\beta}$ and $\dot{e}_{\alpha\beta}$, can be found by using (3.2) together with the associated flow rule (2.8). Thus, the generalized stresses (3.4) can be evaluated to within a certain multiplier whose elimination enables those stresses to be directly related to each other, i.e. to obtain a 'finite relationship' in the form of an interaction surface [3.10]. The set of values $M_{\alpha\beta}, N_{\alpha\beta}$ for all the possible relations between the components $\dot{\kappa}_{\alpha\beta}, \dot{e}_{\alpha\beta}$ constitutes the yield, or interaction surface in the space of generalized stresses,

$$F(M_{\alpha\beta}, N_{\alpha\beta}) = 0. \tag{3.5}$$

General solutions to the problems of finite relationships and properties of the surface (3.5) were considered by Iliushin [3.10], various cases of construction of this surface were elaborated by Shapiro [3.11], Rzhanitzin [3.12, 3.13], Rozhdestvenski [3.14], Onat [3.15], Sawczuk and Rychlewski [3.16], Hodge [3.17, 3.18], Ivlev [3.19], Onat and Prager [3.20] and others.

Approximate methods of building the surface (3.5) were worked out in [3.21]–[3.24]. A different approach to the determination of generalized variables was suggested by Rosenblum [3.25, 3.26]. Finally, a comparison between the surfaces (3.5) corresponding to various yield conditions was made in [3.27, 3.28].

Let us now proceed to the selection of generalized stresses proper

3.1. General method of constructing an introduction surface

to the problems of shakedown. Since, at incremental collapse the distribution of total (per cycle) plastic strain increments $\Delta\epsilon''_{\alpha\beta}$ across the thickness of the plate or shell should be kinematically admissible, we may express it, similarly to (3.2), as

$$\Delta\epsilon''_{\alpha\beta} = z\Delta\kappa_{\alpha\beta} + \Delta e_{\alpha\beta}. \tag{3.6}$$

Taking the integral of the right-hand side of the restated kinematical theorem inequality (2.86) over the thickness, and substituting (3.6) into it, we arrive at

$$\int_{-h}^{h} \sigma^0_{\alpha\beta*}(z\Delta\kappa_{\alpha\beta} + \Delta e_{\alpha\beta})\,\mathrm{d}z = M^0_{\alpha\beta*}\Delta\kappa_{\alpha\beta} + N^0_{\alpha\beta*}\Delta e_{\alpha\beta}, \tag{3.7}$$

where

$$M^0_{\alpha\beta*} = \int_{-h}^{h} \sigma^0_{\alpha\beta*} z\,\mathrm{d}z, \quad N^0_{\alpha\beta*} = \int_{-h}^{h} \sigma^0_{\alpha\beta*}\,\mathrm{d}z. \tag{3.8}$$

Due to the fact that the relationship (2.76)[1] between the fictitious yield surface stresses σ^0_{ij*} and the total (per cycle) plastic strain increments $\Delta\epsilon''_{ij}$ holds and is fully analogous to the law (2.8), the above described procedure of determining the generalized stresses $M^0_{\alpha\beta*}$ and $N^0_{\alpha\beta*}$, adopted from the limit analysis theory, is suitable for the present purpose as well.

The coincidence of (3.3) with (3.7) appears, however, to be merely formal. A peculiarity of the limiting values of the constant generalized stresses $M^0_{\alpha\beta*}, N^0_{\alpha\beta*}$ in general shakedown problems consists in the fact that it is the non-simultaneous plastic flow across the thickness that corresponds to them, because the limiting values of the constant stresses $\sigma^0_{\alpha\beta*}$ at different points of the normal to the middle surface depend on the generally non-isochronous envelope of the distribution of variable 'elastic' stresses.

The construction of the finite relationship between the constant generalized stresses $M^0_{\alpha\beta*}, N^0_{\alpha\beta*}$ usually proves to be more difficult than in corresponding limit analysis problems. This is so because the fictitious yield surface does change its dimensions and even it can change its shape, depending on the coordinates of a point along the normal. As shown in

---

[1] We confine ourselves to the relevant conditions only.

3 Generalized variables in the shakedown analysis

Section 2.5, the number of faces of the fictitious yield surface can be reduced (e.g. from six to four in the plane state) as the variable stress amplitude grows. This number is dependent on the coordinates of a point and the magnitudes of external agencies.

Concerning the integration procedure in connection with (3.7), the determination of generalized stresses in the shakedown problems is comparable with analogous computations encountered in the limit analysis theory of an anisotropic body with temperature-dependent yield stress [3.29, 3.30].

In the two sections to follow some examples of constructions of the fictitious interaction surfaces in terms of generalized stresses will be considered in order to illustrate the proposed method. A beam will be dealt with under two typical, variable repeated loadings generating one- and two-dimensional stress states.

3.2. Fictitious interaction surfaces for a rod under cycling normal stress

Consider the determination of a domain of admissible values of the constant, generalized stress components

$$\phi(M^0_{\alpha\beta*}, N^0_{\alpha\beta*}) = 0 \tag{3.9}$$

by way of the simple example of a bar subjected to variable normal 'elastic' stresses proportional to a certain parameter. For simplicity, the cross-section is assumed to be a rectangle with the depth of $2h$ and with unit width (Fig. 3.2). The constant yield point is denoted by σ_s.

In Fig. 3.2a–3.2e various possible cross-sectional distributions of 'elastic' stresses $\sigma^{(e)}_\tau$ are shown in accordance with the variable components of external actions. During a cycle those 'elastic' variable stresses can vary inside the shaded diagrams. The case (a) represents an action of a constant or a monotonically increasing load alone.

The regions of admissible values of constant normal stresses σ^0_*, corresponding to the diagrams a–e, are shown in Fig. 3.2f–3.2j. They are calculated by means of equation (2.56). If the variable stress components are absent, as in the case (a), the stresses σ^0_* can vary between $-\sigma_s$ and σ_s as in Fig. 3.2f. In all the remaining cases (g–j) the regions of admissible values are suitably smaller. The directions of the longitudinal strain increment vectors are shown by arrows (Fig. 3.2f–j) corresponding

3.2. Fictitious interaction surfaces for a rod

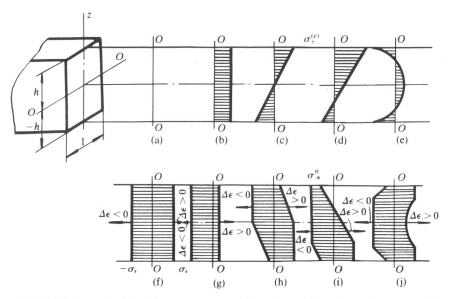

Fig. 3.2. Regions of admissible constant stresses for a beam subjected to various types of variable repeated loading.

to the upper and lower limits of the admissible values of constant stresses. Since the direction of plastic flow is determined by the total stresses $\sigma_*^0 + \sigma_\tau^*$, a positive strain increment can be accompanied by a negative constant stress σ_*^0, as shown, in particular, in Fig. 3.2i.

In the case of incremental collapse the distribution of plastic strain increments $\Delta\epsilon$ (per cycle) over the depth of the cross-section should be governed by the assumption that plane sections must remain flat and plane; hence

$$\Delta\epsilon = z\Delta\kappa + \Delta e. \tag{3.10}$$

In general, the strain increments $\Delta\epsilon$ can develop at different points of the cross-section non-simultaneously.

On proper specification, the expression (3.7) assumes the form

$$\int_{-h}^{h} \sigma_*^0 (z\Delta\kappa + \Delta e)\,dz = M_*^0 \Delta\kappa + N_*^0 \Delta e. \tag{3.11}$$

In order to compute M_*^0 and N_*^0 with the use of definitions of the

97

3 Generalized variables in the shakedown analysis

type (3.8), it is first necessary to know how the stresses σ_*^0 vary over the depth of the section as a function of the relation between $\Delta\kappa$ and Δe. To this end, equality (3.10) should be employed to find the signs of the increments $\Delta\epsilon$ as a function of the coordinate z for each possible ratio of the strain increments. According to condition (2.76), similar to the associated flow rule, to the positive strain increments $\Delta\epsilon$ there corresponds, as seen in Fig. 3.2f–j, the upper limit of the admissible values of constant stresses and vice versa.

For definiteness, consider the case of Fig. 3.2b, g in which the variable components of 'elastic' stresses are prescribed by

$$\sigma_\tau^{(e)} = \sigma_s t(\tau), \quad 0 \leq t(\tau) \leq t_*, \tag{3.12}$$

whereas the bounds on the region of admissible values of constant stresses, in accordance with (3.12), are

$$\sigma_*^0 = -\sigma_s \quad \text{for } \Delta\epsilon < 0,$$
$$\sigma_*^0 = \sigma_s(1 - t_*) \quad \text{for } \Delta\epsilon > 0. \tag{3.13}$$

The arbitrary distribution of normal stresses across the cross-section of the rod can be replaced by an axial force and a bending moment. Now the problem consists in finding the relations between those magnitudes which correspond to those possible distributions of time-independent stresses which, when added to the variable stresses (3.12), lead to the limiting cycle.

Let us calculate the magnitudes entering (3.11) for different ratios of $\Delta\kappa$ to Δe. The neutral axis $z = c$ is located in conformity with (3.10), i.e. from $\Delta\epsilon = c\Delta\kappa + \Delta e = 0$, we have

$$c = -\frac{\Delta e}{\Delta\kappa}. \tag{3.14}$$

For $|c| > h$ the plastic strains have clearly the same sign all over the cross-section and the bending moment is duly absent, $M = 0$. In consequence, it suffices to consider the interval

$$-h \leq c \leq h. \tag{3.15}$$

Let $\Delta\kappa > 0$; then

$$\text{sign } \Delta\epsilon = \text{sign } \Delta\kappa = 1 \quad \text{for } c < z \leq h,$$
$$\text{sign } \Delta\epsilon = -\text{sign } \Delta\kappa = -1 \quad \text{for } -h \leq z < c, \tag{3.16}$$

where $\text{sign } a = a/|a|$ stands for the sign of the magnitude a.

3.2. Fictitious interaction surfaces for a rod

Making use of (3.13), (3.15) and (3.16), we can rewrite the left-hand side of equality (3.11) as

$$\int_c^h \sigma_s(1-t_*)(z\Delta\kappa + \Delta e)\,dz + \int_{-h}^c (-\sigma_s)(z\Delta\kappa + \Delta e)\,dz =$$
$$= \sigma_s[-2c - t_*(h-c)]\Delta e + \sigma_s\left[h^2 - c^2 - \frac{t_*}{2}(h^2 - c^2)\right]\Delta\kappa.$$

On comparing the above with the right-hand side of (3.11) we arrive at

$$N_*^0 = -N_0\left[\frac{c}{h} + \frac{1}{2}t_*\left(1 - \frac{c}{h}\right)\right],$$
$$M_*^0 = M_0\left(1 - \frac{c^2}{h^2}\right)\left(1 - \frac{1}{2}t_*\right),$$
(3.17)

where $N_0 = 2\sigma_s h$, $M_0 = \sigma_s h^2$.

Similar calculations should also be made for $\Delta\kappa < 0$. They lead to the relationships

$$N_*^0 = N_0\left[\frac{c}{h} - \frac{1}{2}t_*\left(1 + \frac{c}{h}\right)\right],$$
$$M_*^0 = -M_0\left(1 - \frac{c^2}{h^2}\right)\left(1 - \frac{1}{2}t_*\right).$$
(3.18)

The expressions (3.17) and (3.18) can be looked upon as the parametric representation with respect to the parameters c, t_* of two families of curves. It is readily seen that these curves are all symmetric with respect to both the horizontal axis N_*^0/N and the vertical axis. Fictitious interaction curves, based on (3.17) and (3.18), are shown in Fig. 3.3.

Let us emphasize that, as the parameter t_* increases, the region of admissible constant load components tends to shrink while undergoing simultaneous translation to the left, i.e. in the direction of the negative normal forces. The latter is due to the action of the variable tensile (in contrast to the compressive) axial force. At $t_* = 2,0$ the region degenerates to a mere point which corresponds to the condition for alternating plastic flow all over the cross-section. At $t_* = 0$ we arrive at the known finite relationship for the limit analysis situation of a beam under simultaneous tension and bending [3.17].

3 Generalized variables in the shakedown analysis

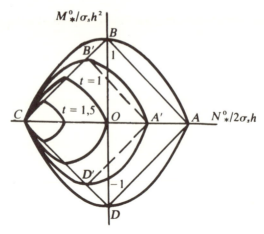

Fig. 3.3. Fictitious interaction curves for a beam (corresponding to Fig. 3.2b).

The interaction curves as shown in Fig. 3.3 are sufficient to predetermine the collapse mechanisms of a non-fixed bar for various relations between the parameters of variable and constant loads. They can be regarded as forming shakedown diagrams with respect to the parameter t_*.

The linearized picture of the situation for the I-beam is shown in Fig. 3.3 by straight lines.

The determination of the fictitious interaction curves for the remainder of the situations indicated in Fig. 3.3 is left to the reader as an instructive exercise.

3.3. Fictitious interaction surfaces for a stretched rod subjected to cycling torque

Consider a circular cylindrical bar (Fig. 3.4) under a time-independent axial force N^0 and torsional moment M^0, subjected to variable action of torsional moment $M_k(\tau)$, changing symmetrically according to

$$-M_* \leq M_k(\tau) \leq M_*. \tag{3.19}$$

We shall find the values of the time-independent torsional moments M^0 and normal forces N^0 that belong to the fictitious interaction surface. The preservation of plane cross-sections leads to axial strain increments

3.3. Fictitious interaction surfaces for a stretched rod

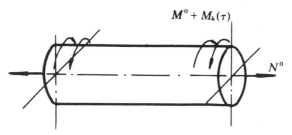

Fig. 3.4. Bar under tension followed by variable torsion.

$\Delta\epsilon$, the same for all the points of the cross-section and to the following linear changes in the shear strain increments $\Delta\gamma$:

$$\Delta\epsilon = \Delta e = \text{const}, \quad \Delta\gamma = r\Delta\theta, \quad \Delta\theta = \text{const}, \quad 0 \leq r \leq R. \tag{3.20}$$

The equation (3.7) acquires now the form

$$N^0 \Delta e + M^0 \Delta\theta = \int_F (\sigma^0 \Delta\epsilon + \tau^0 \Delta\gamma) \, dF, \tag{3.21}$$

where σ^0 and τ^0 stand for the normal and the shearing stresses belonging to the fictitious yield surface and depending on the current radius $0 \leq r \leq R$. The material is assumed to obey the Huber–Mises yield condition (2.9) that takes now the form

$$\sigma^2 + 3\tau^2 = \sigma_s^2. \tag{3.22}$$

Thus the fictitious yield surface for a generic point of the cross-section is, according to its definition (2.56), described by the equation

$$(\sigma^0)^2 + 3(|\tau^0| + \rho\tau_*)^2 = \sigma_s^2, \quad \left(\tau_* = \frac{M_*}{W_p}, \rho = \frac{r}{R}\right). \tag{3.23}$$

This relationship is plotted in Fig. 3.5 for various values of the product $\rho\tau_*$. For $\rho = 0$ the equation (3.23), using specific dimensionless coordinates, is represented by a circle of unit radius and the centre at the origin of coordinates; for $\rho\tau_* > 0$ the fictitious yield curve consists of two circular arcs of the same radius whose centres lie on the τ^0-axis and are at the distance $\rho\tau_*\sqrt{3}/\sigma_s$ from the origin of coordinates.

We shall determine the values of N^0 and M^0, using equation (3.21)

101

3 Generalized variables in the shakedown analysis

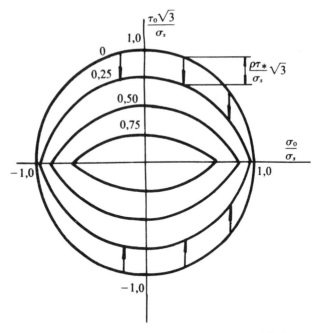

Fig. 3.5. Fictitious yield curve for a generic point of the bar.

and considering different ratios of the generalized strains Δe and $\Delta \theta$. Then, making use of the associated flow rule in the form (2.76), we shall find the corresponding stress components σ^0 and τ^0 belonging to the yield surface.

We shall adopt the notation

$$R \frac{\Delta \theta}{\Delta e} = \psi \tag{3.24}$$

and confine ourselves to the cases in which $\Delta \theta \geqslant 0$, $\Delta e \geqslant 0$.

To a prescribed value of the parameter ψ there corresponds, according to (3.20), the following relationship between the increments of the axial and shear strains:

$$\frac{\Delta \gamma}{\Delta \epsilon} = \frac{r \Delta \theta}{\Delta e} = \rho \psi. \tag{3.25}$$

If the stresses σ^0 and τ^0, related to the given values of ψ and ρ,

3.3. Fictitious interaction surfaces for a stretched rod

belong to a smooth portion of the fictitious yield curve, then, on account of the associated flow rule (2.76), we have

$$\frac{\Delta\gamma}{\Delta\epsilon} = \frac{3(|\tau^0| + \rho\tau_*)}{\sigma^0} \tag{3.26}$$

or, on using the specific dimensionless coordinates of Fig. 3.6, we get

$$\frac{\Delta\gamma}{\Delta\epsilon\sqrt{3}} = \frac{(|\tau^0| + \rho\tau_*)\sqrt{3}}{\sigma^0}.$$

Combining (3.25) and (3.26), we arrive at

$$\frac{3(|\tau^0| + \rho\tau_*)}{\sigma^0} = \rho\psi. \tag{3.27}$$

The solution to the simultaneous equations (3.23) and (3.27) yields the following formulae:

$$\sigma^0 = \frac{\sigma_s}{\sqrt{1 + \frac{1}{3}\rho^2\psi^2}}, \quad |\tau^0| = \left(\frac{\psi\sigma_s}{3\sqrt{1 + \frac{1}{3}\rho^2\psi^2}} - \tau_*\right)\rho. \tag{3.28}$$

They are valid, as seen in Fig. 3.6, for a smooth arc of the fictitious yield curve, where the ratio $\Delta\gamma/\Delta\epsilon$ satisfies

$$\frac{\Delta\gamma}{\Delta\epsilon} \geq \left(\frac{\Delta\gamma}{\Delta\epsilon}\right)_0 \quad \text{and at the corners} \quad 0 \leq \frac{\Delta\gamma}{\Delta\epsilon} \leq \left(\frac{\Delta\gamma}{\Delta\epsilon}\right)_0. \tag{3.29}$$

By making use of (3.26), the yield condition (3.23) and putting $\tau^0 = 0$, we find readily that

$$\left(\frac{\Delta\gamma}{\Delta\epsilon}\right)_0 = \frac{3\rho\tau_*}{\sqrt{\sigma_s^2 - 3\rho^2\tau_*^2}}. \tag{3.30}$$

On rearranging inequality (3.29) by means of (3.30) and (3.25), we obtain the limits for the functions (3.28):

$$0 \leq \rho \leq c, \quad c^2 = \frac{1}{3}\left(\frac{\sigma_s^2}{\tau_*^2} - \frac{9}{\psi^2}\right). \tag{3.31}$$

3 Generalized variables in the shakedown analysis

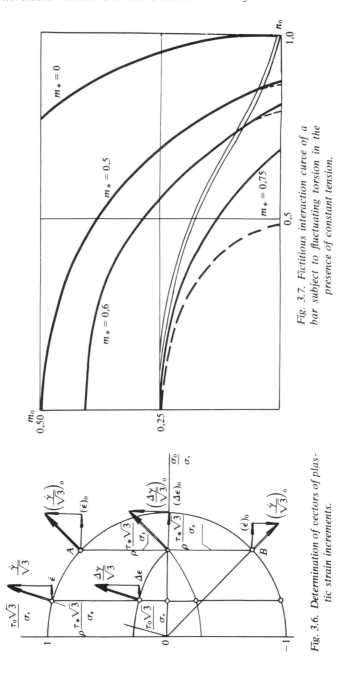

Fig. 3.7. Fictitious interaction curve of a bar subject to fluctuating torsion in the presence of constant tension.

Fig. 3.6. Determination of vectors of plastic strain increments.

3.3. Fictitious interaction surfaces for a stretched rod

When $c \geq 1$, i.e. when the flow régimes at all the points of the cross-section correspond to smooth portions of the fictitious yield curve, the plastic strain rates become non-zero simultaneously and at the largest value of the variable torque. Under those circumstances ('instantaneous' plastic collapse) we can substitute the functions (3.28) into the right-hand side of (3.21) and after suitable integration we get

$$N^0 = 6\pi R^2 \sigma_s \frac{1}{\psi^2}(\sqrt{1+\tfrac{1}{3}\psi^2}-1),$$
$$M_{\max} = M^0 + M_* = 2\pi R^3 \sigma_s \frac{1}{\psi^3}\left[2+\left(\frac{\psi^2}{3}-2\right)\sqrt{1+\tfrac{1}{3}\psi^2}\right]. \quad (3.32)$$

On elimination of the parameter ψ, the equation for the fictitious interaction curve assumes the form

$$(m^0 + m_*)^2 = 1 - (n^0)^2 + \tfrac{1}{4}(1-n^0)(n^0)^2, \quad (3.33)$$

where

$$n^0 = \frac{N^0}{\pi R^2 \sigma_s}, \quad m^0 = \frac{3M^0\sqrt{3}}{2\pi R^3 \sigma_s}, \quad m_* = \frac{3M_*\sqrt{3}}{2\pi R^3 \sigma_s}.$$

For $c < 1$ and $c \leq \rho \leq 1$ (near the periphery of the cross-section) formulae (3.28) cease to hold good and the constant stress components are associated with the corners of the yield curves

$$\tau^0 = 0, \quad \sigma^0 = \sqrt{\sigma_s^2 - 3\rho^2 \tau_*^2} \quad (3.34)$$

constructed for appropriate values of the radius ρ. In this case the vectors of total stresses touch the yield curve simultaneously in all points of the cross-section at maximum value of the variable torque. However, unlike in the solution (3.33), there will be no 'instantaneous' increase in deformation arising. The state of limit equilibrium will not be realized since the directions of the total plastic strain rate vectors obeying (3.30) do not satisfy the compatibility condition (3.25) according to which the ratio of the angular deformation rate to the axial one (as well as the ratio of respective increments) should increase in proportion to the current radius of a considered point.

It is interesting to note that the discussed situation is similar to that encountered in the elemental volume of a micro-nonhomogeneous

3 Generalized variables in the shakedown analysis

medium under a cyclic complicated loading state which is investigated in [3.33]. A deformation, found incompatible under single loading, appears to be acceptable, as can be shown by a suitable analysis, when the elastic deformation begins to vary in the process of cyclic loading (segment AB in Fig. 3.6). Thus, each cycle will result in a finite deformation of the same sign provided the load parameters exceed the prescribed limiting values. The latter can be determined by substituting into the relationship (3.21) the functions (3.28), if $\rho \leq c$, and the functions (3.34), if $c < \rho \leq 1$. The interaction curve is described now in the parametric form by the equations

$$n^0 = \frac{2}{3}\left[\frac{4\sqrt{3}m_*}{\psi^3} - \frac{9}{\psi^2} + \frac{9\sqrt{3}}{4m_*\psi} + \left(1 - \frac{9}{16m_*^2}\right)\sqrt{1 - \frac{16m_*^2}{9}}\right],$$

$$m^0 = \frac{27}{4m_*\psi^2}\left(\frac{\psi^2}{16m_*^2} - 1\right) - m_*\left(\frac{3}{16m_*^2} - \frac{1}{\psi^2}\right)^2$$

(3.35)

which remain valid if

$$m_* > \frac{3\sqrt{1-n^0}}{2(2-n^0)}.$$

The curves (3.35) are shown in Fig. 3.7 by dashed lines; they are all located below the double solid line.

In the above situation, as in the problem discussed in Sec. 3.2, it has been sufficient to consider the current state in a single, arbitrary cross-section of the bar in order to analyse the possible limit states when a non-fixed bar is considered. Hence we may conclude that, in the light of the kinematical theorem as applied to the conditions of incremental collapse, the determination of interaction curves is equivalent to the evaluation of the limiting cycle parameters. In particular, this means that the solution to the problem of the bar under simultaneous tension and torsion (Fig. 3.7) can be treated as the upper bound on the domain of admissible values of constant components of external loading. The above statement also applies to other interaction surfaces as long as they come into being by the preservation of plane sections (for beams) or straight normals (for shells and plates).

A lower bound assessment of the limit equilibrium conditions for a bar under simultaneous tension and torsion was obtained by Nadai in [3.32] on the basis of statical analysis.

3.4. Approximate method of introducing generalized variables

3.4. Approximate method of introducing generalized variables in shakedown problems

The presented method of constructing the fictitious interaction surface consists of two stages: a) determination of the fictitious yield surface in the stress space for every point of the body and b) making use of the obtained surfaces in order to evaluate, by means of procedures similar to those employed in the limit analysis theory, the limiting values of the generalized stresses, i.e. of the fictitious interaction surface.

The calculations necessary in order to construct the fictitious interaction surface are rather cumbersome, especially so for plates and shells acted upon by multi-parameter loading. These complex surfaces, even when exactly calculated, still must usually be replaced in specific practical examples by some approximations (often piecewise linear).

Let us assume that the plastic strain increment $\Delta \epsilon''_{ij0}$ at each point of the body occurs under uniquely determined variable stresses. This amounts to saying that we ignore situations in which certain plastic régimes can be absent in the limiting cycle. As already mentioned in Sec. 2.6, this can result in an overestimation of the admissible constant stresses σ^0_{ij*} at some points of the body. The situation is shown in Fig. 3.8: Under certain variable stresses the plastic régimes II, VI correspond to the given vector of the plastic strain increment, instead of the régimes I, II; to reveal this fact it is necessary to construct the fictitious yield

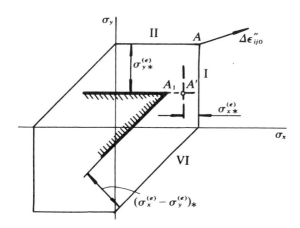

Fig. 3.8. Approximate method of construction of fictitious yield surface.

3 Generalized variables in the shakedown analysis

surface. In practice the error can become appreciable only when the relevant points occupy a substantial part of the body.

The above assumption enables us to make use of a somewhat simpler form of the kinematical theorem. Let us integrate the right-hand side of inequalities (2.80), (2.83) across the thickness,

$$\int_{-h}^{h} \min_{\tau}[(\sigma_{\alpha\beta} - \sigma^{(e)}_{\alpha\beta\tau})\Delta\epsilon''_{\alpha\beta0}] \, dz, \quad (\alpha, \beta = 1, 2), \tag{3.36}$$

and rearrange it as follows:

$$\int_{-h}^{h} \min_{\tau}[(\sigma_{\alpha\beta} - \sigma^{(e)}_{\alpha\beta\tau})(z\Delta\kappa_{\alpha\beta} + \Delta e_{\alpha\beta})] \, dz =$$
$$= (M_{\alpha\beta} - M^{*}_{\alpha\beta\tau})\Delta\kappa_{\alpha\beta} + (N_{\alpha\beta} - N^{*}_{\alpha\beta\tau})\Delta e_{\alpha\beta}, \tag{3.37}$$

$$M_{\alpha\beta} = \int_{-h}^{h} \sigma_{\alpha\beta} z \, dz, \quad N_{\alpha\beta} = \int_{-h}^{h} \sigma_{\alpha\beta} \, dz, \tag{3.38}$$

$$M^{*}_{\alpha\beta\tau} = \int_{-h}^{h} \sigma^{*}_{\alpha\beta\tau} z \, dz, \quad N^{*}_{\alpha\beta\tau} = \int_{-h}^{h} \sigma^{*}_{\alpha\beta\tau} \, dz,$$

where $M_{\alpha\beta}$, $N_{\alpha\beta}$ are the internal forces belonging to the actual yield surface (3.5) and $M^{(e)}_{\alpha\beta *}$, $N^{(e)}_{\alpha\beta *}$ are the generalized stresses expressible in terms of variable actions by means of

$$M^{*}_{\alpha\beta\tau}\Delta\kappa_{\alpha\beta} + N^{*}_{\alpha\beta\tau}\Delta e_{\alpha\beta} = \int_{-h}^{h} \max_{\tau}[\sigma^{(e)}_{\alpha\beta\tau}(z\Delta\kappa_{\alpha\beta} + \Delta e_{\alpha\beta})] \, dz. \tag{3.39}$$

The yield surface $f(\sigma_{\alpha\beta}) = 0$ was assumed to be constant. When the yield stress has to be treated as a temperature- (or temperature–time-) dependent quantity, the variable stress components of $\sigma_{\alpha\beta}$ on the yield surface can be identified with the variable elastic stresses $\sigma^{(e)}_{\alpha\beta\tau}$.

Fig. 3.9 elucidates the determination of the generalized stresses which are the final result of the variable actions in two basic deformation mechanisms of a shell or plate element: bending (a) and tension (b). A number of admissible distributions of one-parameter (varying between zero and a finite value) variable stresses is shown in the figure. The stress distribution satisfying one of the conditions

$$M^{*}_{\tau}\Delta\kappa = \int_{-h}^{h} \max_{\tau}(\sigma^{(e)}_{\tau} z \Delta\kappa) \, dz, \quad (a) \tag{3.40}$$

3.4. Approximate method of introducing generalized variables

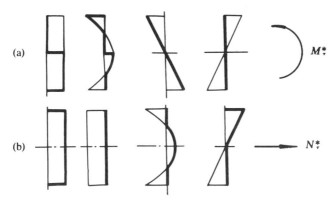

Fig. 3.9. Introduction of generalized forces in the presence of time-dependent stresses.

$$N_\tau^* \Delta e = \int_{-h}^{h} \max_{\tau}(\sigma_\tau^{(e)} \Delta e)\, dz \quad \text{(b)} \tag{3.41}$$

is indicated by heavy line in each of the variants.

The differences appearing in the right-hand side of equation (3.37) represent the approximate expressions for the internal forces belonging to the fictitious interaction surface (bounding the domain of admissible values of constant components $N^0_{\alpha\beta*}$, $M^0_{\alpha\beta*}$). As follows from (3.39), in order to determine them it is necessary to prescribe definite relationships among the components of the generalized strain increment vector ($\Delta\kappa_{\alpha\beta}$, $\Delta e_{\alpha\beta}$). Then we proceed to find values of $\sigma^{(e)}_{\alpha\beta\tau}$ at which the integrand of (3.39) is at the maximum and next we compute the internal forces under variable actions. We assume the relationships between the incremental components $\Delta\kappa_{\alpha\beta}$ and $\Delta e_{\alpha\beta}$ to be exactly the same as the relationships between the rate components $\dot{\kappa}_{\alpha\beta}$ and $\dot{e}_{\alpha\beta}$ pertaining to the actual yield surface (3.5)

$$F(M_{\alpha\beta}, N_{\alpha\beta}) = 0$$

in the space of internal forces. Then, on account of the flow rule associated with equation (3.5), we can find suitable values of $M_{\alpha\beta}$, $N_{\alpha\beta}$ and calculate the differences appearing in the right-hand side of (3.37). Next, different relations between $\Delta\kappa_{\alpha\beta}$, $\Delta e_{\alpha\beta}$ are to be adopted anew and the procedure is repeated, and so on.

Thus, the approximate method of determining the bounds on the

3 Generalized variables in the shakedown analysis

domain of admissible constant internal forces consists in firstly combining, in view of (3.39), the approximate construction of the fictitious yield surface in the physical stress space with the evaluation of limiting forces and, secondly, in employing the established finite relationships (3.5), both known from the limit equilibrium theory. Application of the approximate method allows one to save a considerable amount of computation time in comparison with the exact method.

As mentioned earlier, inaccuracies in the approximate evaluation of the limiting forces are found to influence the results of the shakedown calculations in those cases only in which, during the actual limiting cycle, the work corresponding to the corners of the fictitious yield surface is finite for the considered structure. Those corners become active when more than one flow régime realizes at some points of a shell during the cycle (régimes referring to non-adjacent sides of the yield surface when the Tresca condition applies). Since the assumptions on which the approximate method is hinged lead, in the general case, to the overestimation of the left-hand side of inequality (2.86), the limiting cycle parameters can also turn out to be too high, thus revealing the fact that the predictions of the approximate method are not on the safe side. However, in a number of cases the errors can be irrelevant or even non-existent. The latter circumstance refers to those problems in which only one flow régime is present at each point of the body (can develop during the cycle), or, in particular, to those simpler ones in which both the constant and the time-dependent actions generate internal forces of one type only (mere stretching or mere bending).

The direct application of the relationships (3.37) and (3.39) to the calculation of generalized stresses will be now demonstrated on the example of the beam shown in Fig. 3.2. Let the variable elastic stresses vary per cycle in accordance with the law, Fig. 3.2d,

$$\sigma_\tau^{(e)} = t(\tau)\left(0{,}5 + \frac{z}{h}\right)\sigma_s,$$

$$0 \leq t(\tau) \leq t_*.$$

(3.42)

By (3.6), equation (3.39) assumes the form

$$N_\tau^* \Delta e + M_\tau^* \Delta \kappa = \int_{-h}^{h} \max_\tau \left[t(\tau)\left(0{,}5 + \frac{z}{h}\right)(z\Delta\kappa + \Delta e) \right] dz.$$

(3.43)

The interaction curve for the beam at limit equilibrium is described by

3.4. Approximate method of introducing generalized variables

(3.17) at $t_* = 0$. We get

$$N = \mp 2\sigma_s c, \quad M = \pm \sigma_s(h^2 - c^2), \tag{3.44}$$

where

$$c = -\frac{\Delta e}{\Delta \kappa}, \quad -h \leq c \leq h. \tag{3.45}$$

As seen in Fig. 3.10a, the function $y = (0,5 + z/h)(z\Delta\kappa + \Delta e)$ entering the integrand of (3.43) is positive if

a) $\Delta\kappa > 0$ at $c \leq z \leq h$ and $-h \leq z \leq \tfrac{1}{2}h$ when $h > c > -\tfrac{1}{2}h$,

at $-\tfrac{1}{2}h < z \leq h$ and $-h \leq z < c$ when $-h < c < -\tfrac{1}{2}h$

b) $\Delta\kappa < 0$ at $-\tfrac{1}{2}h < z < c$ when $h > c > -\tfrac{1}{2}h$,

at $c < z < \tfrac{1}{2}h$ when $-h < c < -\tfrac{1}{2}h$.

When $y > 0$, then according to (3.42) the maximum of the integrand in (3.43) is reached for $t(\tau) = t_*$. When $y < 0$, it is reached for $t(\tau) = 0$.

Integrating the right-hand side of the equation (3.43), making use of the above analysis and equating the coefficients at Δe and $\Delta \kappa$ which appear on either side of that equation, we arrive at:

for $\Delta\kappa > 0$, $c > -\tfrac{1}{2}h$,

$$N_\tau^* = \frac{1}{2}\sigma_s h t_* \left[-\left(\frac{c}{h}\right)^2 - \frac{c}{h} + \frac{7}{4}\right],$$

$$N_\tau^* = \sigma_s h^2 t_* \left[-\frac{1}{3}\left(\frac{c}{h}\right)^3 - \frac{1}{4}\left(\frac{c}{h}\right)^2 + \frac{33}{48}\right], \tag{3.46}$$

for $\Delta\kappa > 0$, $c < -\tfrac{1}{2}h$,

$$N_\tau^* = \frac{1}{2}\sigma_s h t_* \left[\left(\frac{c}{h}\right)^2 + \frac{c}{h} + \frac{9}{4}\right],$$

$$M_\tau^* = \sigma_s h^2 t_* \left[\frac{1}{3}\left(\frac{c}{h}\right)^3 + \frac{1}{4}\left(\frac{c}{h}\right)^2 + \frac{31}{48}\right], \tag{3.47}$$

3 Generalized variables in the shakedown analysis

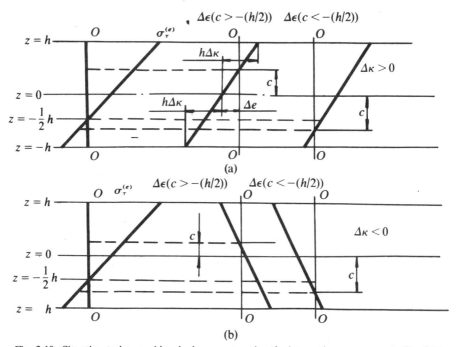

Fig. 3.10. Situation to be considered when constructing the interaction curve acc. to Fig. 3.2d.

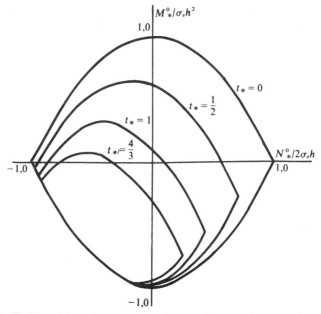

Fig. 3.11. Fictitious interaction curve for a beam to illustrate the approximate method.

3.4. Approximate method of introducing generalized variables

for $\Delta\kappa < 0$,

$$M_\tau^* = \pm \frac{1}{2}\sigma_s h t_* \left(\frac{c}{h} + \frac{1}{2}\right)^2,$$

$$M_\tau^* = \pm \sigma_s h^2 t_* \left[\frac{1}{3}\left(\frac{c}{h}\right)^3 + \frac{1}{4}\left(\frac{c}{h}\right)^2 - \frac{1}{48}\right]. \tag{3.48}$$

In (3.48) the upper sign refers to $c > -\frac{1}{2}h$, the lower sign to $c < -\frac{1}{2}h$.

On subtracting from the internal forces belonging to the interaction curve (3.44), the suitable forces (3.46)–(3.48) induced by variable actions (and calculated for consistent values of c and signs of $\Delta\kappa$), we can evaluate the forces N_*^0, M_*^0 belonging to the fictitious interaction curve that bounds the domain of admissible values of constant force components. Fig. 3.11 illustrates the results obtained for various values of the parameter t_*. At $t_* > \frac{4}{3}$ the beam undergoes alternating plastic flow; the amplitude of 'elastic' stresses during the cycle exceeds at dangerous point the double value of the yield stress. Calculations of the admissible forces caused by constant components of external actions no longer make sense simply because the domain of admissible constant stress components ceases to exist at the relevant points of the body.

The solution obtained (Fig. 3.11) coincides with the exact one since, in the case of linear stress state, only during the alternating plastic flow the plastic strain rates at a given point can become non-zero twice per cycle.

It is also worth noting that the relative simplicity of constructing the fictitious yield surface for a beam results in rather insignificant savings in computational time. In contrast, when plates and shells, to be dealt with further on, are considered, the use of the approximate methods instead of the exact ones is of much greater advantage.

If we employ linear approximations to the actual interaction surfaces (3.5), e.g. a square shown with thin solid lines in Fig. 3.3, we usually obtain discontinuities at the corners of the corresponding fictitious surfaces. This is due to the fact that, according to (3.39), (3.40), (3.41), different internal actions produced by variable load correspond to unique magnitudes of the internal actions at the corner of the yield surface; the ratio of $\Delta\kappa$ to Δe is here relevant, and it changes within the fan bounded by the normals to the adjacent faces of the approximated yield surface. This is a built-in drawback of the approximate method for constructing the fictitious yield surface; it disappears when the above shown exact method is applied.

3 Generalized variables in the shakedown analysis

3.5. Generalized stresses in circular plates under tension and bending

In the chapters to follow, devoted to the solutions of shakedown problems for plates and shells, we shall be faced with the necessity of constructing the interaction surfaces. To demonstrate the procedure, let us determine the domain of admissible values of generalized stresses for circular plates in the simplest situations conceivable, namely at collapse of either purely extensional or purely flexural type. In the case of axially symmetrical loading of circular plates at incremental collapse, the total increments of radial and circumferential strains $\Delta\epsilon_r, \Delta\epsilon_\varphi$ at the limiting cycle are known to be related to the corresponding increments of curvatures $\Delta\kappa_r, \Delta\kappa_\varphi$ and elongations $\Delta e_r, \Delta e_\varphi$ by means of the equations

$$\Delta\epsilon_r = z\Delta\kappa_r + \Delta e_r,$$
$$\Delta\epsilon_\varphi = z\Delta\kappa_\varphi + \Delta e_\varphi, \tag{3.49}$$

following from the preservation of straight normals to the middle surface. Here z denotes the distance from the middle surface of the plate.

We shall give below an analysis of the exact and the approximate methods of solving the considered plate made of either Tresca or Mises material. The yield points will be assumed constant.

A. Let us construct the domain of admissible values of the constant forces in the case of purely extensional collapse mechanism with stretching increments $\Delta e_r, \Delta e_\varphi$, different from zero while $\Delta\kappa_r = \Delta\kappa_\varphi = 0$.

The variable components of the radial and circumferential 'elastic' stresses $\sigma_{rr}^{(e)}, \sigma_{\varphi\tau}^{(e)}$ are assumed to behave during the cycle according to the equations

$$\sigma_{rr}^{(e)} = \varphi(r)\frac{z}{h}\sigma_s t(\tau),$$
$$\sigma_{\varphi\tau}^{(e)} = \psi(r)\frac{z}{h}\sigma_s t(\tau), \tag{3.50}$$

$$-t_* \leq t(\tau) \leq t_*, \quad -h \leq z \leq h, \tag{3.51}$$

where $\varphi(r)$ and $\psi(r)$ are known (from the 'elastic' solution for the particular plate considered) functions of the radius r. For definiteness,

3.5. Generalized stresses in circular plates

these functions are assumed to have the same sign,

$$\operatorname{sign} \varphi(r) = \operatorname{sign} \psi(r) = 1. \tag{3.52}$$

Let us emphasize that in the case here selected the variable action, generates bending moments at any instant of time whereas the assumed collapse mechanism specified by the constant loading is that of stretching, not of bending. The relationship (3.7) assumes now the form

$$\int_{-h}^{h} (\sigma_{r*}^0 \Delta e_r + \sigma_{\varphi*}^0 \Delta e_\varphi) \, dz = N_{r*}^0 \Delta e_r + N_{\varphi*}^0 \Delta e_\varphi. \tag{3.53}$$

Let us find the limiting values of the constant stress components appearing in the integrand. By the Tresca yield condition (2.12) and on account of (2.56) the fictitious yield curve is bounded by the following straight lines:

for $|z| < a$,

$$|\sigma_{r*}^0| = \left(1 - \varphi(r)\frac{|z|}{h} t_*\right) \sigma_s, \tag{3.54}$$

$$|\sigma_{\varphi*}^0| = \left(1 - \psi(r)\frac{|z|}{h} t_*\right) \sigma_s, \tag{3.55}$$

$$|\sigma_{r*}^0 - \sigma_{\varphi*}^0| = \left[1 - |(\varphi(r) - \psi(r))z|\frac{t_*}{h}\right] \sigma_s, \tag{3.56}$$

for $|z| \geq a$,

$$|\sigma_{r*}^0| = \left(1 - \varphi(r)\frac{|z|}{h} t_*\right) \sigma_s, \tag{3.57}$$

$$|\sigma_{\varphi*}^0| = \left(1 - \psi(r)\frac{|z|}{h} t_*\right) \sigma_s. \tag{3.58}$$

In the latter case the fictitious yield curve is a rectangle, in contrast to its former hexagonal shape, Fig. 3.12.

The value of $|z|$, corresponding to the transition from the hexagonal to a rectangular domain of admissible stresses σ_{r*}^0, $\sigma_{\varphi*}^0$ is found to be

$$\frac{a}{h} = \max\left(\frac{1}{2t_*\varphi(r)}, \frac{1}{2t_*\psi(r)}\right), \quad 0 \leq a \leq h. \tag{3.59}$$

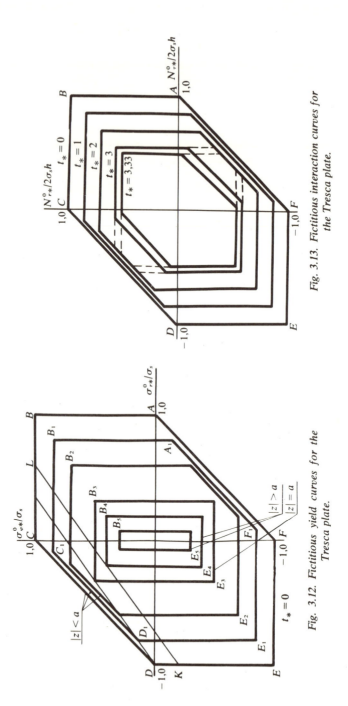

Fig. 3.13. Fictitious interaction curves for the Tresca plate.

Fig. 3.12. Fictitious yield curves for the Tresca plate.

3.5. Generalized stresses in circular plates

The rectangular domains of admissible values of constant stress components (for $|z|>a$) correspond to profiles of the time-dependent stress components at which the plastic flow in the cycle takes place in obedience to two régimes representing two non-adjacent sides of the Tresca hexagon (Fig. 3.12, $ABCDEF$).

In order to construct the curve bounding the domain of admissible values of constant load components it is necessary to find, for each possible ratio of the increments Δe_r and Δe_φ, the corresponding values of $\sigma^0_{r*}, \sigma^0_{\varphi*}$. This is done by making use of (2.76), (3.49), (3.54)–(3.58), and calculating the integral in the left-hand side of the equation (3.53).

When sign $\Delta e_r =$ sign Δe_φ the strain increment vectors fit respectively the corners B_i and E_i, of the fictitious yield curves (Fig. 3.12) or those sides of suitable polygons (given by the equations (3.54), (3.55) and (3.57), (3.58)) that are adjacent to the above corners.

Substituting the latter equations into equality (3.53), we get

$$N^0_{r*} = \pm 2\sigma_s h \left[1 - \frac{1}{2}\varphi(r)t_*\right], \tag{3.60}$$

$$N^0_{\varphi*} = \pm 2\sigma_s h \left[1 - \frac{1}{2}\psi(r)t_*\right]. \tag{3.61}$$

When sign $\Delta e_r = -$ sign Δe_φ, then, for $|z|<a$, the strain increment vector fits the lines (3.56) or the corners which lie at the intersection of the straight lines (3.56) and (3.54) or (3.56) and (3.55), respectively. For $|z| \geq a$ the stresses σ^0_{r*} and $\sigma^0_{\varphi*}$ correspond to the corner created by the intersecting straight lines (3.57) and (3.58). On substituting suitable values of stresses and strain increments into equality (3.53) and integrating, we obtain:

for $\Delta e_\varphi > 0, -\Delta e_\varphi < \Delta e_r < 0$,

$$N^0_{\varphi*}\Delta e_\varphi + N^0_{r*}\Delta e_r = 2\sigma_s h\left[1 - \frac{1}{2}t_*\psi(r)\right]\Delta e_\varphi + 2\sigma_s h \times$$

$$\times \left\{\frac{a}{h} - 1 + \frac{1}{2}t_*\left[\varphi(r)\left(1 - \frac{a^2}{h^2}\right) - \psi(r)\frac{a^2}{h^2} + \right.\right.$$

$$\left.\left. + |\varphi(r) - \psi(r)|\frac{a^2}{h^2}\right]\right\}\Delta e_r, \tag{3.62}$$

3 Generalized variables in the shakedown analysis

for $\Delta e_\varphi > 0, \Delta e_r < -\Delta e_\varphi < 0$,

$$N^0_{\varphi *}\Delta e_\varphi + N^0_{r *}\Delta e_r = 2\sigma_s h \left\{1 - \frac{a}{h} - \frac{1}{2}t_*\left[\psi(r)\left(1-\frac{a^2}{h^2}\right) - \varphi(r)\frac{a^2}{h^2} + \right.\right.$$
$$\left.\left.+ |\varphi(r) - \psi(r)|\frac{a^2}{h^2}\right]\right\}\Delta e_\varphi - 2\sigma_s h\left[1 - \frac{1}{2}\varphi(r)\right]\Delta e_r.$$

(3.63)

Similar equalities for $\Delta e_\varphi < 0$ differ from the above only in the signs of the multipliers appearing in their right-hand sides.

The domains of admissible constant forces, constructed with the use of (3.60)–(3.63), are depicted in Fig. 3.13 by solid lines. It is assumed that

$$\varphi(r) = 0{,}3 \quad \text{and} \quad \psi(r) = 0{,}28. \tag{3.64}$$

For $t_* > \frac{10}{3} \approx 3{,}33$ the alternating plastic flow takes place, as can be concluded from the functions for stresses (3.50) and the known condition (2.86).

Let us now use the approximate method to construct the fictitious interaction curves. In the problem considered equality (3.39) takes the form

$$N^*_{rr}\Delta e_r + N^*_{\varphi r}\Delta e_\varphi = \int_{-h}^{h} \max_\tau (\sigma^{(e)}_{rr}\Delta e_r + \sigma^{(e)}_{\varphi r}\Delta e_\varphi)\,\mathrm{d}z \tag{3.65}$$

and the limits of the admissible region of constant forces are, on account of the equality (3.37), given by

$$N^0_{r*} = N_r - N^*_{rr}, \quad N^0_{\varphi *} = N_\varphi - N^*_{\varphi r} \tag{3.66}$$

where N_r, N_φ are the forces belonging to the yield curve to be determined from the equation [3.17] and (Fig. 3.13, $t_* = 0$)

$$\max(|N_r|, |N_\varphi|, |N_r - N_\varphi|) = 2\sigma_s h. \tag{3.67}$$

Each side of the hexagon (3.67) is referred to a definite force generated by the applied variable actions. For the side AB (Fig. 3.13) we have $N_r = 2\sigma_s h$ and, according to the associated flow rule we have also $\Delta e_\varphi = 0$ and $\Delta e_r = \mu$ ($\mu \geq 0$). In this case equality (3.65), after employing

3.5. Generalized stresses in circular plates

the acting stresses (3.50), (3.51) and reducing μ, takes the form

$$N_{rr}^* = \int_{-h}^{h} \max_{\tau}\left[t(\tau)\varphi(r)\frac{z}{h}\sigma_s\right]dz = \varphi(r)t_*h\sigma_s. \tag{3.68}$$

The integrand of (3.68) attains its maximum at $t(\tau) = t_*$ for $z > 0$ and at $t(\tau) = -t_*$ for $z < 0$.

Results of similar calculations for the remaining sides of the hexagon (3.67) together with the corresponding values of N_{r*}^0 and $N_{\varphi*}^0$ determined from (3.66) are shown in Table 3.1.

According to the approximate method, at the corners of the interaction curve (3.67) the vector of elongation increment is colinear with the rate vector $\dot{e}_r, \dot{e}_\varphi$ and therefore it should be directed along the normal to the suitable side of the hexagon (3.67) whereas the largest value of the integrand in (3.65) should be computed as the sum of maxima for each of its components. The values of N_{r*}^0, $N_{\varphi*}^0$ thus obtained correspond to the intersection points of the straight lines shown in Table 3.1.

Dashed lines in Fig. 3.13 correspond to those values of N_{r*}^0 and $N_{\varphi*}^0$, shown in Table 3.1, that fit the adopted quantities (3.64) and do not coincide with the results of the exact solution.

Let us now make a similar analysis for the Mises yield condition (2.9) which, in the present situation, takes the form

$$\sigma_r^2 - \sigma_r\sigma_\varphi + \sigma_\varphi^2 - \sigma_s^2 = 0. \tag{3.69}$$

The domains of admissible values of constant stresses are bounded by

Table 3.1

	N_r, N_φ	Δe_r	Δe_φ	$N_{r*}^{(e)}/2\sigma_s h, N_{\varphi*}^{(e)}/2\sigma_s h$	$N_{r*}^0/2\sigma_s h, N_{\varphi*}^0/2\sigma_s h$
AB	$N_r = 2\sigma_s h$	μ	0	$\frac{1}{2}t\varphi(r)$	$1 - \frac{1}{2}t\varphi(r)$
BC	$N_\varphi = 2\sigma_s h$	0	μ	$\frac{1}{2}t\psi(r)$	$1 - \frac{1}{2}t\psi(r)$
CD	$N_\varphi - N_r = 2\sigma_s h$	$-\mu$	μ	$\frac{1}{2}t\|\varphi(r) - \psi(r)\|$	$1 - \frac{1}{2}t\|\varphi(r) - \psi(r)\|$
DE	$-N_r = 2\sigma_s h$	$-\mu$	0	$-\frac{1}{2}t\varphi(r)$	$-1 + \frac{1}{2}t\varphi(r)$
EF	$-N_\varphi = 2\sigma_s h$	0	$-\mu$	$-\frac{1}{2}t\psi(r)$	$-1 + \frac{1}{2}t\psi(r)$
FA	$-N_\varphi + N_r = 2\sigma_s h$	μ	$-\mu$	$-\frac{1}{2}t\|\varphi(r) - \psi(r)\|$	$-1 + \frac{1}{2}t\|\varphi(r) - \psi(r)\|$

3 Generalized variables in the shakedown analysis

the curves

$$\left[\sigma_{r*}^0 \pm \varphi(r)\frac{|z|}{h}t_*\sigma_s\right]^2 - \left[\sigma_{r*}^0 \pm \varphi(r)\frac{|z|}{h}t_*\sigma_s\right] \times$$
$$\times \left[\sigma_{\varphi*}^0 \pm \psi(r)\frac{|z|}{h}t_*\sigma_s\right] + \left[\sigma_{\varphi*}^0 \pm \psi(r)\frac{|z|}{h}t_*\sigma_s\right]^2 - \sigma_s^2 = 0 \qquad (3.70)$$

obtained from the expressions (2.56), and (3.50), (3.51). The equations (3.70) should be taken with either all upper or all lower signs read; suitable curves for the selected quantities (3.64) are shown in Fig. 3.14.

Let us discuss certain properties of the corners of admissible domains of constant stresses which will be useful in what follows. Stresses at the corners will be denoted by σ_{rc}^0 and $\sigma_{\varphi c}^0$ as distinct from the stresses $\sigma_{r*}^0, \sigma_{\varphi*}^0$ at the smooth portions of curves. Since both ends (each corresponding to a different instant of time) of the total stress profile crossing the corner $\sigma_{rc}^0, \sigma_{\varphi c}^0$ lie on the yield curve, the corresponding corners of the fictitious surfaces referred to various values of the variable stress parameter are all located on a single straight line

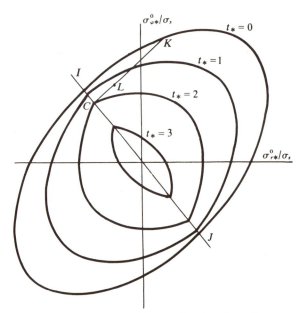

Fig. 3.14. Fictitious yield curves for the Mises plate.

3.5. Generalized stresses in circular plates

(*IJ*, Fig. 3.14)

$$\sigma_{rc}^0(2\varphi - \psi) + \sigma_{\varphi c}^0(2\psi - \varphi) = 0. \tag{3.71}$$

The values σ_{rc}^0 and $\sigma_{\varphi c}^0$ can be computed, as functions of the parameter t_* and the coordinate z from the formulae

$$\sigma_{rc}^0 = \pm \sigma_s \frac{2\psi - \varphi}{\sqrt{3}} \sqrt{\frac{1}{\zeta^2} - t_*^2 \frac{z^2}{h^2}}, \quad \sigma_{\varphi*}^0 = \mp \sigma_s \frac{2\varphi - \psi}{\sqrt{3}} \sqrt{\frac{1}{\zeta^2} - t_*^2 \frac{z^2}{h^2}}, \tag{3.72}$$

where

$$\zeta^2 = \varphi^2 - \varphi\psi + \psi^2, \tag{3.73}$$

and $\varphi = \varphi(r)$, $\psi = \psi(r)$ specify the distribution functions of the stresses (3.50).

From Fig. 3.14 it follows that, given the direction of the increment vector Δe ($\Delta e_r, \Delta e_\varphi$) and the value of t_*, the limiting magnitudes of the constant stresses correspond to the smooth arc of the curve (3.70), provided the sum of ($\sigma_{rc}^0, \sigma_{\varphi c}^0$) and the variable elastic stresses represents the stresses belonging to the interior of the yield curve, i.e. for $z > a$ such that

$$\left[\sigma_{rc}^0 \pm t_* \frac{|z|}{h} \varphi(r)\sigma_s\right]^2 - \left[\sigma_{rc}^0 \pm t_* \frac{|z|}{h} \varphi(r)\sigma_s\right] \times$$

$$\times \left[\sigma_{\varphi c}^0 \pm t_* \frac{|z|}{h} \psi(r)\sigma_s\right] + \left[\sigma_{\varphi c}^0 \pm t_* \frac{|z|}{h} \psi(r)\sigma_s\right]^2 < \sigma_s^2. \tag{3.74}$$

When $a \leq |z| \leq h$, inequality (3.74) is not satisfied and the limiting constant stresses are associated with the corner since they are an intersection of the curves (3.70).

Thus, in the general case when $0 \leq a \leq h$, if we substitute the conditions (3.70) and the formulae (3.72) in equality (3.53), we get

$$N_{r*}^0 \Delta e_r + N_{\varphi*}^0 \Delta e_\varphi = 2 \int_0^a (\sigma_r^0 \Delta e_r + \sigma_\varphi^0 \Delta e_\varphi) \, dz +$$

$$+ 2 \int_a^h (\sigma_{rc}^0 \Delta e_r + \sigma_{\varphi c}^0 \Delta e_\varphi) \, dz =$$

3 Generalized variables in the shakedown analysis

$$= \left[2a\sigma_r \mp \varphi(r) \frac{a^2}{h} t_* \sigma_s \pm \frac{2\varphi - \psi}{\sqrt{3}} \sigma_s \phi(t_*, a) \right] \Delta e_r +$$
$$+ \left[2\sigma_\varphi a \mp \psi(r) \frac{a^2}{h} t_* \sigma_s \mp \frac{2\psi - \varphi}{\sqrt{3}} \sigma_s \phi(t_*, a) \right] \Delta e_\varphi,$$
(3.75)

where

$$\phi(t_*, a) = h \sqrt{\frac{1}{\zeta^2} - t_*^2} - a \sqrt{\frac{1}{\zeta^2} - t_*^2 \frac{a^2}{h^2}} +$$
$$+ \frac{h}{\zeta^2 t_*} \left(\arcsin \zeta t_* - \arcsin \zeta t_* \frac{a}{h} \right),$$
(3.76)

and σ_r, σ_φ are stresses satisfying (3.69).

Equating the coefficients at Δe_r and Δe_φ in the left- and the right-hand sides of equality (3.75), we arrive at

$$N^0_{r*} \pm \varphi(r) \frac{a^2}{h} t_* \sigma_s \mp \frac{2\varphi - \psi}{\sqrt{3}} \sigma_s \phi(t_*, a) = 2a\sigma_r,$$
(3.77)

$$N^0_{\varphi*} \pm \psi(r) \frac{a^2}{h} t_* \sigma_s \pm \frac{2\psi - \varphi}{\sqrt{3}} \sigma_s \phi(t_*, a) = 2a\sigma_\varphi.$$
(3.78)

Next, on eliminating from the above expressions the magnitude a (with the help of (3.74)) as well as the stresses σ_r, σ_φ (on account of (3.69)), we can obtain equations of curves bounding the region of admissible constant forces. Since the general analytical form of those equations is rather cumbersome, we shall only discuss the particular cases $a = h$ and $a = 0$. For the former we have

$$[N^0_{r*} \pm \varphi(r) h t_* \sigma_s]^2 - [N^0_{r*} \pm \varphi(r) h t_* \sigma_s] \times$$
$$\times [N^0_{\varphi*} \pm \psi(r) h t_* \sigma_s] + [N^0_{\varphi*} \pm \psi(r) h t_* \sigma_s]^2 = (2\sigma_s h)^2.$$
(3.79)

The domains of admissible values of constant forces, constructed with the use of (3.77), (3.78) and (3.79) at particular values (3.64) of the functions $\varphi(r), \psi(r)$, are shown in Fig. 3.15 with solid heavy lines, for various values of the parameter t_*.

The construction of the corresponding domains by means of the approximate method is here much simpler. On substituting (3.50), (3.51)

3.5. Generalized stresses in circular plates

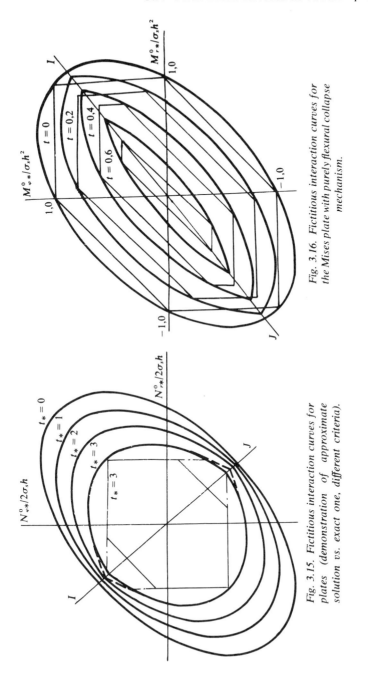

Fig. 3.16. Fictitious interaction curves for the Mises plate with purely flexural collapse mechanism.

Fig. 3.15. Fictitious interaction curves for plates (demonstration of approximate solution vs. exact one, different criteria).

3 Generalized variables in the shakedown analysis

in equality (3.65) we obtain

$$N^*_{rr}\Delta e_r + N^*_{\varphi\tau}\Delta e_\varphi = \int_{-h}^{h} [\mp \varphi(r) t_* \sigma_s \Delta e_r \mp \psi(r) t_* \sigma_s \Delta e_\varphi]\, dz, \tag{3.80}$$

where the upper sign (−) corresponds to the condition

$$\varphi(r)\Delta e_r + \psi(r)\Delta e_\varphi > 0 \tag{3.81}$$

and the max$_\tau$ in the integrand of (3.65) is attained at

$$t(\tau) = t_* \quad \text{if } z > 0 \quad \text{and at} \tag{3.82}$$
$$t(\tau) = -t_* \quad \text{if } z < 0. \tag{3.83}$$

The lower sign (+) yields the condition

$$\varphi(r)\Delta e_r + \psi(r)\Delta e_\varphi < 0. \tag{3.84}$$

By simple rearrangements of the equality (3.80) by means of the finite relationship [3.12]

$$N_r^2 - N_r N_\varphi + N_\varphi^2 = (2\sigma_s h)^2 \tag{3.85}$$

and the use of (3.66), we can obtain a result that practically coincides with that of (3.79). For the values of $\varphi(r)$ and $\psi(r)$ adopted in (3.64) the discrepancies between the approximate and the exact solutions appear insignificant. The approximate results for $t_* = 3$ (where they distinctly differ from the exact ones) are shown in Fig. 3.15 by dashed lines. In the figure there are also shown, by thin lines, the results for $t_* = 3$ in the case of a Tresca plate. Solid thin lines represent the exact solution, dash-dotted thin lines show the approximate one.

B. Let us now construct the appropriate domain of generalized stresses in the case of a purely flexural mechanism of collapse in which both Δe_r and Δe_φ vanish in (3.49).

The variable stresses are assumed to vary in conformity with

$$\sigma_r^{(e)} = t(\tau)\varphi(r)\sigma_s, \quad \sigma_\varphi^{(e)} = t(\tau)\psi(r)\sigma_s, \tag{3.86}$$
$$-t_* \leq t(\tau) \leq t_*, \tag{3.87}$$

i.e. only normal forces are generated at any instant of time.

3.5. Generalized stresses in circular plates

Unlike in the previous analysis, we shall begin by employing the approximate method according to which equality (3.37) takes now the form

$$M^*_{r\tau}\Delta\kappa_r + M^*_{\varphi\tau}\Delta\kappa_\varphi = \int_{-h}^{h} \max_{\tau}\{t(\tau)\sigma_s[\varphi(r)\Delta\kappa_r + \psi(r)\Delta\kappa_\varphi]z\}\,dz, \tag{3.88}$$

hence

$$M^*_{r\tau} = \pm t_*\sigma_s h^2 \varphi(r), \quad M^*_{\varphi\tau} = \pm t_*\sigma_s h^2 \psi(r). \tag{3.89}$$

The upper sign (+) in (3.89) corresponds to

$$\varphi(r)\Delta\kappa_r + \psi(r)\Delta\kappa_\varphi > 0 \tag{3.90}$$

and the integrand in (3.88) reaches its maximum at

$$|t(\tau)| = t_*, \quad \text{sign } t(\tau) = \text{sign } z. \tag{3.91}$$

The limiting values of the constant bending moments M^0_{r*} and $M^0_{\varphi*}$ added to the values (3.89) represent the stress state belonging to the interaction curve

$$M^0_{r*} \pm t_*\sigma_s h^2 \varphi(r) = M_r, \quad M^0_{\varphi*} \pm t_*\sigma_s h^2 \psi(r) = M_\varphi. \tag{3.92}$$

For the Huber–Mises yield condition [3.12] we have

$$M_r^2 - M_r M_\varphi + M_\varphi^2 = (\sigma_s h^2)^2. \tag{3.93}$$

Substituting (3.92) into (3.93), we obtain the sought for equation of the fictitious interaction curve

$$[M^0_{r*} \pm t_*\sigma_s h^2 \varphi(r)]^2 - [M^0_{r*} \pm t_*\sigma_s h^2 \varphi(r)] \times$$
$$\times [M^0_{\varphi*} \pm t_*\sigma_s h^2 \psi(r)] + [M^0_{\varphi*} \pm t_*\sigma_s h^2 \psi(r)]^2 = (\sigma_s h^2)^2. \tag{3.94}$$

The forces generated by variable actions and the limiting values of constant forces in the Tresca plate are displayed in Table 3.2. The actual interaction curve for the Tresca plate is described by [3.12]

$$\max(|M_r|, |M_\varphi|, |M_r - M_\varphi|) = \sigma_s h^2. \tag{3.95}$$

125

3 Generalized variables in the shakedown analysis

Table 3.2

	M_r, M_φ	Δx_r	Δx_φ	$M_{r*}^{(e)}/\sigma_s h^2, M_{\varphi*}^{(e)}/\sigma_s h^2$	$M_{r*}^0/\sigma_s h^2, M_{\varphi*}^0/\sigma_s h^2$
AB	$M_r = \sigma_s h^2$	μ	0	$t\|\varphi(r)\|$	$1 - t\|\varphi(r)\|$
BC	$M_\varphi = \sigma_s h^2$	0	μ	$t\|\psi(r)\|$	$1 - t\|\psi(r)\|$
CD	$-M_r + M_\varphi = \sigma_s h^2$	$-\mu$	μ	$t\|\varphi(r) - \psi(r)\|$	$1 - t\|\varphi(r) - \psi(r)\|$
DE	$-M_r = \sigma_s h^2$	$-\mu$	0	$-t\|\varphi(r)\|$	$-(1 - t\|\varphi(r)\|)$
EF	$-M_\varphi = \sigma_s h^2$	0	$-\mu$	$-t\|\psi(r)\|$	$-(1 - t\|\psi(r)\|)$
FA	$M_r - M_\varphi = \sigma_s h^2$	μ	$-\mu$	$-t\|\varphi(r) - \psi(r)\|$	$-(1 - t\|\varphi(r) - \psi(r)\|)$

The admissible domains of constant bending moments, obeying the conditions (3.93) and (3.95), are shown in Fig. 3.16 at the adopted values of

$$\varphi(r) = -0{,}5 \quad \text{and} \quad \psi(r) = 1. \tag{3.96}$$

Let us now seek the exact solution to the problem of determining the admissible domain of constant forces under the Tresca yield condition. By (2.56) and (3.86), (3.87), the bounds on the admissible region of constant stress components are described as follows:

For $0 \leq t_* \leq t_1$,

$$|\sigma_{r*}^0| = \sigma_s(1 - t_*|\varphi(r)|),$$
$$|\sigma_{\varphi*}^0| = \sigma_s(1 - t_*|\psi(r)|), \tag{3.97}$$
$$|\sigma_{r*}^0 - \sigma_{\varphi*}^0| = \sigma_s(1 - t_*|\varphi(r) - \psi(r)|),$$

where t_1 is the specific value of the parameter t_* given by

$$t_1 = \max\left(\frac{1}{|\varphi(r)|}, \frac{1}{|\psi(r)|}\right). \tag{3.98}$$

For $t_1 \leq t_* \leq t_2$ (t_2 stands for the value of t_* at the onset of alternating plasticity), depending on the signs of $\varphi(r)$ and $\psi(r)$, one of the three equations (3.97) ceases to be a bounding one and thus the hexagon

3.6. Formulation of the statical theorem

degenerates to a parallelogram. It should be noted that, unlike in the case (A), two plastic régimes (within a cycle) corresponding to two non-adjacent sides of the Tresca hexagon will occur either throughout the whole cross-section or nowhere in the latter. This is so because the time-dependent stress components (3.86) are constant over the thickness of the plate.

It is easy to ascertain that, on account of the equality of work done by constant stresses and by the corresponding internal forces

$$M^0_{r*}\Delta\kappa_r + M^0_{\varphi*}\Delta\kappa_\varphi = \int_{-h}^{h} (\sigma^0_{r*}\Delta\kappa_r + \sigma^0_{\varphi*}\Delta\kappa_\varphi)z \, dz, \tag{3.99}$$

which yields values of bending moments perfectly coinciding with those established in the approximate approach (Table 3.3).

A similar coincidence of exact and approximate results is found to exist for the Huber–Mises plate as well.

The above examples of constructing the domains of the admissible constant forces of stretching and bending have indicated that the discrepancies between the exact and approximate results can become substantial only for sufficiently wide ranges of fluctuation of the variable 'elastic' stresses (approaching those for which the alternating plastic flow is supposed to occur) and in case of a non-uniform distribution of these stresses across the thickness of a shell or a plate. Thus for plates and shells it appears reasonable to employ in many a shakedown situation the approximate, time-saving method to evaluate internal forces.

3.6. Formulation of the statical theorem in terms of generalized stresses

As in the limit analysis theory, the formulation of the statical shakedown theorem in terms of generalized variables must be based on special premises whereas the kinematical theorems do not require such prerequisites because the relevant laws of strain distribution across the thickness of the shell are introduced in a direct manner.

Since the determination of the generalized stresses $Q^0_{\alpha\beta*} = (M^0_{\alpha\beta*}, N^0_{\alpha\beta*})$ belonging to the fictitious interaction surface considered above is applicable to the case of incremental collapse only, i.e. when the set of points located inside each of the admissible domains of

3 Generalized variables in the shakedown analysis

constant stress components for the considered body is non-empty, the statical shakedown theorem can be expressed in terms of generalized stresses in the following way:

1. In a shell (or a plate) no incremental collapse takes place if there exists a distribution of time-independent generalized stresses $Q^0_{\alpha\beta}$ which is in equilibrium with the constant components of external actions X^0_i, p^0_i which lie inside the fictitious interaction surfaces

$$\phi(Q^0_{\alpha\beta}) \leq 0 \tag{3.100}$$

constructed in the generalized stress space for the points of the middle surface, according to the prescribed variable actions.

2. On the other hand, the incremental collapse of a shell (or a plate) does take place under variable repeated loading if there does not exist a distribution of the generalized stresses $Q^0_{\alpha\beta}$ which is statically possible, i.e. satisfying both the internal and the boundary equilibrium conditions, and admissible in the sense of the fictitious interaction surfaces $\phi(Q^0_{\alpha\beta *}) = 0$.

The above theorem differs from the statical theorem of the limit analysis theory only in the increased relevance of the fictitious interaction surfaces. These are, in general, varying over the middle surface of the shell and they depend on the ranges (and programs) of fluctuation of the applied variable actions. The fictitious interaction surfaces duly reflect the possible non-simultaneity of the stress vectors, belonging to the same normal to the middle surface, touching the yield surface.

The proof of the above theorem can be constructed along fully analogous lines as in the case of the corresponding limit analysis theorem [3.34]; from all the current relationships one merely has to extract the time-dependent 'elastic' stresses, similarly as this was done in the restated shakedown theorems (Secs. 2.4 and 2.6).

The above theorem refers to the case when another limit state, e.g. the alternating plastic flow, does not take place at any point of the shell, or plate. In consequence, the actual values of the limiting cycle parameters should be determined by using the one of the two limiting vectors of loading parameters whose end is closer to the origin of the coordinate system. These two vectors correspond to the conditions of alternating plasticity and incremental collapse.

References

[3.1] Koiter, W.T., Some remarks on plastic shakedown theorems. *Proc. 8-th Int. Congr. Appl. Mech.* Istanbul, *1*, 1952, 220.
[3.2] Neal, B.G., *The plastic methods of structural analysis*. New York, 1956.
[3.3] Sawczuk, A., On incremental collapse of shells under cyclic loading. *IUTAM Symp. Copenhagen, 1967. Theory of thin shells*. Springer, Berlin, 1969.
[3.4] Sawczuk, A., Evaluation of upper bounds to shakedown loads for shells. *J. Mech. Phys. Solids*, *17*, 1969, 291–301.
[3.5] Franciosi, V., Augusti, G., Sparacio, R., Collapse of arches under repated loading. *Proc. ASCE, J. Struct. Div.*, V. 90 ST 1, Febr., 1964, 165–201.
[3.6] Gokhfeld, D.A., Cherniavsky, O.F., Generalized variables in the shakedown problems for plates and shells, in Russian. *Proc. VII All-Union Conf. on shells and plates, (Dnepropetrovsk, 1969)* Nauka, Moscow, 1970.
[3.7] Gokhfeld, D.A., Cherniavsky, O.F., Methods of solving problems in the shakedown theory of continua. *Proc. Int. Symp. on Found. of Plasticity, Warsaw, 1972*, Noordhoff, Leyden, 1973.
[3.8] Save, M.A., Massonnet, C.E., *Plastic analysis and design of plates, shells and disks*. North-Holland, Amsterdam, 1972.
[3.9] Koiter, W.T., General theorems for elastic-plastic solids. *Progress in solid mechanics*, edited by I.N. Sneddon and R. Hill. North-Holland, Amsterdam, 1960.
[3.10] Iliushin, A.A., *Plasticity*, in Russian, Gostekhizdat, Moscow, 1948.
[3.11] Shapiro, G.S., On yield surfaces for perfectly plastic shells, in Russian, in *Problems of continuum mechanics*, AN SSSR, Moscow, 1961.
[3.12] Rzhanitzin, A.R., Shell solutions by limit equilibrium, in Russian, *Studies in plasticity and strength of building structures*, Gosstroiizdat, Moscow, 1958.
[3.13] Rzhanitzin, A.R., *Structural calculations accounting for plastic properties of materials*, in Russian, Gosstroiizdat, Moscow, 1954.
[3.14] Rozdestvensky, V.V., On limit states in the cross-sections of thin shells, in Russian, in *Studies in structural mechanics and plasticity*, Gosstroiizdat, Moscow, 1956.
[3.15] Onat, E.T., The plastic collapse of cylindrical shells under axially symmetrical loading. *Quart. Appl. Math.*, *13*, 1955, 63.
[3.16] Sawczuk, A., Rychlewski, J., On yield surfaces for plastic shells. *Archiwum mechaniki stosowanej*, Nr 1, 12, 1960.
[3.17] Hodge, Ph.G., *Plastic analysis of structures*, McGraw-Hill, New York, 1959.
[3.18] Hodge, Ph.G., *Limit analysis of rotationally symmetric plates and shells*, Prentice-Hall, Englewood Cliff, N.J., 1963.
[3.19] Ivlev, D.D., *Theory of ideal plasticity*, in Russian, Nauka, Moscow, 1966.
[3.20] Onat, E.T., Prager, W., Limit analysis of shells of revolution. *Konikl. Nederl. Akad. Wetensch.*, Amsterdam, *5*, 1954, 57.
[3.21] Erkhov, M.I., Finite relationships between forces and moments at plastic deformation of shells, in Russian, *Structural mechanics and calculations of structures*, Nr 3, 1959.
[3.22] Erkhov, M.I., Plastic state of shells, plates and beams made of perfectly plastic material, in Russian, *Izv. AN SSSR, OTN, Mechanics and machine-building*, Nr 6, 1960.

3 Generalized variables in the shakedown analysis

[3.23] Mikeladze, M.Sh., Rigid-plastic bending of anisotropic circular discs with non-symmetrical profile, in Russian, *Izv. AN SSSR, OTN*, Nr 2, 1957.

[3.24] Hodge, Ph.G., Yield conditions of rotationally symmetric shells under axisymmetric loading. *J. Appl. Mech.*, 27, Nr 12, 1960.

[3.25] Rosenblum, V.I., Approximate equilibrium theory of technical shells, in Russian, *Applied Math. and Mech.*, 18, Nr 3, 1954.

[3.26] Rosenblum, V.I., On yield condition for thin shells, in Russian, *Applied Math. and Mech.*, 24, Nr 2, 1960.

[3.27] Hodge, Ph.G., Comparison of yield conditions in the theory of plastic shells, in Russian, in *Problems of continuum mechanics*, AN SSSR, 1966.

[3.28] Sawczuk, A., Hodge, Ph.G., Comparison of yield conditions for circular cylindrical shells. *J. Frankl. Inst.*, 269, Nr 5, 1960.

[3.29] Listrova, Yu.P., Rudis, M.A., Some limit equilibrium problems for shells of revolution accounting for temperature effects, in Russian, in *Thermal stresses in structural members*, Nr 5, Naukova dumka, Kiev, 1965.

[3.30] Listrova, Yu.P., Mokashova, P.I., On limit equilibrium of structures made of material with different yield points at tension and compression, in Russian, *Eng. Journ. MTT, AN SSSR*, Nr 6, 1967.

[3.31] Beliakov, A.R., Cherniavsky, O.F., On a certain incremental collapse mechanism under cyclic complex loading, in Russian, in *Problems of strength and design of machinery*, No. 45, Cheliab. Polyt. Inst., Cheliabinsk, 1974.

[3.32] Rabotnov, Yu.N., *Strength of materials*, in Russian, Fizmatgiz, Moskow, 1962.

[3.33] Gokhfeld, D.A., Ivanov, I.A., Sadakov, O.S., Description of complex loading effects on the basis of structural model of continuum, in Russian, in *Developments in mechanics of deformable bodies*, Nauka, Moskow, 1975.

[3.34] Gokhfeld, D.A., Cherniavsky, O.F., Incremental collapse of shells, in Russian, in *Spatial structures*, Nauka, Moscow, 1975.

4

Statical methods of solving shakedown problems

The limiting cycle parameters can be evaluated independently, by means of either the statical or the kinematical shakedown theorem. The computational methods based on the Melan theorem or its modifications will be referred to as statical ones whereas those stemming from the Koiter theorem will be termed kinematical. The present as well as the following chapters are devoted to the systematic exposition of the basic statical and kinematical procedures; the merits and limitations of both methods will be discussed in Chapter 5.

The Melan theorem states the necessary and sufficient conditions for shakedown which essentially consist in the following: A structure will shake down to given external actions if there can be found a time-independent stress distribution σ_{ij}^0 satisfying the equilibrium requirements

$$\sigma_{ij,j}^0 + X_i^0 = 0 \quad \text{in the interior } V,$$
$$\sigma_{ij}^0 n_j = p_i^0 \quad \text{on the surface } S_p \tag{4.1}$$

and belonging to the domain bounded by the fictitious yield surface

$$\varphi(\sigma_{ij}^0) \leq 0 \tag{4.2}$$

determined in advance at each point of the considered body.

Let a structure be subjected to a multi-parameter program of external actions. Parameters of the time-independent actions are introduced in equations (4.1) whereas those of variable actions are appearing in the left-hand side of inequality (4.2), since only the fictitious yield surfaces depend on them. The magnitudes of all parameters but one are assumed to be prescribed. The sole unknown parameter is denoted by p. Now we can formulate, on the basis of the Melan theorem, the problem

4 Statical methods of solving shakedown problems

of the limiting cycle determination as follows: From among all values of the parameter p, satisfying both conditions (4.1) and (4.2), i.e. ensuring shakedown, the maximum one is to be found

$$\max p = ?. \tag{4.3}$$

We can obtain the complete description of the shakedown domain by repeating the computations with various values of fixed external actions parameters.

Differential equations (4.1) admit infinitely many solutions since the number of equations is less than the number of unknown functions. On prescribing the distribution law of three (in the general case) components of the stress state σ_{ij}^0 within the volume of the body, the distribution of the remaining components can be found with the help of the equations (4.1) and then the suitable numerical value of the parameter p can be evaluated by means of condition (4.2). This will clearly constitute a lower bound to the sought for parameter of the limiting cycle and a particular value of p will correspond to each assumed distribution law of stresses σ_{ij}^0 that satisfy equations (4.1). From this point of view the equations (4.1) can be looked upon as the description of a 'control process', whereas the inequalities (4.2) can be treated as imposed constraints. Thus the problem of selecting the maximum value of the parameter p turns out to be a problem within the framework of the mathematical theory of optimal control.

In the general case of a continuous description of the process, and when the differential equations (4.1) together with the constraints (4.2), continuously varying over the volume, are applied, the solution can be based on the well known Pontragin maximum principle [4.1–4.3]. The applications of this principle in the mechanics of deformable bodies were lately discussed by Lepik in his survey [4.4]. Its applications to the shakedown problems were worked our earlier in the paper [4.5].

Using an approximate discrete form of the conditions, obtained by transforming (4.1) and (4.2) into a system of algebraic equations and inequalities, we can apply the tools of mathematical programming, linear or convex, especially with the aid of computers [4.6–4.8].

A number of papers has been published on the application of mathematical programming methods to the solutions of shakedown problems via the statical approach. When adopting the piece-wise linear yield surfaces, especially in the case of bar systems as well as for axisymmetrical plates and shells, the determination of the limiting cycle

4 Statical methods of solving shakedown problems

parameters is usually reduced to solving a problem in linear programming [4.9–4.17]. The non-linear yield surfaces are particularly convenient whenever situations of non-axisymmetric plates and shells are discussed in which the principal planes rotate during the cycle. These yield surfaces lead to problems of convex programming [4.18–4.20].

It is worth noting that in the closely related problems of limit analysis those methods were successfully employed already a decade ago [4.21–4.24]. The reformulated Melan and Koiter theorems, stated in Chapter 2, enable us to use directly suitable computational procedures for solving the limit equilibrium problems in the analysis of shakedown conditions. However, it should be emphasized again that the application of the methods of non-classical variational calculus, such as the maximum principle, and of mathematical programming makes sense for the determination of the progressive deformation conditions, only when the realization of the appropriate situation is probable. This circumstance has not always been taken into account (see, e.g. [4.25]) which has led to a waste of effort since the conditions for alternating plastic flow can be determined in a more elementary manner, as shown in Chapter 2.

Only by using non-classical variational methods, both in the continuous and in the discrete formulations, can we arrive to exact, or nearly exact, solutions to the shakedown problems, on the basis of any one of the two fundamental theorems. Moreover, according to the linear programming theory, the shakedown problems stated via the statical or kinematical approach in a discrete formulation constitute a dual set. The corresponding formulations as well as the very theorems appear to be deducible one from the other by means of purely formal mathematical manipulations, [4.6–4.9] and others.

The approximate methods appear to be much less cumbersome here, particularly in the case of the statical method when a certain stress distribution σ_{ij}^0 is assumed in a reasonable manner, satisfying the conditions (4.1) and (4.2) only. Then the corresponding value of p^- is calculated; it does not, in the general case, coincide with the proper maximum. In this manner we can obtain the lower bound assessments of the limiting cycle parameter sought for, sometimes to within an accuracy of certain coefficients liable to a maximization process.

4 Statical methods of solving shakedown problems

4.1. The Pontragin maximum principle and its application to the problems of limit analysis

Certain definitions will first be considered regarding both the statement and the solution of problems in the optimal control theory [4.1].

Let the state of an object in question be described by a system of differential equations of the type

$$\frac{\mathrm{d}x_i}{\mathrm{d}\rho} = f_i(x_1, x_2, \ldots, x_n; u_1, u_2, \ldots, u_r; p_1, p_2, \ldots, p_s; \rho)$$

$$i = 1, 2, \ldots, n \quad (4.4)$$

where ρ is a single independent variable

$$\rho_0 \leq \rho \leq \rho_1. \quad (4.5)$$

The values of both ρ_0 and ρ_1 are prescribed, and $x_1, x_2, \ldots, x_n$ are the phase coordinates that characterize the state of the object. The vector-valued function $\bar{x}(\rho) = (x_1(\rho), x_2(\rho), \ldots, x_n(\rho))$ determined on the segment (4.5), should belong to a closed region X ($\bar{x} \in X$) whose boundary is a smooth or piece-wise smooth hypersurface. The symbols $u_1, u_2, \ldots, u_r$ stand for the control parameters. It is assumed that the vector-valued function $\bar{u}(\rho) = (u_1(\rho), u_2(\rho), \ldots, u_r(\rho))$, determined over the segment (4.5), admits discontinuities and should belong to a closed region U, $\bar{u} \in U$ where U is the control region. Such a vector-valued function is referred to as an admissible control. The regions X and U are both assumed to be independent of ρ and unrelated to each other. The symbols $p_1, p_2, \ldots, p_s$ denote the ρ-independent control parameters.

The functions f_i are determined for arbitrary values of $x \in X$, $u \in U$ and p and are considered to be continuous will respect to the variables $\bar{x}, \bar{p}, \bar{u}, \rho$ and continuously differentiable with respect to $\bar{x}, \bar{p}, \rho$.

Each combination of $\bar{u}(\rho)$ and $\bar{p}$ corresponds, under the given boundary conditions, to a unique solution of the system (4.4) which is termed a phase trajectory. The problem consists in finding those values of the functions $\bar{u}(\rho)$ and the parameters $\bar{p}$, referred to as the optimum control, which yield the optimum behaviour of the object expressed as the minimum (or maximum) of a certain integral

$$I = \int_{\rho_0}^{\rho_1} f_0(\bar{x}(\rho), \bar{u}(\rho), \bar{p}, \rho) \, \mathrm{d}\rho, \quad (4.6)$$

4.1. The Pontragin maximum principle

where f_0 is a given function, positive for any values of its arguments.

The obtained solution to the system (4.4), thus the optimum trajectory, should satisfy the given boundary conditions

$$\bar{x}(\rho_0) = \bar{x}^0, \quad \bar{x}(\rho_1) = \bar{x}^1. \tag{4.7}$$

In the general case, the necessary optimality conditions are stated within the theory of optimum control. If a solution satisfying those conditions turns out to be unique, then the necessary conditions are at the same time sufficient.

Let us assume that the optimum trajectory has been found and that it consists of a finite number of portions each of which either lies on a smooth part of the boundary of the region X or belongs (except, possibly, for the ends) to the interior of that region.

The necessary optimality conditions for those portions of the trajectory which correspond to the interior of the region X are obtainable from the following theorem: Let $\bar{u}(\rho)$, $\rho_0 \leq \rho \leq \rho_1$, be an admissible control which moves (for a specified value of ρ) a phase point from the location $\bar{x}^0$ into $\bar{x}^1$ and let $\bar{x}(\rho)$ be a corresponding trajectory (see (4.4)). Then a necessary condition which must be satisfied by $\bar{u}(\rho)$ and p when such a solution to the stated problem exists is that there should exist a non-vanishing continuous vector-valued function $\bar{\psi}(\rho) = (\psi_0(\rho), \psi_1(\rho), \ldots, \psi_n(\rho))$ which satisfies the following requirements:

a)
$$\frac{d\psi_i}{d\rho} = -\frac{\partial H}{\partial x_i}, \quad i = 0, 1, \ldots, n \tag{4.8}$$

where

$$H = \sum_{k=0}^{n} \psi_k f_k(\bar{x}, \bar{u}, \bar{p}, \rho), \tag{4.9}$$

b) for all values of $\rho_0 \leq \rho \leq \rho_1$ the function

$$H(\bar{\psi}(\rho), \bar{x}(\rho), \bar{u}, \bar{p}, \rho), \quad u \in U, \tag{4.10}$$

attains its maximum (minimum) at the point $u = u(\rho)$ for any fixed functions

$$\bar{\psi}(\rho), \bar{x}(\rho), \tag{4.11}$$

4 Statical methods of solving shakedown problems

c) the function $\psi_0(\rho) = $ const. is non-positive,

d)
$$\int_{\beta_i}^{\beta_{i+1}} \sum_{k=0}^{n} \psi_k \frac{\partial f_k(\bar{x}, \bar{u}, \bar{p}, \rho)}{\partial p_j} \, d\rho = 0, \quad j = 1, 2, \ldots, s. \tag{4.12}$$

By β_i, β_{i+1} are here denoted the segments on which the optimum trajectory $x(\rho)$ belongs to the interior of the region X.

The necessary optimality conditions for the portion of the optimum trajectory lying on the boundary of the region X are formulated in the monograph [4.1]; for lack of space they will not be given here. They will not be employed in what follows, since in the examples to be considered the whole optimum trajectory will belong to the interior of the region X.

In some situations the second boundary condition (4.7) is not prescribed; in such cases it has to be determined from the solution of the problem of a free end of the trajectory. Let $\bar{x}(\rho_1)$ belong to a closed convex region determined by the condition that

$$\text{for } \rho = \rho_1, \quad F(\bar{x}) \leq 0 \tag{4.13}$$

if the cost function is subject to minimization (and $F(\bar{x}) > 0$ if otherwise).

A corresponding formulation of the optimum control problem for the case in which the cost function (optimality criterion) has the form

$$I = \sum_{1}^{n} c_i \bar{x}_i(\rho_1) \tag{4.14}$$

was given in [4.30]. Following this paper, we write now the boundary conditions for the coupled system (4.8). Two cases should be here distinguished:

a) the end of the phase trajectory belongs to an open region corresponding to the condition (4.13). Thus we have

$$\psi_i(\rho_1) = -c_i, \tag{4.15}$$

b) the end of the phase trajectory belongs to the boundary of the region corresponding to the condition (4.13), i.e. $F(\bar{x}) = 0$. Hence

$$\psi_i(\rho_1) = -\lambda c_i - \mu b_i[\bar{x}(\rho_1)], \quad b_i = \frac{\partial F}{\partial x_i}, \tag{4.16}$$

4.1. The Pontragin maximum principle

where $\lambda \geq 0$, $\mu \geq 0$ are non-negative numbers that cannot vanish simultaneously. Without loss of generality we can write $\lambda = 1$ if $\lambda > 0$ or $\mu = 1$ if $\mu > 0$.

Thus the problem of optimum control described by $2n$ differential equations has $2n$ boundary conditions.

The formulation of the optimum control problems allows for not more than one independent variable ρ. Hence we can apply it only in the case of one-dimensional situations of limit equilibrium and shakedown in which the conditions (4.1) are formulated in terms of ordinary differential equations. A somewhat more general formulation is presented in the monograph [4.2].

The applications of the theory of optimum control to the determination of the limit actions on the basis of the statical theorem will now be elucidated by means of examples. We shall deal with circular and annular plates subject to axisymmetrical loading.

The equilibrium equation for such plates has the following one-dimensional form [4.26]:

$$\frac{dm_r^0}{d\rho} = f_1(m_r^0, m_\varphi^0, \bar{p}, \rho) \equiv \frac{1}{\rho}[m_\varphi^0 - m_r^0 - \bar{p}f(\rho)], \qquad (4.17)$$

where we have denoted by $\rho = r/R$ the current radius, $k \leq \rho \leq 1$, by m_r^0, m_φ^0 the time-independent radial and circumferential bending moments referred to the fixed value of the ultimate bending moment M_0 (it will be assumed, for simplicity, that $M_0(\rho) = $ const.); by $\bar{p}$ the dimensionless load factor, and by $f(\rho)$ a function specified by the variation of the applied transverse forces.

Equation (4.17) describes the control process (see (4.4)) in which the continuous and differentiable function m_r^0 plays the role of the phase coordinate whereas the function m_φ^0, admitting discontinuities, is a control parameter (in the presence of such discontinuities the force transmitted across the discontinuity surface remains constant). The other control parameter, independent of ρ, is $\bar{p}$.

Both the phase coordinate m_r^0 and the control parameter m_φ^0 should stay within the boundaries of the fictitious yield surface. Under these circumstances an admissible control m_φ^0 must be found which satisfies (4.17) and suitable boundary conditions and for which the parameter $\bar{p}$ attains its maximum value $\bar{p} = \bar{p}_0$.

Let us reduce the relevant function, which coincides here with the control parameter, to the form (4.6),

4 Statical methods of solving shakedown problems

$$I = \int_k^1 \frac{\tilde{p}}{1-k} \, d\rho, \tag{4.18}$$

and let us consider a number of situations corresponding to the assumed form of the yield surface as well as to the support conditions of the plate. We are here confining ourselves to the limit analysis problems simply because they are sufficient to illustrate the computational procedure while they differ from the shakedown problems only in admitting a somewhat simpler description of the region of admissible values of internal forces. In the shakedown situation the fictitious interaction surface has to be constructed whereas here the actual one will do. Examples of the determination of the incremental collapse conditions for plates and shells by means of the Pontragin maximum principle will be given in Chapters 7 and 8.

Example 1. An annular plate as shown in Fig. 4.1 is simply supported on the outer periphery and subjected to a ring force P applied around the central opening. Let us evaluate the ultimate load carrying capacity P if the yield condition is given by the constraints

$$|m_r| \leq 1, \tag{4.19}$$

$$|m_\varphi| \leq 1, \tag{4.20}$$

visualized in Fig. 4.2 by a square AB_1DE_1 with $\tilde{p} = P/2\pi M_0$ and $f(\rho) \equiv 1$. The function (4.9) and the system (4.8) take the form

$$H = \psi_0 \frac{\tilde{p}}{1-k} + \frac{\psi_1}{\rho}(m_\varphi - m_r - \tilde{p}), \tag{4.21}$$

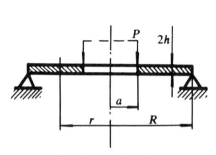

Fig. 4.1. Annular plate.

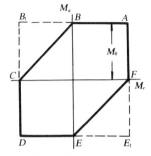

Fig. 4.2. Yield conditions for circular plates.

4.1. The Pontragin maximum principle

$$\frac{d\psi_0}{d\rho} = -\frac{\partial H}{\partial x_0} = 0, \quad \frac{d\psi_1}{d\rho} = -\frac{\partial H}{\partial m_r} = \frac{\psi_1}{\rho}.$$

These equations imply that

$$\psi_0 = C_0, \quad \psi_1 = C_1\rho,$$

where C_0 and C_1 are constants.

It follows from the expression (4.21) and the conditions (4.10) and (4.20) that

$$m_\varphi = -\operatorname{sign} \psi_1. \tag{4.22}$$

The minimum of H ensuring the maximum of the objective cost is obtained when the signs of ψ_1 and m_φ differ while the absolute value of m_φ is at its maximum, i.e. $|m_\varphi| = 1$.

Then the conditions (4.12) take the form

$$\int_{\beta_i}^{\beta_{i+1}} \left[\frac{\psi_0}{1-k} - \frac{\psi_1}{\rho}\right] d\rho = \int_{\beta_i}^{\beta_{i+1}} \left(\frac{C_0}{1-k} - C_1\right) d\rho = 0.$$

This result implies that $C_1 < 0$, since $C_0 < 0$ according to (4.11). It is necessary to point out that $C_0 \neq 0$, otherwise the above equation would yield $C_1 = 0$, whereas we seek the non-vanishing solution to the system (4.8). Therefore sign $\psi = -1$ as $\psi_1 = C_1\rho$, $\rho > 0$.

Substituting the obtained results into (4.22), we arrive at $m_\varphi \equiv 1$.

Thus the equilibrium equation assumes the form

$$\frac{d}{d\rho}(\rho m_r) = 1 - \bar{p}_0. \tag{4.23}$$

The boundary conditions

$$m_r(k) = 0, \quad m_r(1) = 0$$

enable us to find the integration constant together with the value of the collapse load multiplier. The latter turns out to be

$$\bar{p} = 1. \tag{4.24}$$

The ultimate loads for other support conditions can be evaluated in

4 Statical methods of solving shakedown problems

a similar manner. For instance, it is easy to show that for an annular built-in plate under uniformly distributed load p the condition for the limit equilibrium is given by

$$\frac{d}{d\rho}(\rho m_r) = 1 - \frac{p_0 R^2}{2M_0}(\rho^2 - k^2) \qquad (4.25)$$

which leads to

$$\frac{p_0 R^2}{2M_0} = 3\frac{1 - k - m_r(1)}{1 - 3k^2 + 2k^3} = \frac{3(2-k)}{1 - 3k^2 + 2k^3}. \qquad (4.26)$$

The magnitude of $m_r(1)$ has been selected in such a way as to ensure, within the constraint (4.19), the largest value of p_0.

It can readily be shown that in the considered situations the whole optimum trajectory $m_r(\rho)$, excluding perhaps its end points, belongs to the interior of the region, given by the inequalities (4.19).

In the cases when the extremum of the phase coordinate belongs to the given closed region, the optimum control theory provides special conditions referred to as the transversality conditions. They enable us to find out if the optimum trajectory passes through a point of that region. Those conditions are here inoperative since the selection of $m_r(1)$ is obvious enough.

Example 2. Let us take the same type of annular plate but obeying the Tresca yield condition as shown in Fig. 4.2 by the hexagon *ABCDEF*. Thus we have now that

$$\max(|m_r|, |m_\varphi|, |m_r - m_\varphi|) = 1. \qquad (4.27)$$

In order to reduce the corresponding limit analysis problem to an optimum control problem of the considered type it is necessary to exclude the relationships between the admissible domains of m_r and m_φ as given by (4.27). Let us rewrite (4.27) in the form

$$|m_r| \leq 1, \qquad (4.28)$$

$$\left|m_\varphi - \frac{1}{2}m_r\right| + \left|\frac{1}{2}m_r\right| \leq 1 \qquad (4.29)$$

and let us introduce two new control parameters u_1 and u_2 by which m_φ can be uniquely determined with the use of (4.17). Those parameters

4.1. The Pontragin maximum principle

belong to the domains which are independent of m_r,

$$u_1 = \left|m_\varphi - \frac{1}{2}m_r\right| + \left|\frac{1}{2}m_r\right|, \quad 0 \leq u_1 \leq 1, \tag{4.30}$$

$$u_2 = \text{sign}\left(m_\varphi - \frac{1}{2}m_r\right), \quad \text{sign } x = \frac{x}{|x|}. \tag{4.31}$$

Let us now express m_φ in terms of u_1, u_2 and m_r from (4.30), (4.31). We get

$$u_1 u_2 = m_\varphi - \frac{1}{2}m_r + \left|\frac{1}{2}m_r\right| u_2,$$

and substitution into equation (4.17) results in

$$\frac{dm_r}{d\rho} = \frac{1}{\rho}\left[u_1 u_2 - \frac{1}{2}|m_r|u_2 - \frac{1}{2}m_r - \bar{p}f(\rho)\right]. \tag{4.32}$$

This equation describes the optimum control process; the control region is given by the expressions (4.30) and (4.31); the region X by the inequality (4.28). The objective function remains unchanged. Using the theorem on the necessary optimality conditions we obtain

$$H = \psi_0 \frac{\bar{p}}{1-k} + \frac{\psi_1}{\rho}\left[\left(u_1 - \frac{1}{2}|m_r|\right)u_2 - \frac{1}{2}m_r - \bar{p}f(\rho)\right], \tag{4.33}$$

$$\frac{d\psi_0}{d\rho} = 0, \quad \frac{d\psi_1}{d\rho} = \frac{1}{2}\cdot\frac{\psi_1}{\rho}u_2\,\text{sign } m_r + \frac{1}{2}\cdot\frac{\psi_1}{\rho}. \tag{4.34}$$

It follows from (4.30), (4.31) and (4.28) that the minimum value of H is attained at

$$u_1 \equiv 1, \quad u_2 = -\text{sign } \psi_1. \tag{4.35}$$

The second equation (4.34) is linear on any interval in which $u_2 \text{ sign } m_r = \text{const}$:

$$\psi_1 = C_1, \quad \text{for } u_2 \text{ sign } m_r = -1,$$
$$\psi_1 = C_2 \rho^{1/2}, \quad \text{for } u_2 \text{ sign } m_r = 0, \tag{4.36}$$

4 Statical methods of solving shakedown problems

$\psi_1 = C_3\rho,$ for u_2 sign $m_r = 1$,

where C_1, C_2, C_3 denote arbitrary constants. Since ψ_1 is a continuous function, it follows from the expression (4.36) that the function does not change sign at any $\rho > 0$.

Conditions (4.12) assume the form

$$\int_{\beta_i}^{\beta_{i+1}} \left[\frac{\psi_0}{1-k} - \frac{\psi_1}{\rho} f(\rho) \right] d\rho = 0. \tag{4.37}$$

If $\int_{\beta_i}^{\beta_{i+1}} \rho^{-\alpha} f(\rho) \, d\rho > 0$, ($\alpha = 0, \frac{1}{2}, 1$) then, similarly as in the previous example, sign $\psi_1 = -1$. Keeping in mind this result, let us substitute the values u_1, u_2 from (4.35) into the equation (4.32). We obtain that

$$\frac{dm_r}{d\rho} = \frac{1}{\rho} \left[1 - \frac{1}{2} |m_r| - \frac{1}{2} m_r - \bar{p}_0 f(\rho) \right]. \tag{4.38}$$

For continuous $m_r(\rho)$ this equation together with the boundary conditions describes the optimum trajectory and the value of the parameter $\bar{p} = p_0$.

Equation (4.37) is linear in each interval (β_i, β_{i+1}) in which sign $m_r = $ const:

$$\frac{d}{d\rho}(\rho m_r) = 1 - \bar{p}_0 f(\rho) \quad \text{for } m_r > 0, \tag{4.39}$$

$$\rho \frac{dm_r}{d\rho} = 1 - \bar{p}_0 f(\rho) \quad \text{for } m_r \leq 0. \tag{4.40}$$

The number of such intervals is not known beforehand, thus the sign of m_r should first be specified in a zone neighbouring to one of the boundaries of the plate, and next one should estimate the limits of the intervals (β_i, β_{i+1}) with the help of the equations (4.39) and (4.40) and the continuity condition of $m_r(\rho)$. In this way one obtains two different phase trajectories. However, since the distribution of the internal forces at collapse is known to be unique [4.27], it is only one of them that constitutes the optimum solution under given boundary conditions.

Consider, for instance, an annular plate under uniformly distributed load of intensity p, simply supported around the outer circumference. We have

$$f(\rho) = \rho^2, \quad m_r(k) = 0, \quad m_r(1) = 0, \quad \bar{p} = pR^2/2M_0.$$

4.1. The Pontragin maximum principle

Assume that $m_r < 0$ over the segment (k, β_1). Since $m_r(k) = 0$ and $m_r(\beta_1) = 0$, equation (4.40) yields

$$\tilde{p}_0 = \frac{2 \ln \frac{\beta_1}{k}}{\beta_1^2 - k^2}. \tag{4.41}$$

Substituting this result into (4.40), we get

$$\left(\frac{dm_r}{d\rho}\right)_{\rho=k} = \frac{1}{k} - \frac{2k \ln \frac{\beta_1}{k}}{\beta_1^2 - k^2} > 0 \tag{4.42}$$

which contradicts our assumption that $m_r < 0$.

Thus we conclude that $m_r > 0$ over (k, β_1). Equation (4.39) clearly yields the result

$$\tilde{p}_0 = \frac{3}{\beta_1^2 + \beta_1 k + k^2}. \tag{4.43}$$

Let β_1 be such that $k \leq \beta_1 < 1$. Then equation (4.40) should be satisfied in (β_1, β_2), $\beta_1 \leq \beta_2 \leq 1$, under the conditions that $m_r(\beta_1) = 0$ and $m_r(\beta_2) = 0$. But this is clearly impossible.

We conclude finally that $\beta_1 = 1$ and thus

$$\tilde{p}_0 = \frac{p_0 R^2}{2M_0} = \frac{3}{1 + k + k^2}. \tag{4.44}$$

It can now be readily seen that $|m_r| < 1$ holds everywhere.

When the above considered plate is clamped around its outer circumference instead of being simply supported, then a similar argument leads to a phase trajectory for which

$$m_r \geq 0 \quad \text{when} \quad k \leq \rho \leq \beta, \quad m_r \leq 0 \quad \text{when} \quad \beta \leq \rho \leq 1. \tag{4.45}$$

The requirement that m_r is continuous and that the load parameter attains its maximum leads to a system of equations in $\tilde{p}_0$ and β:

$$(\beta - k) - \frac{1}{3}\tilde{p}_0(\beta^3 - k^3) = \ln \beta + \frac{1}{2}(1 - \beta^2) - 1 = 0. \tag{4.46}$$

4 Statical methods of solving shakedown problems

For a circular plate without any central opening we get

$$\tilde{p}_0 = \frac{p_0 R^2}{2M_0} = \frac{3}{\beta^2}, \quad \beta^2 e^{(3/\beta^2)-5} = 1, \tag{4.47}$$

which fully coincides with the known result [4.28] (see also Chapter 7).

There are also other rotationally symmetric problems of limit equilibrium and shakedown that can be conveniently reduced to belong to the mathematical optimum control theory. For example, the equilibrium equation for a cylindrical shell under edge loading assumes the form [4.26]

$$\frac{d^2 m}{d\rho^2} + n = 0, \tag{4.48}$$

where $m = M_x/M_0$, $n = N_\varphi/N_0$, $M_0 = \sigma_s h^2$, $N_0 = 2\sigma_s h$, $\rho = x\sqrt{2/ah}$ and a is the shell radius. This equation can now be looked upon as one describing the optimum control process with the phase coordinates m and $dm/d\rho$. Under the constraints corresponding to the 'square' yield condition, namely

$$|m| \leq 1, \quad |n| \leq 1 \tag{4.49}$$

n plays the role of the control parameter; when the 'hexagonal' yield condition is selected ([4.29], see also Chapter 8),

$$|m| \leq 1, \quad \left|n \pm \frac{1}{2}m\right| \leq 1, \tag{4.50}$$

then one must perform manipulations similar to those pertinent to the Tresca plate in order to identify the control parameters.

Let us mention that in the latter case the load parameter, instead of entering the equilibrium equation, appears in the accompanying boundary conditions. Therefore it is advisable to formulate the plate problem somewhat differently, as in [4.30].

It has been established that neither the presence of variable thickness of the plate nor the variability of the yield point of the material (in rotationally symmetrical situations) lead to any substantial difficulties. A suitable change of variables can always be made such that both the regions U and X become independent of ρ. This usually

4.2. Application of the mathematical programming methods

happens at the cost of a somewhat more complicated form of the equilibrium equation.

The optimum control theory enables us virtually to solve the limit analysis problems not only for piece-wise linear yield conditions but also for non-linear ones. The problem is then described by a system of non-linear differential equations.

4.2. Application of the mathematical programming methods to problems of incremental collapse

The solutions to limit analysis problems by means of the Pontragin maximum principle, as discussed in the preceding section, have been based on the direct use of the differential equations of equilibrium together with some constraints, continuous with respect to the coordinates and imposed on the stresses. However, certain difficulties arise when one attempts to solve those problems by means of the above method using a universal mathematical procedure (computer program). This is due to the fact that once we try to deal with practical problems, various equilibrium conditions as well as various shapes of the fictitious yield surface must be employed. Though, the difficulties can be overcome if the differential equations (4.1) are replaced by a system of simultaneous algebraic equations and the constraints (4.2) are observed at only a finite number of points within the body. The discrete system thus obtained consisting of constraints together with the objective function (4.3) constitute a problem in the mathematical programming [4.6-4.8, and others]. There exists a vast library of programs belonging to convex and, in particular to linear programming. This, together with the simplicity of discretization of conditions (4.1) and (4.2) makes such an approach a better tool than the direct use of the maximum principle.

Let us discuss the linear programming technique as applied to the shakedown problems. The linear programming problem can be formulated in general terms as follows [4.7]:

We seek the maximum (or minimum[1]) of a linear function of the variables x_j, $(j = 1, 2, \ldots, n)$

$$z = \sum_{j=1}^{n} p_j x_j, \qquad (4.51)$$

[1] Minimization of a linear function can clearly be replaced by its maximization on account of obvious equality $\min z = -\max(-z)$.

4 Statical methods of solving shakedown problems

where the variables are related to each other by given inequalities

$$y_i = -\sum_{j=1}^{n} a_{ij}x_j + a_i \geq 0, \quad (i = 1, 2, \ldots, m) \tag{4.52}$$

and by the equations

$$y_i = -\sum_{j=1}^{n} a_{ij}x_j + a_i = 0, \quad (i = m+1, m+2, \ldots, k). \tag{4.53}$$

Equations (4.53) can be solved with respect to any $k - m$ variables, provided the values of the remaining x_j's are known. Thus it is clear that, with no loss of generality, the system of relevant conditions consists of the inequalities (4.52) only.

It is worth noting that by excluding equations from the system of constraints, we have decreased the number of variables and the computations to follow became suitably simpler. This elimination can either be made upon the formulation of the problem or, should this be more convenient, it can be treated as an independent initial stage of the solution to the linear programming problem [4.7].

The variables whose signs are not restrained are termed free variables while those suffering from such restraints ($x_j \geq 0$) are termed bounded ones.

The conditions (4.52) together with the linear form of (4.51) are shown in Table 4.1.

The linear programming problem can be interpreted in geometrical terms. The constraints (4.52) are easily visualized, in the multi-dimen-

Table 4.1

	$-x_1$	$-x_2$	...	$-x_n$	1
$y_1 =$	a_{11}	a_{12}	...	a_{1n}	a_1
...	...	...	...	...	...
$y_m =$	a_{m1}	a_{m2}	...	a_{mn}	a_m
$z =$	$-p_1$	$-p_2$	...	$-p_n$	0

4.2. Application of the mathematical programming methods

sional Euclidean x_j-space, as a certain convex polyhedron Ω. The magnitude of the cost function (4.51) at a generic point can be considered to be a distance of this point from the hyperplane

$$\sum_{j=1}^{n} p_j x_j = 0. \tag{4.54}$$

Thus the geometrical sense of the programming problem consists in finding the point on the polyhedron Ω whose distance from the hyperplane (4.54) is the largest (or the smallest). This point clearly coincides with one of the polyhedron vertices.

Nowadays the simplex method is most commonly used to solve linear programming problems. The method consists of an algorithm either to find a starting solution to the system of linear constraints (4.52) corresponding to a certain vertex of the polyhedron Ω, or to state the fact of incompatibility of the system. Another algorithm is then used to pass from the given starting solution to a new starting solution for which the cost function (4.51) is not smaller than before (or not greater than before, depending on the formulation of the problem). The process is to be repeated until the optimal solution is reached.

Let us now outline, following [4.7], some basic rules of solving problems by means of the simplex method, based on the modified Jordan elimination. The rules will suffice for solving the examples that follow.

A step in the modified Jordan elimination with the pivotal element a_{rs} that transforms the Table 4.1 into a new table is the following:
  a) the pivotal element is replaced by unity,
  b) the remaining elements of the row remain unchanged,
  c) the remaining elements of the columns change signs,
  d) 'ordinary' elements b_{ij} ($i \neq r, j \neq s$) and free terms of the new table are calculated by means of

$$b_{ij} = a_{ij}a_{rs} - a_{is}a_{rj}, \quad b_i = a_i a_{rs} - a_{is}a_r, \tag{4.55}$$

  e) all the elements of the new table are divided by a_{rs}.

Before embarking on finding the starting solution, the free variables should be eliminated from the top row of Table 4.1. Let, for instance, the variables x_j, ($j = 1, \ldots, n$) be free. Thus, using further steps of the modified Jordan elimination procedure, all the free variables x_j from the top row can be shifted to the left-hand column and replaced by suitable

4 Statical methods of solving shakedown problems

y_i's. The expressions replacing the x_j's will be needed to express the results in terms of the original coordinates after the calculations are completed.

Thus certain rows in Table 4.2 are left empty.

As to the selection of the starting solution, we proceed as follows: If all the free terms in Table 4.2 are non-negative, the starting solution has the form

$$y_i = 0, \quad (i = 1, 2, \ldots, n). \tag{4.56}$$

If some of the free terms are negative, a sufficient number of steps in the modified elimination procedure must be taken to cause all the free terms to become non-negative.

To select the pivotal element suitable for the current elimination step to be made, the following rules apply:

a) A row should be chosen having a negative free term (let, for instance, $b_r < 0$). If there are no negative coefficients in this row, then the system (4.52) is incompatible.

b) If there are some negative coefficients in the considered row, then any of them should be selected (let $b_{rs} < 0$) and the corresponding column is adopted as the pivotal one.

c) The selection of the pivotal row must be made as follows: First all the non-negative ratios, $b_i/b_{is} \geq 0$, of the free terms to the appropriate non-vanishing coefficients of the pivotal column should be calculated and the smallest one should be chosen, corresponding to $i = i_0$. Then the row i_0 is taken as the pivotal one since $b_{i_0 s}$ is a pivotal element.

In the degenerate case in which $\min_i (b_i/b_{is} \geq 0) = b_{i_0}/b_{i_0 s} = 0$, the

Table 4.2

	$-y_1$	$-y_s$	...	$-y_n$	1
$y_{n+1} =$	$b_{n+1,1}$	$b_{n+1,s}$	...	$b_{n+1,n}$	b_{n+1}
...	...	...	...	...	...
$y_r =$	b_{r1}	b_{rs}	...	b_{rn}	b_r
...	...	...	...	...	...
$y_m =$	b_{m1}	b_{ms}	...	b_{mn}	b_m
$Z =$	q_1	q_s	...	q_n	Q

4.2. Application of the mathematical programming methods

element b_{i_0s} is adopted as the pivotal one only when $b_{i_0s} > 0$.

After a finite number of steps in the modified Jordan elimination procedure we either find out that the system (4.52) is incompatible or we obtain a table (similar to Table 4.2) with no negative free terms and we arrive at the starting solution to the system by equating to zero all the y_i's which appear in the top row of the table.

The optimum solution to the linear programming problem can be obtained in the following manner:

If all the coefficients of the z-row are non-negative, the linear programming problem has the solution

$$\max z = Q,$$
$$y_1 = y_2 = \cdots = y_n = 0. \tag{4.57}$$

If there is a negative coefficient in the z-row (for example, $q_s < 0$), a step in the modified elimination procedure should be made; the pivotal element should be selected according to the following rules:

a) The column containing the negative element of the z-row (s-th column) should be taken as the pivotal column.

b) Suitable free terms are divided by all the positive coefficients, if any, of the pivotal column, and the smallest value occurring is selected, say, this is for $i = i_0$. Then the i_0-row is taken as the pivotal one, the pivotal element being b_{i_0s}.

When, after the above described elimination step is done, all the coefficients in the z-row become non-negative, the solution is terminated, see (4.57). But when in the z-row of the modified table the negative free terms continue to appear, the next elimination steps should be made until the solution (4.57) is finally arrived at.

If there exist no positive coefficients in the column that contains the negative element of the z-row, the objective function z is not bounded from above.

It should be mentioned that the optimum solution to the problem corresponds to the maximum (or minimum) value of the cost function (specifically, of the load parameter) at which the system of constraints becomes incompatible. This is in full agreement with the second statement of the statical shakedown theorem, as well as with its limit analysis counterpart, according to which the purely elastic behaviour of the structure (or its equilibrium) is impossible under the circumstances.

The solution to the discussed problem enables us to state what type of constraints under suitable conditions are violated, i.e. for

4 Statical methods of solving shakedown problems

shakedown problems, what type of cyclic plastic deformation (alternating plastic flow or incremental collapse) is specifically taking place in the limiting cycle.

Let us now proceed to some simple examples whose aim is to illustrate the basic peculiarities associated with the application of the simplex method; to this end all the intermediate computational steps will be shown in detail. Somewhat more complex and computerized examples dealing with plates and shells will be given in Chapters 7 and 8 in which some special features of applying computers will also be presented together with a discussion of their numerical accuracy.

4.3. Simple examples of solving the shakedown problems via the statical approach by using the simplex method

Example 1. The first, simplest structure to be considered is the two-parameter bar system [4.10] shown in Fig. 1.14. Let the system be subjected to cyclic actions of a force $P = P(\tau)$ and temperature (heating and cooling of element 1). Both load and temperature can vary arbitrarily and independently of each other, provided they stay within the bounds shown in Fig. 4.3 by means of a rectangle in the P, t-plane.

The 'elastic' stresses σ_{pi} and σ_{ti}, ($i = 1, 2, 3$) in particular bars, generated by mechanical loading and by heating have already been found. They will now be written in the form

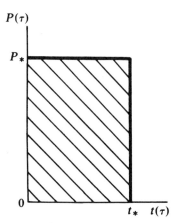

Fig. 4.3. Domain of load factors.

4.3. Simple examples of solving the shakedown problems

$$\sigma_{pi}(\tau) = p(\tau)R_i,$$
$$\sigma_{ti}(\tau) = q(\tau)Q_i, \tag{4.58}$$

where $p(\tau) = P(\tau)$ is the load factor, $q(\tau) = \alpha E t(\tau)$ is the temperature factor, α is the linear expansion coefficient, E is Young's modulus. The product αE is found to be nearly independent of the temperature, hence it is assumed constant. The intervals are

$$0 \le p(\tau) \le p_*, \quad 0 \le q(\tau) \le q_*. \tag{4.59}$$

The quantities R_i and Q_i represent the distribution of 'elastic' stresses at $p(\tau) = 1$ and $q(\tau) = 1$. The following inequalities hold:

$$R_i > 0, \quad (i = 1, 2, 3),$$
$$Q_1 < 0, Q_i > 0 \quad \text{for } i = 2, 3. \tag{4.60}$$

Let the yield point σ_{si} of each bar be temperature-independent. Then the inequality constraints of the rearranged statical theorem take the form

$$\max_\tau [|\sigma_{pi}(\tau) + \sigma_{ti}(\tau) + \sigma_i^0| - \sigma_{si}] \le 0, \tag{4.61}$$

where σ_i^0 are the self-equilibrating stresses.

On substituting the expressions (4.58) into inequality (4.61) and computing $\max_\tau$ with the aid of conditions (4.59) and (4.60), we get

$$\begin{aligned}
&p_* R_1 + \sigma_1^0 - \sigma_{s1} \le 0, & &-q_* Q_1 - \sigma_1^0 - \sigma_{s1} \le 0, \\
&-\sigma_2^0 - \sigma_{s2} \le 0, & &p_* R_2 + q_* Q_2 + \sigma_2^0 - \sigma_{s2} \le 0, \\
&p_* R_3 + q_* Q_3 + \sigma_3^0 - \sigma_{s3} \le 0, & &-\sigma_3^0 - \sigma_{s3} \le 0,
\end{aligned} \tag{4.62}$$

or, in terms of forces,

$$\begin{aligned}
&p_* R_1^* + N_1^0 - Y_1 \le 0, & &-q_* Q_1^* - N_1^0 - Y_1 \le 0, \\
&-N_2^0 - Y_2 \le 0, & &p_* R_2^* + q_* Q_2^* + N_2^0 - Y_2 \le 0, \\
&p_* R_3^* + q_* Q_3^* + N_3^0 - Y_3 \le 0, & &-N_3^0 - Y_3 \le 0,
\end{aligned} \tag{4.63}$$

4 Statical methods of solving shakedown problems

where

$$N_i^0 = \sigma_i^0 F_i, \quad Y_i = \sigma_{si} F_i, \quad R_i^* = R_i F_i, \quad Q_i^* = Q_i F_i, \tag{4.64}$$

and F_i is the cross-sectional area of the i-th bar. The self-equilibrating forces N_i^0 satisfy the equation

$$\sum_{i=1}^{3} N_i^0 = 0. \tag{4.65}$$

Let us find the limit value of the parameter p_*,

$$\max_{N_i^0} p_* = ?, \tag{4.66}$$

treating q_* as a known quantity. The constraints (4.63), (4.65) together with the objective function (4.66) constitute a problem in linear programming. Since the structure considered is actually a discrete one, the formulation is directly derivable from the statical shakedown theorem.

In order to solve the problem (4.63)–(4.66) by means of the simplex method, the following specific values are adopted for the ratios of the ultimate forces Y_i and the stiffness coefficients $c_i = (EF/l)_i$:

$$\begin{aligned} Y_1 = Y_2 = \frac{1}{2} Y_3 = Y, \\ c_1 = c_2 = 2c_3 = 2c. \end{aligned} \tag{4.67}$$

The quantities R_i^*, Q_i^* amount now to

$$R_1^* = R_2^* = 0{,}4, \quad R_3^* = 0{,}2, \quad Q_1^* = -1{,}2c, \quad Q_2^* = 0{,}8c, \quad Q_3^* = 0{,}4c.$$

Putting those values of R_i^* and Q_i^* into the inequalities (4.63) and expressing N_3^0 in terms of N_1^0, N_2^0 (using (4.65)), we arrive at a system of constraints consisting solely of inequalities. This is shown in Table 4.3 which has been prepared according to the rules given before. The unknown forces N_i^0 as well as the sought for load factor p are shown in the top row of the table as multipliers, their signs being reversed, in accordance with the inequalities (4.52). Recalling the constraints system resulting from (4.63), we adopt here as free terms those which are

4.3. Simple examples of solving the shakedown problems

Table 4.3

	$-N_1^0$	$-N_2^0$	$-p$	1
y_1	①	0	0,4	Y
y_2	-1	0	0	$Y - 1,2qc$
y_3	0	1	0,4	$Y - 0,8qc$
y_4	0	-1	0	Y
y_5	-1	-1	0,2	$2Y - 0,4qc$
y_6	1	1	0	$2Y$
Z	0	0	-1	0

constraining the yield force Y as well as the specified temperature factor q (for simplicity, the asterisks are omitted again).

The computational procedure is shown in Tables 4.4–4.7. Tables 4.4 and 4.5 illustrate two steps in the modified Jordan elimination scheme which result in the shift of the free variables N_1^0, N_2^0 from the top row to

Table 4.4

	$-y_1$	$-N_2^0$	$-p$	1
N_1	1	0	0,4	Y
y_2	1	0	0,4	$2Y - 1,2qc$
y_3	0	①	0,4	$Y - 0,8qc$
y_4	0	-1	0	Y
y_5	1	-1	0,6	$3Y - 0,4qc$
y_6	-1	-1	0,4	Y
Z	0	0	-1	0

4 Statical methods of solving shakedown problems

Table 4.5

	$-y_1$	$-y_3$	$-p$	1
N_1^0	1	0	0,4	Y
N_2^0	0	1	0,4	$Y - 0,8qc$
y_2	1	0	⓪,④	$2Y - 1,2qc$
y_4	0	1	0,4	$2Y - 0,8qc$
y_5	1	1	(1̇)	$4Y - 1,2qc$
y_6	-1	-1	$-0,8$	$0,8qc$
Z	0	0	-1	0

occupy the left-hand column. The variable p_* is non-negative on account of (4.59). The pivotal elements are encircled in each table.

At each elimination step we obtain the elements of the tables according to the rules given in Sec. 4.2, making special use of formulae (4.55). For instance, the elements of Table 4.4 are calculated as follows:

$$b_{23} = (a_{23}a_{11} - a_{21}a_{13})/a_{11} = 0 \cdot 1 - (-1,0)0,4 = 0,4,$$

$$b_2 = (a_2 a_{11} - a_{21}a_1)/a_{11} = (Y - 1,2q_*c) \cdot 1 - (-1)Y = 2Y - 1,2q_*c,$$

the common divisor being $a_{rs} = a_{11} = 1$.

After elimination of the free variables (Table 4.5), all the free terms appear to be non-negative at $q_*c \leq \frac{5}{3}Y$. Consequently we find the starting solution from that condition. The next elimination step, made according to the rules of Sec. 4.2, leads to the optimum solution (Table 4.6):

$$p_* = 5Y - 3q_*c, \quad \begin{array}{l} N_1^0 = -Y + 1,2q_*c, \\ N_2^0 = -Y + 0,4q_*c \end{array} \tag{4.68}$$

which holds good as long as $q_*c \geq \frac{5}{9}Y$ (because $q_*c < \frac{5}{9}Y$ corresponds to another pivotal element shown in Table 4.5 by dashed circle). From

4.3. Simple examples of solving the shakedown problems

Table 4.6

	$-y_1$	$-y_3$	$-y_2$	1
N_1^0				$-Y + 1,2qc$
N_2^0				$-Y + 0,4qc$
p	2,5	0	2,5	$5Y - 3qc$
y_4				$0,4qc$
y_5				$-Y + 1,8qc$
y_6				$4Y - 1,6qc$
Z	2,5	0	2,5	$5Y - 3qc$

Table 4.6 it clearly follows that at $q_*c > \tfrac{5}{3}Y$ the system of constraints becomes incompatible: The negative free term is not accompanied by any negative coefficient in the same row.

As follows from Table 4.7 (p. 160), we have at $0 \leq q_*c \leq \tfrac{5}{9}Y$ that

$$p_* = 4Y - 1,2q_*c, \qquad \begin{aligned} N_1^0 &= -0,6Y + 0,48q_*c, \\ N_2^0 &= -0,6Y - 0,32q_*c. \end{aligned} \qquad (4.69)$$

The results (4.68) and (4.69) can be easily explained in kinematical terms. Table 4.6 clearly indicates that the inequalities y_1 and y_2 corresponding to the state of one bar at different instants of time, i.e. at different relations of external actions become equalities, as seen from the first two inequalities (4.63). Thus plastic deformation develops in this bar twice per cycle which means that alternating plastic flow takes place. This type of behaviour, corresponding to the conditions (4.68) is shown in Fig. 4.4 by the straight segment AB bounding a part of the shakedown diagram.

In another situation, as Table 4.7 shows, in none of the bars are the shakedown conditions violated twice per cycle. Comparison with the inequalities (4.63) reveals that there is occurring consecutively in each of the three bars an elongation. This means that incremental collapse takes

4 Statical methods of solving shakedown problems

place according to the conditions (4.69). This is shown in Fig. 4.4 by the segment *BC*.

The results (4.68) and (4.69) fully coincide with those obtained in Chapter 1.

Example 2. Let us consider some ways of constructing a discrete model to study the shakedown of a continuous body and let us evaluate the limiting cycle parameters. Take a plane disc rotating with constant angular velocity and subjected to repeated cycles of heating and cooling with linear distribution of temperature $t(\tau, \rho)$ along the radius, Fig. 4.5,

$$t(\tau, \rho) = t_0(\tau) + \rho t_1(\tau), \qquad (4.70)$$

where $\rho = r/R$ stands for the dimensionless current radius. For simplicity of constructing the fictitious yield surface, the yield point σ_s is assumed to be temperature-independent. Thus the function $t_0(\tau)$ describing the uniform part of heating no longer makes sense here and it will not enter the description.

Given the interval of temperature fluctuations per cycle

$$-t_* \leq t_1(\tau) \leq t_*, \qquad (4.71)$$

let us find the limit value of the angular velocity parameter ω. The variable components σ_{rt} and $\sigma_{\varphi t}$ of the radial and of the circumferential, thermo-elastic stresses were described in [4.31] for the case of the temperature distribution (4.70). They take, after referring to the magnitude σ_s, the form

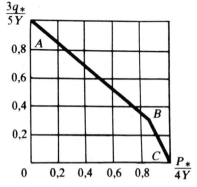
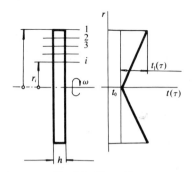

Fig. 4.4. Shakedown diagram for annular plate.   Fig. 4.5. Plane disc.

4.3. Simple examples of solving the shakedown problems

$$\sigma_{rt}(\tau) = q(\tau)(1-\rho), \quad \sigma_{\varphi t}(\tau) = q(\tau)(1-2\rho), \tag{4.72}$$

$$-q_* \leq q(\tau) \leq q_*, \quad q(\tau) = \frac{\alpha E t_1(\tau)}{3\sigma_s}. \tag{4.73}$$

For the Tresca yield condition (2.10) the domain of the admissible constant stresses $\sigma_r^0, \sigma_\varphi^0$ is determined by the inequalities (2.61) which, on account of the expressions (4.72), (4.73), assume the form

$$-1 + q_*(1-\rho) \leq \sigma_r^0 \leq 1 - q_*(1-\rho),$$
$$-1 + q_*|1-2\rho| \leq \sigma_\varphi^0 \leq 1 - q_*|1-2\rho|, \tag{4.74}$$
$$-1 + q_*\rho \leq \sigma_r^0 - \sigma_\varphi^0 \leq 1 - q_*\rho.$$

Fig. 4.6 depicts the domains (4.74) for $\rho = 0{,}5$ and various values of the thermal parameter q_*.

The constraints (4.74) are active for all the radii $0 \leq \rho \leq 1$. Moreover, the constant components of the radial and circumferential stresses should be in equilibrium with the time-independent centrifugal forces [4.31],

$$\frac{d}{d\rho}(\rho\sigma_r^0) - \sigma_\varphi^0 + 3p\rho^2 = 0, \tag{4.75}$$

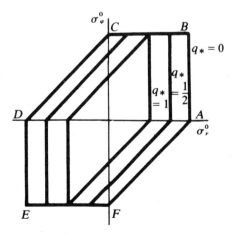

Fig. 4.6. Domains of admissible values of constant stress components.

4 Statical methods of solving shakedown problems

$$\sigma_r^0 = 0 \quad \text{for } \rho = 1, \tag{4.76}$$

where

$$p = \frac{1}{3}\frac{\gamma\omega^2}{g\sigma_s}R^2.$$

The equilibrium condition can also be given a different form; by integrating (4.75) under the boundary condition (4.76), we get

$$\rho\sigma_r^0 = -\int_\rho^1 \sigma_\varphi^0 \, d\rho + p(1-\rho^3). \tag{4.77}$$

Let us now discretize the system (4.74), (4.77). The integration in (4.77) will be replaced by finite summation with constant step of the numerical integration $\Delta\rho = \rho_i - \rho_{i-1}$. The function $\sigma_\varphi^0(\rho)$ is continuous and piece-wise linear, and

$$\rho_i\sigma_{ri}^0 = \frac{1}{-2}\sum_{k=1}^{i-1}[(\sigma_\varphi^0)_k + (\sigma_\varphi^0)_{k+1}] + p(1-\rho_i^3), \tag{4.78}$$

where $\sigma_{ri}^0, \sigma_{\varphi i}^0$ are the constant stress components at $\rho = \rho_i$, and $i = 1, 2, \ldots, n$ is the numbering of sections running towards the centre of the disc.

Such an approach clearly results in a replacement of the discontinuities of $\sigma_\varphi^0(\rho)$, if there are any, by certain zones of more or less rapid continuous change of this function, depending on the magnitude of $\Delta\rho$. The stresses $\sigma_r^0(\rho)$ must be continuous, otherwise the magnitude of σ_φ^0 entering (4.75) cannot be finite as required by the inequalities (4.74).

The constraints (4.74) must now be satisfied not at any ρ but only at the interfaces of zones within which the function $\sigma_\varphi^0(\rho)$ is linear, i.e. at $\rho = \rho_i$. To reduce the dimension of the matrix of the constraints system, let us substitute (4.78) into the inequalities (4.74). This done, the number of unknowns (excluding σ_{ri}^0) is cut down by half and the system of constraints is reduced by the n equations (4.78), thus reducing to inequalities only:

$$-1 + q_*|1 - 2\rho_i| \leq \sigma_{\varphi i}^0 \leq 1 - q_*|1 - 2\rho_i|, \tag{4.79}$$

$$\rho_i[-1 + q_*(1-\rho_i)] \leq -\frac{1}{2}\Delta\rho\sum_{k=1}^{i-1}[(\sigma_\varphi^0)_k + (\sigma_\varphi^0)_{k+1}] + p(1-\rho_i^3)$$

4.3. Simple examples of solving the shakedown problems

$$\leq \rho_i[1 - q_*(1 - \rho_i)], \tag{4.80}$$

$$\rho_i(-1 + q_*\rho_i) \leq -\frac{1}{2}\Delta\rho \sum_{k=1}^{i-1}[(\sigma_\varphi^0)_k + (\sigma_\varphi^0)_{k+1}] + p(1 - \rho_i^3) - \rho_i\sigma_{\varphi i}^0$$

$$\leq \rho_i[1 - t(1 - \rho_i)]. \tag{4.81}$$

The above system together with the optimality criterion

$$\max p = ? \tag{4.82}$$

constitutes a linear programming problem.

Let us solve the problem for $\Delta\rho = 0.5$, which means that only three sections of the disc are relevant: $\rho_1 = 1$, $\rho_2 = 0.5$ and $\rho_3 = 0$. At $\rho_1 = 1$ the inequalities (4.80) are then trivial on account of (4.76), whereas the inequalities (4.79) and (4.81) coincide. At $\rho_3 = 0$ the constraints (4.80) and (4.81) coincide and yield the equality

$$-\frac{1}{4}(\sigma_{\varphi 1}^0 + 2\sigma_{\varphi 2}^0 + \sigma_{\varphi 3}^0) + p = 0 \tag{4.83}$$

from which the magnitude $\sigma_{\varphi 3}^0$ can be expressed in terms of $\sigma_{\varphi 1}^0, \sigma_{\varphi 2}^0$. Finally, the constraints system assumes the form

$$-1 + q_* \leq \sigma_{\varphi 1}^0 \leq 1 - q_*, \quad -1 \leq \sigma_{\varphi 2}^0 \leq 1,$$

$$-\frac{1}{2} + \frac{1}{4}q_* \leq -\frac{1}{4}\sigma_{\varphi 1}^0 - \frac{1}{4}\sigma_{\varphi 2}^0 + \frac{7}{8}p \leq \frac{1}{2} - \frac{1}{4}q_*,$$

$$\tag{4.84}$$

$$-\frac{1}{2} + \frac{1}{4}q_* \leq -\frac{1}{4}\sigma_{\varphi 1}^0 - \frac{3}{4}\sigma_{\varphi 2}^0 + \frac{7}{8}p \leq \frac{1}{2} - \frac{1}{4}q_*,$$

$$1 - q_* \leq \sigma_{\varphi 1}^0 - 2\sigma_{\varphi 2}^0 + 4p \leq 1 - q_*.$$

The solution to the extremum problem (4.82), (4.84) is shown in Tables 4.8–4.11. The first two steps of the modified elimination leading to Table 4.10 are executed to get rid of the free variables $\sigma_{\varphi 1}^0, \sigma_{\varphi 2}^0$. The next step, made according to the rules of seeking the starting solution, leads directly to the optimum solution, Table 4.11, valid for $q_* < \frac{6}{7}$:

$$p = 1 - \frac{1}{2}q_*,$$

$$\sigma_{\varphi 1}^0 = \sigma_{\varphi 3}^0 = 1 - q_*, \quad \sigma_{\varphi 2}^0 = 1. \tag{4.85}$$

4 Statical methods of solving shakedown problems

Table 4.7

	$-y_1$	$-y_3$	$-y_5$	1
N_1^0				$-0{,}6Y + 0{,}48qc$
N_2^0				$-0{,}6Y - 0{,}32qc$
p				$4Y - 1{,}2qc$
y_2				$0{,}4Y - 0{,}72qc$
y_4				$0{,}4Y - 0{,}32qc$
y_6				$5Y - 0{,}7qc$
Z	1	1	1	$4Y - 1{,}2qc$

Table 4.8

	$-\sigma_{\varphi 1}^0$	$-\sigma_{\varphi 2}^0$	$-p$	1
y_1	1	0	0	$1-q_*$
y_2	-1	0	0	$1-q_*$
y_3	0	1	0	1
y_4	0	-1	0	1
y_5	$\frac{1}{4}$	$\frac{1}{4}$	$-\frac{7}{8}$	$\frac{1}{2}-\frac{1}{4}q_*$
y_6	$-\frac{1}{4}$	$-\frac{1}{4}$	$\frac{7}{8}$	$\frac{1}{2}-\frac{1}{4}q_*$
y_7	$\frac{1}{4}$	$\frac{3}{4}$	$-\frac{7}{8}$	$\frac{1}{2}-\frac{1}{4}q_*$
y_8	$-\frac{1}{4}$	$-\frac{3}{4}$	$\frac{7}{8}$	$\frac{1}{2}-\frac{1}{4}q_*$
y_9	1	2	-4	$1-q_*$
y_{10}	-1	-2	4	$1-q_*$
Z	0	0	-1	0

4.3. Simple examples of solving the shakedown problems

Table 4.9

	$-y_1$	$-\sigma_{\varphi 2}^0$	$-p$	1
$\sigma_{\varphi 1}^0$	1	0	0	$1-q_*$
y_2	1	0	0	$2(1-q_*)$
y_3	0	1	0	1
y_4	0	-1	0	1
y_5	$-\frac{1}{4}$	$\frac{1}{4}$	$-\frac{7}{8}$	$\frac{1}{4}$
y_6	$\frac{1}{4}$	$-\frac{1}{4}$	$\frac{7}{8}$	$\frac{3}{4}-\frac{1}{2}q_*$
y_7	$-\frac{1}{4}$	$\frac{3}{4}$	$-\frac{7}{8}$	$\frac{1}{4}$
y_8	$\frac{1}{4}$	$-\frac{3}{4}$	$\frac{7}{8}$	$\frac{3}{4}-\frac{1}{2}q_*$
y_9	-1	2	-4	0
y_{10}	1	-2	4	$2(1-q_*)$
Z	0	0	-1	0

Table 4.10

	$-y_1$	$-y_3$	$-p$	1
$\sigma_{\varphi 1}^0$	1	0	0	$1-q_*$
y_2	1	0	0	$2(1-q_*)$
$\sigma_{\varphi 2}^0$	0	1	0	1
y_4	0	1	0	2
y_5	$-\frac{1}{4}$	$-\frac{1}{4}$	$-\frac{7}{8}$	0
y_6	$\frac{1}{4}$	$\frac{1}{4}$	$\frac{7}{8}$	$1-\frac{1}{2}q_*$
y_7	$-\frac{1}{4}$	$-\frac{3}{4}$	$-\frac{7}{8}$	$-\frac{1}{2}$

4 Statical methods of solving shakedown problems

Table 4.10 (continued)

y_8	$\frac{1}{4}$	$\frac{3}{4}$	$\frac{7}{8}$	$\frac{6}{4}-\frac{1}{2}q_*$
y_9	-1	-2	-4	-2
y_{10}	1	2	4	$4-2q_*$
Z	0	0	-1	0

Table 4.11

	$-y_1$	$-y_3$	$-y_{10}$	1
$\sigma_{\varphi 1}^0$				$1-q_*$
y_2				$2(1-q_*)$
$\sigma_{\varphi 2}^0$				1
y_4				2
y_5				$\frac{7}{8}(1-\frac{1}{2}q_*)$
y_6				$\frac{1}{8}(1-\frac{1}{2}q_*)$
y_7				$\frac{3}{8}-\frac{7}{16}q_*$
y_8				$\frac{5}{8}-\frac{1}{16}q_*$
y_9				$2(1-q_*)$
p				$1-\frac{1}{2}q_*$
Z	$\frac{1}{4}$	$\frac{1}{2}$	$\frac{1}{4}$	$1-\frac{1}{2}q_*$

When $q_* > \frac{6}{7}$, a negative free term appears in the row y_7.

Having determined p, $\sigma_{\varphi 1}^0$, $\sigma_{\varphi 2}^0$, we can find the magnitude of $\sigma_{\varphi 3}^0$ from (4.83). The solution (4.85) can readily be shown to describe the incremental collapse. It should be mentioned that neither the obtained result nor, in particular, the equation (4.85) which is represented in Fig. 4.7 by the

4.3. Simple examples of solving the shakedown problems

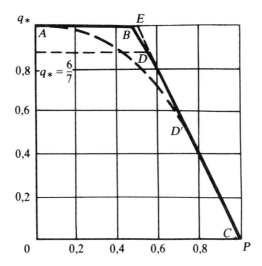

Fig. 4.7. Shakedown diagram for plane disc.

straight line *CD*, change when the integration step $\Delta\rho$ is getting smaller. The result coincides with the exact one; we shall show in Chapter 5 that the latter can be arrived at by simpler means than demonstrated here.

By expanding the solution of the considered problem, we can find that also for $\frac{6}{7} \leq q_* \leq 1$ the incremental collapse takes place, however, the associated mechanism turns out to be different. Whereas in the case of the solution (4.85) the circumferential stresses reached their limit values (and, according to the associated flow rule, the circumferential strain rates differed from zero), now, i.e. for $q_* > \frac{6}{7}$ we have in the central part of the disc the yielding régime

$$\sigma_\varphi - \sigma_r = \sigma_s \tag{4.86}$$

corresponding to the segment *CD*, Fig. 4.6. Both circumferential and radial strain rates are non-vanishing in an appropriate zone of the disc. However, this zone can be shown to be rather small and thus leading to only negligible discrepancies between the solution (4.85) and the present one. The latter is shown in the shakedown diagram of Fig. 4.7 by the straight segment *DB*, branching to the left of the straight line *CDE*. Unlike the solution of (4.85), the present one depends on the magnitude of the integration step.

4 Statical methods of solving shakedown problems

For $q_* > 1$ the system of constraints becomes incompatible. A detailed analysis shows that a condition of the type (2.90) is here violated which means that alternating plastic flow takes place. Thus the exact solution, obtained by assuming sufficiently small steps $\Delta\rho$, can be drawn in Fig. 4.7 as the line $AD'C$ bounding the shakedown domain (recall that the temperature distribution is linear with respect to the current radius ρ).

4.4. Discretization of the continuum in the shakedown problems. Finite element method

In the example considered above the finite summation together with the piece-wise linear approximation of the stress distribution $\sigma_\varphi^0(\rho)$ enabled us to construct the discrete model and thus to reduce the shakedown problem to that of linear programming. There exist, however, other methods to build such a model. In particular, the differential equation (4.75) can be replaced by a system of linear algebraic equations by using finite differences and a piece-wise linear approximation of the function $\sigma_r^0(\rho)$, or better still, of the function $\rho\sigma_r^0(\rho)$. In the latter case we get instead of equation (4.75), the following one:

$$\frac{1}{\Delta\rho}[\rho_i\sigma_{ri}^0 - \rho_{i-1}\sigma_{r(i-1)}^0] - \sigma_{\varphi i}^0 + 3p\rho_i^2 = 0. \tag{4.87}$$

Further manipulations are similar to the previous ones. This approach was employed in the problems of limit analysis by Kupman and Lance [4.22], who dealt with the limit equilibrium of plates by means of the simplex method.

The inaccuracies due to the discrete finite difference model (4.87) resulting from numerical differentiation are usually larger than those due to the discrete finite summation model (4.78) resulting from numerical integration.

In order to replace the differential equilibrium equations by a system of linear algebraic equations, one can also employ wave discretization similarly as it was done by Mirzabekjan and Reitman [4.32] in the evaluation of the limit load for shells. For instance, it can be assumed that in the interval $\rho_i \leq \rho \leq \rho_{i+1}$ the radial stresses take the form of a series

4.4. Discretization of the continuum in the shakedown problems

$$\sigma_r^0 = \sum_{n=1}^{k} (a_n \sin n\pi\rho + b_n \cos n\pi\rho). \tag{4.88}$$

A substitution of the above series into the differential equation (4.75) leads to a system of algebraic equations, linear with respect to both a_n and b_n. Some of those coefficients can be found from the boundary conditions, the rest from the condition of the load factor maximization.

In cases in which the actual stress distributions in the limiting cycle are of undulating character, the wave discretization leads to a reduction in the number of intervals and thus also in the number of constraints. Such a situation takes place, for example when analysing short shells. However, it should be mentioned that in the presence of discontinuities in the actual stress distribution the advantages of wave discretization, using smooth functions, become less clearly pronounced.

The above considered methods of obtaining the discrete models of the shakedown problem and in particular, of the limit analysis problem, possess certain general features. First, the solutions obtained can be arbitrarily close to the exact ones, provided the number of intervals into which a structure is split is large enough. Secondly, the stress discontinuities can conveniently be tackled either directly or by introducing zones of rapid change for the relevant functions.

The last decade has witnessed broad applications of the finite element method (FEM).

Some time ago mechanical models, analogous to that of FEM, were employed in the limit analysis of elastic-plastic structures. In one of them a structure is treated as consisting of rigid elements and deforming (inelastically) along certain rectilinear interfaces referred to as plastic hinges. In applications to the limit equilibrium of plates and especially of reinforced concrete slabs the main concept of such an approach was outlined by Gvozdev. His work was followed by Rzhanitzin [4.33], who also extended the method to cover shells [4.24]. The method remains applicable to the shakedown problems provided the actual yield surface is suitably replaced by the fictitious one determined beforehand according to the relationships presented in Chapters 2 and 3.

The finite element method was applied to the shakedown problems by the Italian school of mechanics (Maier, Corradi, Donato, Vitello, Capurso and others). FEM was used together with the theory of mathematical programming to formulate the fundamental theorems, also to offer a broader interpretation (allowance for strain-hardening, geometrical non-linearity etc.), and to estimate the displacements and strains

4 Statical methods of solving shakedown problems

accumulated in the process of shakedown. Maier's lecture notes [4.34] should be here mentioned; other references were given in Chapter 2.

The general solvability of the finite element procedure, applied to shakedown situations, was first considered in [4.18]. With the use of a non-linear yield condition (2.9) the problem of a plate with an aperture under two-parameter in-plane load was shown equivalent to a problem in convex quadratic programming.

4.5. Approximate statical methods

As in the limit equilibrium problems, the approximate methods of solution are also frequently used in the shakedown problems. Thus the two fundamental theorems can be used to assess from above and from below the limiting cycle parameters. The various modifications of the statical lower bound method have been most commonly employed during a certain period [4.35–4.38 and others]. All of them are based on an a priori assumption of self-equilibrating stress fields (the results of various authors differ in the specific manners in which these stress fields are chosen) whereupon a selection of load factors is made which ensures that all the conditions of the Melan theorem are satisfied.

In other words, some admissible static, but not optimal, solution to the problem (4.1)–(4.3) is constructed, yielding the lower bound to the limiting cycle parameters. The discrepancy between the obtained assessment and the exact solution depends on how good the assumed stress distribution σ_{ij}^0 has been. This selection escapes formalization, leaving room for experience and intuition of the engineer. A certain improvement in the accuracy of assessment can be made by introducing into the distribution σ_{ij}^0 a restricted number of multipliers that can be determined by maximizing the sought factor of the applied load.

We shall consider two variants of the approximate statical method, differing in the way in which in the distributions of time-independent stresses are presented. Some other methods do also exist, however these two variants illustrate adequately the application of the statical method.

1. Let the 'elastic' stresses generated by a variable load and temperature field be expressed as a linear function

$$\sigma_{ij}^{(e)} = \sum_{k=1}^{\alpha} p_k P_{ij}^{(k)} + \sum_{r=1}^{\beta} q_r Q_{ij}^{(r)}, \qquad (4.89)$$

4.5. Approximate statical methods

where p_k, q_r are load and temperature factors respectively, both depending on time; and $P_{ij}^{(k)}$, $Q_{ij}^{(r)}$ are 'unit' tensors of 'elastic' stresses, both functions of spatial coordinates.

The time-independent residual stresses that, according to the Melan theorem, should develop as a result of plastic flow over the first loading stages will be assumed in the form

$$\bar{\rho}_{ij} = \sum_{n=1}^{\gamma} m_n M_{ij}^{(n)}, \qquad (4.90)$$

where m_n is the self-stress factor; $M_{ij}^{(n)}$ are linearly independent self-equilibrating 'unit' stress systems satisfying the boundary conditions and γ is the statical redundancy of the system, or equivalently, the number of degrees of freedom for residual stresses, i.e. for a continuum $\gamma = \infty$.

This kind of stress representation is most convenient for systems with a finite number of degrees of freedom. Unit residual stress states can be selected in such a way as to satisfy the conditions of orthogonality and normalization [4.33], [4.35]. So far as the limit equilibrium problems are concerned, it was stated that 'if it is possible to approximate the stress field by a linear sum of orthogonal functions representing the whole system (as, for instance, it is the case with the Fourier method) then the load factor can virtually be maximized with the use of the standard methods of analysis' [4.22].

However, in order to obtain the approximate lower bound one can limit the number of terms in the series (4.90); the easiest result will be reached by assuming $\gamma = 1$. Solutions of this kind were given for certain shakedown problems which were characterized by a one-parameter load ($\alpha = 1$) and by a thermal régime for which the thermal stresses were all proportional to a single parameter ($\beta = 1$). It is here convenient to use the geometrical representation in the form of the admissible states diagram, as introduced by A.R. Rzhanitzin [4.33].

Following [4.39], we can assume that

$$M_{ij} = Q_{ij}^{(r)}, \qquad (4.91)$$

since $Q_{ij}^{(r)}$ is also a self-equilibrating stress distribution satisfying the statical conditions. Thus, for $\alpha = \beta = 1$, the admissible states diagram can be shown to lie in the plane with coordinates $p, q' = q + m$. When the necessity arises to consider the yield stress as a temperature-

167

4 Statical methods of solving shakedown problems

dependent quantity, the admissible states constitute a certain domain in the three-dimensional space. The shapes of the bounding surfaces (corresponding to those values of the load factors at which stresses at a given point of the body reach the yield point) are determined by the adopted yield condition as well as by the temperature-dependent yield stress function. The domain of admissible states is bounded by those portions of the surfaces that are closer to the coordinate origin or by their envelopes. The maximum value of the parameter p at which the relevant point will still lie on the boundary of the domain (limit equilibrium) can be found analytically. One can also establish the relationships between the limiting amplitudes of load and temperature parameters (in the shakedown problems).

To a certain extent, the admissible states diagram can be used for the geometrical interpretation of the basic problem in linear programming. This appears to be particularly apparent in two- and three-dimensional situations.

The original formulation of the shakedown problem as based on the Melan theorem in the form (2.16) is known to fail to distinguish between the two limit states: alternating plastic flow and incremental collapse. However, the results thus obtained can be interpreted from this standpoint. It is interesting to note that the equations for the limit resistance against the alternating plastic flow resulting from the approximate method practically coincide with the corresponding exact solutions, whereas for the conditions of incremental collapse it is only the lower bound to the load parameters that can be found in this manner. The reasons for such a situation become obvious when we recall the problem of stress uniqueness in the limiting cycle as considered in Chapter 2.

We can use the representation of the residual stresses in the form (4.90) for $1 < \gamma < \infty$ in connection with the mathematical programming methods, whenever the general formulation of the problem leads to a matrix of constraints which is too large to be dealt with by available computational means.

2. The second variant of the approximate statical method stems from the reformulated Melan theorem (Sec. 2.4) and from certain kinematical notions.

Let a certain suitable collapse mechanism be assumed, i.e. we assume a kinematically admissible distribution of plastic strain increments $\Delta\epsilon_{ij0}''$. The associated flow rule (2.76) enables us to determine the corresponding magnitudes of some time-independent stress components

4.6. Examples illustrating the use of approximate statical methods

σ_{ij}^0 which fit the fictitious yield surface at each point of the body. It can now be assumed that the components thus found differ from those developing in the actual limiting cycle by a constant multiplier C. Thus the distribution law is found to within a multiplier.

The preceding calculations are based on the conditions of the Melan theorem. The equilibrium equations (4.1) supply the remaining stress components and the parameters of external actions. All those quantities are functions of the multiplier C; its value has to be computed in such a way as to yield the maximum magnitude of the unknown parameter of external actions and to satisfy the conditions (4.2).

4.6. Examples illustrating the use of approximate statical methods in shakedown problems

Example 1. Following [4.40], let us consider the shakedown conditions for a thick-walled tube as shown in Fig. 4.8, subjected to a cycling internal pressure, $0 \leq p_a \leq p_a^*$, and to a temperature field

$$t(\tau, \rho) = t_b(\tau) + [t_a(\tau) - t_b(\tau)] \frac{\ln \rho}{\ln k}, \qquad (4.92)$$

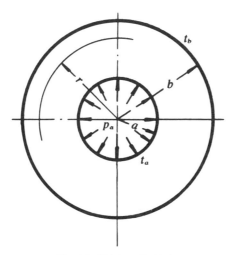

Fig. 4.8. Thick-walled tube.

4 Statical methods of solving shakedown problems

where τ denotes time, $\rho = r/b$, $k = a/b$, $t_a(\tau) - t_b(\tau) = t_1(\tau)$, $0 \leq t_1(\tau) \leq t_*$.
The temperature field (4.92) takes place when heating and cooling are sufficiently slow.

'Elastic' stresses generated by internal pressure in a long closed tube are equal to [4.31][1]

$$\sigma_{rp}^{(e)} = p\left(1 - \frac{1}{\rho^2}\right), \quad \sigma_{\varphi p}^{(e)} = p\left(1 + \frac{1}{\rho^2}\right), \quad \sigma_{zp}^{(e)} = p, \qquad (4.93)$$

where

$$p = \frac{p_a}{\sigma_s} \cdot \frac{k^2}{1-k^2}$$

is the load factor.

The temperature field (4.92) generates thermo-elastic stresses [4.31] which are expressible in the form

$$\sigma_{rq}^{(e)} = -q\left(1 - \frac{1}{\rho^2} + 2\delta \ln \rho\right),$$

$$\sigma_{\varphi q}^{(e)} = -q[1 + \frac{1}{\rho^2} + 2\delta(1 + \ln \rho)], \qquad (4.94)$$

$$\sigma_{zq}^{(e)} = -2q[1 + \delta(1 + 2 \ln \rho)],$$

where

$$q = \frac{\alpha E t_1(\tau)}{2\sigma_s(1-\mu)} \cdot \frac{k^2}{1-k^2}$$

is the temperature field factor and $\delta = (1 - k^2)/2k^2 \ln k$, $0 \leq q \leq q_*$.

As mentioned before, the residual stress state in a solid body is a function of an unlimited number of parameters. However, when calculating the approximate solution to the considered problem, we shall assume that only one of them (m) is non-vanishing and that the corresponding 'unit' state M_{ij} is similar to the thermal stress state (4.94).

Therefore the resultant stresses are

---

[1] Throughout this section stresses in the tube are referred to the yield point σ_s at room temperature.

4.6. Examples illustrating the use of approximate statical methods

$$\sigma_r = p\left(1 - \frac{1}{\rho^2}\right) + (m - q)\left(1 - \frac{1}{\rho^2} + 2\delta \ln \rho\right),$$

$$\sigma_\varphi = p\left(1 + \frac{1}{\rho^2}\right) + (m - q)\left[1 + \frac{1}{\rho^2} + 2\delta(1 + \ln \rho)\right], \quad (4.95)$$

$$\sigma_z = p + 2(m - q)[1 + \delta(1 + 2 \ln \rho)].$$

The yield point is assumed to decrease linearly in the interval $t_b \leq t \leq t_a$, t_b being adopted as fixed value, so that

$$\sigma_{st} = \sigma_s[1 - n(t - t_b)] = \sigma_s(1 - 2\lambda q\delta \ln \rho), \quad (4.96)$$

where

$$\lambda = \frac{2(1 - \mu)n\sigma_s}{\alpha E}.$$

Substituting the stresses (4.95) into the Huber–Mises yield condition (2.9) in which the yield point (4.96) is a temperature-dependent quantity, we arrive at a set of limit surfaces bounding the domains of variation of the parameters p, q, m:

$$a_{11}q^2 + a_{22}m^2 + a_{33}p^2 + 2a_{12}qm + 2a_{23}mp + 2a_{13}qp + 2a_{14}q + a_{44} = 0. \quad (4.97)$$

Inside those domains the strains due to the resultant stresses (4.95) corresponding to appropriate radii remain elastic. Indeed, the coefficients in (4.97) are functions of the current radius,

$$a_{11} = a_{22} - a_{14}^2, \quad a_{13} = -a_{23} = -\frac{3}{\rho^2}\left(\frac{1}{\rho^2} + \delta\right),$$

$$a_{14} = 2\lambda\delta \ln \rho, \quad a_{22} = -a_{12} = [1 + \delta(1 + 2 \ln \rho)]^2 + 3\left(\frac{1}{\rho^2} + \delta\right)^2, \quad (4.98)$$

$$a_{33} = \frac{3}{\rho^4}, \quad a_{44} = -1.$$

An analysis [4.40] of (4.98) shows that the equation (4.97) represents in the p, q, m-space a set of elliptical cones whose common axis runs through the middle of the coordinate angle in the plane $p = 0$.

4 Statical methods of solving shakedown problems

The coordinates of the apices of the cones are

$$q' = m' = \frac{1}{a_{14}} = \frac{1}{2\lambda\delta \ln \rho}. \qquad (4.99)$$

For the outermost points of the tube, $\rho = 1$, the apex is removed to infinity and the expressions (4.97), (4.98) yield the equation of a cylinder

$$4(1 + \delta)^2(q - m) - 6(1 + \delta)(q - m)p + 3p^2 - 1 = 0. \qquad (4.100)$$

The domain of admissible (i.e. elastic) states is shown in Fig. 4.9 as bounded by surfaces of the cone for $\rho = k$ and the cylinder (4.100). The surfaces (4.97) corresponding to other values of the current radius are located outside the shown one.

Similar calculations can readily be obtained for the Tresca yield condition (2.12). Elliptical cones will be replaced by hexagonal pyramids, cylinders by prisms.

Considering various cross-sections of the admissible states domain, we can find the relationships describing shakedown conditions for different loading programs. In particular, intersection by the plane $p =$ const. yields

$$q_1^0 = \frac{(1 + k^2\delta)\sqrt{4k^4 - 3p^2} - \lambda k^2(1 - k^2)}{2(1 + k^2\delta)^2 - 0{,}5\lambda^2(1 - k^2)^2} \qquad (4.101)$$

for relatively low pressures and

$$q_2^0 = \frac{3p - \sqrt{4 - 3p^2}}{4(1 + \delta)} - \frac{3p - \sqrt{4k^4 - 3p^2}}{4(1 + k^2\delta)} \qquad (4.102)$$

for higher pressures.

The expressions obtained supply the lower bounds to the limit values of temperature drops at which the thermal fluctuations under constant internal pressure will lead to shakedown of the tube.

The intersections of the admissible domains corresponding to (4.101) and (4.102) are shown in Fig. 4.10a, b. In the first case both ends of the shakedown trajectory 1–2 are lying on the limit surface (4.97) plotted for $\rho = k$. This means that for $q > q_1^0$ the thermal fluctuations will result in alternating plastic flow in the neighbourhood of the innermost points of the tube, the trajectory being 1–2–3–4–1. In the second case the tra-

4.6. Examples illustrating the use of approximate statical methods

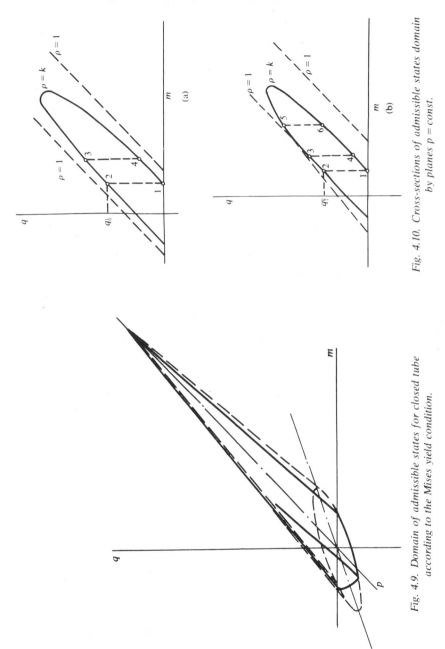

Fig. 4.10. Cross-sections of admissible states domain by planes $p = const.$

Fig. 4.9. Domain of admissible states for closed tube according to the Mises yield condition.

4 Statical methods of solving shakedown problems

jectory touches different limit surfaces ($\rho = 1$ at heating and $\rho = k$ at cooling) at different stages of the cycle and thus the conditions for strain accumulation are roughly imitated; for $q > q_2^0$ the incremental collapse should take place.

The admissible states domain permits us to study also somewhat more complicated loading programs in which not only the temperature but also the internal pressure is variable. On analysing the inclined intersections $q = \beta p$, β being a numerical coefficient, the limit relationships can be established in the case of proportional changes in load and temperature parameters,

$$q_1^0 = \frac{(2k^2 + 3p)(1 + k^2\delta) - \lambda k^2(1 - k^2)}{4(1 + k^2\delta)^2 - \lambda^2(1 - k^2)^2} +$$
$$+ \frac{\{[(2k^2 + 3p)(1 + k^2\delta) - \lambda k^2(1 - k^2)]^2 - 3p(k^2 + p)[4(1 + k^2\delta)^2 - \lambda(1 - k^2)^2]\}^{1/2}}{4(1 + k^2\delta)^2 - \lambda^2(1 - k^2)^2},$$

(4.103)

$$q_2^0 = \frac{3p - \sqrt{4 - 3p^2}}{4(1 + \delta)} + \frac{k^2}{2(1 + k^2\delta)}.$$

(4.104)

for relatively low and relatively high pressures, respectively.

As to a cycle which represents the general formulation of the shakedown problem, i.e. in which both the pressure and the temperature can vary arbitrarily inside prescribed intervals, the solution is obtainable from the analysis of the intersections of the admissible domain by the planes $m = $ const., Fig. 4.9. One such section is plotted in Fig. 4.11 for low pressure. A difference can easily be seen between the ordinate of the inscribed rectangle corresponding to an arbitrary sequence of loading (q_3^0, point A) and the limit ordinate corresponding to the program $p = $ const. (point B). The former amounts to

$$q_3^0 = \frac{2k^2 - 3p + \sqrt{4k^4 - 3p^2}}{2[2(1 + k^2\delta) + \lambda(1 - k^2)]}.$$

(4.105)

A detailed analysis shows that, after a certain increase in the maximum value of the load factor p, the curve $\rho = 1$ shifts in such a way that both points A and B coincide. Then $q_3^0 = q_2^0$ and the formula (4.102) applies.

We can estimate from the admissible states diagram the lower bound to the limit values of $p = p_0$ at which the load-carrying capacity of

4.6. Examples illustrating the use of approximate statical methods

the tube is of an instantaneous type, Fig. 4.12. Account can also be taken here of the change in the ultimate pressure due to temperature rise, as affecting the magnitude of yield stress.

The shakedown diagram of the tube under the considered loading conditions for $k = 0{,}8$, $\lambda = 0{,}17$ is presented in Fig. 4.13. For the cycle $p = $ const. the shakedown domain is bounded by the lines 1 and 2 corresponding to the equations (4.101) and (4.102), respectively. For a proportional loading cycle, the lines 4, 5, 3 are the bounding ones corresponding, respectively, to (4.103), (4.104) and to the change in ultimate pressure owing to temperature rise. Finally, for an arbitrary sequence of load and temperature variations the shakedown domain is bounded by the lines 6 and 2; it appears to be clearly the most confined one.

In Fig. 4.13 the location of side 6 is also indicated by a dashed line corresponding to the temperature-independent yield point, $\lambda = 0$. The location of the other side, 2, already known to correspond to the incremental collapse, is not influenced by dependence of the yield point on temperature. The reasons for this are clearly seen in Fig. 4.11 since the line $\rho = 1$ is now passing lower than before and therefore the points A and B coincide.

The exact magnitude of the ultimate pressure [4.42],

$$p_a^2 = -\frac{2}{\sqrt{3}} \sigma_s \ln k \qquad (4.106)$$

and its approximate magnitude, as found by means of the admissible states diagram, are both shown in Fig. 4.14 as functions of the ratio a/b.

For the adopted distribution of the residual stresses, similar to that of thermal stresses, the discrepancies appear to be rather insignificant; they are getting smaller as the tube becomes thinner.

Example 2. Let us now consider the plane rotating disc as analysed in Sec. 4.3 by means of the simplex method. The lower bound solution to the progressive deformation process is sought. Recalling the known limit equilibrium solution to the problem [4.43], we shall assume the kinematically admissible collapse mechanism in the form

$$\Delta\epsilon_r \equiv 0, \quad \Delta\epsilon_\varphi = \frac{A}{r}, \quad A = \text{const.}, \quad (A > 0), \qquad (4.107)$$

where $\Delta\epsilon_r$ and $\Delta\epsilon_\varphi$ are respectively the radial and circumferential plastic

4 Statical methods of solving shakedown problems

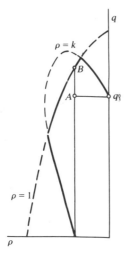

Fig. 4.11. Cross-sections of admissible states domain by the plane $m = $ const.

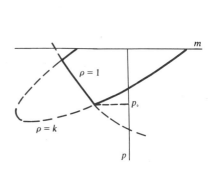

Fig. 4.12. Cross-sections of admissible states domain by the plane $q = 0$.

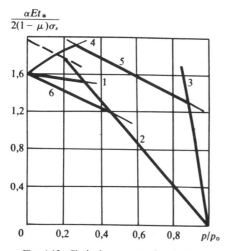

Fig. 4.13. Shakedown range for the tube.

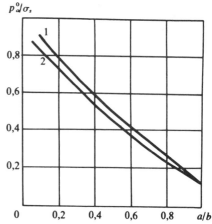

Fig. 4.14. Ultimate loads (instantaneous plastic collapse); exact (1) and approximate (2) methods.

4.6. Examples illustrating the use of approximate statical methods

strain increments per cycle. The stresses

$$\sigma_{\varphi *}^0 = 1 - q_*|1 - 2\rho| \qquad (4.108)$$

corresponding to them belong to the fictitious yield surface by which the domain (4.74) is bounded. Let us emphasize that $1 - 2\rho \geq 0$ for $0 \leq \rho \leq 0{,}5$ and $1 - 2\rho \leq 0$ for $0{,}5 \leq \rho \leq 1$.

The time-independent loop stresses at limiting cycle will now be assumed proportional to the stresses (4.108),

$$\sigma_\varphi^0 = C(1 - q_*|1 - 2\rho|), \quad C = \text{const.} \qquad (4.109)$$

Substituting the above into the equilibrium equation (4.75), we shall calculate the radial stresses σ_r^0. The integration constant is found from the condition prevailing at the centre of the disc,

$$\sigma_r^0 = \sigma_\varphi^0 \quad \text{at } \rho = 0, \qquad (4.110)$$

whereas the angular velocity parameter p can be evaluated from the boundary condition (4.76) at $\rho = 1$. Thus, keeping in mind the signs of $1 - 2\rho$, we get:

$$\sigma_r^0 = C[1 - q_*(1 - \rho)] - p\rho^2 \quad \text{for } 0 \leq \rho \leq 0{,}5, \qquad (4.111)$$

$$\sigma_r^0 = C\left[1 + q_*(1 - \rho) - \frac{q_*}{2\rho}\right] - p\rho^2 \quad \text{for } 0{,}5 \leq \rho \leq 1, \qquad (4.112)$$

$$p = C\left(1 - \frac{1}{2}q_*\right). \qquad (4.113)$$

Substituting the expressions (4.109), (4.111)–(4.113) into the inequalities (4.74), we conclude that

$$|C| \leq 1, \qquad (4.114)$$

for $0 \leq \rho \leq 0{,}5$

$$-1 + q_*(1 - \rho) \leq C\left[1 - \rho^2 - q_*\left(1 - \rho - \frac{1}{2}\rho^2\right)\right] \leq 1 - q_*(1 - \rho),$$
$$-1 + q_*\rho \leq C\left[-\rho^2 - q_*\rho\left(1 - \frac{1}{2}\rho\right)\right] \leq 1 - q_*\rho, \qquad (4.115)$$

4 *Statical methods of solving shakedown problems*

and for $0.5 \leq \rho \leq 1$

$$-1 + q_*(1-\rho) \leq C\left[1 - \rho^2 - t\left(-1 + \rho + \frac{1}{2\rho} - \frac{1}{2}\rho^2\right)\right] \leq 1 - q_*(1-\rho),$$

$$-1 + q_*\rho \leq C\left[-\rho^2 - q_*\rho\left(-1 - \frac{1}{2}\rho + \frac{1}{2\rho^2}\right)\right] \leq 1 - q_*\rho.$$
(4.116)

In order to obtain from (4.113) a lower bound estimate of the limiting cycle parameters, the maximum value of the constant C must first be found for which the inequalities (4.114)–(4.116) hold true. Calculating C from (4.115) and (4.116) we readily get

$$\max C = 1 \quad \text{for } q_* < 0.85. \tag{4.117}$$

Therefore (4.114) bounds the magnitude of C. On combining (4.117) and (4.113), we have

$$p = 1 - \frac{1}{2}q_*. \tag{4.118}$$

For $q_* > 0.85$ the left-hand side inequality in the last row of (4.116) is the bounding one. After rearrangements we obtain

$$C \leq \frac{1 - q_*\rho}{\rho^2 + q_*\rho\left(-1 - \frac{1}{2}\rho + \frac{1}{2\rho^2}\right)}. \tag{4.119}$$

The expression (4.113) assumes, on account of (4.119), the form

$$p = \left(1 - \frac{1}{2}q_*\right)\min_{\rho}\left[\frac{1 - q_*\rho}{\rho^2 + q_*\rho\left(-1 - \frac{1}{2}\rho + \frac{1}{2\rho^2}\right)}\right], \quad (0.5 \leq \rho \leq 1). \tag{4.120}$$

The lower bound estimates (4.118) and (4.120) are found to be practically identical with the exact solution obtained in the previous section with the help of the simplex method (Fig. 4.7).

Let us finally point out that the application of the above variant of the approximate statical method is most favourable when the approximate kinematical method is used for the upper bound assessment simultaneously, see Chapter 5. Indeed, in both cases the same collapse mechanism can be adopted as a basis for the calculations. This permits us to get upper and lower bonds for the parameters characterizing the limiting cycle.

A characteristic feature of the considered method is that in the analysis of plates and shells, as a rule, there is no need to employ the entire interaction surface in terms of the generalized stresses. This considerably simplifies the computations.

References

[4.1] Pontragin, L.S., *Mathematical theory of optimal control*, in Russian, Fizmatgiz, Moscow, 1961.
[4.2] Butkovsky, A.G., *Methods of systems control with distributed parameters*, in Russian, Nauka, Moscow, 1975.
[4.3] Boltyansky, W.G., *Optimum control of discrete systems*, in Russian, Nauka, Moscow, 1973.
[4.4] Lepik, Yu.R., Application of Pontragin maximum principle to strength calculations of shells and plates, *IX All-Union Conf. on shell and plate theory*, in Russian, Leningrad, 1973.
[4.5] Cherniavsky, O.F., On solving limit equilibrium and shakedown problems with the help of Pontragin's maximum principle, in Russian, Izvestiya AN SSSR, *Solid Body Mechanics*, 4, 1970.
[4.6] Gass, S., *Linear programming. Methods and applications*, McGraw-Hill, New York, 1958.
[4.7] Zukhovitsky, S.P., Avdeeva, L.I., *Linear and convex programming*, in Russian, Nauka, Moscow, 1967.
[4.8] Himmelblau, D.M., *Applied nonlinear programming*, McGraw-Hill, New York, 1972.
[4.9] Chiras, A.A., *Linear programming methods in analysis of elastic-plastic systems*, in Russian, Strojizdat, Leningrad, 1969.
[4.10] Gokhfeld, D.A., Cherniavsky, O.F., Simplex method in shakedown problems, in Russian, in *Thermal stresses in structural members*, V. 6, Naukova Dumka, Kiev, 1966.
[4.11] Gokhfeld, D.A., Cherniavsky, O.F., *On incremental collapse conditions in axisymmetric shakedown problems by means of linear programming*, in Russian, Izv. AN SSSR, Mechanica Tverd. Tela, Nr 3, 1970.
[4.12] König, J.A., A method of shakedown analysis of frames and arches. *Int. J. Solid. Struct.*, 7, 1971, 327–344.
[4.13] Gokhfeld, D.A., Cherniavsky, O.F., Methods of solving problems in the shakedown theory of continua. *Proc. Int. Symp. Found. of Plasticity. Warsaw, 1972* Noordhoff, Leiden, 1973.
[4.14] Corradi, L., Zavelani, A., A linear programming approach to shakedown analysis of structures. *Comput. Math. Appl. Mech. and Eng.*, 3, Nr 1, 1974.

4 Statical methods of solving shakedown problems

[4.15] Fox, J.D., Kraus, H., Penny, R.K., Shakedown of pressure vessels with ellipsoidal heads. *Pap. ASME*, NPVP-34, 1971.
[4.16] Cohn, M.Z., Ghosh, S.K., Parimi, S.R., Unified approach to theory of plastic structures. *J. Eng. Mech. Div. Proc. ASCE*, *98*, Nr 5, 1972, 1133–1158.
[4.17] Guerlement, G., Application de la theorie de la stabilisation plastique au dimensionnement des reservoirs. *Rev. Meth.*, *20*, Nr 1, 43–48, 1974.
[4.18] Belytschko, T., Plane stress shakedown analysis by finite elements. *Int. J. Mech. Sci.*, *14*, 1972.
[4.19] Chiras, A.A., Atkochunas, Yu.Yu., *Mathematical models for analysis of elastic-plastic body under repeated load*, in Russian, Lit. Mekh, Sb., Nr 2(7), 1970.
[4.20] Lubarov, B.I., On elastic-plastic analysis under variable repeated loading, in Russian, *Stroit. Mekch. i Razch. Soor.*, Nr 2, 1974, 28–32.
[4.21] Wolfensberger, R., *Traglast und optimale Bemessung von Platten*, Diss. Doct. techn. Wiss., Eidgenos Techn. Hochschule, Zurich, 1964.
[4.22] Koopman, D.C.A., Lance, R.H., On linear programming and plastic limit analysis, *Journal of the Mechanics and Physics of Solids*, *13*, Nr 2, 1965, 77–87.
[4.23] Ceradini, G., Gavarini, C., Calcolo a vottura e programmazion lineare, *Giorn. genio civile*, *103*, Nr 1–2, 1965.
[4.24] Rzhanitzin, A.R., Limit equilibrium of shells by linear programming, in Russian, in *Proc. VI All-Union Conf. on shell and plate theory*, (Baku, 1966), Nauka, Moscow, 1966.
[4.25] Muspratt, M.A., Shakedown of steel plates. *Trans. ASME*, *38*, Nr 4, 1971.
[4.26] Timoshenko, S.P., Woinowsky-Kriger, *Theory of plates and shells*, McGraw-Hill, New York, 1959.
[4.27] Koiter, W.T., General theorems for elastic-plastic solids, *Progress in solid mechanics*, v. I, edited by I.N. Sneddon and R. Hill, North-Holland, Amsterdam, 1960.
[4.28] Hopkins, H.G., Prager, W., The load-carrying capacities of circular plates. *J. Mech. Phys. Solids*, *2*, 1953, 1–13.
[4.29] Hodge, Ph.G., Jr., *Limit analysis of rotationally symmetric plates and shells*. Prentice-Hall, Englewood Cliff, N.J., 1963.
[4.30] Rozonoer, L.I., Pontragin's maximum principle in optimum systems theory, in Russian, *Automatika i telemekhanika*, *20*, Nr 10–12, 1959.
[4.31] Ponomarev, S.D., *Strength calculation in machinery*, in Russian. vol. II, Mashgiz, Moscow, 1959.
[4.32] Mirzabekian, Yu.Yu., Rejtman, M.I., *Load-carrying capacity of shells with the use of linear programming*, in Russian, Inzh. Zhurn. Mechanica Tverd. Tela, Nr 1, 1968, 122–124.
[4.33] Rzhanitzin, A.R., *Structural analysis taking account of plastic properties*, in Russian, Gosstroiizdat, Moscow, 1954.
[4.34] Maier, G., Matrix theory of shakedown allowing workhardening and second-order geometric effects. *Proc. Int. Symp. Found. of Plasticity. Warsaw, 1972*, Noordhoff, Leiden, 1973.
[4.35] Prager, W., *Problemen der Plastizitätstheorie*, Birkhäuser, Basel-Stuttgart, 1955.
[4.36] Rozenblum, W.I., On shakedown theory of elastic-plastic bodies, in Russian, Izv. AN SSSR, OTN, Nr 6, 1958.
[4.37] Leckie, F.A., Penny, R.K., Shakedown pressures for flush cylinder-sphere shell intersections. *J. Mech. Eng. Sci.*, *7*, NT, 1965, 367–371.

References

[4.38] Macfarlane, W.A., Findley, G.E., A simple technique for calculating shakedown loads in pressure vessels. *Proc. Inst. Mech. Eng.*, *186*, Nr 4, 1972, 45–52.

[4.39] Prager, W., Plastic design and thermal stresses, *British Welding Journal*, *3*, Nr 8, 1956, 355–359.

[4.40] Gokhfeld, D.A., Ermakov, P.I., Shakedown of thick-walled tubes under non-uniform heating, in Russian, *Zhurn. Prikl. Mekh. Tekh. Fiz.*, Nr 3, 1963.

[4.41] Muskhelishvili, N.I., *Analytical geometry*, in Russian, Gostekhizdat, Moscow, 1947.

[4.42] Ilushin, A.A., *Plasticity*, Gostekhizdat, Moscow, 1948.

[4.43] Gokhfeld, D.A., On rotating discs at limit state, in Russian, *Mashinostroenie*, Nr 5, 1965.

5

Kinematical methods of solving the shakedown problems

The problem of determination of the limiting cycle parameters is a non-classical variational problem, also when based on the kinematical shakedown theorem.

Let one of the factors of external loads (constant or variable) be unknown. We denote it by p. The values of the remaining load factors at limiting cycle are assumed to be given. Then, according to the reformulated kinematical theorem (Sec. 2.6), the problem consists in finding the distribution of plastic strain increments $\Delta\epsilon''_{ij0}$ per cycle, i.e. the collapse mechanism, which leads to the minimum value of the unknown,

$$\min_{\Delta\epsilon''_{ij0}} p = ?, \qquad (5.1)$$

under the following conditions:

$$\int X_i^0 \Delta u_{i0}\, dv + \int_{S_p} p_i^0 \Delta u_{i0}\, dS \geq \int \sigma^0_{ij*} \Delta\epsilon''_{ij0}\, dv + \sum_v \int_{S_v} \sigma^0_{ij*} n_j \Delta u'_{i0}\, dS, \qquad (5.2)$$

$$\Delta\epsilon''_{ij0} = \frac{1}{2}(\Delta u_{i0,j} + \Delta u_{j0,i}), \qquad (5.3)$$

$$\Delta\epsilon''_{ij0} = \sum_n \mu \frac{\partial \varphi_n(\sigma^0_{ij*})}{\partial \sigma^0_{ij*}}, \quad \mu \geq 0. \qquad (5.4)$$

It should be remembered that the stresses σ^0_{ij*} in (5.2) and (5.4) belonging to the fictitious yield surface are functions of the 'elastic' stresses generated by variable external actions. The inequality (5.2) allows for the possibility of discontinuities in the displacement field; as it follows from the sum of integrals on its right-hand side. The associated flow rule, linking the vector of plastic strain increment per cycle with the fictitious yield surface $\varphi(\sigma^0_{ij*}) = 0$ (see Sec. 2.5) is here adopted for the

5 Kinematical methods of solving the shakedown problems

particular case in which the actual yield surface and/or the enveloping hodograph of variable stresses are piece-wise linear. In the general case, in which no such condition applies, it is a somewhat more awkward relationship (2.77) that must be satisfied.

The differential equations (5.3) describing the increments of residual displacements allow for an infinite number of solutions even discontinuous ones. They can be looked upon as a description of the 'control process'. The components of the residual deflection increments Δu_{i0} and of the plastic strain increments $\Delta \epsilon''_{ij0}$ play the roles of phase coordinates and controls, respectively. Conditions (5.2) and (5.4) specify the constraints imposed on those phase coordinates and controls; the process itself is to obey the optimality criterion (5.1).

Thus the problem (5.1)–(5.4) can be considered as one of the mathematical optimum control theory. From this statement it follows that Pontragin's maximum principle [5.1] can be applied in some simpler (one-dimensional) situations whereas in more complex cases one should employ its suitable generalizations [5.2, 5.3 and others].

However it should be emphasized that the above outlined solution to the problem (5.1)–(5.4) via the kinematical theorem turns out to be considerably more difficult than that based on the statical theorem (Sec. 4.1). This is due to the fact that the integral inequality (5.2) imposes some relationships between magnitudes of both phase coordinates and controls whereas in the statical approach the corresponding relationships are exclusively of a local character. For extremum problems of the discussed type in the continuous formulation, there still have to be worked out convenient computational methods.

The discrete form of the problem (5.1)–(5.4) can be obtained as follows: We replace the differential equations (5.3) by a system of algebraic equations, we require that the constraints (5.4) are satisfied at a finite number of points, and we give an algebraic representation of the basic inequality (5.2) where we replace integration by finite summation. Then the problem becomes one of mathematical programming. Furthermore, when a piece-wise linear yield condition is employed together with the assumption that both strains and displacements remain small, we have a linear programming problem. Since there is available a variety of methods of discretization, including the finite element technique, the linear programming method has been broadly applied in the limit analysis, both in the statical and kinematical approach. In this chapter the peculiarities of its application will be discussed via the latter approach and some simple examples will be demonstrated.

5.1. *Application of the mathematical programming methods*

A determination of the shakedown conditions with the use of the powerful mathematical techniques which are available for solving the non-classical variational problem (5.1)–(5.4) is found to be rather cumbersome. It is worthwhile doing this only when the incremental collapse situation occurs and when there exists no other possibility, except by calculations, to make a reasonable guess as to the type of expected collapse mechanism. The clarity of the geometrical representation of the kinematical approach is the reason for greater potentialities of this method as compared with the statical one, as in the latter the statically admissible stress fields have to be guessed.

An additional loading method (or overload method) can be also classified as belonging to the approximate kinematical methods since it is based on kinematical notions and it leads, in general, to an upper bound assessment of the limiting cycle parameters.

We devote this chapter to various computational procedures based on the kinematical theorem. The procedures will be illustrated by suitable examples.

5.1. Application of the mathematical programming methods to shakedown problems in the kinematical formulation

The application of linear programming methods to the solution of shakedown problems via the kinematical theorem is due to Chyras [5.4, 5.5], who studied one-dimensional bar systems under variable mechanical loading. He showed that the linear programming problems formulated on the basis of the fundamental theorems constituted the so-called dual pair. Because of this fact, one can employ certain formal mathematical rules to obtain, without reference to any mechanical terms, one discrete formulation of the shakedown problem directly from the other. It should be mentioned here that the latter fact was recently pointed out by the authors of [5.6]; using the reformulated Melan theorem in the form similar to that given in Sec. 2.4 and allowing for the duality of linear programming problems, they arrived at the discrete form of the kinematical theorem (5.1)–(5.4) which does not contain the current time similarly as in the considerations of Sec. 2.6.

On the other hand, the simultaneous consideration of the pair of dual problems of linear programming opens new avenues to devise suitable computational procedures [5.7] of the kind which seems to have been used until now rather rarely.

5 Kinematical methods of solving the shakedown problems

As to isothermal as well as nonisothermal shakedown problems, the linear programming methods, particularly the simplex method, were first employed in [5.8, 5.9] to deal with continuous bodies within the framework of the kinematical theorem. Their applications to the limit analysis of shells and plates were worked out earlier in [5.10–5.12].

Let us now illustrate the solution to (5.1)–(5.4) by means of a simple example.

Example. A bar system shown in Fig. 5.1 is subjected to a constant force P and cyclic variation in the temperature of bar 3:

$$0 \leqslant t_3(\tau) \leqslant t_*, \quad t_1(\tau) = t_2(\tau) \equiv 0. \tag{5.5}$$

The transverse beams are assumed to be perfectly rigid, whereas the material properties of all the bars are the same and totally independent of the temperature. The cross-sectional areas are such that

$$F_1 : F_2 : F_3 = 1 : 1{,}8 : 2. \tag{5.6}$$

Let us find the magnitude of the force P and of the temperature t_*, both corresponding to the limiting cycle.

In order to specify the relationships (5.2)–(5.4) it is first necessary to compute the thermo-elastic stresses and to determine the domains of admissible values of time-independent axial forces in each of the constituent bars, i.e. to specify the fictitious yield surfaces. The appropriate results are shown in Table 5.1, in which $N_{ti}^{(e)}$ stand for the forces caused by non-uniform heating, under the assumption that the material is ideally elastic, and $q(\tau)$ is a temperature parameter,

$$q(\tau) = \frac{\alpha E t_3(\tau)}{9{,}5\sigma_s}, \quad 0 \leqslant q(\tau) \leqslant q_*. \tag{5.7}$$

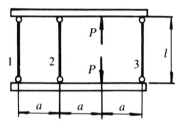

Fig. 5.1. A bar system.

5.1. Application of the mathematical programming methods

Table 5.1

N bar	$N_{ii}^{(e)}/\sigma_s F_1$	$N_{i*}^0/\sigma_s F_1$	$\Delta\epsilon_i''$
1	$-2q(\tau)$	$\dfrac{1}{-1+2q_*}$	$\Delta\epsilon_1 > 0$ $\Delta\epsilon_1 < 0$
2	$3q(\tau)$	$\dfrac{1{,}8-3q_*}{-1{,}8}$	$\Delta\epsilon_2 > 0$ $\Delta\epsilon_2 < 0$
3	$-q(\tau)$	$\dfrac{2}{-2+q_*}$	$\Delta\epsilon_3 > 0$ $\Delta\epsilon_3 < 0$

The largest and the smallest admissible values of time-independent axial forces are shown, in accordance with (2.56), in the third column of Table 5.1. The corresponding signs of plastic strain increments $\Delta\epsilon_i''$, obeying (5.4), are given in the last column.

Since we expect that various yielding regimes may generally take place over a cycle, we assume the strain increments to consist of an elongation ($\Delta\epsilon_i^+$) and of a contraction ($-\Delta\epsilon_i^-$) part,

$$\Delta\epsilon_i'' = \Delta\epsilon_i^+ - \Delta\epsilon_i^-, \quad i = 1, 2, 3, \tag{5.8}$$

$$\Delta\epsilon_i^+ \geq 0, \quad \Delta\epsilon_i^- \geq 0. \tag{5.9}$$

Thus, noting the data in Table 5.1, we see that the condition (5.2) for the shakedown not to occur takes the form

$$p\delta \geq [\Delta\epsilon_1^+ + (1-2q_*)\Delta\epsilon_1^- + (1{,}8-3q_*)\Delta\epsilon_2^+ + 1{,}8\Delta\epsilon_2^- + 2\Delta\epsilon_3^+ \\ + (2-q_*)\Delta\epsilon_3^-]l, \tag{5.10}$$

where δ is a displacement increment per cycle at the points of application of the forces P, Fig. 5.1, and p denotes a dimensionless load factor,

$$p = P/\sigma_s F_1. \tag{5.11}$$

The strain increments $\Delta\epsilon_i''$ are related to the displacement incre-

5 Kinematical methods of solving the shakedown problems

ments δ by the compatibility conditions

$$\Delta\epsilon_3'' = 3\Delta\epsilon_2'' - 2\Delta\epsilon_1'', \quad \frac{\delta}{l} = 2\Delta\epsilon_2'' - \Delta\epsilon_1''. \qquad (5.12)$$

The formulation of the problem of finding the limiting cycle parameters is thus as follows:
Values of the variables $\Delta\epsilon_i^+, \Delta\epsilon_i^-$ should be found for which the conditions (5.8), (5.9), (5.10), (5.12) hold and the factor p, considered as a function of the factor q attains its minimum (or vice versa, i.e. q attains its minimum as a function of p).

The problem is peculiar in that there appears in the system of constraints (5.10) a product of the unknowns δ and p. However the incremental collapse problems can be reduced to the form at which the algorithms of Sec. 4.2 are usable. Taking into account that the constraining system (5.8)–(5.12) allows for an evaluation of both the strain and displacement increments only to within a multiplier and that the external work done on the displacement increments is always positive, we can assume that

$$\delta = 1. \qquad (5.13)$$

Let us now substitute this value into the inequality (5.10), at the same time simplifying the system of constraints by excluding the equalities. To this end, we solve (5.12), using (5.8) and (5.13) for any two of the unknowns, for instance $\Delta\epsilon_1^+$ and $\Delta\epsilon_3^+$.

After some rearrangements the system of constraints takes the form

$$p - 2(1 - q_*)\Delta\epsilon_1^- - (1{,}8 - 3q_*)\Delta\epsilon_2^+ - 1{,}8\Delta\epsilon_2^- - (4 - q_*)\Delta\epsilon_3^- - 3 \geq 0, \qquad (5.14)$$

$$\Delta\epsilon_1^- + 2\Delta\epsilon_2^+ - 2\Delta\epsilon_2^- - 1 \geq 0, \quad -\Delta\epsilon_2^+ + \Delta\epsilon_2^- + \Delta\epsilon_3^- + 2 \geq 0, \qquad (5.15)$$

$$\Delta\epsilon_1^- \geq 0, \quad \Delta\epsilon_2^+ \geq 0, \quad \Delta\epsilon_2^- \geq 0, \quad \Delta\epsilon_3^- \geq 0, \quad p \geq 0. \qquad (5.16)$$

The magnitudes $\Delta\epsilon_1^+, \Delta\epsilon_3^+$ can be readily found from

$$\Delta\epsilon_1^+ = \Delta\epsilon_1^- + 2\Delta\epsilon_2^+ - 2\Delta\epsilon_2^- - 1, \quad \Delta\epsilon_3^+ = -\Delta\epsilon_2^+ + \Delta\epsilon_2^- + \Delta\epsilon_3^- + 2. \qquad (5.17)$$

In order to employ the algorithm to solve the maximum problem

5.1. Application of the mathematical programming methods

(Sec. 4.2), the optimality criterion can be conveniently rewritten as

$$\min p = -\max(-p) = ?. \tag{5.18}$$

The problem (5.14)–(5.16), (5.18) is presented numerically in Table 5.2. The variables in the top row are non-negative by condition (5.16), and therefore the calculations are initiated, once the starting solution is found. This can be accomplished by performing two steps in the modified Jordan elimination procedure shown in Tables 5.3 and 5.4. The pivotal elements are those circled in Tables 5.2 and 5.3. Table 5.5 gives the results of the first step towards finding the optimum solution; the corresponding pivotal element is circled in Table 5.4. At $q_* < 0{,}6$ the z-row of Table 5.5 contains no negative coefficients which means that the optimum solution has just been arrived at; it is given by

$$p = 3{,}9 - 1{,}5 q_*, \tag{5.19}$$
$$\Delta\epsilon_1^- = \Delta\epsilon_2^- = \Delta\epsilon_3^- = 0, \quad \Delta\epsilon_2^+ = 0{,}5.$$

Substituting the above found quantities into (5.17) and (5.8), we get

$$\Delta\epsilon_1'' = 0, \quad \Delta\epsilon_2'' = 0{,}5, \quad \Delta\epsilon_3'' = 1{,}5. \tag{5.20}$$

At $q_* < 0{,}6$ it is only the first coefficient in the z-row of Table 5.5 that is negative and therefore one more step is necessary in order to obtain the optimum solution. This is shown in Table 5.6. For $0{,}6 \leq q_* \leq 1$ the optimum solution is found to be

$$p = 6{,}6 - 6 q_*, \tag{5.21}$$

Table 5.2

	$-\Delta\epsilon_1^-$	$-\Delta\epsilon_2^+$	$-\Delta\epsilon_2^-$	$-\Delta\epsilon_3^-$	$-p$	1
y_1	$2(1-q_*)$	$1{,}8 - 3q_*$	$1{,}8$	$4 - q_*$	(-1)	-3
y_2	-1	-2	2	0	0	-1
y_3	0	1	-1	-1	0	2
z	0	0	0	0	1	0

5 Kinematical methods of solving the shakedown problems

Table 5.3

	$\Delta\epsilon_1^-$	$-\Delta\epsilon_2^+$	$-\Delta\epsilon_2^-$	$-\Delta\epsilon_3^-$	$-y_1$	1
p	$-2(1-q_*)$	$-1{,}8+3q_*$	$-1{,}8$	$-4+q_*$	-1	3
y_2	$\boxed{-1}$	-2	2	0	0	-1
y_3	0	1	-1	-1	0	2
Z	$2(1-q_*)$	$1{,}8-3q_*$	$1{,}8$	$4-q_*$	1	-3

whereas

$$\Delta\epsilon_1^- = \Delta\epsilon_2^- = \Delta\epsilon_3^- = 0,$$

and consequently,

$$\Delta\epsilon_2'' = 2, \quad \Delta\epsilon_1'' = 3, \quad \Delta\epsilon_3'' = 0. \tag{5.22}$$

The computational results are shown diagrammatically in Fig. 5.2. The lines AB and BC correspond to the incremental collapse conditions (5.19), (5.20) and (5.21), (5.22), respectively; the corresponding collapse mechanisms are also shown. The line DC refers to alternating plastic flow in bar 1, $q_* = 1$. This fact can readily be seen by inspecting the data in Table 5.1: At $q_* = 1$ the domain of the admissible force in bar 1 starts

Table 5.4

	$-y_2$	$-\Delta\epsilon_2^+$	$-\Delta\epsilon_2^-$	$-\Delta\epsilon_3^-$	$-y_1$	1
p	$-2(1-q_*)$	$2{,}2-q_*$	$-5{,}8+4q_*$	$-4+q_*$	-1	$5-2q_*$
$\Delta\epsilon_1^-$	-1	$\boxed{2}$	-2	0	0	1
y_3	0	1	-1	-1	0	2
Z	$2(1-q_*)$	$-2{,}2+q_*$	$5{,}8-4q_*$	$4-q_*$	1	$-5+2q_*$

5.1. Application of the mathematical programming methods

Table 5.5

	$-y_2$	$-\Delta\epsilon_1^-$	$-\Delta\epsilon_2^-$	$-\Delta\epsilon_3^-$	$-y_1$	1
p	$-0{,}9+1{,}5q_*$	$-1{,}1+0{,}5q_*$	$-3{,}6+3q_*$	$-4+q_*$	-1	$3{,}9-1{,}5q_*$
$\Delta\epsilon_2^+$	$-0{,}5$	$0{,}5$	-1	0	0	$0{,}5$
y_3	$0{,}5$	$-0{,}5$	0	-1	0	$1{,}5$
Z	$0{,}9-1{,}5q_*$	$1{,}1-0{,}5q_*$	$3{,}9-3q_*$	$4-q_*$	1	$-3{,}9+1{,}5q_*$

Table 5.6

	$-y_3$	$-\Delta\epsilon_1^-$	$-\Delta\epsilon_2^-$	$-\Delta\epsilon_3^-$	$-y_1$	1
p	$1{,}8-3q_*$	$-2+2q_*$				$6{,}6-6q_*$
$\Delta\epsilon_2^+$	1	0				2
y_2	2	-1				3
Z	$-1{,}8+3q_*$	$2-2q_*$	$3{,}9-3q_*$	$5{,}8-4q_*$	1	$-6{,}6+6q_*$

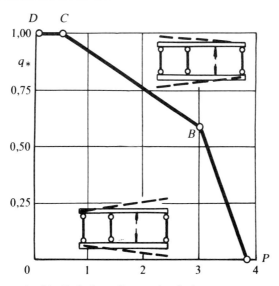

Fig. 5.2. Shakedown diagram for the bar system.

5 Kinematical methods of solving the shakedown problems

to degenerate whereas the admissible regions for the remaining bars remain non-degenerate.

5.2. Solution to the problem of incremental collapse of a plane disc by means of the simplex method

We shall consider an example which will adequately illustrate the application of linear programming methods to the solution of shakedown problems for solids via the kinematical approach.

Let a plane solid disc rotating with constant velocity be subjected to thermal changes associated with a most simple linear distribution of temperature along its radius, Fig. 4.5. By using the statical approach, we have evaluated in Sec. 4.3 the limiting cycle parameters for the disc.

Now we can formulate as follows the problem of finding the limits of shakedown domain by using the kinematical theorem.

We have to find the minimum value of the centrifugal force factor

$$\min_{\Delta \epsilon_r'', \Delta \epsilon_\varphi''} p = ?, \quad p = \frac{\gamma \omega^2 R^2}{3 g \sigma_s}, \tag{5.23}$$

under the conditions (5.2), which for the plane disc take the form

$$3p \int_0^1 \Delta u \rho^2 \, d\rho \geq \int_0^1 (\sigma_{r*}^0 \Delta \epsilon_r'' + \sigma_{\varphi*}^0 \Delta \epsilon_\varphi'') \rho \, d\rho + \sum_\nu (\sigma_{r*}^0 \Delta u' \rho)_\nu, \tag{5.24}$$

as well as under the compatibility conditions (5.3), in terms of the plastic strain increments

$$\Delta \epsilon_r'' = \frac{d(\Delta u)}{d\rho}, \quad \Delta \epsilon_\varphi'' = \frac{\Delta u}{\rho}. \tag{5.25}$$

There also have to be satisfied relationships of the type (5.4), resulting from the flow rule (2.76) associated with the fictitious (piece-wise linear) yield surface,

$$\Delta \epsilon_r'' = \sum_i \mu_i \frac{\partial \varphi_i(\sigma_{r*}^0, \sigma_{\varphi*}^0)}{\partial \sigma_{r*}^0}, \quad \Delta \epsilon_\varphi'' = \sum_i \mu_i \frac{\partial \varphi_i(\sigma_{r*}^0, \sigma_{\varphi*}^0)}{\partial \sigma_{\varphi*}^0}, \tag{5.26}$$

where

$$\mu_i \geq 0. \tag{5.27}$$

5.2. Solution to the problem of incremental collapse

We shall denote by σ_{r*}^0, $\sigma_{\varphi*}^0$ the radial and hoop stresses belonging to the fictitious yield curve $\varphi(\sigma_{r*}^0, \sigma_{\varphi*}^0) = 0$ and conveniently referred to the yield point σ_s, by $\Delta\epsilon_r''$, $\Delta\epsilon_\varphi''$ the increments of radial and hoop plastic strains per cycle, by Δu the radial displacement increment referred to the outer radius of the disc, and by ρ_ν the radius (also referred to the outer radius of the disc) at which a jump $\Delta u_\nu'$ takes place in the radial displacement field.

The fictitious yield locus for the points of the considered disc are described by (see Sec. 4.3)

$$\max[|\sigma_{r*}^0| + q_*(1-\rho), |\sigma_{\varphi*}^0| + q_*|1-2\rho|, |\sigma_{r*}^0 - \sigma_{\varphi*}^0| + q_*\rho] = 1, \quad (5.28)$$

$$\left(q = \frac{\alpha E}{3\sigma_s} t_1(\tau), -q_* \leq q(\tau) \leq q_*\right),$$

and are shown in Fig. 5.3a–e for various points of the disc and for different values of the factor q_*. Recalling condition (5.28), we easily prove that at

$$0 \leq q_* \leq \frac{1}{2|1-2\rho|} \quad (5.29)$$

the yield locus is a hexagon for all the points of the disc. At

$$\frac{1}{2|1-2\rho|} \leq q_* \leq 1 \quad (5.30)$$

the fictitious yield curve is a rectangle in the central part of the disc, $\rho < \frac{1}{2}$, Fig. 5.3a, b, whereas it becomes a parallelogram at the periphery, $\frac{1}{2} < \rho \leq 1$, Fig. 5.3d, e. At $\rho = \frac{1}{2}$ we have that the circumferential thermo-elastic stress $\sigma_{\varphi q}^{(e)} = 0$ and the fictitious yield curves depend on the radial thermo-elastic stresses only; for $q_* \geq 1$ they are like the parallelogram shown in Fig. 5.3c.

Let us now rearrange inequality (5.24) so that it takes a form which is more convenient in linear programming situations. As in the previous example, the multiplier $\int_0^1 \Delta u \rho^2 \, d\rho$ appearing in the left-hand side of the inequality at p, will be assumed to be equal to unity, namely

$$\int_0^1 \Delta u \rho^2 \, d\rho = 1. \quad (5.31)$$

This means that the collapse mechanism in the problem (5.1)–(5.4) will

5 Kinematical methods of solving the shakedown problems

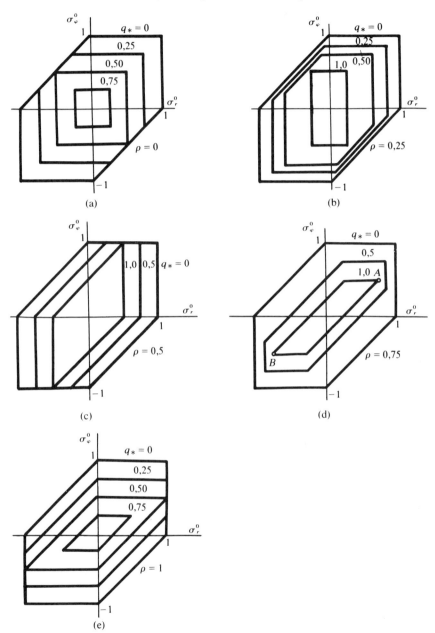

Fig. 5.3. Fictitious yield loci for a plane disc.

5.2. Solution to the problem of incremental collapse

be chosen to within an accuracy of a multiplier. It should be remembered that the equality (5.3) holds only in an incremental collapse situation in which the displacement increments per cycle Δu are non-vanishing at least at some points of the disc.

Let us express the strain increments per cycle $\Delta \epsilon_r''$, $\Delta \epsilon_\varphi''$ as sums of components, each corresponding to one of the six possible yielding régimes as shown in Fig. 5.4a,

$$\Delta \epsilon_r'' = \Delta \epsilon_r^+ - \Delta \epsilon_r^- + \Delta \epsilon_{r(\varphi)}^+ - \Delta \epsilon_{r(\varphi)}^-,$$
$$\Delta \epsilon_\varphi'' = \Delta \epsilon_\varphi^+ - \Delta \epsilon_\varphi^- - \Delta \epsilon_{r(\varphi)}^+ + \Delta \epsilon_{r(\varphi)}^-. \tag{5.32}$$

The fictitious yield locus stresses are also depicted in Fig. 5.4a, as related to the strain increments of (5.32) by the conditions (5.26). Now the work done by the stresses $\sigma_{r*}^0, \sigma_{\varphi*}^0$, appearing in the integral on the right-hand side of the inequality (5.24), can be expressed as

$$\sigma_{r*}^0 \Delta \epsilon_r + \sigma_{\varphi*}^0 \Delta \epsilon_\varphi = [1 - q_*(1 - \rho)](\Delta \epsilon_r^+ + \Delta \epsilon_r^-) +$$
$$+ (1 - q_*\rho)(\Delta \epsilon_{r(\varphi)}^+ + \Delta \epsilon_{r(\varphi)}^-) + [1 - q_*|1 - 2\rho|](\Delta \epsilon_\varphi^+ + \Delta \epsilon_\varphi^-), \tag{5.33}$$

where

$$\Delta \epsilon_r^+ \geq 0, \quad \Delta \epsilon_r^- \geq 0, \quad \Delta \epsilon_\varphi^+ \geq 0, \quad \Delta \epsilon_\varphi^- \geq 0, \quad \Delta \epsilon_{r(\varphi)}^+ \geq 0, \quad \Delta \epsilon_{r(\varphi)}^- \geq 0. \tag{5.34}$$

The expression (5.33) for work done by the stresses belonging to the fictitious yield locus, although written here for the case in which the locus is a hexagon, remains also valid in cases when the occurrence of particular yielding régimes in the limiting cycle is impossible. In this situation, the corresponding sides of the fictitious yield polygon disappear, as for instance in Fig. 5.4b. This follows from the fact that the requirement (5.23) on the minimum of the parameter p with respect to the strain increments, i.e. with respect to the possible collapse mechanism, implies under the assumption of both the inequality (5.24) and the condition (5.31) that in (5.33) only those terms will remain non-zero that correspond to the actual yielding régimes. In particular, it can be readily seen that the expression (5.33) remains correct also in the case presented in Fig. 5.4b, provided some of the terms in the general formulae (5.32) are assumed to vanish.

A similar procedure can be applied to rewrite the second term on

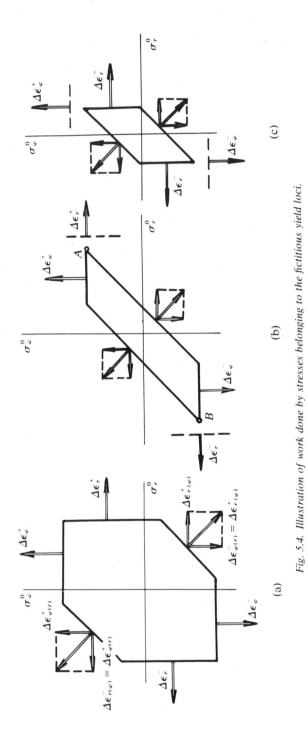

Fig. 5.4. Illustration of work done by stresses belonging to the fictitious yield loci.

196

5.2. Solution to the problem of incremental collapse

the right-hand side of inequality (5.24). Assuming that the jumps in radial displacements constitute the limiting case of rapid but continuous changes, the stresses σ_{r*}^0 corresponding to a certain prescribed displacement jump $\Delta u'_\nu$ can be determined by means of (5.26) under the following conditions, derived from (5.25):

for $\quad \Delta u'_\nu > 0, \quad \Delta \epsilon''_r > 0, \quad \Delta \epsilon''_r \gg |\Delta \epsilon''_\varphi|,$ \hfill (5.35)

for $\quad \Delta u'_\nu < 0, \quad \Delta \epsilon''_r < 0, \quad |\Delta \epsilon''_r| \gg |\Delta \epsilon''_\varphi|.$ \hfill (5.36)

Since the considered disc displays rotational symmetry, the radial displacement discontinuities are clearly the only possible ones. In plane stress such jumps are supposed to be admissible due to the generally accepted notion of an ideal plastic hinge. Thus the jump can be understood as localized deformation of a plastic layer, the thickness of which tends to zero.[1]

If the fictitious yield locus turns out to be a hexagon, as on Fig. 5.4a, or a rectangle of the type shown in Fig. 5.3a, b, then in accordance with the expression (5.26), the stresses corresponding to the displacement discontinuities (5.35) and (5.36) are

$$\sigma_{r*}^0 = [1 - q_*(1 - \rho_\nu)] \operatorname{sign} \Delta u'_\nu. \tag{5.37}$$

For a fictitious yield locus having the shape shown in Fig. 5.4b, the jumps (5.35) and (5.36) correspond to the corners A and B, i.e. to the stresses

$$\sigma_{r*}^0 = [2 - q_* \rho_\nu - q_* |1 - 2\rho_\nu|] \operatorname{sign} \Delta u'_\nu. \tag{5.38}$$

Representing work done by stresses σ_{r*}^0 on the jump $\Delta u'_\nu$ as the corresponding linear combination of works, we get

$$\sigma_{r*}^0 \Delta u'_\nu = [1 - q_*(1 - \rho_\nu)] \Delta u_\nu^{(1)} + [2 - q_* \rho_\nu - q_* |1 - 2\rho_\nu|] \Delta u_\nu^{(2)}, \tag{5.39}$$

where

$$\Delta u_\nu^{(1)} \geq 0 \quad \text{and} \quad \Delta u_\nu^{(2)} \geq 0 \tag{5.40}$$

---

[1] The problem of discontinuity of normal velocity (or displacement increment) in plane stress is considered, for instance, in the monograph [5.13] in which the formation of a neck or a bulge is discussed in this context.

5 Kinematical methods of solving the shakedown problems

are related to the jump $\Delta u'_\nu$ in the radial displacement increment by means of

$$|\Delta u'_\nu| = \Delta u_\nu^{(1)} + \Delta u_\nu^{(2)}. \tag{5.41}$$

In fact, at each jump exactly one out of the possible two yielding regimes takes place. Again the minimum requirement on the parameter p, as in the case of the work (5.33), leads to the conclusion that only some of the values $\Delta u_\nu^{(1)}$, $\Delta u_\nu^{(2)}$ are non-vanishing and at each section only one of them can differ from zero.

We shall not dwell here any longer on discontinuities as they somewhat complicate the calculations in comparison with the statical formulation. Moreover, discontinuities can be replaced by zones of rapid change of the respective functions.

Now, substituting (5.33), (5.39) into the inequality (5.24) and making use of condition (5.31), we arrive at

$$3p \geq \int_0^1 \{[1 - q_*(1-\rho)](\Delta\epsilon_r^+ + \Delta\epsilon_r^-) + (1 - q_*\rho)(\Delta\epsilon_{r(\varphi)}^+ + \Delta\epsilon_{r(\varphi)}^-) +$$
$$+ [1 - q_*|1 - 2\rho|](\Delta\epsilon_\varphi^+ + \Delta\epsilon_\varphi^-)\} \rho \, d\rho +$$
$$+ \sum_\nu \{[1 - q_*(1 - \rho_\nu)]\Delta u_\nu^{(1)} + [2 - q_*\rho_\nu - q_*|1 - 2\rho_\nu|]\Delta u_\nu^{(2)}\}\rho_\nu. \tag{5.42}$$

The equations (5.31) and (5.25) can be rewritten with the help of (5.32) as

$$\int_0^1 (\Delta\epsilon_\varphi^+ - \Delta\epsilon_\varphi^- - \Delta\epsilon_{r(\varphi)}^+ + \Delta\epsilon_{r(\varphi)}^-)\rho^3 \, d\rho = 1, \tag{5.43}$$

$$(\Delta\epsilon_r^+ - \Delta\epsilon_r^- + \Delta\epsilon_{r(\varphi)}^+ - \Delta\epsilon_{r(\varphi)}^-) = \frac{d}{d\rho}[\rho(\Delta\epsilon_\varphi^+ - \Delta\epsilon_\varphi^- - \Delta\epsilon_{r(\varphi)}^+ + \Delta\epsilon_{r(\varphi)}^-)]. \tag{5.44}$$

The last expression can be integrated which leads to

$$\rho(\Delta\epsilon_\varphi^+ - \Delta\epsilon_\varphi^- - \Delta\epsilon_{r(\varphi)}^+ + \Delta\epsilon_{r(\varphi)}^-) =$$
$$= \int_0^\rho (\Delta\epsilon_r^+ - \Delta\epsilon_r^- + \Delta\epsilon_{r(\varphi)}^+ - \Delta\epsilon_{r(\varphi)}^-) \, d\rho + C_0, \tag{5.45}$$

where C_0 is an integration constant which evidently should vanish, $C_0 = 0$.

5.2. Solution to the problem of incremental collapse

In order to reformulate the differential and integral relationships (5.42)–(5.45) in an algebraic manner, let us adopt a piece-wise approximation of the unknown functions of the current radius (such as $\Delta\epsilon_r^+$, $\Delta\epsilon_r^-$ and so on) such that the extremum values of those functions can be attained at the junctions of intervals only, i.e. at nodal points. If the constraints (5.34), (5.40) are satisfied at a finite number of nodal points, then those constraints will hold at all points of the body. One of the simplest approximations satisfying the above requirements is a piece-wise linear one.

The introduction of approximating functions clearly entails some additional constraints apart from those formulated in the problem (5.1)–(5.4). Thus the class of admissible collapse mechanisms becomes necessarily narrower. According to the formulation of the kinematical theorems, as given in Sec. 2.6, this fact can result in obtaining upper bounds to the limiting cycle parameters instead of exact solutions. However, the inaccuracies will diminish as the number of nodal points increases.

Let, in a certain interval $\rho_i \leq \rho \leq \rho_{i+1}$, the strain increment components vary according to the following rule:

$$\rho\Delta\epsilon_\varphi^+ = A^+ + B^+\rho, \quad \rho\Delta\epsilon_{r(\varphi)}^+ = C^+ + D^+\rho,$$
$$\rho\Delta\epsilon_\varphi^- = A^- + B^-\rho, \quad \rho\Delta\epsilon_{r(\varphi)}^- = C^- + D^-\rho, \qquad (5.46)$$
$$\rho\Delta\epsilon_r^- = E^- + F^-\rho.$$

The last component, $\Delta\epsilon_r^+$, can be determined from (5.44) and therefore it is not shown above.

Within the given interval all the coefficients A, B, C, D, E, F appearing in (5.46) are constant. Since $0 \leq \rho \leq 1$, the conditions (5.34) are satisfied if and only if the quantities on the right-hand sides of the equalities (5.46) are non-negative. This condition is clearly met in the whole interval (ρ_i, ρ_{i+1}) once it holds good at its end points.

At an interface of adjacent segments of the disc the strain increments $\Delta\epsilon_r''$, $\Delta\epsilon_\varphi''$, and consequently their components (5.32), (5.46), can suffer jumps, since the conditions (5.25) allow for discontinuities in the radial displacement increments. Therefore the coefficients A, B, C, D, E, F constitute an independent system of quantities for different intervals $\rho_i \leq \rho \leq \rho_{i+1}$. Only one of the coefficients can be eliminated by means of (5.43) from the whole system for the entire disc.

The number of displacement jumps in an actual collapse mechanism is usually small (and in discs it is not more than one). Hence the dimen-

5 Kinematical methods of solving the shakedown problems

sionality of the matrix of the constraints system can be reduced by imposing the additional requirement of strain increment continuity for the whole disc or for its particular parts consisting of certain segments. In this case the displacement discontinuities inside such parts are clearly replaced by zones of more or less rapid, smooth changes depending on $\rho_{i+1} - \rho_i$.

Let, for instance, $\rho_i = 0$ and $\rho_{i+1} = 1$, which means that the most rough, linear, approximation is adopted for the whole disc. On substituting the values of the strain increments (5.46) into (5.44), we get:

$$\rho \Delta \epsilon_r^+ = E^- + F^- \rho - C^+ - 2D^+ \rho + C^- + 2D^- \rho + B^+ \rho - B^- \rho. \qquad (5.47)$$

Placing (5.46) in the equality (5.43) leads to

$$A^+ = 3 + A^- + C^+ - C^- - \frac{3}{4}(B^+ - B^- - D^+ + D^-), \qquad (5.48)$$

which enables us to eliminate one of the elements from the simplex-table.

On substituting (5.46)–(5.48) into the inequalities (5.34), (5.42) and by simple rearrangements, we arrive at the linear programming problem shown in Table 5.7. The variables A^-, C^+, C^-, E^- should be non-negative; the variables B^+, B^-, D^+, D^-, F^- are free. Applying the procedure of the simplex method, we find that the solution to the problem formulated in Table 5.7 can be written as follows:

$$p = 1 - \frac{1}{2} q_*, \quad \rho \Delta \epsilon_\varphi = A^+ = 3. \qquad (5.49)$$

The obtained result must generally be considered as an upper bound assessment of the limiting cycle parameters. Comparison of (5.49) and (4.118) shows that for small values of the parameter q_*, the now obtained result coincides with the lower bound answer, thus being an exact solution.

In order to make the kinematical solution more precise at larger values of the parameter q_*, see (4.120), it appears necessary to split up the disc into a number of parts so that the coefficients appearing in (5.46) have different values for each of the parts. One should study possibilities of displacement field discontinuities taking place at the interfaces of the parts according to the inequality (5.42), in the presence of the constraints (5.40).

5.2. Solution to the problem of incremental collapse

Table 5.7

	$-A^-$	$-C^+$	$-C^-$	$-E^-$	$-B^-$	$-D^+$	$-D^-$	$-F^-$	$-B^+$	1
y_1	-1	0	0	0	-1	0	0	0	0	0
y_2	0	-1	0	0	0	-1	0	0	0	0
y_3	0	0	-1	0	0	0	-1	0	0	0
y_4	0	0	0	-1	0	0	0	-1	0	0
y_5	-1	-1	$+1$	0	$-\frac{3}{4}$	$-\frac{3}{4}$	$+\frac{3}{4}$	0	$\frac{3}{4}$	3
y_6	-1	-1	$+1$	0	$-\frac{3}{4}$	$-\frac{3}{4}$	$+\frac{3}{4}$	0	$-\frac{1}{4}$	3
y_7	0	$+1$	-1	-1	1	0	0	0	0	0
y_8	0	$+1$	-1	-1	$+1$	2	-2	-1	-1	0
$3Z$	$2(1-\frac{1}{2}q_*)$	$1-\frac{1}{2}q_*$	$1-\frac{1}{2}q_*$	$2(1-\frac{1}{2}q_*)$	$\frac{3}{4}(1-\frac{11}{18}q_*)$	$\frac{1}{4}(1-\frac{3}{2}q_*)$	$\frac{3}{4}(1-\frac{7}{18}q_*)$	$1-\frac{1}{3}q_*$	$\frac{1}{4}(1-\frac{1}{24}q_*)$	$-3(1-\frac{1}{2}q_*)$

5 Kinematical methods of solving the shakedown problems

5.3. Approximate kinematical method of determining parameters of the limiting cycle. Simple examples

In the strong methods we seek the actual collapse mechanism in accordance with the minimization condition for the unknown external action factor and under definite constraints. In contrast to this, the approximate kinematical methods consist in assuming a certain suitable, kinematically admissible velocity cycle as in the original formulation of the Koiter theorem, Sec. 2.3, or a kinematically possible distribution of plastic strain increments as in the reformulated theorem, Sec. 2.6. Such an approach enables us to estimate from above the sought for parameters of external actions since the actual collapse mechanism corresponds to their minimum values.

There exist broad classes of practically important situations in which we are able to make a sufficiently reasonable guess as to the expected form of the collapse mechanism. One of them is an axially symmetric situation encountered in plates, shells, thick-walled reservoirs and similar structures. The collapse mechanism is here characterized by a certain displacement which is a function of a single argument. The relative simplicity of application of the approximate kinematical method [5.14] in the restated formulation permits one to enhance accuracy, however at the cost of only gradually approaching the actual collapse mechanism. Thus, by simultaneously employing both the kinematical and the statical methods, the exact solution might often be approached well enough by using very simple analytical means.

The sequence of calculations, using the reformulated kinematical theorem (5.2)–(5.4), can be the following: After a likely displacement distribution Δu_{i0} has been assumed in accordance with the expected collapse mechanism, the plastic strain increments are computed with the help of (5.3); then, using the associated flow rule in the form (5.4), the corresponding stresses σ^0_{ij*} are found belonging to the fictitious yield surface. The obtained results are substituted into the basic relationship (5.2) in which the equality sign is taken in order to reach the best upper bound possible on the sought for parameter of external actions.

The calculations are made more convenient by the fact that the above method does not lead to the necessity of constructing the whole fictitious surface beforehand at each point of the body. Indeed, since the directions of the vector of plastic strain increment per cycle (resulting from the assumed collapse mechanism) are given, we can find from (see

5.3. Approximate kinematical method of determining parameters

Secs. 2.3, 2.6)

$$\sigma^0_{ij*}\Delta\epsilon''_{ij0} = \min_{\tau}[(\sigma_{ij} - \sigma^{(e)}_{ij\tau})\Delta\epsilon''_{ij0}] \tag{5.50}$$

the determining (enveloping) values of the variable stresses corresponding at each point of the body to a certain single point of the fictitious yield locus.

It must be emphasized that condition (5.50) is written for the case in which there occurs only one yielding regime at a point of the body and therefore only one determining value of the actual variable stresses exists at that point. It has already been mentioned in Sec. 2.6 that such a situation can lead in specific cases to too large a value of the sought for load factor. In problems which easily lend themselves to approximate treatment the inaccuracies are usually found to be rather insignificant; an initial analysis of variable stresses related to the assumed collapse mechanism leads often to improved accuracy.

When the yield point is sensitive to temperature, the magnitude of the difference $\sigma_{ij} - \sigma^{(e)}_{ij\tau}$ must be found for which the right-hand side of the equality (5.50) attains its minimum. This should be followed by the determination of the values of variable stresses. In the case of temperature-independent yield point, the equality (5.50) can be rewritten to become

$$\sigma^0_{ij*}\Delta\epsilon''_{ij0} = \sigma_{ij}\Delta\epsilon''_{ij0} - \max_{\tau}(\sigma^{(e)}_{ij\tau}\Delta\epsilon''_{ij0}). \tag{5.51}$$

This renders directly the sought for stationary values of the stresses $\sigma^*_{ij\tau}$ characterizing the variable actions.

When we wish to employ the expression (2.14) for the plastic dissipation energy stemming from the Tresca yield condition (2.10), the last equation can be expressed in a very convenient form

$$\sigma^0_{ij*}\Delta\epsilon''_{ij0} = \sigma_s|\Delta\epsilon''_{ij0}|_{\max} - \sigma^*_{ij\tau}\Delta\epsilon''_{ij0}. \tag{5.52}$$

The part of the body in which

$$\max_{\tau}(\sigma^{(e)}_{ij\tau}\Delta\epsilon''_{ij0}) > 0 \tag{5.53}$$

will from now on be called the volume of overload (additional loading)

5 Kinematical methods of solving the shakedown problems

since the application of variable stresses brings it closer to the plastic state. In the remainder of the body either unloading occurs or the values of the global stresses remain unchanged.

While dealing with plates and shells by means of the approximate kinematical method we don't need to introduce, as in Chapter 3, any special generalized variables that would have to reflect the peculiarities of time-dependent loading. The Love–Kirchhoff hypothesis is used only to derive strain increments from displacement increments; the work done by time-dependent stresses on the admissible strain increments can be directly computed in the course of calculations. This circumstance, coupled with no necessity to completely construct the yield surfaces for each point of the body, lends considerable simplicity to the procedure as compared with other known methods of shakedown analysis.

We shall illustrate now the approximate kinematical method by applying it to some simple bar systems subject to variable repeated mechanical load. For the first example we shall also provide, for the sake of comparison, a solution based on the original Koiter theorem (Sec. 2.3) requiring a detailed analysis of both time-dependent loading and unloading processes.

Since the determination of the limiting cycle parameters under the conditions of alternating plasticity is straightforward, our attention will again be focused on the incremental collapse situations.

Example 1. Let us analyse the incremental collapse of a two-span beam loaded in turn by a single force $0 \leq P_1 \leq P_{1*}$ and two simultaneous forces $0 \leq P_1 \leq P_{1*}$, $0 \leq P_2 \leq P_{2*}$, Fig. 5.5.

The postulated collapse mechanism is shown at the bottom (it is assumed here that $P_{2*} = P_{1*}$). The per cycle increments of linear and angular displacements clearly obey the following relationships:

$$\Delta \varphi_b = \frac{1}{2}\Delta \varphi_a, \quad \Delta u_1 = \frac{l}{2}\Delta \varphi_b, \quad \Delta u_2 = 0. \tag{5.54}$$

It is reasonable to assume that at the first stage of a cycle, under the action of force P_1 only, a plastic hinge forms in the first midspan, i.e. at the point of application of the load. At the second stage, when two forces act at the same time, a plastic hinge forms over the middle support. This corresponds to the bending moment fields at each of the stages.

We shall first use the original formulation (2.39) of the kinematical theorem. In the present situation of a bar system with concentrated

5.3. Approximate kinematical method of determining parameters

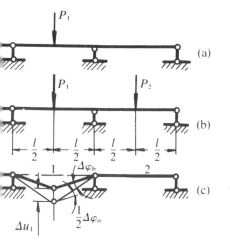

Fig. 5.5. Variable repeated loading of two-span beam.

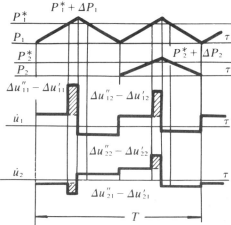

Fig. 5.6. Variations in generalized rates in accordance with the loading program.

plastic hinges the equation (2.39) takes the form

$$\int_0^T P_i \dot{u}_{i0} \, d\tau = \int_0^T M_{0i} \dot{\varphi}_{i0} \, d\tau. \tag{5.55}$$

The summation convention applies as usual; $M_{0i} = M_0 = \tfrac{1}{4}\sigma_s bh^2$ is the ultimate bending moment, $\dot{\varphi}_{i0}$ stands for the residual rotation rate at the i-th plastic hinge, $\dot{u}_{i0}$ denotes the residual deflection rate of a point of application of the force P_i. Those deflection rates can be computed as differences of rates under elastic-plastic deformation (taking into account suitable plastic hinges) and under purely elastic deformation in the presence of the same load as before.

In Fig. 5.6 the loading program is shown together with the actual per cycle rates of deflection of the points at which the loads are applied. The maximum loads are assumed to exceed their limiting cycle values by small quantities ΔP_1 and ΔP_2. The residual deflection increments per cycle amount to

$$\Delta u_1 = (\Delta u''_{11} - \Delta u'_{11}) + (\Delta u''_{12} - \Delta u'_{12}),$$
$$\Delta u_2 = -(\Delta u''_{21} - \Delta u'_{21}) + (\Delta u''_{22} - \Delta u'_{22}) = 0 \tag{5.56}$$

5 Kinematical methods of solving the shakedown problems

and these correspond to shaded areas in Fig. 5.6. The first subscript labels the point of load application, the second one denotes the stage of loading. Elastic deflection increments are primed whereas the elastic-plastic increments (plastic hinge inclusive) are double primed.

On account of (5.56), the left-hand side of the equation (5.55), describing the work of external load on the residual displacements, assumes the form

$$A = P_1(\Delta u''_{11} - \Delta u'_{11}) + P_1(\Delta u''_{12} - \Delta u'_{12}) + P_2(\Delta u''_{22} - \Delta u'_{22})$$

$$= P_1 \Delta u_1 + P_2 \Delta u''_{21} \left(1 - \frac{\Delta u'_{21}}{\Delta u''_{21}}\right) \tag{5.57}$$

and its right-hand side, that of plastic energy dissipation, is equal to

$$D = M_0 \Delta \varphi_a + M_0 \Delta \varphi_b = M_0(\Delta \varphi_a + \Delta \varphi_b). \tag{5.58}$$

The further calculations require now the use of (5.54) and the knowledge of the ratio $\Delta u'_{21}/\Delta u''_{21}$ as well as of the relationship between $\Delta u''_{21}$ and $\Delta \varphi_a$. This can be done by the methods of structural mechanics and the result is depicted in Fig. 5.7. We have:

$$\frac{\Delta u'_{21}}{\Delta u''_{21}} = \frac{3}{16}, \quad \frac{\Delta u''_{21}}{l \Delta \varphi''_{11}} = \frac{3}{52}, \quad (\Delta \varphi''_{11} = \Delta \varphi_a). \tag{5.59}$$

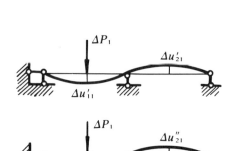

Fig. 5.7. Deflections at elastic and elastic-plastic deformation.

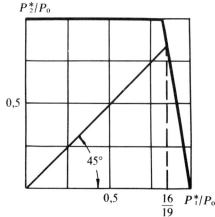

Fig. 5.8. Shakedown diagram of the two-span beam shown in Fig. 5.5.

5.3. Approximate kinematical method of determining parameters

On substituting (5.57), (5.58) into the equation (5.55) and making use of (5.59), we arrive at the conditions of incremental collapse in the form

$$\frac{P_{1*}}{P_{01}} + \frac{3}{16}\frac{P_{2*}}{P_{02}} = 1, \tag{5.60}$$

where $P_{01} = P_{02} = P_0 = 6M_0/l$ is the load carrying capacity. The obtained relationship is shown in Fig. 5.8. When $P_{1*} = P_{2*} = P_*$, we get

$$P_* = \frac{16}{19}P_0 \approx 5{,}05\frac{M_0}{l} \tag{5.61}$$

which agrees perfectly with the statical solution to the problem [5.15]. Let us now solve the problem on the basis of the restated kinematical theorem, Sec. 2.6, i.e. with the help of the inequality (5.2). As we know, the time dependence of the loading process need not be considered. When applied to the bending of beams, the relevant equality (5.2) takes the form

$$P_i^0 \Delta u_{i0} = (M_{0i} - M_{ir}^*)\Delta\varphi_{i0}, \tag{5.62}$$

where M_{ir}^* is the envelope of the 'elastic' bending moments satisfying the condition

$$M_{ir}^*\Delta\varphi_{i0} = \max_{\tau}(M_{ir}^{(e)}\Delta\varphi_{i0}). \tag{5.63}$$

Both bending moments generated by the concentrated load P_1 do positive work in the plastic hinges located according to the assumed collapse mechanism. Therefore the situation is equivalent to that caused by the constant load $P_1 = P_{1*}$. Condition (5.63) and Fig. 5.9 show that the distribution of the envelope of the moments generated by the concentrated force P_2 is as follows: At the plastic hinge in the left-hand side midspan $M_{a\tau}^* = 0$; at the plastic hinge over middle support $M_{b\tau}^* = \tfrac{3}{32}P_{2*}l$. Substituting those values into (5.62), we get

$$P_{1*}\Delta u_1 = M_0\Delta\varphi_a + \left(M_0 - \frac{3}{32}P_{2*}l\right)\Delta\varphi_b. \tag{5.64}$$

Making use of (5.54), we arrive at the already known solution (5.60).

5 Kinematical methods of solving the shakedown problems

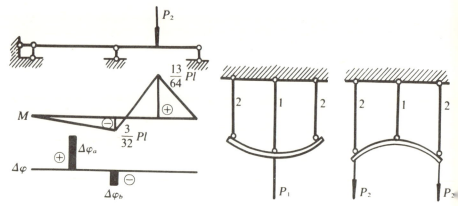

Fig. 5.9. "Enveloping" bending moments caused by force P_2.

Fig. 5.10. Variable repeated loading of a system of bars connected by a flexible beam.

As readily seen, the solution provided by the restated theorem is simpler, even in such an elementary situation. Moreover, the conditions necessary for incremental collapse to take place are better exhibited.

Example 2. Consider a system consisting of three vertical bars and a flexible beam, Fig. 5.10. The concentrated forces $0 \leq P_1 \leq P_{1*}$, $0 \leq P_2 \leq P_{2*}$ are applied by turns. This example is given in many monographs and textbooks to serve as a qualitative illustration of strain accumulation. However, to the best of the author's knowledge in no one of the books the suitable limiting condition has been totally determined.

In application to the stretched system the equation (5.2) assumes the form

$$P_i^0 \Delta u_{i0} = (Y_i - N_{i\tau}^*)\Delta l_{i0}, \qquad (5.65)$$

where $Y_i = Y = \sigma_s F$ are the ultimate axial forces, assumed to be the same for all the bars and Δl_{i0} are the elongation increments per cycle, where $i = 1, 2, 3$ numbers the bars.

In the considered example the constant load is absent; $P_i^0 = 0$. The collapse mechanism (admissible cycle of plastic strain rates) is described by

$$\Delta u_0 = \Delta l_{10} = \Delta l_{20} = \Delta l_{30}. \qquad (5.66)$$

It is now necessary to find the largest tensile forces $N_{i\tau}^*$ satisfying

5.3. Approximate kinematical method of determining parameters

the condition

$$N^{(e)}_{i*}\Delta l_{i0} = \max_{\tau}(N^{(e)}_{i\tau}\Delta l_{i0}). \tag{5.67}$$

The overload volume contains all the bars. The largest force in the middle (first) bar is at $P_1 = P_{1*}$, and according to the loading program, accompanied by $P_2 = 0$. The largest force in the side bars will occur at $P_2 = P_{2*}$ while $P_1 = 0$.

From basic structural relationships we have

$$N^*_{1\tau} = \frac{1+2k}{3+2k} P_{1*}, \quad N^*_{2\tau} = \frac{2(1+k)}{3+2k} P_{2*}, \tag{5.68}$$

where

$$k = \frac{c_a}{c_b}, \quad c_a = \frac{EF}{l}, \quad c_b = \frac{48EI}{a^3},$$

E is Young's modulus, F, l are respectively the cross-sectional area and the length of each bar, and I, a are the moment of inertia and the length of the flexible beam.

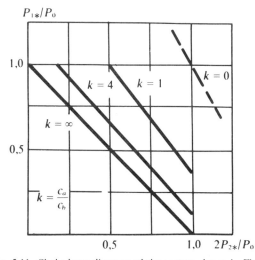

Fig. 5.11. Shakedown diagram of the system shown in Fig. 5.10.

5 Kinematical methods of solving the shakedown problems

Combining the above results with equation (5.65), we obtain the following condition for the incremental collapse to commence:

$$\frac{1+2k}{3+2k} \cdot \frac{P_{1*}}{P_0} + \frac{4(1+k)}{3+2k} \cdot \frac{P_{2*}}{P_0} = 1, \qquad (5.69)$$

where $P_0 = 3\sigma_s F$ stands for the ultimate load. The situation is presented in Fig. 5.11. We can find from it the limiting cycle parameters for various ratios of bar stiffness to beam stiffness. At $k = 0$ (perfectly rigid beam) the incremental collapse is impossible: The limiting cycle parameters reach, or even exceed, the magnitudes of instantaneous ultimate loads.

5.4. Incremental collapse of a thick-walled tube subjected to thermal cycling and internal pressure

Consider the situation of Sec. 4.6 but with the difference that the yield stress is now assumed, for simplicity, to be temperature-independent. A section of the tube is shown in Fig. 4.8.

The principal stresses accompanying the plastic behaviour of the tube are assumed to satisfy the inequalities

$$\sigma_\varphi > \sigma_z > \sigma_r. \qquad (5.70)$$

According to the flow rule associated with the Tresca yield condition (2.12) we have

$$\Delta\epsilon''_{\varphi 0} = -\Delta\epsilon''_{r 0}, \quad \Delta\epsilon''_{z 0} = 0. \qquad (5.71)$$

On account of the relationships (5.3) which take now the form

$$\Delta\epsilon''_{\varphi 0} = \frac{\Delta u_0}{b\rho}, \quad \Delta\epsilon''_{r 0} = \frac{\mathrm{d}(\Delta u_0)}{b\,\mathrm{d}\rho} \qquad (5.72)$$

(where $\rho = r/b$ and b is the outer radius), we obtain the differential equation

$$\frac{\mathrm{d}(\Delta u_0)}{\mathrm{d}\rho} + \frac{\Delta u_0}{\rho} = 0 \qquad (5.73)$$

5.4. Incremental collapse of a thick-walled tube

which yields

$$\Delta u_0 = C/\rho, \quad \Delta\epsilon''_{\varphi 0} = -\Delta\epsilon''_{r0} = C/b\rho^2. \tag{5.74}$$

These relationships are fully analogous to those obtained in the case of limit analysis of the tube [5.16].

If the diameter of the tube is assumed to grow under cyclic actions, then $C > 0$. Hence (5.50) implies that the determining values of the stresses $\sigma^*_{\varphi p\tau}, \sigma^*_{rp\tau}$, see (4.93), generated by internal pressure correspond at all points to its largest value, $p_a = p_*$. This is so because the difference $\sigma^{(e)}_{\varphi p\tau} - \sigma^{(e)}_{rp\tau}$ is positive everywhere. Consequently, the loading program in which the pressure remains constant and equal to its maximum value, $p_a = p_{a*}$, is the most disadvantageous as regards the incremental collapse.

The values of the one-parameter thermo-elastic stresses (4.94) that correspond to condition (5.50) are the following:

$$\max_\tau[(\sigma^{(e)}_{\varphi t\tau} - \sigma^{(e)}_{rt\tau})\Delta\epsilon''_{\varphi 0}] = 0 \quad \text{for } k \leq \rho \leq \gamma, \tag{5.75}$$

$$\max_\tau[(\sigma^{(e)}_{\varphi t\tau} - \sigma^{(e)}_{rt\tau})\Delta\epsilon''_{\varphi 0}] = (\sigma^*_{\varphi t\tau} - \sigma^*_{rt\tau})\Delta\epsilon''_{\varphi 0} \quad \text{for } \gamma \leq \rho \leq 1, \tag{5.76}$$

where $\sigma^*_{\varphi t\tau}, \sigma^*_{rt\tau}$ are the stresses in the overload zone $\gamma \leq \rho \leq 1$, corresponding to the largest magnitude of the temperature interval. In the region $k \leq \rho \leq 1$ in which the thermo-elastic stresses result in unloading the plastic state can be reached only at those instants of time when these stresses are absent.

The overloading and unloading zones are shown in Fig. 5.12 in which there are also depicted the thermo-elastic stresses and hoop strain increments in accordance with the collapse mechanism (5.74). The interface has the radius $\rho = \gamma$ on which the difference between the hoop and the radial stresses vanishes. It should be remembered that γ is constant at any instant of time inside the cycle only in the case of one-parameter thermo-elastic stresses.

The expressions (4.94) yield

$$\gamma^2 = -1/\delta. \tag{5.77}$$

Thus the equation (5.2) assumes, in the presence of (5.52), (5.74)–

5 Kinematical methods of solving the shakedown problems

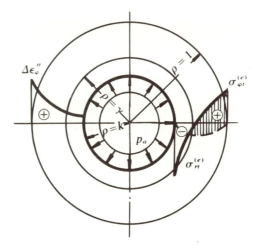

Fig. 5.12. Limiting cycle situation for a hollow cylinder.

(5.76), the form

$$p_{a*} \frac{C}{k} kb = \sigma_s b^2 \int_k^1 \frac{C}{b\rho^2} \rho \, d\rho - |\sigma^*_{n\gamma\tau}| \frac{C}{\gamma} \gamma b \sigma_s. \tag{5.78}$$

We have employed here the identity

$$\int_{V_l} \sigma^*_{ij\tau} \Delta \epsilon''_{ij0} \, dV = \int_{S_l} \sigma^*_{ij\tau} n_j \Delta u_{i0} \, dS, \tag{5.79}$$

valid because of (2.15), for one-parameter actions in the absence of body forces. S_l denotes a surface enclosing the overload region V_l in which the integrand differs from zero on account of (5.75), (5.76).

The first expression (4.94) together with (5.77) yields

$$|\sigma^{(e)}_{n\gamma\tau}| = q_* \left(1 - \frac{1}{\gamma^2} + 2\delta \ln \gamma \right) = q_*[1 + \delta(1 - \ln|\delta|)]. \tag{5.80}$$

Hence the equation (5.78) can be rewritten to become

$$\frac{p_*}{p_0} + D \frac{q_*}{q^0} = 1, \tag{5.81}$$

where $p_0 = -\sigma_s \ln k$ is the ultimate pressure, $q^0 = k^2/(1 + k^2\delta)$ is the largest value of the parameter q_* at which, owing to the known criterion (2.91), the thermal changes for p_a = const. are not supposed to bring about the alternating plastic flow. Lastly

$$D = -\frac{[1 + \delta(1 - \ln|\delta|)]k^2}{(k^2\delta + 1)\ln k}. \tag{5.82}$$

In order to compare the above solution with that obtained by means of the approximate statical method, Sec. 4.5, the yield condition employed in each case must first be taken into account. For the Tresca material the kinematical result (5.81), (5.82) turns out to be the exact one since at all stages of the cycle it is associated with statically admissible stress fields [5.17].

5.5. Overload method. Conditions for incremental collapse of a spherical vessel under variable repeated temperature

A suitable generalization of the results obtained in Chapter 1, where some simple bar systems were investigated, makes it possible to work out more variations of the approximate approach to the determination of the incremental collapse conditions.

It must again be borne in mind that what constitutes the main features of a given limiting state is the consecutive appearance of plastic zones spreading inside a cycle over either the whole body or a part of it in such a way that, on concentration in time of all the regions in which the yield condition is satisfied, the state of the body corresponds to that of instantaneous plastic collapse. In particular, it follows from this that in many incremental collapse situations and in the limit analysis as well it may be convenient to employ the standard equilibrium equations instead of their generalization in form of the virtual work principle. Such an approach is especially instructive in some simpler cases, such as axisymmetrical ones, in which it is possible to show the direct correspondence of the stress diagrams to different stages of the limiting cycle.

On one hand, because of the associated flow law, the stress fields should be related to a certain kinematically admissible collapse mechanism that forms inside the cycle and, on the other, they should be

5 Kinematical methods of solving the shakedown problems

statically admissible. When both these requirements hold good, the obtained solution is exact; if the statical admissibility prevails, the solution is an approximate statical one and the external actions parameters constitute the lower bound on the true ones, see Sec. 4.5. Finally, when the existence of a collapse mechanism is postulated as a necessary condition, the method belongs to the kinematical approach irrespectively of the fact that the basic variables are those of stresses at the limiting cycle.

In the conditions of limiting cycle it is the imposition of suitable 'elastic' stresses that gives rise to a transition from one state to the other, characterized by new values of the parameters of external variable actions. The addition of stresses results in preloading some regions and, at the same time, unloading others. Overloading is associated with the next stage of development of the collapse mechanism. Thus the discussed approach will be called the method of constructing stress diagrams for stages of the limiting cycle stresses or, for brevity, the overload method.

In the stress analysis it is assumed that in the limiting cycle the steady stage is already over, that is the stage at which a certain advantageous distribution of self-equilibrating stresses develops at the cost of confined plastic deformation. Although the residual stresses do not have to be explicitly stated in the overload method, their presence is tacitly assumed since the stress fields are constructed merely on the basis of the equilibrium equations and the yield condition.

It is proper to mention that the basic idea behind the discussed approach to the shakedown problems was expressed in the monograph [5.18] by Rzhanitzin in connection with simultaneous bending and stretching of bars. It was also formulated in [5.19] and other related papers which will be quoted later. An analogous approach to bar systems was employed by Symonds and Neal [5.20]. The overload method was first applied to the shakedown analysis of solid bodies transmitting all the three principal stresses in [5.17, 5.21].

Let us investigate the shakedown of a spherical vessel under variable repeated action of pressure and temperature generated by a working medium inside the vessel. Both heating and cooling are assumed to be slow enough and the temperature to be distributed across the thickness according to the function

$$t(\tau, \rho) = t_b(\tau) + [t_a(\tau) - t_b(\tau)] \frac{k(1-\rho)}{\rho(1-k)}, \tag{5.83}$$

where

$$\rho = r/b, \quad k = a/b, \quad 0 \leq t_a(\tau) - t_b(\tau) \leq t_{1*},$$

and a, b and r are the inner, outer and current radii respectively. Subscripts show where the relevant temperatures appear.

The global elastic stresses in the hollow sphere, generated by the internal pressure p_a and the temperature field (5.83), and referred for convenience to the yield point, can be expressed after [5.22] in the form

$$\begin{aligned}\sigma_{rr}^{(e)} &= p\left(1 - \frac{1}{\rho^3}\right) + q\left(\frac{1+k}{k^2}\delta - \frac{3}{\rho} + \frac{\delta}{\rho^3}\right), \\ \sigma_{\varphi\tau}^{(e)} &= p\left(1 + \frac{1}{2\rho^3}\right) + q\left(\frac{1+k}{k^2}\delta - \frac{3}{2\rho} - \frac{\delta}{2\rho^3}\right),\end{aligned} \quad (5.84)$$

where:

$p = \dfrac{p_a}{\sigma_s} \cdot \dfrac{k^3}{1-k^3}$ is the load factor, $0 \leq p \leq p_*$,

$q = \dfrac{\alpha E(t_a - t_b)k}{3\sigma_s(1-\mu)(1-k)}$ is the temperature field factor, $0 \leq q \leq q_*$,

$\delta = 3k^2/(1 + k + k^2)$ is an auxiliary coefficient.

In the conditions of point symmetry the Huber–Mises (2.9) and the Tresca–Saint Venant (2.10) yield conditions are known to coincide and take the form

$$\sigma_\varphi - \sigma_r = \sigma_s. \quad (5.85)$$

The biggest changes of stresses during the cycle occur on the inner surface of the hollow sphere and it is there that the alternating plastic flow can take place first. On account of the criterion (2.91), the alternating flow under arbitrary program of temperature and pressure variations will be prevented provided we have the inequality

$$\frac{3p_*}{2k^3} + \frac{3q_*}{2k}\left(\frac{\delta}{k^2} - 1\right) \leq 2, \quad (5.86)$$

derived from (5.84) and (5.85). The equality sign corresponds to the

5 Kinematical methods of solving the shakedown problems

limiting condition which after simple rearrangements can be written as

$$A\frac{p_*}{p_0} + B\frac{q_*}{q^0} = 1, \qquad (5.87)$$

where:

$p_0 = \dfrac{2k^3 \ln k}{k^3 - 1}$ is the ultimate load factor at the limit equilibrium of the vessel [5.13],

$q^0 = \dfrac{4k^3}{3(\delta - k^2)}$ is the ultimate temperature field factor at which the thermal cycles are not accompanied by plastic yielding in the absence of pressure,

$A = \dfrac{3 \ln k}{2(k^3 - 1)}, \quad B = 1.$

Similar conditions can also be established for particular, fixed programs of temperature and pressure fluctuations. For instance, thermal cycles at a constant internal pressure of the hollow sphere lead to the limiting condition

$$q_*/q_0 = 1. \qquad (5.88)$$

The equations (5.87) and (5.88) are represented in the shakedown diagrams of Fig. 5.13 as lines 1 and 2 and shown for two different ratios of inner to outer radius of the vessel.

Let us now proceed to the condition for incremental collapse. Due to central symmetry in the situation considered, a unique collapse mechanism is possible; it consists in an unlimited increase in the diameter of the vessel. The limit state will be approached as the internal pressure grows. Therefore, the most unfavourable program will be that of temperature cycles in the presence of a constant, maximum pressure. It can readily be found using the condition (5.50).

Thermal changes are responsible for the redistribution of stresses between the inner and the outer parts of the vessel. In the state just preceding the incremental collapse the strains will still be elastic but the region in which the stresses related to the yield condition (5.85) reach

5.5. Overload method

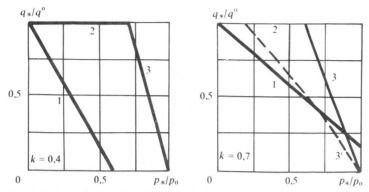

Fig. 5.13. Shakedown diagram for a hollow sphere at cyclic variations in internal pressure and temperature (for various relative thicknesses).

the yield values at various stages of the cycle, is supposed to spread over the entire thickness of the wall. A spherical interface separating overloading and unloading zones will have the radius $\rho = \gamma$ which, on account of (5.85), can be found from the equation

$$\sigma_{\varphi t}^{(e)} - \sigma_{rt}^{(e)} = 0, \tag{5.89}$$

where the subscript t indicates thermal stresses.

Substituting the relevant terms of (5.84) into (5.89), we get

$$\gamma = \sqrt{\delta}. \tag{5.90}$$

The stress field can be constructed by assuming that at the beginning of a semi-cycle, when pressure is acting in the absence of heating, the stresses reach the yield point in the inner part of the vessel, $k \leq \rho \leq \gamma$; at the end of the semi-cycle, when pressure is acting together with non-uniform heating of the wall, the stresses are at yield in the outer part, $\gamma \leq \rho \leq 1$.

The stress distribution should satisfy during each stage the statical requirements such as the internal and boundary equilibrium conditions. Making use of the differential equilibrium equations [5.13]

$$\rho \frac{d\sigma_r}{d\rho} = 2(\sigma_\varphi - \sigma_r), \tag{5.91}$$

5 Kinematical methods of solving the shakedown problems

the yield condition (5.85) and the boundary condition

$$\sigma_{ra} = -\frac{p_{a*}}{\sigma_s} = \frac{2p_*}{p_0} \ln k \tag{5.92}$$

on the inner surface of the vessel, we arrive at the formulae for the stresses in the inner part $k \leq \rho \leq \gamma$,

$$\sigma_r = 2\left(\ln\frac{\rho}{k} + \frac{p_*}{p_0}\ln k\right),$$

$$\sigma_\varphi = 1 + 2\left(\ln\frac{\rho}{k} + \frac{p_*}{p_0}\ln k\right). \tag{5.93}$$

The stresses in the outer part $\gamma \leq \rho \leq 1$, at the beginning of a semi-cycle, when pressure alone is acting, differ from the yield point by the magnitude of the thermo-elastic stresses,

$$\sigma_\varphi - \sigma_r = 1 - (\sigma_{\varphi t}^{(e)} - \sigma_{rt}^{(e)}) = 1 + \frac{2k^3(\delta - \rho^2)}{\rho^3(\delta - k^2)} \cdot \frac{q_*}{q_0}. \tag{5.94}$$

Equation (5.91) leads, under the assumption of the outer boundary condition $\sigma_{rb} = 0$, to the formulae

$$\sigma_r = 2\ln\rho + \frac{4k^3}{\delta - k^2}\left[\frac{\delta}{3}\left(1 - \frac{1}{\rho^3}\right) - \left(1 - \frac{1}{\rho}\right)\right]\frac{q_*}{q_0},$$

$$\sigma_\varphi = 1 + 2\ln\rho + \frac{4k^3}{\delta - k^2}\left[\frac{\delta}{3}\left(1 + \frac{1}{2\rho^3}\right) - \left(1 - \frac{1}{2\rho}\right)\right]\frac{q_*}{q_0}. \tag{5.95}$$

Insisting on stress continuity across the interface $\rho = \gamma$, we get the relationship

$$\frac{p_*}{p_0} + D\frac{q_*}{q_0} = 1, \tag{5.96}$$

where

$$D = \frac{2k^3(\delta^2 - 3\delta + 2\sqrt{\delta})}{3\delta(k^2 - \delta)\ln k}. \tag{5.97}$$

5.5. Overload method

The expression (5.96) constitutes the sought for condition for incremental collapse of a thin hollow sphere under thermal fluctuations. It is represented by line 3 in Fig. 5.13. The shakedown diagrams are different: The shakedown domain is bounded by the line 1 for low ratios of inner to outer radius and by the lines 1 and 3 for higher ratios, if the program of temperature and pressure fluctuations is not prescribed. When the vessel responds to thermal changes under constant pressure, the shakedown domain is bounded by the lines 2 and 3 irrespectively of the value $k = a/b$.

Figs. 5.14a, b show the limiting cycle stress distributions in two particular states which differ from each other by the thermo-elastic stresses depicted in Fig. 5.14d. The residual stresses whose presence over a number of initial cycles ensures that the vessel will shake down, can also be found in a known manner. The residual hoop stresses are shown in Fig. 5.14c; the radial ones are so minute that they cannot be seen on the same scale. It is worth observing that the obtained distribution of residual stresses is found dissimilar to that of thermo-elastic stresses, Fig. 5.14d. The similarity is assumed to take place sometimes when the approximate statical method is used, see Chap. 4.

The stresses were calculated for $k = 0{,}7$; $p_*/p_0 = 0{,}7$; $q_*/q^0 = 0{,}77$; The stress fields were obtained by assuming that the kinematically admissible collapse mechanism takes place due to an arbitrary finite increase in any of the external actions. The stress fields turned out to be statically admissible and thus the solution (5.96) can be looked upon as exact within the system of assumptions accepted in the shakedown theory.

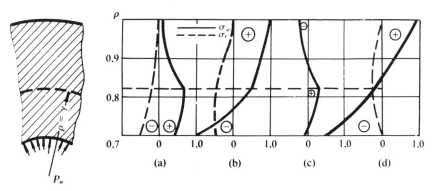

Fig. 5.14. Stress distributions in the hollow sphere at the limiting cycle.

5 Kinematical methods of solving the shakedown problems

The dashed line 3′ in Fig. 5.13 shows the approximate statical solution to the considered problem. The incremental collapse parameters are clearly evaluated from below. In the alternating plasticity situation both estimates, i.e. the one obtained on the basis of the approximate statical method and the one based on the criterion (2.91), coincide as expected (line 1).

5.6. Full shakedown analysis of a beam under simultaneous steady tension and variable repeated bending

The overload method is applicable, in some simpler situations, to the analysis of steady stress cycles not only at shakedown but also at cyclic plastic deformation. Let us demonstrate this by way of an example.

Consider a rectangular beam under constant tensile axial load P and a symmetrically variable bending moment M, Fig. 5.15.

The limit equilibrium of such a beam received due attention in the literature [5.18]. It is described by

$$\left(\frac{P}{P_0}\right)^2 + \frac{M_*}{M_0} = 1 \tag{5.98}$$

which is the parabola 5 in Fig. 5.16. Two ultimate internal forces are here involved; $P_0 = \sigma_s b h$ and $M_0 = \frac{1}{4}\sigma_s b h^2$.

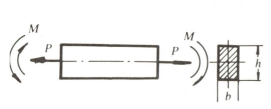

Fig. 5.15. Cyclic bending of a pretensioned beam.

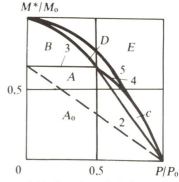

Fig. 5.16. Complete shakedown diagram for the beam under constant tension and variable bending.

5.6. Full shakedown analysis of a beam

According to (2.91), the condition for alternating plastic flow under the prescribed loading program $P = \text{const.}$, $-M_* \leq M \leq M_*$ is $|\sigma_{M_*}^{(e)}| = \sigma_s$. Hence we get

$$M_*/M^0 = 1, \tag{5.99}$$

where $M^0 = \frac{2}{3}M_0$. This condition is depicted by the line 1 in Fig. 5.16.

Cyclic variations of the bending moment lead to increases in tensile stresses in consecutive parts of the cross-section and thus they can lead to incremental strain accumulation. It is easy enough to construct suitable stress diagrams at main stages of the limiting cycle, Fig. 5.17. To this end, one must start from imposing 'elastic' stresses (b) generated by bending. These, together with the stresses (a) equilibrating the longitudinal force P, should bring about plastic yielding in the upper part of the cross-section (c) at the end of one semi-cycle, i.e. when $M = M_*$, as represented by the solid line in Fig. 5.17b, and in the lower part (d) at the end of the next semi-cycle, represented by the dashed line in Fig. 5.17b. A change in stresses per semi-cycle, Fig. 5.17c, d, corresponds to $2M_*$.

In order to establish a relationship between the magnitude of the force P and the maximum bending moment M_* it is necessary to account for the fact that the beam will behave as if at limit equilibrium, provided we neglect unloading occurring in one of the parts of the cross-section as a result of 'elastic' stresses generated by $2M_*$. From this fact it follows that

$$P + \Delta P = P_0, \tag{5.100}$$

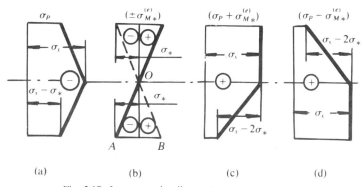

Fig. 5.17. Incremental collapse situation for a beam.

5 Kinematical methods of solving the shakedown problems

where

$$\Delta P = \frac{3\Delta M}{2h} = \frac{3M_*}{h}. \quad (5.101)$$

ΔP represents the longitudinal force equivalent to the stresses in the overload region generated by the bending moment $\Delta M = 2M_*$. The force ΔP corresponds to the area OAB in Fig. 5.17b, its arm being $\frac{2}{3}h$.

The equation of the corresponding line 2 bounding the shakedown domain, Fig. 5.16, can be obtained from (5.100). After simple rearrangements we get

$$\frac{P}{P_0} + \frac{1}{2} \cdot \frac{M_*}{M_0} = 1. \quad (5.102)$$

Thus the whole shakedown domain is determined by (5.99) and (5.102). It is depicted in Fig. 5.16 by lines 1, 2. The dashed line separates the region of solely elastic behaviour of the beam.

The magnitudes of external actions, corresponding to the portion of the plane which is bounded by the lines 1, 2 and the curve 5, are such that on repeated loading they result in cyclic plastic deformation. In the considered example, based on the overload method, we can make a further discrimination: two regions can be separated; one corresponding to the alternating plastic flow and the other to the incremental collapse.

Fig. 5.18 illustrates a limiting cycle in which the one-sign strain accumulation commences in the conditions of advanced alternating plasticity. The thin line in Fig. 5.18b shows the elastic stress distribution ($\sigma_{M_*}^{(e)} > \sigma_s$) generated by the moment $\pm M_*$. The thick line corresponds to

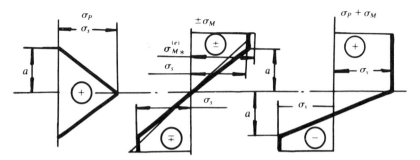

Fig. 5.18. Alternating plastic flow followed by strain accumulation.

5.6. Full shakedown analysis of a beam

the actual distribution of stresses which are in equilibrium with the same magnitude of bending moment; plastic flow is seen to suppress stresses in the outermost fibres.

Making use of (5.100) and of a suitable expression for ΔP, easily derivable from Fig. 5.18b, we arrive at the equation

$$\frac{4}{3}\left(\frac{P}{P_0}\right)^2 + \frac{M_*}{M_0} = 1. \tag{5.103}$$

Similarly we can find the condition describing a limiting cycle at which an advanced one-sign yielding is about to be accompanied by alternating deformation, Fig. 5.19. We get

$$\frac{3}{4}\frac{M_*}{M_0} = \left(1 - \frac{P}{P_0}\right)\left(\frac{1}{2} + \frac{P}{P_0}\right). \tag{5.104}$$

As in Fig. 5.18, in Fig. 5.19c the stress distribution is shown only at the end of one semi-cycle. A mirror image of the diagram with respect to the horizontal axis would supply the proper stress distribution at the end of the next semi-cycle.

In Fig. 5.16 the equations (5.103), (5.104), are represented by the lines 3 and 4, respectively. This can now be called the full diagram of the behaviour of the beam. The external actions plane is finally divided into six regions: A_0 which corresponds to the initially elastic state, A concerning shakedown, B relating to the alternating plastic flow, C corresponding to one-sign deformation mounting with each cycle, i.e. the

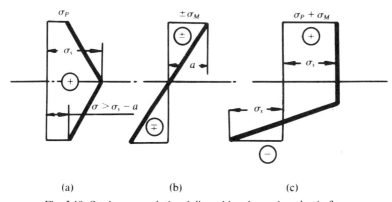

(a)   (b)   (c)

Fig. 5.19. Strain accumulation followed by alternating plastic flow.

5 Kinematical methods of solving the shakedown problems

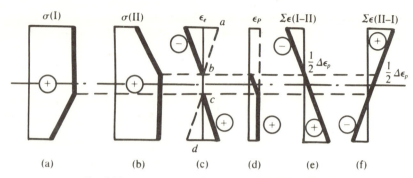

Fig. 5.20. Increase in elongation of the bar per cycle.

incremental collapse, D representing a mixed response when both types of cyclic deformation are possible, and E relating to the case when no equilibrium is possible, since the limit equilibrium corresponds to line 5.

Also when the shakedown conditions are violated the strains per cycle can be evaluated relatively easily in the above situation. Let us show this for the régime corresponding to the domain C. In Fig. 5.20a, b the stress diagrams are given at two particular instants of time bounding the semi-cycle, namely when the bending moment attains its extremum values. Unlike in Fig. 5.17, the magnitudes of loading exceed here those limiting ones given by the equation (5.102). Consequently, we obtain an overlapping of the regions in which the stresses reach the yield point at different stages of the cycle.

The plastic strain increment per cycle can be evaluated from the following conditions:

1. On transition from state a to state b the elastic strains will develop only in the portions ab and cd, Fig. 5.20c; their values are proportional to the stress changes at respective points; elastic strains over reverse transition are shown by dashed lines.

2. Plastic strains in this semi-cycle will develop in the portion bd, Fig. 5.20d, in which the stresses at the end of semi-cycle reach the yield point; plastic strains in the next semi-cycle are shown by dashed line.

3. Plastic strain distribution in each semi-cycle is such that the totally strained cross-sections remain plane.

The global strains associated with the transition from state I to state II are shown in Fig. 5.20e; the reverse situation is shown in Fig. 5.20f.

The axial strain increments in each semi-cycle are clearly equal to each other and thus to the half of the whole increment per cycle.

The above statements permit us to evaluate the increment of the elongation. This takes the form

$$\frac{\Delta\epsilon_p}{\epsilon_s} = \frac{16\left(1-\frac{P}{P_0}\right)^2\left(\frac{P}{P_0}+\frac{3}{4}\frac{M_*}{M_0}-1\right)}{\left(1-\frac{P}{P_0}-\frac{M_*}{2M_0}\right)^2}. \qquad (5.105)$$

The formulae for strain increments per cycle (in the case of alternating plasticity, per semi-cycle) can be found in a similar manner at loads corresponding to the regions B and D, see Fig. 5.16.

The above example seems to adequately illustrate the convenience of application of the overload method. The analysis has been made in terms of stresses. By using strong statical or kinematical methods we could determine the incremental collapse conditions, corresponding to line 2, Fig. 5.16 in terms of the internal forces, according to the concept introduced in Chapter 3. The analysed example appears to be very characteristic: If the non-simultaneity of plastic flow across the depth of the beam is ignored and ordinary representation of generalized stresses is used, then the variable load is found to have no influence whatever on the axial elongation. This is so because at no instant of time the given alternating load in the global sense generates any axial force in the beam, $N_\tau^{(e)} = 0$.

The applications of linear programming were demonstrated earlier in this chapter on the examples of the bar system and the plane rotating disc. We leave to the reader the recalculation of these examples with the use of the overload method.

References

[5.1] Pontryagin, L.S., *Mathematical theory of optimum control*, in Russian, Fizmatgiz, Moscow, 1961.
[5.2] Boltyansky, W.G., *Mathematical methods of optimum control*, in Russian, Nauka, Moscow, 1966.
[5.3] Lurie, K.A., *Optimum control in the mathematical physics*, in Russian, Nauka, Moscow, 1975.
[5.4] Chyras, A.A., *Linear programming methods in analysis of elastic-plastic systems*, in Russian, Gosstroiizdat, Leningrad, 1969.
[5.5] Chyras, A.A., *Optimization theory in limit analysis of solid deformable bodies*, in Russian, Mintis, Vilnius, 1971.

5 Kinematical methods of solving the shakedown problems

[5.6] Corradi, L., Zavelani, A.A., A linear programming approach to shakedown analysis of structures. *Comput. Math. Appl. Mech. and Eng. 3*, Nr 1, 1974.

[5.7] Zukhovitzky, S.I., Avdeeva, L.I., *Linear and covnex programming*, in Russian, Nauka, Moscow, 1967.

[5.8] Gokhfeld, D.A., Cherniavsky, O.F., Application of linear programming methods to kinematically formulated shakedown problems, in Russian, in *Thermal stresses in structural members, 9*, Naukova Dumka, Kiev, 1970.

[5.9] Gokhfeld, D.A., Cherniavsky, O.F., Incremental collapse conditions in axi-symmetric shakedown problems dealt with linear programming methods, in Russian, Izv. AN SSSR, *Mekhanika tverdogo tela*, Nr 3, 1970.

[5.10] Rzhanitzin, A.R., Analysis of cylindrical shells by linear programming, in Russian, Stroj. Mekh. Raschot Soor., Nr 4, 1966.

[5.11] Rzhanitzin, A.R., Shell analysis at limit equilibrium by linear programming, in Russian, in *Shells and plates theory*, (Proceed. VI All-Union Conference, Baku, 1966), Nauka, Moscow, 1966.

[5.12] Lance, R.H., On automatic construction of velocity fields in plastic limit analysis. *Acta Mech., 3*, Nr 1, 1967.

[5.13] Kachanov, L.M., *Foundations of the theory of plasticity*, North-Holland, Amsterdam, 1971.

[5.14] Gokhfeld, D.A., Some problems in shakedown theory of plates and shells, in Russian, in *Shells and plates theory*, (Proceed. VI All-Union Conference, Baku, 1966), Nauka, Moscow, 1966.

[5.15] Freundenthal, A., Geiringer, H., *The mathematical theories of the inelastic continuum*, Springer, Berlin, 1958.

[5.16] Prager, W., Hodge, Ph.G., Jr., *Theory of perfectly plastic solids*. John Wiley and Sons, New York; Chapman and Hall, London, 1951.

[5.17] Gokhfeld, D.A., *Incremental collapse at thermal cycling*, in Russian, Prikladnaya Mekhanika, *1*, Nr 6, 1965.

[5.18] Rzhanitzin, A.R., *Structural analysis taking account of plastic properties of materials*, in Russian, Gosstrojizdat, Moscow, 1954.

[5.19] Miller, D.R., Thermal stress ratchet mechanism in pressure vessels. *Trans. ASME*, Ser. D, *81*, Nr 2, 1959.

[5.20] Symonds, P.S., Neal, B.G., Recent progress in plastic methods of structural analysis. *J. Franklin Inst., 252*, 1951, 387–407, 469–492.

[5.21] Gokhfeld, D.A., *Carrying capacity of turbine discs in nonstationary conditions*, Mashinostrojenie, Nr 6, 1965.

[5.22] *Handbook on machinery*, in Russian, (in 6 vol.) Vol. 3, Mashgiz, Moscow, 1962.

6

Carrying capacity of a turbine disc under single and repeated loading

A vast literature has been devoted to the strength calculations of turbine discs. In particular, a considerable number of methods has been devised to evaluate stresses and deformations in thin revolving discs under non-uniform temperature (see for instance [6.1–6.5]). Thanks to the current computational facilities we can take easily into account such effects as the influence of temperature on the physical-mechanical properties of materials, the inelastic deformations and the creep phenomena. Moreover, the computational procedures dealing with the above listed effects do not differ much from each other, especially if based on similar approaches.

But the determination of the stress and the strain states in structures under external actions appears to be merely the first step in the strength calculations. It should be immediately followed by a suitable assessment of the global and the local strength based on properly selected assumptions as to the nature of structural failure. Let us trace the evolution of those assumptions while analysing the turbine disc.

At the initial stage of development of the turbine industry the strength of rotating discs was being evaluated under the assumption of ideal elasticity, i.e. only the maximum elastic stresses were computed as sums of centrifugal and thermal effects and then checked against the critical stresses. The latter were usually identified with the strength limit or with the yield point for metals and alloys used to build turbine rotors.

Soon it became quite apparent that such a conservative approach was not sufficient to uncover the strength reserves stored in the ductile materials. For instance, experimental evidence [6.6] showed that, owing to its plastic properties, the ultimate velocity of a disc with a small central hole was practically equal to that of a solid disc of the same dimensions irrespective of the fact that the maximum 'elastic' stresses in the first case are known to be twice as large as in the second. It thus

6 Carrying capacity of a turbine disc

became increasingly clear that a single application of thermal stresses to a sufficiently ductile material could only result in localized plastic strains and was not critical to the overall strength of the disc. At that time some researchers were even suggesting to completely ignore the thermal stresses while evaluating the strength on the mere basis of the stresses resulting from the centrifugal forces.

At the next stage, directly related to the experimental evidence obtained, there was worked out a method of analysing the limit equilibrium of discs [6.7–6.9]. This method makes it possible to easily evaluate the limit angular velocity at which an unconfined increase in deformation is about to commence, under the assumption of perfect plasticity. The concept of limit equilibrium, based on the postulate that a possible mechanism of either global or partial plastic collapse does exist is sometimes employed to assess the failure conditions in the original sense, i.e. in the case of fracture. This is then the case when, due to the plastic properties of the material, the local failure is impossible. Collapse conditions are arrived at on replacing the yield point by the strength limit where necessary [see 6.6 and other papers].

In a heated disc creep is found to equalise stresses as the time progresses. A suitable application of the approximating isochronous creep curves [6.10] enables us here to employ the limit analysis methods. Depending on the way in which the isochronous curve is approximated, i.e. which characteristic stress value is adopted as a 'yield point' (either the creep limit corresponding to a certain preassigned inelastic strain or the long-term strength), the realization of the 'limit equilibrium' may indicate that either excessive deformation has taken place or actual collapse occurred for the given duration of loading and temperature effects. In the Soviet Union Birger [6.3] initiated extensive practical applications of the long-term strength analysis to the design of gas turbines. Experimental evidence showed that discrepancies between the actual burst velocities, measured in the long-term tests on heated discs, and the calculated ones, based on a suitable interpretation of the limit analysis theory, did not exceed 2–10 per cent [6.11]. Bigger discrepancies, rarely reaching 15 per cent, were encountered for discs made from low ductility steels or alloys. As known, in this case the limit analysis leads to the upper bound on the burst velocity.

In a number of investigations attempts were made to experimentally verify the accuracy of ultimate velocity computations for discs made of various steels and revolving at room temperature. It must not be here forgotten that a good agreement of the computational and the experimental results depends to a high degree on the choice of the approximation to the actual stress–strain diagram as well as on the magnitude of plastic strain or displacement assumed in the tests to be the measure of transition into the limit state.

Since the ductility of real metal alloys is found to be confined, the limit state analysis is usually accompanied by the determination of the maximum global stresses generated by centrifugal forces and the temperature field. Thus, two types of safety factors are being recommended in the design of discs [6.3]: the ultimate load safety factor, or a related factor of burst velocity, and the local strength safety factor. Thermal stresses are accounted for in the latter case only. Other methods of estimating the strength of turbine discs are also known.

Further development of technology together with ever increasing thermal régimes and growing complexity of structures has induced the investigators to pose anew the problem of thermal stress effects. Both in the actual structures and in the tests on steel specimens it has been observed that structural collapse can be caused by thermal fatigue or, in some cases, by unacceptable changes in the geometry of discs mounting with the number of starts [6.12, 6.13 and others]. This problem has become increasingly important in the field of gas turbine engines whose working régimes suffer frequent and rapid changes.

In this chapter we shall not deal with the numerous step-by-step procedures of disc analysis under variable repeated actions and we shall not employ a plasticity and creep theory based on a particular physical model of continuum. Instead, our attention will be focused on the methods of limit analysis which will be shown to be capable of qualitatively describing the basic unsafe states. A precise distinction of the fundamental effects leading to a suitable choice of safety factors will pave the way for establishing the proper design procedures.

6.1. Limit analysis of rotating disc. Global and partial collapse mechanisms

The limit analysis theory is based on the fundamental concept of a certain state that just precedes the structural collapse. This state is associated with the possibility of kinematical changes that manifest themselves by an unlimited increase in deformations unaccompanied by any increments in loading. The limit analysis theory is usually associated with the hypothesis of perfect plasticity, since due to the plastic proper-

6 Carrying capacity of a turbine disc

ties one finds that local failure is nearly absent at loads lower than the ultimate ones. It is in principle immaterial whether the limit state will be followed by the unconfined plastic flow, i.e. plastic collapse, or by failure of another kind. Therefore, for example, the accepted computational techniques for dimension welded and riveted connections in which the cross-sectional stress distributions at collapse are uniform, are based on the limit analysis theory. Similarly, reinforced concrete structures are analysed according to the limit states theory, although concrete alone is known to fracture in a rather brittle manner.

As mentioned before, the limit analysis philosophy has widely been applied in the practical design of turbine discs. The methodology, as recommended in a number of design guides, is based on the assumption that the collapse will occur in the radial cross-section of the disc and that the plastic zone at the instant of collapse will spread to the whole disc, provided that elastic-perfectly plastic behaviour of the material is assumed. Making use of the Tresca–Saint Venant yield condition (2.10) and insisting that the largest stresses should be the circumferential ones, we arrive at the equality

$$\sigma_\varphi = \sigma_s. \tag{6.1}$$

This must hold at collapse along the whole radial cross-section. Thus, on considering equilibrium on one half of the disc, Fig. 6.1, we get the

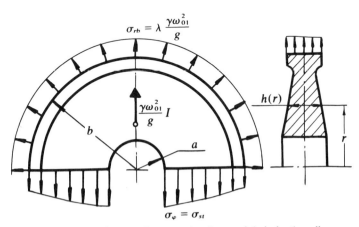

Fig. 6.1. Limit equilibrium diagram of a disc at global plastic collapse.

6.1. Limit analysis of rotating disc

known formula for the ultimate angular velocity [6.9]

$$\omega_{01}^2 = \frac{g\sigma_s F}{\gamma(I + \lambda F_b)}, \tag{6.2}$$

where $F = \int_a^b h \, dr$ and $I = \int_a^b hr^2 \, dr$ designate respectively the area and the moment of inertia of the radial section of the disc with respect to the axis of rotation, λ stands for the proportionality factor in the expression

$$\sigma_{rb} = \lambda \frac{\gamma \omega^2}{g}, \tag{6.3}$$

$F_b = h_b b$, and a, b, r denote the inner, outer and current radius, respectively. Moreover, $h = h(r)$ is the thickness of the disc and the subscript b refers to quantities that correspond to the outer radius.

In the case of non-uniform heating the yield point will be assumed to depend on the temperature t,

$$\sigma_{st} = \sigma_{st}(t).$$

Then the expression (6.2), see Fig. 6.1, takes the form

$$\omega_{01}^2 = \frac{g \int_a^b \sigma_{st} h \, dr}{\gamma(I + \lambda F_b)}. \tag{6.4}$$

In the beginning it was believed that the considered collapse mechanism was the only one possible. But another type was later discovered in which an unlimited increase in deformation was associated not with the entire disc but with its peripheral part between a certain intermediate radius $r = c$ and the outer radius $r = b$. Such a situation will be referred to as a partial collapse in contrast to the formerly discussed global collapse.

The occurrence of partial collapse will be explained by way of a stepped disc consisting of two parts, each of constant thickness. Both the elasto-plastic and the fully plastic states of such a disc were investigated by Sokolovsky [6.8] under the assumption that nowhere in the disc the radial stresses exceeded the hoop ones. This assumption fails to hold when the thicknesses of the two parts differ much from each other and it turns out that then an increase in angular velocity leads

6 Carrying capacity of a turbine disc

to the situation in which the radial stresses first reach the yield point,

$$\sigma_{rc} = \sigma_s. \tag{6.5}$$

This occurs when approaching the radius $r = c$ from the thinner part of the disc where the radial stress suffers a jump. As the angular velocity increases, the radial stresses at that radius remain fixed, due to ideal plasticity, and therefore the increase in circumferential stresses in the peripheral part of the disc will speed up, whereas it will slow down in the central part. The plastic zone will gradually grow to eventually spread over the whole outer part $c \leq r \leq b$. Once this situation sets in, Fig. 6.2, the radial displacements in this part cease to be finite and the load-carrying capacity of the disc will be exhausted in spite of the fact that the whole central part remains either purely elastic or elasto-plastic.

When condition (6.5) is satisfied, the radial rates suffer jump at $r = c_0$. In the present analysis of limit equilibrium the body could have been as well assumed to be rigid-perfectly plastic. However, such an approach would have made it impossible to consider the situation to be discussed next, namely that of repeated application of actions, as in that case the elastic strains are known to play a definite role. This situation can also be dealt with by assuming that the plastic annular zone in which

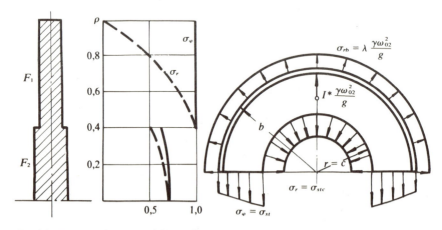

Fig. 6.2. Dimensionless stress field in a disc at partial plastic collapse.

Fig. 6.3. Limit equilibrium diagram of a disc at partial plastic collapse.

6.1. Limit analysis of rotating disc

(6.5) holds has a finite, even if quite small, thickness. This idea is somewhat similar to the generally accepted ideal plastic hinge concept.[1]

In the case of a disc of a complex profile the possibility of partial collapse should necessarily be taken into account. Similarly as before, the ultimate velocity can be derived from the equilibrium requirements, but referring now to the peripheral part of the disc as shown in Fig. 6.3. Accounting for non-uniform heating, we obtain [6.15]

$$\omega_{02}^2 = \frac{g\left[\int_c^b \sigma_{st} h \, dr + \sigma_{stc} F_c\right]}{\gamma(I^* + \lambda F_b)}, \qquad (6.6)$$

where σ_{stc} is the yield point corresponding to the temperature at the radius $r = c$, $F_c = h_c c$ and $I^* = \int_c^b h r^2 \, dr$ is the moment of inertia of this part of the disc profile that entered the limit equilibrium.

At constant temperature we get

$$\omega_{02}^2 = \frac{g\sigma_s(F_c^* + F_c)}{\gamma(I^* + \lambda F_b)}, \quad \text{where } F_c^* = \int_c^b h \, dr. \qquad (6.7)$$

The radius $r = c$ at which the discontinuity of the disc with complex profile occurs is generally unknown. The expression (6.6) is used to obtain the relationship

$$\omega_{02} = \omega_{02}(c). \qquad (6.8)$$

As follows from the kinematical theorem of limit analysis (see Chapter 2) the actual collapse mechanism is that corresponding to the smallest ultimate velocity

$$\omega_0 = \min_c [\omega_{01}, \omega_{02}(c)]. \qquad (6.9)$$

Depending on the disc profile, the magnitude of radial load, the temperature distribution and the type of function $\sigma_{st} = \sigma_{st}(t)$, this minimum velocity refers to either global or partial collapse.

Possible variants of the curve (6.8) for a solid disc and for a disc with central hole are shown in Fig. 6.4 in which suitable ultimate

---
[1] As mentioned in Chapter 5, the problem of discontinuity in the normal velocity and of creation of a neck under plane stress was discussed in [6.14].

6 Carrying capacity of a turbine disc

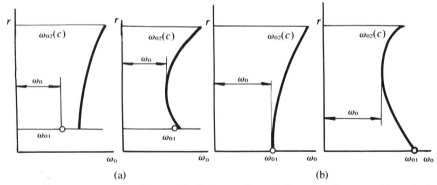

Fig. 6.4. Diagrams for evaluating the limit angular velocity of a disc; a – with central aperture, b – without.

velocities are also indicated. For the solid disc the formulae (6.4) and (6.7) clearly coincide when we take $c = 0$ in the latter.

Partial collapse of discs was investigated experimentally at the engine test benches [6.16]. Discs transmitting large radial stresses were found to undergo failure in the following manner: The peripheral part of the disc first detached from the central one and then burst into pieces; the central part continued to stay solid.

The computational analysis of numerous turbine discs in real structures has shown that the actual burst velocities often correspond to the conditions of partial collapse. The main reason for this is a certain drop in the yield point towards the periphery of the disc as a result of increase in the temperature. For a disc with very thin rim we get a low actual safety factor, substantially lower than that determined on the assumption of global collapse.

As mentioned before, the limit analysis theory is used in the design of discs in order to approximately estimate their long-term strength [6.3]. The whole methodology remains unchanged except for the yield point σ_{st} which has to be replaced by the long-term strength σ_{bt}.

The long-term strength safety factor is a known function of the temperature and the material and is defined by

$$n_{bt} = \left(\frac{\omega_0}{\omega}\right)^2, \qquad (6.10)$$

where ω_0 designates the smallest ultimate angular velocity obtained by assuming (6.9) and replacing in (6.4) and (6.6) the yield strength by the

6.2. The evaluation of carrying capacity of discs

long-term strength. Sometimes the disc strength is characterized by the 'burst velocity safety factor'

$$k_{bt} = \sqrt{n_{bt}} = \frac{\omega_0}{\omega}. \tag{6.11}$$

6.2. Some additional remarks on the evaluation of carrying capacity of discs

The described methodology can also be used for rough strength assessments of discs in radial-flow turbines and centrifugal compressors. If we restrict ourselves to the case when bending action is absent,[1] the relevant formulae are (6.4) and (6.7). We take $\lambda = 0$, signifying that the outer periphery of the disc is free from load, and we modify the expressions for the inertia moments I, I^* so that the influence of blade mass is incorporated. This means,

$$\omega_{01}^2 = \frac{g \int_a^b \sigma_{st} h \, dr}{\gamma I_{eq}}, \tag{6.12}$$

$$\omega_{02}^2 = \frac{g \left[\int_c^b \sigma_{st} h \, dr + \sigma_{stc} F_{ce} \right]}{\gamma I_{eq}^*},$$

$$I_{eq} = \int_a^b (1+\zeta) h r^2 \, dr, \quad I_{eq}^* = \int_c^b (1+\zeta) h r^2 \, dr, \quad \zeta = \frac{zF(r)}{2\pi rh},$$

$$F_{ce} = h_c c (1+\zeta_c), \tag{6.13}$$

where z is the number of blades and $F(r)$ denotes the cross-sectional area of a blade with cylindrical shape of radius r.

The influence of openings in the disc web can be taken into account in a similar manner. It is a recognized fact that eccentric holes can be a sufficient reason for the partial collapse of a disc [6.16]. Indeed, it can be readily shown by using suitable formulae that partial collapse will occur for a disc of constant thickness having same number of small holes with diameter d spaced evenly around the radius $r = c$. The critical number of

---

[1] The influence of bending on the ultimate behaviour of a bladed wheel will be discussed later.

6 Carrying capacity of a turbine disc

holes is found to be

$$n \geq \pi \left[1 + \frac{2(b-d)}{d} \left(\frac{c}{b}\right)^3 \right].$$ (6.14)

For $c = 0,5b$, $d = 0,1b$, we get $n \geq 10$. This means that an increase in the number of holes from 2, lying on one diameter, to 10 does not lead to a weakening of the disc. However a further increase does result in a drop of the ultimate angular velocity since a transition from one to the other collapse mechanism takes place.

In a similar way some other types of weakening of a disc can be considered. It is only necessary to assume a reasonable collapse mechanism in each particular case. Needless to say that only those discs are here discussed that display sufficient ductility to be considered approximately perfectly plastic.

The notion of partial collapse may be used for the proper design of disc profiles. Indeed, the disc can be termed to be one of equal strength if $\omega_{02} = \omega_{01} = $ const., where ω_{02} is determined for all the values of the transition radius, $a \leq c \leq b$. Since no tensile stress can be applied at the contour of the central hole, the equal strength disc should be solid, $a = 0$. Equal strength of a disc with central hole can be roughly secured at the cost of forming a fairly thick hub near the hole.

Making use of the expression (6.6) in which the yield point is replaced by the long-term strength, and of the formula (6.11), we arrive at

$$h_c = \frac{1}{c\sigma_{btc}} \left[\frac{\gamma}{g} k_{bt}^2 \omega^2 (I^* + \lambda F_b) - \int_c^b \sigma_{bt} h \, dr \right],$$ (6.15)

where k_{bt} is the safety margin with respect to the burst velocity. The calculations start from the outer contour of the disc. Its thickness is found to be

$$h_b = \frac{k_{bt}^2}{\sigma_{btb}} (\sigma_{rb} h_b).$$ (6.16)

The product $\sigma_{rb} h_b$ is determined by the centrifugal forces coming from the blades that are in some way fixed to the peripheral rim [6.3].

In actual turbine discs both the shape and the dimensions of the rim are designed in such a way as to facilitate the connection between

6.3. Shakedown of plane disc

the blades and the disc. Thus it is only the web which can be of equal strength. The dimensioning commences with the thickness h_d at the periphery of the web, found from (6.15) by putting $c = d$ and $I^* = I$, where I is the moment of inertia of the rim alone. When the thicknesses h_d and h_b differ too much from each other, Fig. 6.5, the radius d and the thickness h_d can be adjusted in such a manner as to make the transition from the rim to the disc a gradual one.

When some protrusions like lips and brackets have to be provided we can use again the expression (6.15) for designing a disc as close to the equal strength one as possible. It can be readily shown that under the conditions of uniform heating the expression (6.15) leads to the well-known profile of exponential type,

$$h = h \exp \frac{b^2 - r^2}{2\lambda}. \qquad (6.17)$$

6.3. Shakedown of plane disc at cyclic variation of angular velocity and temperature

To approach the problem of the ultimate behaviour of turbine discs in unsteady conditions, it appears expedient to begin with the analysis of a plane disc.

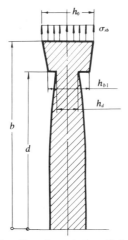

Fig. 6.5. Turbine disc of equal strength (excluding rim).

6 Carrying capacity of a turbine disc

Let the disc rotate with angular velocity $0 \leq \omega \leq \omega_*$ and let the cyclic variations in the temperature field, caused by non-uniform heat exchange between the peripheral surface and the disc faces, be described by a simple series

$$t(\tau, \rho) = t_0(\tau) + \rho t_1(\tau) + \rho^2 t_2(\tau) + \cdots + \rho^n t_n(\tau), \qquad (6.18)$$

where $\rho = r/b$.

Next, let the functions of current time, entering (6.18), vary independently within the prescribed intervals

$$0 \leq t_i(\tau) \leq t_i^*. \qquad (6.19)$$

The thermal stresses in a plane solid disc under arbitrary radial distribution of the temperature are found to be [6.17]

$$\sigma_{rt} = \alpha E \left[\int_0^1 t(\tau, \rho) \rho \, d\rho - \frac{1}{\rho^2} \int_0^\rho t(\tau, \rho) \rho \, d\rho \right],$$
$$\sigma_{\varphi t} = \alpha E \left[\int_0^1 t(\tau, \rho) \rho \, d\rho + \frac{1}{\rho^2} \int_0^\rho t(\tau, \rho) \rho \, d\rho - t(\tau, \rho) \right]. \qquad (6.20)$$

Using the adopted temperature distribution (6.18), we get

$$\sigma_{rt} = \alpha E \sum_{i=1}^n \frac{t_i(\tau)}{i+2} (1 - \rho^i),$$
$$\sigma_{\varphi t} = \alpha E \sum_{i=1}^n \frac{t_i(\tau)}{i+2} [1 - (i+1)\rho^i]. \qquad (6.21)$$

We shall establish presently the conditions for incremental collapse of a disc with the help of the approximate kinematical method. The radial displacement increments at the limiting cycle are assumed to be equal for all points of the disc,

$$\Delta u_0 = c, \quad (c > 0). \qquad (6.22)$$

The increments of radial and circumferential plastic strains follow from the conditions of the type (5.25) and take the form

$$\Delta \epsilon_{r0}'' = \frac{d(\Delta u_0)}{dr} = 0, \quad \Delta \epsilon_{\varphi 0}'' = \frac{\Delta u_0}{r} = \frac{c}{b\rho} > 0. \qquad (6.23)$$

6.3. Shakedown of plane disc

The fictitious yield surface stresses appearing in the right-hand side of the equation (5.2) are related to the strain increments by expressions of the type (5.4). In particular, according to the relations (6.23) and the flow law associated with the Tresca yield condition (2.10), the determining stresses are the hoop stresses

$$\sigma^0_{\varphi *}(\rho) = \min_\tau(\sigma_s - \sigma^{(e)}_{\varphi \tau}). \tag{6.24}$$

When the yield point σ_s remains constant as the cycle proceeds, condition (6.24) takes the form

$$\sigma^0_{\varphi *}(\rho) = \sigma_s - \max_\tau \sigma^{(e)}_{\varphi \tau}. \tag{6.25}$$

From this it follows that the most unfavourable is the loading program during which the angular velocity stays constant over the whole cycle and equals its maximum, $\omega = \omega_*$. Because of this fact and due to the absence of the periphery loading we can write

$$X^0 = \frac{\gamma \omega^2_* b\rho}{g}, \quad p^0_i = 0. \tag{6.26}$$

Therefore it is the thermo-elastic stresses (6.20) that should be identified with the time-dependent stresses appearing in the formulae (6.24), (6.25). The determination of their 'enveloping' values for the adopted temperature distribution is illustrated in Fig. 6.6.

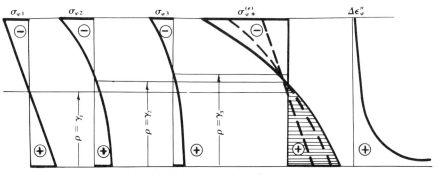

Fig. 6.6. Envelope of thermal stresses.

6 Carrying capacity of a turbine disc

We take here only the first three terms of (6.18), disregarding the function $t_0(\tau)$ which describes the general heating of the disc and does not affect the thermal stresses. Using (6.21)–(6.26), we can express equation (5.2) in the form

$$\frac{\gamma\omega_*^2 b^2}{g}\int_0^1 \rho^2\,d\rho = \int_0^1 \sigma_s\,d\rho - \alpha E \sum_{i=1}^n \frac{t_i^*}{i+2}\int_0^{\gamma_i}[1-(i+1)\rho^i]\,d\rho, \quad (6.27)$$

where γ_i can be obtained from the conditions

$$1-(i+1)\gamma_i^i = 0. \quad (6.28)$$

It must be borne in mind that the acceptance of equality sign in the inequality (5.2) corresponds to the best upper bound on the sought for parameters of the limiting cycle.

On integrating (6.27) we obtain

$$\frac{p^*}{p_0} + \sum_{i=1}^n D_i \frac{t_i^*}{t_i^0} = 1, \quad (6.29)$$

where $p^*/p_0 = (\omega_*/\omega_0)^2$, $\omega_0^2 = 3g\sigma_s/\gamma b^2$, $D_i = 2(i+1)^{-\frac{i+1}{i}}$, and $t_i = 2\sigma_s(i+2)/\alpha E i$ denotes the maximum value of t_i^* at which the cyclic fluctuations of the temperature field $t_i(\tau, \rho) = \rho^i t_i(\tau)$, $0 \leq t_i(\tau) \leq t_i^*$ do not result in the alternating plastic flow of a disc at rest.

Taking the first three terms in the series (6.21) we obtain the incremental collapse condition (6.29) in the form

$$\frac{p^*}{p_0} + \frac{1}{2}\cdot\frac{t_1^*}{t_1^0} + \frac{2\sqrt{3}}{9}\cdot\frac{t_2^*}{t_2^0} + \frac{\sqrt[3]{2}}{4}\cdot\frac{t_3^*}{t_3^0} = 1. \quad (6.30)$$

Exactly the same result was derived by using the overload method [6.18] and assuming the quadratic law of temperature distribution in which only t_2^* is non-vanishing; we obtained then

$$\frac{p^*}{p_0} + \frac{2\sqrt{3}}{9}\cdot\frac{t_2^*}{t_2^0} = 1. \quad (6.31)$$

The stress fields at all the stages of the cycle described by (6.31) can be verified to be statically admissible and thus the obtained solution is complete.

The above statement can also be made concerning the solutions

6.3. Shakedown of plane disc

obtained by means of the kinematical method and based on the distribution of residual displacement increments (6.22) for more general temperature fields described by a series (6.18). Interestingly enough, the linear law of temperature distribution ($t_i^* \neq 0$ for $i = 1$ only) is here excluded since at the load intensities approaching those causing the alternating plastic flow, when t_1^*/t_1^0 differs but little from unity, the solution (6.30) constitutes an upper bound on the limiting cycle parameters and not the exact answer. The statical conditions for shakedown are then not fulfilled.[1] The conclusion is that the linear temperature field generates a collapse mechanism that will be different from that given by the relationships (6.22), (6.23), and more complex.

The shakedown solutions for a disc under linear radial distribution of temperature, obtained with the help of both the approximate and the exact statical methods, were shown in Chapter 4, Fig. 4.7, as the lines BD and CDE respectively. In order to compare those solutions with the just obtained upper bound let us rewrite (6.30) for $i = 1$. We get

$$\frac{p^*}{p_0} = 1 - \frac{1}{2} \cdot \frac{t_1^*}{t_1^0}. \tag{6.32}$$

This equation is represented in Fig. 4.7 by the straight line CDE. For $t_1^*/t_1^0 < 0{,}85$ the kinematical and the statical solutions coincide; this means that formula (6.32) yields the exact result. For $t_1^*/t_1^0 > 0{,}85$ the lower and the upper estimates of the progressive collapse drift apart although not very significantly.

Let us consider again the shakedown conditions assuming the quadratic law of temperature distribution. In addition to the incremental collapse conditions (6.31) let us find the alternating plasticity condition under arbitrary loading program. The most unfavourable program appears to be that under which the angular velocity of the disc varies and therefore we have to employ here the expressions for the determination of 'elastic' stresses generated by a centrifugal load [6.17]. The stresses can be written as

$$\sigma_r = \frac{3\sigma_s}{3-\kappa} \cdot \frac{p}{p_0}(1-\rho^2),$$
$$\sigma_\varphi = \frac{3\sigma_s}{3-\kappa} \cdot \frac{p}{p_0}(1-\kappa\rho^2), \tag{6.33}$$

where $\kappa = (1+3\mu)/(3+\mu)$ and μ is Poisson's ratio.

---

[1] The actual reason for this is the appearance of substantial compressive stresses in the central part of the disc at cooling.

6 Carrying capacity of a turbine disc

Remembering the criterion (2.90) of the alternating plastic flow, we obtain the condition

$$\frac{3(1-\kappa)}{2(3-\kappa)} \cdot \frac{p^*}{p_0} + \frac{t_2^*}{t_2^0} = 1. \tag{6.34}$$

As in condition (6.31), the temperature field is here described by only one term of the series (6.18), namely the one corresponding to $i = 2$; the present condition refers to the radius $\rho = 1$ and concerns the sequence of actions which is most disadvantageous at this particular radius: starting – heating – stoppage – cooling. For constant velocity we get

$$\frac{t_2^*}{t_2^0} = 1. \tag{6.35}$$

It can be readily shown that at the radius $\rho = 1$ it is the alternating flow that will take place first.

The shakedown diagram is shown in Fig. 6.7. Line 1 corresponds to the alternating plasticity condition (6.34), for arbitrary loading program; line 2 corresponds to the same condition but at constant angular velocity; line 3 indicates the incremental collapse condition (6.31). Depending on the sequence of actions, the shakedown domain is bounded by either the lines 1, 3 representing an arbitrary program, or by the lines 2, 3 corresponding to thermal actions at constant velocity.

The solution to the shakedown problem of a plane disc was first

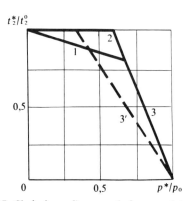

Fig. 6.7. Shakedown diagram of plane revolving disc.

6.3. Shakedown of plane disc

arrived at in an approximate manner, namely by means of the statical theorem [6.19]. The shakedown domain corresponding to an arbitrary program of actions is bounded by lines 1, 3^1. However their interpretation was not offered because in the framework of statical methods it is difficult to distinguish between the various types of cyclic plastic deformation. It is interesting to note that in the case of alternating flow the approximate solution coincides with the exact one, whereas in the case of incremental collapse the statically determined parameters of the limiting cycle are those estimated from below. The reasons for coincidence in the former case were explained in Sec. 2.7.

The shakedown analysis of discs, based on the kinematical method as considered in this section, can be extended to cover real temperature fields described either by using the analytical tools of the heat conduction theory, as exposed for instance in [6.20], or by employing the experimental results. Most essential is here the proper choice of a collapse mechanism and the construction of the envelope of thermal (or total) time-dependent stresses, as shown in Fig. 6.6.

The fundamental mechanisms of incremental collapse can conveniently be examined by considering the stress fields at various stages of the limiting cycle. In those cases in which the strain accumulation in the peripheral part of the disc is associated with the yielding régimes

$$\sigma_\varphi = \sigma_s \ (c \leq r \leq b) \quad \text{and} \quad \sigma_r = \sigma_s \ (r = c), \tag{6.36}$$

where global collapse corresponds to $c = a$ and partial collapse to $c > a$, the virtual work equation reduces to the standard equilibrium equation on one half of the disc.

To illustrate the results let us consider the stress fields in the characteristic states of the limiting cycle for two types of discs. In both cases the angular velocity, accompanied by cyclic variations in the thermo-elastic stresses is assumed to be constant and the stresses themselves to be generated by a one-parameter temperature field. The uniform part of heating will not be incorporated since no influence is taken into account of arbitrary thermal conditions upon the physical-mechanical properties of the material.

In Fig. 6.8, corresponding to a plane disc with a central hole and to the values $p^*/p_0 = 0.85$, $q^*/q_0 = 0.31$ there are shown successively: the fictitious 'elastic' stresses resulting from the rotation and the temperature field (a); the actual stresses in two states bounding the semi-cycle, i.e. maximum heating (b) and full cooling (c), and differing from

6 Carrying capacity of a turbine disc

one another by the thermo-elastic stresses (d); and finally the residual stresses (e). The latter were computed as differences between the actual stresses (b) and the fictitious elastic ones (a) in a heated disc.

It can be readily ascertained that in the limiting cycle, i.e. in the state prior to incremental collapse, the condition formulated in Chapter 5 is satisfied. In the present application this condition is as follows: If the stress distribution in the unheated disc (Fig. 6.8, area of the hoop stress diagram is equal to the centrifugal forces resultant to within a constant multiplier) is imposed on the thermo-elastic stresses (c) and the unloading which is bound to occur in a part of the profile is disregarded, then the situation arises in which the load-carrying capacity has just been exhausted, i.e. we have a case of instantaneous collapse.

In Fig. 6.9 we can see a similar situation concerning incremental partial collapse. The actual stresses in the two states bounding the semi-cycle are shown at maximum heating (a) and at cooling (b). Diagram (c) shows the residual stresses computed in the standard manner. They are a result of the plastic flow of the thin part of the disc which occurs while its central part undergoes elastic deformations only under radial load.

We can easily deduce from the figures the mechanism of incremental collapse occurring in each case considered, provided one of

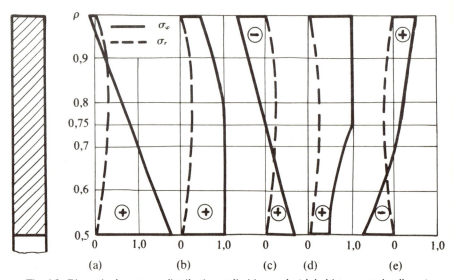

Fig. 6.8. Dimensionless stress distribution at limiting cycle (global incremental collapse).

6.4. Influence of creep on the shakedown conditions

the external action parameters, either the angular velocity or the maximum temperature interval, undergoes a finite increment above its limiting cycle value. For instance, in the disc of stepped profile the strain accumulation in the thin part will take place as a result of the hoop stresses attaining the yield point at the end of heating in the hub part and at the end of cooling in the peripheral part. In the former case the radial stresses, at the connection of the two parts, reach the yield point as well. The stress fields shown in Fig. 6.9 correspond to the limiting cycle; plastic zones at various stages will overlap.

6.4. Influence of creep on the shakedown conditions

We shall again discuss on the example of the plane disc the peculiarities of the shakedown solutions in the case of alternation of long-term, i.e. stationary and short-term, i.e. transient loading régimes. We shall use the most simple formulation permitting the display of the qualitative creep effects.

Let the temperature field in the disc be

$$t(\tau, \rho) = t_0(\tau) + \rho^2 t_1(\tau), \tag{6.37}$$

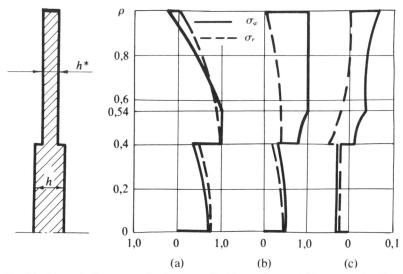

Fig. 6.9. Dimensionless stress distribution at limiting cycle (partial incremental collapse).

6 Carrying capacity of a turbine disc

where $0 \leq t_0(\tau) \leq t_0^*$, $0 \leq t_1(\tau) \leq t_1^*$. Let us assume that the disc is revolving with constant angular velocity ω and that, at the first stage, an increase in the uniform temperature $t_0(\tau)$ is accompanied by a rapid rise of the temperature $t_1(\tau)$ from zero to its maximum value, the latter being reached simultaneously with t_0^*. Then $t_1(\tau)$ begins to drop and the temperature field in the whole disc becomes uniform, equal to t_0^*, Fig. 6.10. Thus a stationary regime of definite duration sets in, to be followed by cooling. For the sake of simplicity, the temperature gradients as well as the thermal stresses at the last stage are both assumed to be negligibly small.

The properties of the material at heating stages of short duration are assumed not to depend on the temperature nor on the number of cycles (creep effects at these stages can be incorporated by a suitable selection of the loading rate when obtaining the stress–strain diagrams). The material deformation characteristics in stationary régimes depend upon the global time of N preceding cycles and can be determined using the corresponding isochronous stress–strain curve, Fig. 6.11, whose parameters such as the yield point under creep conditions and long-term strength decrease as the number of cycles increases. In shakedown analysis the real short-term and long-term stress–strain diagrams are to be approximated by the elastic-ideally plastic ones and only the difference in corresponding yield stresses is taken into account. Let the yield stress due to stationary regimes be $k\sigma_s$, where σ_s denotes the yield point on the approximated short-term curve and $k < 1$ depends on the

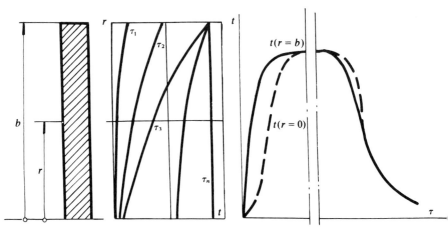

Fig. 6.10. A cycle including the steady state régime.

6.4. Influence of creep on the shakedown conditions

temperature and the total life-time of the structure, i.e. on the global duration of the stationary regimes $\tau_\Sigma = TN$.

It is worth mentioning that by means of the above introduced simplifications we can emphasize the way in which the duration of loading is accounted for in the shakedown problems. It is also easy to interpret the obtained results. Some of the simplifications are not obligatory and can be dismissed once the necessary information is available.

Let us analyse the conditions for incremental collapse of the disc. Making use of the approximate kinematical method, we adopt the kinematically admissible distribution of the radial displacement increments along the radius, $0 \leq \rho \leq 1$, according to (6.22). In other words, we assume that a global collapse of the disc will take place. Using the condition (6.24) and the expression (6.21), we arrive at the condition for incremental collapse

$$\frac{p}{p_0} = \int_0^1 \min_\tau \left[k - \frac{t_1(\tau)}{t_1^0} (1 - 3\rho^2) \right] d\rho. \tag{6.38}$$

For the transient régimes we adopt $k = k_0 = 1$, and for the stationary ones the value $k \leq 1$ should be selected, in accordance with the available data on the material properties.

It is easy to see that, depending on the radial distribution of temperature, the integrand in (6.38) attains its minimum for the following magnitudes of k and $t_1(\tau)$:

a) in the region $\rho \geq \rho_* \geq 0$ at the stationary régime

$$k < 1, \quad t_1(\tau) = 0; \tag{6.39}$$

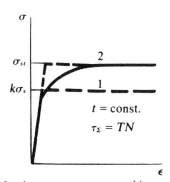

Fig. 6.11. Isochronous creep curve and its approximation.

6 Carrying capacity of a turbine disc

b) in the region $0 \leq \rho \leq \rho_*$ at the transient régime

$$k = k_0 = 1, \quad t_1(\tau) = t_1^*, \tag{6.40}$$

where

$$\rho_*^2 = \frac{1}{3}\left[1 - (1-k)\frac{t_1^0}{t_1^*}\right]. \tag{6.41}$$

The expressions (6.39) and (6.40) show that the inelastic strain rates in the central part of the disc at the limiting cycle cease to be zero as soon as the temperature interval $t_1(\tau)$ attains its maximum on heating while the thermal stresses in this part of the disc are tensile. In the peripheral part the inelastic strain rates are non-vanishing in the stationary régime, i.e. due to creep. As the process continues, especially at lower temperature gradients during the heating period, the radius ρ_* decreases and can even, under certain conditions, shrink to zero. In the latter case the minimum of the integrand in (6.38) corresponds to a single régime for all the points of the disc and to the rise of compatible creep strains at stationary stages of the limiting cycle.

The results obtained by the analysis of conditions for strain accumulation with the help of equation (6.38) and expressions (6.39) and (6.40) are shown diagrammatically in Fig. 6.12 for $k = 1,0; 0,75; 0,5$. Such a sequence of values corresponds to an increase of the life-time of the considered structure. Dashed lines indicate the occurrence of alternating plastic flow and are found in the known manner according to the accepted approximation of the material properties at various stages of the cycle of the same duration (see (2.91)).

Each shakedown curve in Fig. 6.12 is bounding a domain of the limiting cycle parameters within which, for the prescribed duration, the disc will deform only elastically at all the stages of the cycle, excluding a certain initial finite deformation. If a point whose location indicates the relevant cycle parameters, lies outside the shakedown domain constructed for $k_0 = 1$, the repeated loading will at once result in the cyclic plastic deformation, either alternating or one-sign or a combination of both. If a point lies inside the shakedown domain, the inelastic strain will set in not earlier than after a number of cycles such that the duration of the process on them is sufficient for the shakedown diagram, constructed with a suitable value of k, to meet the point in question. Depending on which side of the shakedown domain comes in contact with this point,

6.5. Shakedown analysis of rotating discs

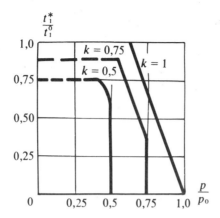

Fig. 6.12. Shakedown diagrams for various periods of steady régime.

the dashed, the solid inclined, or solid vertical, we have to do, respectively, with a visco-plastic alternating flow, a one-sign strain of the same nature accumulating with each cycle and, finally, a viscous flow at the stationary régime, i.e. creep strain. As the process continues, or the temperature of the stationary régime increases, the vertical side of the shakedown boundary is getting longer. This means that the corresponding interval of the values of cycle parameters extends; inside this interval the carrying capacity of the structure is limited by the conditions of the stationary régime.

Thus the construction of the shakedown diagram accounting for the duration of the cycle is found to be free from any further difficulties and it falls within the framework of the already described methods. However, this is clearly so only when the short-term and the long-term stress–strain diagrams are known for various loading histories. More complex situations of interaction between the short- and the long-term deformation effects will not be discussed here. To investigate this question, different models of continuum are needed, as given for instance in [6.21, 6.22].

6.5. Shakedown analysis of rotating, non-uniformly heated discs of arbitrary profiles

Let us determine the incremental collapse conditions for a disc with arbitrarily variable thickness $h = h(r)$. It is assumed that the disc profile

6 Carrying capacity of a turbine disc

is symmetric with respect to its middle plane and the temperature field is axially symmetric and constant across the thickness. If the elastic moduli E, μ and the thermal expansion coefficient α are kept temperature-independent, the 'elastic' stresses generated by the centrifugal forces and non-uniform heating can be expressed by the linear relationships

$$\sigma_{rr}^{(e)} = p(\tau)\varphi_1(r) + q(\tau)\psi_1(r, \tau),$$
$$\sigma_{\varphi\tau}^{(e)} = p(\tau)\varphi_2(r) + q(\tau)\psi_2(r, \tau), \qquad (6.42)$$
$$0 \leq p(\tau) \leq p_*, \quad q^- \leq q(\tau) \leq q^+,$$

where

$$p = \frac{\gamma \omega^2 b^2}{g\sigma_s^0}, \quad q(\tau) = \frac{\alpha E[t(b, \tau) - t_0(\tau)]}{\sigma_s^0}$$

are the centrifugal and the temperature parameters. The latter is seen to depend on the difference of temperatures between the periphery $t(b)$ and the centre $t_0(\tau)$, or at the contour of the central hole. The stresses (6.42) are conveniently referred to the yield point at room temperature σ_s^0. The functions $\varphi_1(r), \varphi_2(r)$ are given by the disc profile; the functions $\psi_1(r, \tau), \psi_2(r, \tau)$ depend on the profile and also on the radial distribution of temperature. In the thermal regimes with proportional temperature gradients

$$t(r, \tau) = t_0(\tau) + t_1(\tau) \cdot f(r) \qquad (6.43)$$

the functions ψ_1, ψ_2 cease to be time-dependent and the only remaining argument is that of the current radius.

The dependence of the yield point on temperature is assumed to be given by

$$\sigma_{st} = \sigma_s(t)/\sigma_s^0 = 1 - \phi(t), \qquad (6.44)$$

where $\phi(t)$ is to be determined experimentally. Because the temperature of the disc depends on the current radius and time only, we can write $\phi(t) \equiv \phi(r, \tau)$. Recalling (6.43), we note that the function $t_0(\tau)$ describing the uniform heating of the disc is here involved.

Let the global collapse mechanism (6.22) be pertinent to the limiting

6.5. Shakedown analysis of rotating discs

cycle situation. After performing similar rearrangements to those made in Sec. 6.3 and using condition (5.2) we may write the equation of the limiting cycle in the form

$$\int_a^b \sigma_{\varphi*}^0 h(r)\,dr = 0. \tag{6.45}$$

In the case of partial incremental collapse, when strains accumulate in the outer part with each cycle, we obtain that

$$\int_c^b \sigma_{\varphi*}^0 h(r)\,dr + ch_c \sigma_{r*}^0(c) = 0. \tag{6.46}$$

This results from considering the work done by the fictitious surface stresses at the displacement discontinuity $r = c$.

Let us recall the notation used in the equations (6.45), (6.46): a, b are the inner and the outer radii (for a solid disc $a = 0$); c is the inner radius of the peripheral part in which strains accumulate; $\sigma_{\varphi*}^0, \sigma_{r*}^0$ denote the hoop and the radial stresses belonging to the fictitious yield curve and $h_c = h(c)$.

It is reasonable to expect that in the plastic part of the disc the flow régime will be determined by hoop stresses, whereas at the radius $r = c$ partial collapse occurs when the radial stress reaches the yield point at a certain instant of time.

Hence we can write

$$\sigma_{\varphi*}^0 = \min_\tau[\sigma_{st}(\tau) - \sigma_{\varphi\tau}^{(e)}],$$
$$\sigma_{r*}^0 = \min_\tau[\sigma_{st}(\tau) - \sigma_{rr}^{(e)}]. \tag{6.47}$$

Substituting (6.42) and (6.44), we get

$$\sigma_{\varphi*}^0 = \min_\tau[1 - \phi(r,\tau) - p(\tau)\varphi_2(r) - q(\tau)\psi_2(r,\tau)],$$
$$\sigma_{r*}^0 = \min_\tau[1 - \phi(r,\tau) - p(\tau)\varphi_1(r) - q(\tau)\psi_1(r,\tau)]. \tag{6.48}$$

The limiting cycle equations (6.45), (6.46) for the global and the partial

6 Carrying capacity of a turbine disc

incremental collapse respectively assume now the form

$$\int_a^b h\{1 - \max_\tau[\phi(r,\tau) + p(\tau)\varphi_2(r) + q(\tau)\psi_2(r,\tau)]\}\,dr = 0, \tag{6.49}$$

$$\int_c^b h\{1 - \max_\tau[\phi(r,\tau) + p(\tau)\varphi_2(r) + q(\tau)\psi_2(r,\tau)]\}\,dr +$$

$$+ ch_c\{1 - \max_\tau[\phi(c,\tau) + p(\tau)\varphi_1(c) + q(\tau)\psi_1(c,\tau)]\}. \tag{6.50}$$

A better upper bound on the load factors obeying (6.49), (6.50) is usually that corresponding to the exact solution, i.e. simultaneously satisfying the statical requirements. However, at higher intensities of thermal actions, particularly in the region close to the limit state of local alternating flow, other, more complex collapse mechanisms may be critical. On the other hand, the upper bound on the load factors obtained by minimization of (6.49), (6.50) appears often to be not far from the exact answer. Certain refinements can be obtained by accepting in the conditions of partial collapse the occurrence of radial displacement at $r = c$, i.e. in the zone of very rapid change abolishing the discontinuity, under flow régimes different from those leading to equation (6.50). This circumstance is shown in Fig. 6.13b by means of the fictitious yield locus for such points of the disc at which the time-dependent hoop thermal stresses attain certain larger negative (compression) values during the cycle. One side of the fictitious yield polygon can disappear at the periphery of the disc provided the condition

$$\min_\tau[\sigma_{st}(\tau) - \sigma_{rr}^{(e)}] > \min_\tau[\sigma_{st}(\tau) - \sigma_{rr}^{(e)} + \sigma_{\varphi\tau}^{(e)}] + \min_\tau[\sigma_{st}(\tau) - \sigma_{\varphi\tau}^{(e)}] \tag{6.51}$$

is satisfied.

Now the fictitious stresses curve can be determined from the formulae

$$\sigma_{\varphi*}^0 = \min_\tau[\sigma_{st}(\tau) - \sigma_{\varphi\tau}^{(e)}],$$

$$\sigma_{r*}^0 = \min_\tau[\sigma_{st}(\tau) - \sigma_{rr}^{(e)} + \sigma_{\varphi\tau}^{(e)}] + \min_\tau[\sigma_{st}(\tau) - \sigma_{\varphi\tau}^{(e)}], \tag{6.52}$$

which differ from (6.47).

When we assume that the inequality (6.51) holds good at the radius

6.5. Shakedown analysis of rotating discs

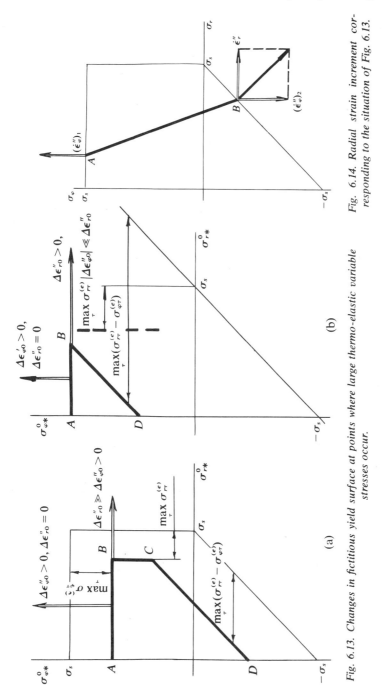

Fig. 6.13. Changes in fictitious yield surface at points where large thermo-elastic variable stresses occur.

Fig. 6.14. Radial strain increment corresponding to the situation of Fig. 6.13.

6 Carrying capacity of a turbine disc

$r = c$, the condition for the partial incremental collapse takes the form

$$\int_c^b h\{1 - \max_\tau[\phi(r,\tau) + p(\tau)\varphi_2(r) + q(\tau)\psi_2(r,\tau)]\}\,\mathrm{d}r +$$

$$+ ch_c\{2 - \max_\tau[\phi(c,\tau) + p(\tau)\varphi_1(c) + q(\tau)\psi_1(c,\tau) - p(\tau)\varphi_2(c) -$$

$$- q(\tau)\psi_2(c,\tau)] - \max_\tau[\phi(c,\tau) + p(\tau)\varphi_2(c) + q(\tau)\psi_2(c,\tau)]\} = 0.$$

(6.53)

Fig. 6.14 is provided to elucidate the generation of strains at the radius $r = c$: The flow regimes A ($\sigma_\varphi = \sigma_{st}$) at one stage, corresponding to maximum angular velocity, and B ($\sigma_r - \sigma_\varphi = \sigma_{st}$) at another stage, associated with the maximum temperature interval, are accompanied by non-vanishing radial strain increments in each cycle of the alternating plastic flow in the circumferential direction. Thus the hoop strains must also accumulate then, since $(\dot\epsilon''_\varphi)_1 > (\dot\epsilon''_\varphi)_2$. This inequality turns out to be the necessary condition for incremental collapse.

In order to find the maximum values of the functions entering (6.49), (6.50), (6.53), we need to have a relationship between the functions $p(\tau)$ and $q(\tau)$. This largely depends on the usage of the disc and can be either deterministic or stochastic. In the extreme case the values of angular velocity and temperature can be completely independent from each other varying within the prescribed intervals. If this is so, the equations (6.49), (6.50), (6.53) can be rearranged with the help of the virtual work principle to assume a more convenient form. Taking into account that the maximums with respect to τ on the left-hand sides are all attained at $p(\tau) = p_*$, because the functions $\varphi_1(\rho)$, $\varphi_2(\rho)$ are positive-valued, we finally arrive at

$$\frac{\gamma\omega_*^2}{g\sigma_s^0}(I + \lambda F_b) + \int_a^b h \max_\tau[\phi(r,\tau) + q(\tau)\psi_2(r,\tau)]\,\mathrm{d}r = F_a^*,\qquad (6.54)$$

$$\frac{\gamma\omega_*^2}{g\sigma_s^0}(I^* + \lambda F_b) + \int_c^b h \max_\tau[\phi(r,\tau) + q(\tau)\psi_2(r,\tau)]\,\mathrm{d}r +$$

$$+ F_c \max_\tau[\phi(c,\tau) + q(\tau)\psi_1(c,\tau)] = F_c^* + F_c,\qquad (6.55)$$

6.5. Shakedown analysis of rotating discs

$$\frac{\gamma\omega_*^2}{g\sigma_s^0}(I^* + \lambda F_b) + \int_c^b h \max_\tau[\phi(r,\tau) + q(\tau)\psi_2(r,\tau)]\,dr +$$

$$+ F_c\{\max_\tau[\phi(c,\tau) + q(\tau)(\psi_1(c,\tau) - \psi_2(c,\tau))] +$$

$$+ \max_\tau[\phi(c,\tau) + q(\tau)\psi_2(c,\tau)]\} = F_c^* + 2F_c, \tag{6.56}$$

where

$$I^* = \int_c^b hr^2\,dr, \quad F_c^* = \int_c^b h\,dr, \quad F_a^* = \int_a^b h\,dr, \quad F_c = ch_c. \tag{6.57}$$

The first equation corresponds to global incremental collapse, the second and the third ones to partial collapse. In the latter case the alternating plastic yielding in the circumferential direction occurs at the radius $r = c$ together with the radial strain accumulation.

If the disc temperature is assumed to obey the rule (6.43), the equations (6.54)–(6.56) can be further simplified. Let us generally agree that one of the states at which the maximums with respect to τ of the expressions appearing in (6.54)–(6.56) are reached is associated with that instant of heating when the function $t_1(\tau)$, describing the temperature intervals and consequently the thermal stresses, attains the largest value

$$t' = t_0' + t_1' \cdot f(r), \tag{6.58}$$

where

$$t_1' = t_1(\tau_1) = [t_1(\tau)]_{\max}, \quad t_0' = t_0(\tau_1).$$

For the other state we shall take that instant of cooling when the temperature field becomes uniform

$$t'' = t_0'' + t_1'' \cdot f(r), \tag{6.59}$$

where

$$t_1'' = t_1(\tau_2) = 0, \quad t_0'' = t_0(\tau_2).$$

To avoid a large number of parameters, let $t_0'' = t_0' = t_0$. When dealing with specific structures and having the necessary data, we can suitably

6 Carrying capacity of a turbine disc

refine the relevant conditions. For instance, assuming that $t_1'' < 0$, we can account for the so-called 'reversed' temperature gradients encountered in the transient working regimes of turbine discs.

In what follows p^*, q^* will denote the largest values of the respective factors, reached per cycle and corresponding to ω_{max} and t_1'.

According to Fig. 6.13a, the adopted collapse mechanisms (6.22), (6.23) lead in this part of the disc where the one-sign strains accumulate, to a flow regime at which the yield point is reached by the hoop stresses. It should be expected that at the end of heating the above condition will be fulfilled in the central part in which the thermal stresses generated by the temperature field (6.58) give rise to overloading. At the periphery, where the thermal stresses are negative, the yield point can be reached only when condition (6.59) is fulfilled.

The overload region is shown in Fig. 6.15 where the variations of the yield point during heating are accounted for. After imposing the extremum thermo-elastic stresses $\sigma_{\varphi q}^{(e)}$ on the segment $e_0 \leq r \leq e$, the decrease in total stresses (resulting from both rotation and temperature) is slower than the decrease in the yield point at the termination of heating. Thus the overload region suitably expands and its radius can be found from the equation

$$-q\psi_2(e) = \phi(t_e') - \phi(t_0), \qquad (6.60)$$

where $t_e' = t_0 + t_1' \cdot f(e)$.

Given t_0, the condition $\sigma_\varphi = \sigma_{st}$ makes it possible to establish the relationship

$$e = e(q). \qquad (6.61)$$

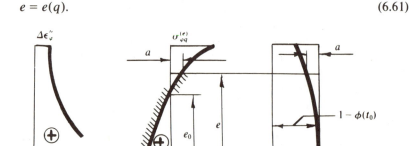

Fig. 6.15. Determination of overload region caused by thermal stresses (for temperature-dependent yield point).

6.5. Shakedown analysis of rotating discs

Here $e \geq e_0$, where e_0 can be derived from the condition

$$\psi_2(e_0) = 0 \tag{6.62}$$

corresponding to the temperature-independent yield point.

The constraint (6.61) is rather weak, thus a sufficiently small change in e has little influence on the conditions for incremental collapse. Therefore the boundary of the overload region can be found from (6.62) to within an acceptable accuracy.

We can obtain the global incremental collapse condition from (6.54) by making simple rearrangements and bearing in mind that the exhaustion of the load-carrying capacity in the overloading ($a \leq r \leq e$) and in the unloading ($e \leq r \leq b$) regions corresponds to different thermal conditions:

$$\frac{\gamma \omega_*^2}{g\sigma_s^0}(I + \lambda F_b) + q_* \int_a^e \psi_2(r) h \, dr = \int_a^e \sigma_{st1} h \, dr + \int_e^b \sigma_{st2} h \, dr. \tag{6.63}$$

Here $\sigma_{st1}, \sigma_{st2}$ stand for the yield points, according to (6.44) and corresponding to the thermal conditions (6.58) and (6.59) respectively.

Making use of the identity

$$\int_a^e \psi_2(r) h \, dr = -\int_a^b \psi_2(r) h \, dr = F_e \psi_1(e), \tag{6.64}$$

which follows from the equilibrium of thermal stresses, we may write the equation (6.45) in the form

$$\frac{p_*}{p_0} + D(q_*)\frac{q_*}{q^0} = 1 - \frac{1}{F}\left[\int_a^e \phi(t')h \, dr + F_e^* \phi(t_0)\right], \tag{6.65}$$

where

$$p_0 = \frac{\gamma \omega_0^2 b^2}{g\sigma_s^0}, \quad q^0 = \frac{2}{|\psi_2(b)|}, \tag{6.66}$$

$$D(q^*) = \frac{F_e}{F} q^0 \psi_1(e), \quad \text{see (6.61),}$$

$$F_e^* = \int_e^b h \, dr, \quad F_e = h_e e, \quad h_e = h(e), \tag{6.67}$$

6 Carrying capacity of a turbine disc

$\omega_0 = \omega_{01}$ is obtained from formula (6.2) at $\phi(t) \equiv 0$, and the factor q^0 is associated with the limit intensity of thermal fluctuations in a disc at rest in which the alternating flow takes place at the periphery.

On insisting that $e = e_0$, the equation (6.65) becomes linear in p_* and q_*, since the coefficient D no longer depends on q_*,

$$D = \frac{F_{e0}}{F} q^0 \psi_1(e_0). \tag{6.68}$$

One would expect a partial incremental mechanism to form in a disc with rather large radial stresses. Let us first, however, consider the mechanism described by the equation (6.54). At $c < e$ the situation will differ from the one discussed before only by the fact that at the end of heating (6.58), apart from the condition $\sigma_\varphi = \sigma_{st}$ at $c \leq r \leq e$, the condition

$$\sigma_{rc} = \sigma_{stc} \tag{6.69}$$

will additionally be satisfied with

$$\sigma_{stc} = 1 - \phi(t'_c), \quad t'_c = t_0 + t'_1 \cdot f(c).$$

The equation (6.60) remains valid and the suitable limiting condition can be derived from the equation

$$\frac{\gamma \omega_*^2}{g \sigma_s^0}(I^* + \lambda F_b) + q_* \int_a^e \psi_2(r) h \, dr = \sigma_{stc} F_c + \int_c^e \sigma_{st1} h \, dr + \int_e^b \sigma_{st2} h \, dr. \tag{6.70}$$

Recalling that the ultimate velocity ω_{02} of partial collapse at $\phi(t) \equiv 0$ is determined by (6.7), we obtain after routine rearrangements that

$$C^*(c) \frac{p_*}{p_0} + D^*(c, q_*) \frac{q_*}{q^0} =$$
$$= 1 - \frac{1}{F_c^* - F_c} \left[\int_c^e \phi(t') h \, dr + F_c \phi(t'_c) + F_c^* \phi(t_0) \right], \tag{6.71}$$

where

$$C^*(c) = \frac{p_0}{p_{02}} = \left(\frac{\omega_{01}}{\omega_{02}}\right)^2, \quad D^*(c, q_*) = \frac{F_e}{F_c^* + F_c} q^0 \psi_1(e). \tag{6.72}$$

6.5. Shakedown analysis of rotating discs

Similarly as before, putting $e = e_0$, we get

$$D^*(c) = \frac{F_{e0}}{F_c^* + F_c} q^0 \psi_1(e). \tag{6.73}$$

At $c \geq e$ in the limiting cycle preceding the partial incremental collapse, Fig. 6.16, it is condition (6.69) that is satisfied at the end of heating (I), whereas condition $\sigma_\varphi = \sigma_{st}$ holds good on the segment $c \leq r \leq b$ at the equalized temperature during cooling (II). The limiting relationship can be obtained from (6.71) on substituting there $e = c$ as well as into (6.72). The coefficient D^* in (6.72) will not depend here on q_*.

In the case of partial collapse at variable repeated flow under the limit condition (6.56), the maximum of the expression $\phi(r, \tau) + q(\tau)\psi_2(r)$ under the assumption of (6.58), (6.59) and $c \geq e$, will be equal to $\phi(t_0)$. The maximum is reached at $q(\tau) = 0$ (since $\psi_2(\rho) < 0$), which describes the instant of equalized temperature. Under these conditions the rate of plastic hoop strain is positive in the region $c \leq r \leq b$ and the maximum of the second expression appearing on the left-hand side of the equation (6.56), i.e.

$$\phi(c, \tau) + q(\tau)[\psi_1(c) - \psi_2(c)],$$

is attained at the maximum temperature gradient when $q(\tau) = q_*$ and $\phi(c, \tau) = \phi(t_c')$, because $\psi_1(c) - \psi_2(c) > 0$. At the radius $r = c$ we have $\sigma_r - \sigma_\varphi = \sigma_{st}$, and according to the associated flow law, the rate of the plastic hoop strain is negative while the rate of the plastic radial strain is positive. The condition (6.56) for the limiting cycle assumes now the

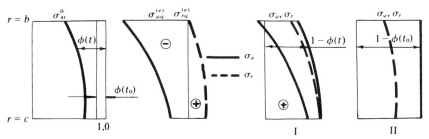

Fig. 6.16. Stress distributions at two stages of the limiting cycle resulting in partial incremental collapse; case $c > e$.

6 Carrying capacity of a turbine disc

form

$$C^*(c)\frac{p_*}{p_0} + D^{**}(c)\frac{q_*}{q^0} = 1 - \phi(t_0) + \frac{F_c}{F_c^* + F_c}[1 - \phi(t'_c)], \tag{6.74}$$

where

$$D^{**}(c) = \frac{F_c}{F_c^* + F_c} q^0[\psi_1(c) - \psi_2(c)]. \tag{6.75}$$

Aiming at full description of the shakedown domain for the considered disc, we should also find the limiting values of load factors under the conditions of alternating flow. A suitable limiting condition is associated, as shown in Chapter 2, with the degeneration into a line or a point of the fictitious yield surface constructed for unsafe radius of the disc. A more detailed analysis has shown that at arbitrary loading program the stress state generated by the variable actions, i.e. revolutions and temperature, is such that the alternating flow condition takes usually the form

$$p_* \varphi_i(r) + q_* |\psi_i(r)| = \sigma_{st1} + \sigma_{st2}, \tag{6.76}$$

where i is equal to either 1 or 2, depending on which sum (of hoop or radial stresses) appears to be larger. The temperature conditions at the two stages relevant to the determination of the yield points σ_{st1} and σ_{st2} are here accounted for.

By simply rearranging condition (6.76) and using the temperature conditions (6.58), (6.59) together with the notation (6.66), we obtain

$$A(r)\frac{p_*}{p_0} + B(r)\frac{q_*}{q^0} = 1 - \frac{1}{2}[\phi(t') + \phi(t_0)], \tag{6.77}$$

where

$$A(r) = \frac{1}{2} p_0 \varphi_i(r), \quad B(r) = \frac{1}{2} q^0 |\psi_i(r)|. \tag{6.78}$$

In the plane of dimensionless coordinates $p_*/p_0, q_*/q^0$ the equation (6.77), at t_0 kept constant, describes a family of straight lines parametrized by r. The segments of those lines lying nearer to the

6.6. Peculiarities of incremental collapse analysis

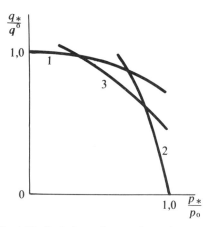

Fig. 6.17. Shakedown diagram for turbine disc.

origin of the coordinates, or their envelopes, form one of the limiting curves, i.e. the boundaries of the shakedown domain, Fig. 6.17, curve 1. The second curve, labelled 2, can be constructed by using the lines represented by the equations (6.65) and (6.71); this curve will correspond to incremental collapse, partial or global. Finally, when a special mechanism of partial incremental collapse is formed, involving the variable repeated flow at the radius separating the two regions of the disc, equation (6.74) leads to the curve 3 that can be critical for certain values of the load parameters.

The domain of Fig. 6.17 bounded by the curves 1, 2, 3 and containing the origin of coordinates indicates to what values of the load factors p_*, q_* the disc will shake down.

6.6. Peculiarities of incremental collapse analysis of rotor discs of radial-flow turbine

The working wheels of radially axial turbines and compressors can be either symmetric with respect to their middle planes, e.g. with blades on both sides of the disc, or non-symmetric, e.g., with blades on one side, Fig. 6.18. In the former case and also for a symmetric loading pattern the collapse mechanism is clearly that of rupture, of no-bending type both under single (Sec. 6.2) and repeated loading.

It can be assumed that roughly the same collapse mechanism is

6 Carrying capacity of a turbine disc

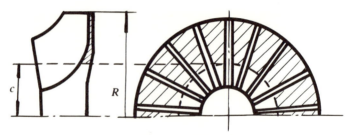

Fig. 6.18. Working wheel of the radial-axial-flow turbine.

formed in the latter case due to the centrifugal forces. Let us remember that the actual disc has usually a much thicker central part forming a hub; a partial collapse is therefore most probable. Indeed, experiments show that the peripheral part of the disc does often tear away from the central part. However, this is not to say that the bending stresses generated by the lack of symmetry, the configuration of blades or radial drop in the temperature are all irrelevant; their influence in a similar situation was considered in Sec. 5.6 where a beam was discussed.

The determination of elastic stresses in a rotating hot disc with blades on one side is rather difficult. The roughest approximating idealization for computation purposes consists in neglecting both the bending action and the rigidity of blades; the latter are replaced by distributed mass applied to the middle surface of the disc. Since the disc profile is assumed to be symmetric, the stress analysis in the disc with side blades is reduced to that for the symmetric disc with density of its material so adjusted as to account for the mass of the blades [6.23].

Other computational procedures have also been developed accounting for the rigidity of blades [6.24, 6.25] and the bending [6.25-6.27 and others]. The problem is usually reduced to that of an orthotropic plate or shell, the blades being assumed to be uniformly smeared out on the face of the disc. As the number of blades decreases, each having larger mass, the above model ceases to be adequate and it becomes natural to treat blades as discrete lumps. Formerly, the problem thus stated was tackled by using a trigonometric series but subsequently the rapid development of the finite element technique has provided a very efficient and powerful tool to deal with the situation.

The incremental collapse condition for a working wheel of a radial-flow turbine will now be established on the assumption that the 'elastic' stresses in the disc and the blades are both determined as certain

6.6. Peculiarities of incremental collapse analysis

functions of current radius with due allowance for bending. We assume the incremental collapse mechanism

$$\Delta u = \text{const.} > 0 \quad \text{for} \quad c \leq r \leq R, \quad \Delta u = 0 \quad \text{for} \quad r < c$$

associated with the accumulation, per each cycle, of the radial displacements Δu in the peripheral part of the disc as shaded in Fig. 6.18. No off-plane deflections are admitted in this no-bending mechanism of partial collapse. The components of strain rates in the peripheral part, except on the radius $r = c$, are described by the formulae (6.23); in all cross-sections of the blades, except on the cylindrical surface $r = c$, the plastic strain rates are assumed to vanish. The fact that $\Delta\epsilon_r > 0$ at $r = c$ both on the disc and the blades results in a jump on that radius of the displacement rates.

The above assumptions and the equation (5.2) for arbitrary loading program lead to the following equation describing the limiting cycle situation under the conditions of incremental collapse:

$$\int_c^R dr \int_{-h}^h \min_\tau [\sigma_{st}(\tau, z) - \sigma_{\varphi p\tau}^{(e)}(z) - \sigma_{\varphi q\tau}^{(e)}(z)] \, dz +$$

$$+ c \int_{-h_c}^{h_c} \min_\tau [\sigma_{st}(\tau, c) - \sigma_{rp\tau}^{(e)}(c) - \sigma_{rq\tau}^{(e)}(c)] \, dz +$$

$$+ \frac{\nu}{2\pi} \int_{F_b(c)} \min_\tau [\sigma_{st}(\tau, c) - \sigma_{p\tau}^{(e)}(c) - \sigma_{q\tau}^{(e)}(c)] \, dz = 0, \qquad (6.79)$$

where $2h = 2h(r)$ is the thickness of the disc, $h_c = h(c)$, $F_b(c)$ is the area of one blade cut out by the cylindrical surface $r = c$, ν denotes the number of blades, and the subscripts p and q denote the effects of rotation and temperature, respectively. The yield point is treated as a temperature-dependent quantity and depends, as the stresses do, on the coordinates r, z of the disc or the blade and on the current time τ,

$$\sigma_{st} = \sigma_{st}(\tau, r, z), \quad \sigma_{\varphi p\tau}^{(e)} = \sigma_{\varphi p\tau}^{(e)}(\tau, r, z), \text{ etc.}$$

The relationship (6.79) has been formulated for the case in which the thermal stresses are reasonably low and the flow regime, corresponding to Fig. 6.13a takes place. When the hoop thermal stresses in compression are of considerable magnitude for the radius $r = c$, different flow regimes can set in corresponding to the corner of the fictitious

6 Carrying capacity of a turbine disc

yield locus as shown in Fig. 6.13b (see also Fig. 6.14). The suitable incremental collapse condition differs from that given by (6.79) as much as the equation (6.56) differs from (6.55).

In order to establish the condition for the alternating plastic flow in a wheel of the radial-flow turbine, we need only to make the local analysis of 'elastic' stress variations at each point.

6.7. Using the shakedown diagram to evaluate the load-carrying capacity of structures under variable repeated loading

In general the shakedown surface bounds the space of the values of the loading factors of the disc at which the plastic deformation process will not continue with every subsequent cycle and thus adequate strength is secured depending on the properties of material used. However, to ensure strength in real situations it is necessary to adopt some definite coefficients of safety to compensate for the inaccuracies both in the initial data and in the computational methods and procedures. Therefore we face immediately the problem of finding a relationship between the given working cycle and the limiting cycle.

In what follows the temperature field is assumed to be expressed by the simple function (6.43). Thus at the fixed level of general heating temperature, determining the yield point, the shakedown domain can be diagrammatically shown in the plane of dimensionless coordinates p_*/p_0, q_*/q^0, Fig. 6.19. The curves 1 and 2 correspond to the basic ultimate situations, the alternating plastic flow and the incremental collapse. For clarity, we shall ignore more complex mechanisms in which for the whole set of values of the load parameters there can occur an unsafe state as a combination of strain accumulation in a certain portion of the disc and alternating flow at a certain radius. The safety margins for the disc as regards the two described states, possibility of the low-cycle fatigue and the appearance of one-sign strains or excessive displacements, will be chosen by comparing the working cycle, defined by the coordinates of the working point, with a certain limiting cycle.

Let us focus our attention on an analogy between the shakedown diagram of Fig. 6.19 and the known diagram of limiting stress amplitudes due to fatigue; the analogy is most complete when the shakedown domain is constructed for thermal cycling at constant angular velocity. Under these circumstances we can use certain notions and

6.7. Load-carrying capacity of structures

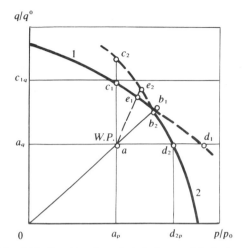

Fig. 6.19. Consideration of shakedown safety factors.

procedures from the analysis of life of members made of ductile materials and transmitting variable stresses.

The first problem encountered in determining the strength margins is to find the limiting cycle corresponding to the considered working cycle. Let, for instance, the extreme angular velocity and the temperature interval change in such a manner that the working point, denoted by W.T. in Fig. 6.19, moves in the direction ae_1e_2. The strength margins can here be understood as ratios between the relevant coordinates of the limiting points e_1, e_2 and of the working point.[1]

However, the actual direction in which the working point moves during the process is usually unknown. It therefore appears useful to consider three characteristic loading paths: ab_2b_1 indicating the similarity between the working and the limiting cycles, ac_1c_2 which corresponds to the constant maximum angular velocity, and ad_2d_1 corresponding to the constant maximum intervals of temperature. Provided no special data have to be accounted for, it is clearly sufficient to determine the two safety margins with respect to the

1) alternating flow under $p_* = \text{const.}$,

$$n_q^{(a)} = \overline{0c_{1q}}/\overline{0a_q}, \tag{6.80}$$

[1] The limiting curves 1, 2 are here continued beyond the point of intersection for conceptual reasons. The complete shakedown diagram should be constructed similarly as for a beam in Sec. 5.6. This problem will be discussed in more detail in Chapter 11.

6 Carrying capacity of a turbine disc

2) incremental collapse under $q_* = \text{const.}$,

$$n_p^{(i)} = \overline{0d}_{2p}/\overline{0a}_p. \tag{6.81}$$

The first safety margin assures against the local loss of strength, the second one secures against the exhaustion of global load-carrying capacity of the disc subjected to thermal cycling and repeated variations in angular velocity.

The factors n_p and n_q by which the angular velocity or the temperature must be multiplied to bring about the corresponding limit state can be introduced directly in the construction of the shakedown diagram, i.e. into equations (6.65), (6.71), (6.77) or analogous ones. This circumstance makes it possible to estimate the probability of occurrence of the limit states in relation to the peculiarities of turbine behaviour. It must be remembered that one more parameter enters the picture, namely the uniform heating temperature t_0 which affects the mechanical properties of the material in general, and the yield point in particular.

It is proper to stress that the variety of consequences that can arise as a result of each of the basic types of cyclic plastic deformation, not to mention the effects of certain factors ignored in the shakedown analysis, call for a thorough study of the issue of safety factors separately for the alternating flow (6.80) and the incremental collapse (6.81).

In the turbine disc situation it is important for how long the loads are applied. Creep has already been shown to make the shakedown domain shrink. The approximate analysis was made in Sec. 6.4 for a disc with constant thickness; no difficulties arise in generalizing the solution to a disc with a more complex profile. However, the problem as a whole requires further and deeper studies, especially concerning the effects due to interaction of the short-term and the long-term inelastic deformation processes.

In practical design [6.3] the following method of evaluating the disc strength has found broad applications. It consists in the determination of two safety factors, one with respect to local statical strength (allowable stress method) and the other with respect to the limit analysis (6.10), or to the burst angular velocity (6.11). This is done by using formulae of the type (6.4), (6.6). It is of interest to compare this practical approach with that based on the shakedown theory. In such a comparison the fundamental mechanical properties of the material should be assumed the same. Depending on the working conditions and the adopted failure criterion, the relevant property can be one of the following: the yield

6.7. Load-carrying capacity of structures

point, the limit creep stress, the short-term strength or the long-term strength. In the framework of limit analysis, i.e. the instantaneous or incremental collapse, the problem consists in a proper choice of the way in which the short-term or the long-term stress–strain relationships are to be approximated.

Starting from such premises, the long-term strength safety factor with respect to the limit equilibrium (6.10) would have, as its counterpart, the incremental collapse safety factor (6.81); the local strength safety factor, as a ratio of the long-term strength to the maximum current stress in the disc, would correspond to the alternating flow safety factor (6.80).

This analogy does not insist on exact coincidence. The incremental collapse safety factor incorporates the possibility of repeated load applications and reflects a drop in the load-carrying capacity of the disc as a result of thermal stress cycling. As shown before, this decrease can be substantial and different for different structures since the intensity of thermo-elastic stresses depends on a number of factors, especially on the cooling system of the disc and on its profile.

The thermal stresses are known to exert no influence on the conditions of instantaneous collapse and they do not enter the procedure for evaluating the respective safety factor. This is why a drop in the temperature of any portion of the disc always results in an increase of this safety margin, as readily seen in (6.4), (6.6), (6.10). It has been experimentally confirmed that an extensive cooling of the central part of the disc in some cases can result in a decrease in its load-carrying capacity and finally leads to an increase in its outer diameter [6.12]. Such facts fail to fit in the framework of the limit analysis theory, whereas they are well incorporated in the shakedown theory, reflecting on the magnitude of the incremental collapse safety margin, as defined by (6.81).

Returning to the problem of local strength safety factors, it is necessary to mention that the alternating flow safety margin has a clearer mechanical meaning than the former one because a real danger of failure due to low-cycle fatigue is here considered. When evaluating the factor, the maximum variation in stresses is determined at the points of the disc during the cycle, unlike in the previously adopted procedure in which only the absolute maximum enters the calculations. Thus, for instance, the 'reversal' temperature gradients can be accounted for leading to the reversal of sign of the total stresses in the rim. The collapse in the presence of reverse temperature intervals was discussed in [6.13].

6 Carrying capacity of a turbine disc

Finally, a methodical advantage is associated with evaluating both the safety margins with respect to the global and the local strength on the basis of a single shakedown diagram constructed for a particular disc. On the other hand, it must not be forgotten that the shakedown analysis, although reflecting the basic quantitative effects rather well, is an approximation and so are the estimates of the load-carrying capacity. Thus the magnitudes of safety factors should be standardized, taking account of the performance and failures of working structures as well as the results of suitable tests.

6.8. Analysis of some test results and adoption of shakedown safety factors

The above considered types of failure of the turbine discs have been encountered both under working conditions of actual structures and in specially designed tests. The occurrence of thermo-fatigue cracks under unsteady conditions was described, for instance, in [6.12, 6.13]. Partial 'instantaneous' collapse was frequently observed on the engine test benches [6.3, 6.16, 6.28 and others]. Global and partial incremental collapse was registered on repeated starting.

A survey of tests on the collapse of discs in both the stationary and nonstationary working regimes is given in the monograph [6.16]. In particular, strain accumulation is attributed to repeated starts of a test gas turbine. Although the qualitative discussion of results offered by the authors is not directly based on the shakedown theory, it fully corresponds to the incremental collapse behaviour as shown in Fig. 6.8.

Let us review some tests on repeatedly started discs, reported in [6.29, 6.30], in the light of the predictions on the performance given by the shakedown analyses.

Subject to tests were plane and conical discs made of pearlitic and austenitic steel. Some test results on plane discs with central holes, $k = a/b = 0{,}156$, are shown in Table 6.1 in which i_c stands for the number of cycles prior to collapse, i_d denotes the number of cycles prior to the deformation measurements taken at the inner $(r = a)$ and at the outer $(r = b)$ contours of the disc and shown in the neighbouring columns.

Discs N-1, N-8, M-4 were subject to a series of repeated cycles of temperature and angular velocity, whereas for discs N-7, N-9, M-5 each cycle was separated by one hour of steady working regime. The program

6.8. Analysis of some test results

Table 6.1
The results after [6.29, 6.30]

Disc	Steel	i_c	$\dfrac{\Delta a}{a}\%$	$\dfrac{\Delta b}{b}\%$	i_d
N-1	aust.	20	5,2	0,5	15
N-7	aust.	7	–	–	–
N-8	aust.	90	5,6	0,6	83
N-9	aust.	45	3,8	0,4	14
M-4	pearl.	150	0,92	0,1	115
M-5	pearl.	43	7,1	1,8	25

of thermal ($\Delta t = t_b - t_a$) and velocity variations is roughly shown in Fig. 6.20. The shakedown analysis should be suitably based on this prescribed loading program.

Let the thermal régime correspond to the equation (6.43) and let the temperature field (function $t_1(\tau)$) be expressed by the quadratic parabola. The total stresses generated by the temperature field and the rotations of a plane disc with central hole can be computed from (6.21) at $i = 2$ and

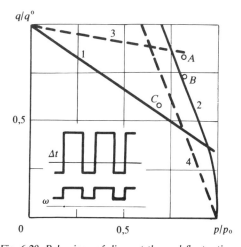

Fig. 6.20. Behaviour of discs at thermal fluctuations.

6 Carrying capacity of a turbine disc

from (6.33). The total elastic stresses are found to be:

$$\sigma_r^{(e)} = \frac{3\sigma_s}{3-\kappa} \cdot \frac{1}{1+k+k^2} \cdot \frac{p}{p_0}\left(1-\frac{k^2}{\rho^2}\right)(l-\rho^2) +$$

$$+ \frac{\sigma_s}{1-k^2} \cdot \frac{t_2}{t_2^0}(1-\rho^2)\left(1-\frac{k^2}{\rho^2}\right),$$

$$\sigma_\varphi^{(e)} = \frac{3\sigma_s}{3-\kappa} \cdot \frac{1}{1+k+k^2} \cdot \frac{p}{p_0}\left(1+k^2-\kappa\rho^2+\frac{k^2}{\rho^2}\right) +$$

$$+ \frac{\sigma_s}{1-k^2} \cdot \frac{t_2}{t_2^0}\left(1+k^2-3\rho^2+\frac{k^2}{\rho^2}\right), \quad (6.82)$$

where

$$\frac{p}{p_0} = \left(\frac{\omega}{\omega_0}\right)^2, \quad \omega_0^2 = \frac{3g\sigma_s}{\gamma b^2} \cdot \frac{1}{1+k+k^2}, \quad t_2^0 = \frac{4\sigma_s}{\alpha E} \cdot \frac{1}{1-k^2}. \quad (6.83)$$

Additional 'elastic' stresses imposed in the course of a semi-cycle can be determined from (6.82) on replacing the parameters p, t_2 by their appropriate maximum increments per cycle.

The boundary of the overload region $\rho = \gamma$ [1] associated with global collapse (6.22) is obtained from, see Fig. 6.21,

$$\frac{3}{3-\kappa} \cdot \frac{1}{1+k+k^2} \cdot \frac{\Delta p_*}{p_0}\left(1+k^2-\kappa\gamma^2+\frac{k^2}{\gamma^2}\right) +$$

$$+ \frac{1}{1-k^2} \cdot \frac{\Delta t_2^*}{t_2^0}\left(1+k^2-3\gamma^2+\frac{k^2}{\gamma^2}\right) = 0. \quad (6.84)$$

All the tests were performed for

$$p_1/p_2 = (\omega_1/\omega_2)^2 \approx (10000/16000)^2 = 0{,}39;$$

hence

$$\Delta p_* = p_2 - p_1 = p_2\left(1-\frac{p_1}{p_2}\right) = 0{,}61 p_2 = 0{,}61 p_*. \quad (6.85)$$

---

[1] This γ should not be confused with the one in the second expression (6.83) where it stands for density.

6.8. Analysis of some test results

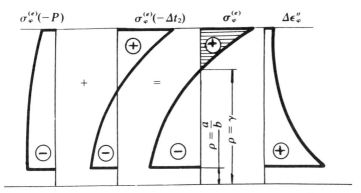

Fig. 6.21. Overload region at the programmed loading of a disc.

Further, we recall that

$$\frac{\Delta t_2}{t_2^0} = \frac{\alpha E}{4\sigma_s}(1-k^2)\Delta t_2 = \frac{\alpha E}{4\sigma_s}\Delta t_{ba}, \qquad (6.86)$$

where Δt_{ba} is the variation in the temperature interval. The radius $r = \gamma$ can be found from (6.84) and (6.85) as a certain function of the ratio

$$m = \frac{p_*}{p_0} : \frac{\Delta t_2^*}{t_2^0}. \qquad (6.87)$$

To establish the condition for incremental collapse we use the equation of the overload method, see Sec. 5.5,

$$I\frac{\gamma\omega_2^2}{g} + \frac{F}{1-k}\int_\gamma^1 \sigma_\varphi^* \, d\rho = F\sigma_s, \qquad (6.88)$$

where F, I designate the area of the disc profile and its second moment, respectively.

Identifying σ_φ^* in the overload region with the maximum additional stresses per semi-cycle (Fig. 6.21) expressed by the formulae (6.82) at Δp and Δt_{ba} as supplied by (6.85) and (6.86), respectively, we obtain the sought for condition in the form

$$C(m)\frac{p_*}{p_0} + D(m)\frac{\Delta t_2^*}{t_2^0} = 1, \qquad (6.89)$$

6 Carrying capacity of a turbine disc

where

$$C(m) = 1 + \frac{1,83}{3-\kappa} \cdot \frac{1}{1-k^3} \int_\gamma^1 \left(1 + k^2 - \kappa\rho^2 + \frac{k^2}{\rho^2}\right) d\rho, \tag{6.90}$$

$$D(m) = \frac{1}{(1-k)(1-k^2)} \int_\gamma^1 \left(1 + k^2 - 3\rho^2 + \frac{k^2}{\rho^2}\right) d\rho. \tag{6.91}$$

The above calculations are valid for the constant ratio $\Delta p/p_* = 0{,}61$ obtained in (6.85).

The boundary of the shakedown domain (Fig. 6.20, line 2) corresponding to the condition (6.89) can be constructed as follows: Selecting consecutive values of the radius $\rho = \gamma$ from the range $\gamma_0 \leq \gamma \leq 1$, where γ_0 follows from (6.84) at $\Delta p_* = 0$, one finds first the magnitudes of the ratio m. Then the coefficients of the equation (6.89) are computed from (6.90), (6.91). The required point of the boundary will be found by intersecting the ray (6.87) starting from the origin of coordinates with the line (6.89).

For the sake of comparison the line 4 is also shown in Fig. 6.20, constructed for the situation in which thermal fluctuations take place at a constant angular velocity ($\Delta p = 0$), i.e. in less favourable loading conditions than those created in the tests.

The condition for alternating plastic flow can be obtained from (2.91) and (6.82). It takes the form

$$A\frac{p_*}{p_0} + \frac{\Delta t_2^*}{t_2^0} = 1. \tag{6.92}$$

The critical points are those on the radius $\rho = k$. We get

$$A = \frac{0{,}92}{3-\kappa} \cdot \frac{2 + k^2(1-\kappa)}{1 + k + k^2}.$$

At $k = 0{,}156$ and $\kappa = 0{,}6$ the above formula yields $A = 0{,}65$.

Condition (6.92) is represented in Fig. 6.20 by line 1. Line 3 corresponds to the alternating flow at the outer contour of the disc; it is constructed under the assumption that the changes in temperature and angular velocity are not synchronous.

Both conditions (6.89) and (6.92) have been established specially for a material with a temperature-independent yield point in order to obtain

6.8. Analysis of some test results

one single shakedown diagram for all the tested discs. As the discs are made of two different grades of steel, variation in the yield point can roughly be incorporated by suitable adjustment of the working point coordinates. At the same time, the effect of reverse temperature intervals on cooling can be approximated by simply adding them to those on heating. Thus Δt_{ba} in the expression (6.86) should be understood as a sum of 'direct' and 'reverse' intervals.

Let us now determine the coordinates of the working points in the plane of the shakedown diagram, associated with the actual condition under which the discs of Table 6.1 were revolving.

The discs N-1, N-7 (austenitic steel) had the following data: $\sigma_s = 5700$ kG/cm² (the yield point was almost constant within the temperature interval 20–700°C), $\alpha E = 28,5$ kG/cm² · °C (mean value for the given temperature interval), $\gamma = 8,2 \times 10^{-3}$ kG/cm³, $b = 22,5$ cm, $\Delta t_{ba} = 570° - (-105°) = 675°$C.

The working point coordinates were

$$\frac{p_*}{p_0} = \left(\frac{\omega_2}{\omega_0}\right)^2 = 0{,}81, \quad \frac{\Delta t_2^*}{t_2^0} = \frac{\alpha E}{4\sigma_s} \Delta t_{ba} = 0{,}84.$$

This point is labelled A in Fig. 6.20.

The discs N-8, N-9 (austenitic steel) had data differing from the above only in that the total temperature gradient was

$$\Delta t_{ba} = 480 - (-100) = 580°\text{C}.$$

The coordinates were

$$\frac{p_*}{p_0} = 0{,}81, \quad \frac{\Delta t_2^*}{t_2^0} = 0{,}72,$$

indicating the location of point B.

The discs M-4, M-5 were made of pearlitic steel.

The yield point of this steel substantially changes at elevated temperatures. In connection with the alternating flow at the contour of the central hole where temperature varies from 120° to 350°C per cycle, we adopted the mean value of the yield point, equal for this interval to $\sigma_s = 6200$ kG/cm². The dimensions of the discs were the same as of the previous ones. Maximum frequency of revolutions was $n_2 = 15750$ rev/min (for the other discs $n_2 = 16000$ rev/min). Given those data,

6 Carrying capacity of a turbine disc

we obtain

$$\frac{p_*}{p_0} = 0{,}69, \quad \frac{\Delta t_2^*}{t_2^0} = 0{,}56.$$

These are the coordinates of point C.

Comparison of the locations of working points on the shakedown diagram, Fig. 6.20, with the test results, Table 6.1, leads to certain definite conclusions. The discs were collapsing due to thermal fatigue; the first flaw started from the edge of the central hole and its propagation resulted in bursting the disc. Only in the discs N-8, N-9 the first flaws opened at the periphery where small apertures were present (the stress concentration was not taken into account in the calculation). In all the tests strain accumulation was recorded to have led to the expansion of discs.

The tests have also shown a considerable influence of creep on the strength of discs under given conditions. For the discs N-9 and M-5 whose programs included steady periods of one hour duration, the number of cycles prior to collapse was found to be considerably lower than that for discs having no hold-time periods within the loading program. Those plateaus turned out to affect mainly the discs made of pearlitic steel which is a material with less pronounced relaxation resistance. This was the reason for not only the lesser number of revolutions prior to collapse but also for the appearance of strain accumulation.

6.9. Tests and analysis of the rotor disc of a radial-axial-flow turbine

The radial-axial-flow turbines of diesel engine jet-superchargers work in highly unsteady-state conditions. This is why in servicing them one finds that thermo-fatigue cracks are apt to form in the thinner parts of the working wheels. The strain accumulation is also occasionally encountered leading to an increase in the outer radius of the wheel accompanied by warping of its rim.

Some test results will now be surveyed including the temperature effects in a wheel working under non-stationary conditions, suitable stress analysis and tests on the strength of real discs [6.31, 6.32].

The temperature measurements were made on a special bench whose

6.9. Tests and analysis of the rotor disc

wheel was either at rest or, for checking purposes, was rotating with a service frequency reaching about 45000 rev/min.

The temperature distributions at various instants of heating are shown in Fig. 6.22, and of cooling in Fig. 6.23. The temperatures on the blade side (1) of the disc are shown by solid lines, on the opposite side (2), by dashed lines. The wheel was heated by a flow of gas whose temperature T_{∂} is indicated in the figure captions, and it was cooled by air of temperature T_a between 20° and 30°. To bring the test conditions as close as possible to the service conditions (especially in winter time when cooling of the exposed parts can be quite abrupt) the gas duct was provided with special water jets. The corresponding results are shown in Fig. 6.24. Considerable reverse temperature gradients can here be noticed along the radius. A special check was made to ensure that the temperature régimes in tests were sufficiently similar to those in service.

In order to analyse the shakedown conditions, the thermo-elastic stresses generated in the wheel at various instants of starting and cooling were determined by means of temperature measurements. Both the rigidity of blades and the disc bending were taken into account [6.27]. The obtained thermoelastic stress distributions are shown in Figs. 6.25 and 6.26 for the heating and the cooling stages, respectively. The stresses induced by rotations are shown in Fig. 6.27. In all three figures the number 1 indicates stress diagrams on the blade side, the number 2 on the opposite smooth side of the wheel. The bending action was found to be of importance only near the hub; this is particularly true for stresses generated by centrifugal forces.

To evaluate the limiting cycle parameters we have assumed that the yield points σ_{st} in the working and in the limiting cycles are both equal and that the corresponding 'elastic' stresses from centrifugal forces and non-uniform heating differ from the working cycle stresses by the multipliers n_p and n_q respectively. On prescribing one of the multipliers and using equation (6.79), we can find the other multiplier as a certain function of the radius c corresponding to the peripheral part of the disc undergoing partial incremental collapse. In Fig. 6.28 the function of the incremental collapse safety margin $n_q = n_q(c)$ is shown at $n_p = 1,0$, curve 1. Two branches of curve 2 correspond to the condition for the alternating plastic flow. In accordance with the kinematical theorem the lowest ordinates of curves shown give the best upper bounds on the limiting cycle parameters.

The calculated shakedown diagram is shown in Fig. 6.29. Line 1 corresponds to the best upper bound assessment of partial incremental collapse according to (6.79) and fits the radii: $c = 24$ mm, $b = 55$ mm. Line

6 Carrying capacity of a turbine disc

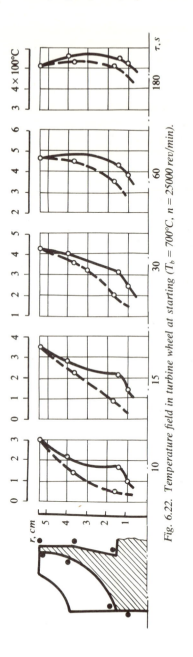

Fig. 6.22. Temperature field in turbine wheel at starting ($T_b = 700°C$, $n = 25000\ rev/min$).

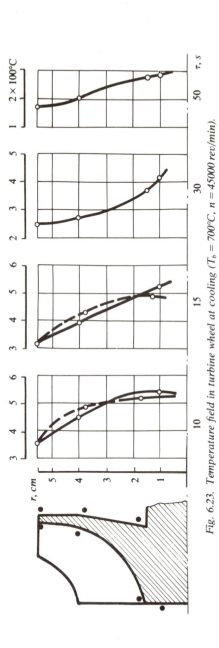

Fig. 6.23. Temperature field in turbine wheel at cooling ($T_b = 700°C$, $n = 45000\ rev/min$).

6.9. Tests and analysis of the rotor disc

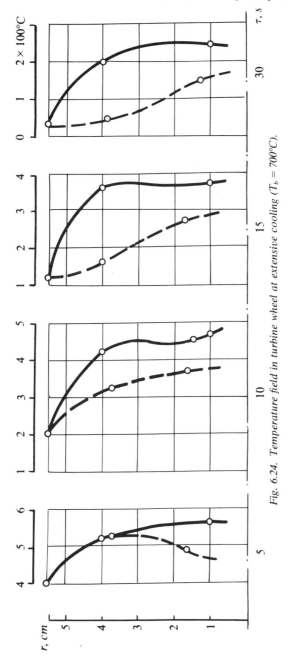

Fig. 6.24. Temperature field in turbine wheel at extensive cooling ($T_b = 700°C$).

6 Carrying capacity of a turbine disc

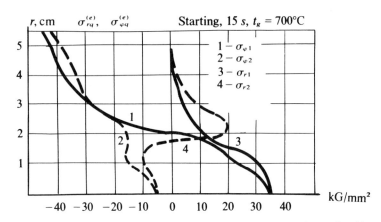

Fig. 6.25. Instantaneous distribution of thermal stresses on main surfaces of turbine wheel (heating).

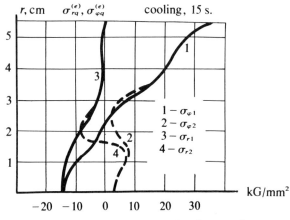

Fig. 6.26. Instantaneous distribution of thermal stresses on main surfaces of turbine wheel (cooling).

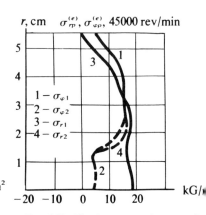

Fig. 6.27. Elastic stresses in a revolving wheel.

6.9. Tests and analysis of the rotor disc

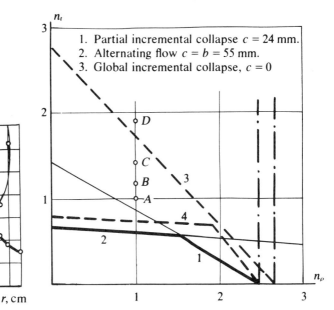

Fig. 6.28. Consideration of shakedown safety factors.

Fig. 6.29. Shakedown diagram of working wheel of a radial-axial-flow turbine.

2 visualizes the alternating flow situation, the best assessment being for $c = b = 55$ mm. Global incremental collapse ($c = 0$) is depicted by line 3; the estimates of the safety factors n_q and n_p are here substantially worse. As a background, the dashed line 4 in Fig. 6.29 represents results obtained by using elastic stresses which were calculated without taking account of bending action or the rigidity of blades.

The points A, B, C, D correspond to those working régimes at which the temperature fields were investigated, corresponding to the temperatures 500, 600, 700°C and, for the point D, to 700°C followed by rapid cooling.

The results show that at repeated starts and sudden changes of working conditions, the alternating plastic deformation should occur in turbine wheels. There are also good reasons for maintaining that, under severe conditions such as a high temperature of the gas and extensive cooling, the combined effect can take place consisting of alternating flow and strain accumulation.[1] According to calculations, partial in-

---

[1] As pointed out earlier (see the footnote on p. 265) we have formally assumed that the two limiting conditions are independent.

6 Carrying capacity of a turbine disc

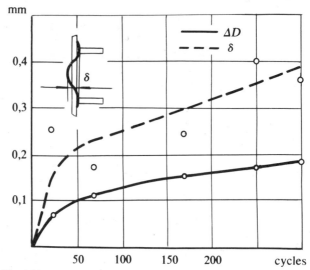

Fig. 6.30. Strain accumulation in turbine wheel at repeated starting.

cremental collapse should take place here in the form of a gradual expansion of the thin part of the disc.

The discussed results are found to be in good agreement, at least qualitatively, with the data gathered in field tests formed on the same bench as used to measure the temperature distributions. The applied cycle consisted of starting accompanied by rapid increase of temperature, a 30 min long period of operating at steady regime, cutting out of the combustion chamber, and stopping the wheel with its simultaneous cooling by the air blow. The cycle lasted about 45 min.

In the course of experiments the turbine was periodically dismantled to be inspected for flaws and to be supplied with a wheel of different dimensions. The results are shown diagrammatically in Fig. 6.30. Repeated starts were found to lead to a gradual increase in the diameter of the wheel as indicated by the solid line. Progressive warping of the rim[1] was also encountered; it is shown by a dashed line. The test conditions for the first 300 cycles correspond to the point C, for the following 30 – to the point D. After 334 cycles a crack appeared at the periphery and began to penetrate through the whole thickness of the disc.

[1] Buckling at thermal variations will be dealt with in Chapter 10.

References

[6.1] Kovalenko, A.D., *Plates and shells in turbines rotors*, in Russian, AN UkSSR, Kiev, 1955.
[6.2] Ponomarev, S.D., et al., *Strength analysis in machinery*, in Russian, Vol. 3, Mashgiz, Moscow, 1959.
[6.3] *Thermal strength of machine parts*, in Russian, ed. I.A. Birger, B.F. Shorr, Mashinostroyenye, Moscow, 1975.
[6.4] Chubb, E.J., An investigation into the non-elastic behaviour of turbine rotor discs at high temperatures. *J. Strain. Analysis*, Vol. 7, Nr 2, 1972.
[6.5] Amada, Sh., Mashida, A., *Elasto-plastic stress analysis of rotating discs*. Papers of Research Institute, Tokyo, 1973.
[6.6] Robinson, E., Bursting tests of steam-turbine disc wheels. *Trans. ASME*, 66, Nr 5, 1944.
[6.7] Nadai, A., Donnel, L., Stress distribution in rotating discs of ductile material after the yield point has been reached. *Trans. of the ASME*, 51, Part 1, 1929.
[6.8] Sokolovsky, W.W., *Plasticity theory*, in Russian, Gostekhnizdat, Moscow, 1950.
[6.9] Gokhfeld, D.A., On strength analysis of turbine discs, in Russian, *Proceed. of Moscow Aircraft Institute*, 17, 1952.
[6.10] Rabotnov, Yu.N., *Creep problems in structural members*. North-Holland, Amsterdam, 1969.
[6.11] Rabotnov, Yu.N., Rabinovitsh, V.P., On creep strength of discs, in Russian, *Izvestia AN SSSR, Mekh. i Mashinostrojenie*, Nr 4, 1959.
[6.12] Moltshanov, E.I., Plotkin, E.R., Temperatures and stresses in the disc TWD IT-700 at starting, in Russian, in *Thermal stresses in structural members*, Vol. 6, Naukova Dumka, Kiev, 1966.
[6.13] Potemkina, A.M., Shwarzman, P.I., Muslin, E.S., On the collapse of turbine discs at reverse temperature gradient, in Russian, in *Thermal stresses in structural members*, 1 AN SSSR, Kiev, 1961.
[6.14] Kachanov, L.M., *Foundations of the theory of plasticity*. North-Holland, Amsterdam, 1971.
[6.15] Gokhfeld, D.A., *On limit analysis of rotating discs*, in Russian, Mashinostrojenie, Moscow, Nr 5, 1965.
[6.16] Rabinovitsh, V.P., *Strength of turbine discs*, in Russian, Mashinostrojenie, Moscow, 1966.
[6.17] Timoshenko, S.P., *Strength of materials*. Part II Advance theory and problems. van Nostrand, New York, 1956.
[6.18] Gokhfeld, D.A., Shakedown of multi-parameter systems at nonuniform heating, in Russian, in *Thermal stresses in structural members*, Vol. 4, Naukova Dumka, Kiev, 1964.
[6.19] Rosenblum, W.I., On shakedown theory of elastic-plastic bodies, in Russian, *Izvestia AN SSSR*, OTN, Nr 6, 1958.
[6.20] Kostuk, A.G., Temperature and thermal stresses in cooled gas turbine discs at unsteady thermal regimes, in Russian, *Izvestia AN SSSR, Mekhanika i Mashinostrojenie*, Nr 4, 1962.
[6.21] Sadakov, O.S., Stress and strain analysis of structural members at cyclic nonisothermal loading based on the structural model of a medium, in Russian, *Proc.*

6 Carrying capacity of a turbine disc

All-Union Symposium on low-cycle fatigue at elevated temperatures, Vol. 3, Cheliabinsk, 1974.

[6.22] Gokhfeld, D.A., Sadakov, O.S., A mathematical model of medium for analysing the inelastic deforming of structures subjected to repeated actions of load and temperature. *Trans. of the 3rd Intern. Conf. on Struct. Mech. in React. Technol.*, Vol. 5, L 5/7, London, 1975.

[6.23] Birger, I.A., Shorr, B.F., Shneiderovitsh, R.M., *Strength analysis of machine parts*, in Russian, Mashinostrojenie, Moscow, 1966.

[6.24] Traupel, W., *Thermische Turbomaschinen*. Zweiter Band: Regelverhalten, Festigkeit und Dynamische Probleme. Springer, 1960.

[6.25] Birger, I.A., *Plates and shells of revolution*, in Russian, Oborongiz, Moscow, 1961.

[6.26] Demianushko, I.V., Strength analysis of working wheels of centrifugal compressor, in Russian, in *Strength and Dynamics of aircraft engines*, Vol. 5, Mashinostrojenie, 1969.

[6.27] Zavarceva, N.A., *Strength analysis of radial-flow turbine discs*, in Russian, Trudy NAMI, Nr 55, 1963.

[6.28] Kozlov, I.A. Bazhenov, W.G., Leschenko, W.M., Investigations on stress state and load-carrying capacity of gas turbine discs, in Russian, in *Thermal strength of materials and structural members*, AN SSSR, Naukova Dumka, Kiev, 1965.

[6.29] Dinerman, A.P., Influence of starts on the service of turbine discs, in Russian, in *Experimental studies on strength of turbine discs*, Trudy CNIITMASh, Nr 12, 1960.

[6.30] Dinerman, A.P., On influence of rapid starting on the serviceability of turbine discs, in Russian, in *Thermal stresses in turbine parts*, Nr 2, AN SSSR, Kiev, 1962.

[6.31] Kononov, K.M., Strength of centripetal gas turbines at repeated starting, in Russian, in *Thermal stress of materials and structural members*, AN SSSR, Naukova Dumka, Kiev, 1965.

[6.32] Gokhfeld, D.A., Kononov, K.M., Temperatures and thermal stresses in the working wheel of centripetal gas turbine at unsteady thermal regimes, in Russian, in *Thermal stresses in structural members*, Nr 5, Naukova Dumka, Kiev, 1965.

7

Shakedown of plates

In many branches of technology it is important to investigate the behaviour of plates and shells (the latter will be dealt with in Chapter 8) beyond the elastic limits as well as their load-carrying capacity. Vast literature has been accumulated on the limit analysis of plates and shells alone. Let us mention that the fundamental statical and kinematical theorems were first formulated in the Soviet Union as early as in the thirties, Gvozdiev [7.1]. A number of problems in the limit analysis of plates was later solved by Sokolovsky [7.2], Ilyushin [7.3], Fejnberg [7.4] and others. An important contribution to the theory of elastic-plastic shells and plates was made, among others, in the papers of Rabotnov [7.7], Shapiro, Rosenblum, Erkhov, Hodge, Sawczuk and Onat. Valuable surveys are given by Hodge [7.8] and Shapiro [7.9].

Investigations of the behaviour of plates and shells under repeated loading are also of great importance. The first solutions to non-trivial shakedown problems of plates were arrived at a decade ago [7.10–7.12] with the help of approximate kinematical and statical methods; in particular, the reformulated Koiter theorem (see Chapter 2) was then employed. Later on, the circular plates were tackled under multi-parameter loading with the use of similar methods [7.13].

The simplex method of linear programming was first applied in [7.14] in combination with the statical theorem to analyse the incremental deformation conditions in circular plates. Soon after this, methods were worked out based on the restated kinematical theorem [7.15]. Pontragin's maximum principle found its application in [7.16] where circular plates were studied.

It must be emphasized that the formal mathematical tools such as linear programming and optimal control theory often turn out to be practically usable only when generalized variables are suitably introduced. In [7.14–7.16] those variables were used to employ the procedure described in Chapter 3 which was meaningful only in the incremental collapse situation. The analysis of alternating plastic flow consists in

7. Shakedown of plates

studying stresses at each point of the body and no special mathematical means are required. Another approach to the use of generalized variables was taken by König [7.17]; some solutions are presented in [7.18].

To apply the shakedown theory in practical design it is necessary to develop suitable computational procedures for the plotting of arbitrary contours. This has been done for instance in [7.19], where the incremental deformations of perforated plates exerted by thermal and mechanical cycling were studied.. However, the most powerful tool for dealing with complex situations is provided by the finite element method. Its application to the shakedown problems received due attention in [7.20]. Two more papers will be quoted here, both devoted to some special problems in the shakedown theory of plates. The first one [7.21] deals with the shakedown conditions for plates with stochastically distributed mechanical properties; the second one [7.22] is devoted to the estimation of the deflections accumulating as a plate shakes down.

In the present chapter we shall be concerned with the most typical shakedown situations encountered in practical design and in the service of structures subject to mechanical and thermal repeated actions. Some problems will be considered to illustrate the corresponding mathematical methods.

7.1. Alternating mechanical loading of a circular plate

Strain accumulation under repeated mechanical loading has so far been illustrated by means of a limited number of examples, mainly on the bar systems. We shall analyse structures of various types by means of necessary conditions for incremental collapse derived from (2.84).

Consider a circular simply supported plate, Fig. 7.1, loaded in turn by uniform pressure $0 \leq p(\tau) \leq p_*$ at $M(\tau) \equiv 0$ and applied bending moments $0 \leq M(\tau) \leq M_*$ at $p(\tau) \equiv 0$. Making use of a suitable limit equilibrium analysis [7.6], we shall adopt for the kinematically admissible displacement increment field the expression[1]

$$\Delta w = \Delta w_0(1-\rho), \quad \left(\rho = \frac{r}{R}\right) \tag{7.1}$$

as describing the admissible plastic deflections per cycle.

[1] For brevity in what follows, some indices will be suppressed.

7.1. Alternating mechanical loading of a circular plate

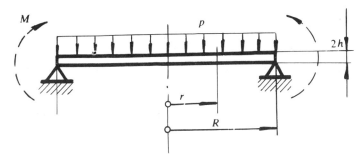

Fig. 7.1. Variable repeated loading of circular plate.

Recalling the known relationships [7.23]

$$\kappa_r = \frac{1}{R^2} \cdot \frac{d^2 w}{d\rho^2}, \quad \kappa_\varphi = -\frac{1}{R^2} \cdot \frac{dw}{\rho \, d\rho}, \qquad (7.2)$$

we get the increments of plastic principal curvatures in the form

$$\Delta \kappa_r = 0, \quad \Delta \kappa_\varphi = \frac{\Delta w_0}{R^2 \rho}. \qquad (7.3)$$

Using generalized variables (3.37)–(3.40) and (7.3) we may write equation (2.88) in the form

$$\int_0^1 M_{\varphi *}^0 \Delta \kappa_\varphi \rho \, d\rho = \int_0^1 \min_\tau [(M_0 - M_{\varphi\tau}^{(e)}) \Delta \kappa_\varphi] \rho \, d\rho =$$
$$= \int_0^1 M_0 \Delta \kappa_\varphi \rho \, d\rho - \int_0^1 \max_\tau [M_{\varphi\tau}^{(e)} \Delta \kappa_\varphi] \rho \, d\rho =$$
$$= \int_0^1 M_0 \Delta \kappa_\varphi \rho \, d\rho - \int_0^1 M_{\varphi\tau}^* \Delta \kappa_\varphi \rho \, d\rho = 0. \qquad (7.4)$$

Here $M_0 = \sigma_s h^2$, $M_{\varphi\tau}^*$ is the envelope of distribution of circumferential bending moments satisfying the condition (3.39). At non-vanishing plastic curvature increment the inequality

$$M_{\varphi\tau}^* \Delta \kappa_\varphi = \max_\tau [M_{\varphi\tau}^{(e)} \Delta \kappa_\varphi] > 0. \qquad (7.5)$$

holds in the additional loading region.

285

7. Shakedown of plates

In the considered case the determination of bending moments generated by variable loading is a trivial task since a purely flexural collapse mechanism forms, for which stresses at points belonging to the same normal to the middle surface reach the yield surface simultaneously.

Incremental collapse is possible when the ordinates of the enveloping diagram correspond to more than one instant of time, i.e. when the bending moment distribution is not isochronous. Using the known results [7.23]

$$M^{(e)}_{\varphi p} = \frac{pR^2}{16}[3 + \mu - \rho^2(1 + 3\mu)], \qquad (7.6)$$

$$M^{(e)}_{\varphi M} = M, \qquad (7.7)$$

we find the following relation leading to the radius $\rho = \gamma$ at which the bending moment diagrams (7.6) and (7.7) intersect at maximum loads:

$$\gamma^2 = \frac{1}{3}\left(5 - 4\frac{M_*}{M_0} \cdot \frac{p_0}{p_*}\right), \quad \left(p_0 = \frac{6M_0}{R^2}\right). \qquad (7.8)$$

The Poisson ratio is taken to be $\mu = \frac{1}{3}$.

The necessary condition for incremental collapse holds for

$$0 < \gamma < 1. \qquad (7.9)$$

This corresponds to the following relationship between the loads p_* and M_*:

$$\frac{1}{2} < \frac{M_*}{M_0} \cdot \frac{p_0}{p_*} < \frac{5}{4}. \qquad (7.10)$$

The following relationship between the limiting cycle parameters at incremental collapse follows from the equation (7.4):

$$\frac{M_*}{M_0}(1 - \gamma) + \frac{1}{4}\frac{p_*}{p_0}(5 - \gamma^2)\gamma = 1. \qquad (7.11)$$

The obtained results are shown in Fig. 7.2. Strains will mount with each cycle of loading when the maximum values of loads represent a point lying between the curve and the lines $M_*/M_0 = 1$, $p_*/p_0 = 1$. Attention is drawn here to the fact that the values of loads whose alternating

7.1. Alternating mechanical loading of a circular plate

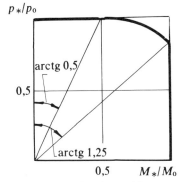

Fig. 7.2. Shakedown diagram for circular plate subject to variable repeated loading.

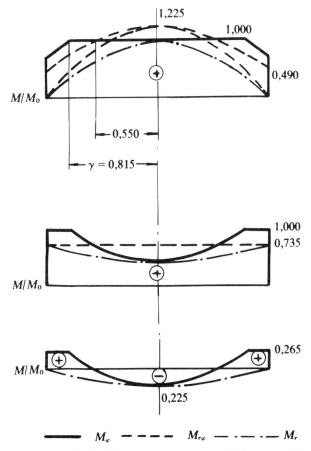

Fig. 7.3. Limiting cycle distribution of moments at variable repeated loading.

7. Shakedown of plates

actions result in strain accumulation, differ but little from their ultimate values in the sense of instantaneous collapse under separate loads. This circumstance, as has been pointed out earlier, is a characteristic feature of structural behaviour at purely mechanical load cycling. However there exist exceptions such as the action of moving loads to be discussed later on.

In Fig. 7.3 are shown the bending moment fields in the limiting cycle at two characteristic stages and the residual moment field corresponding to $M_*/M_0 : p_*/p_0 = \frac{3}{4}$. The moment distributions at all the stages are statically admissible which means that the solution, obtained via the kinematical approach, is complete. A suitable check shows that at $p_* < p_0$ the repeated loading will not create alternating plastic flow (at $M_* < M_0$ the above statement is obvious).

7.2. Shakedown of a simply supported circular plate at thermal cycling

Consider the conditions for progressive flexure of a simply supported plate loaded with uniform pressure $0 \leq p \leq p_*$ at variable temperature field[1]

$$t(\tau, \zeta) = t_0(\tau) + t_1(\tau)\left(\zeta + \frac{1}{2}\zeta^2\right), \tag{7.12}$$

where $0 \leq t_1(\tau) \leq t_*$, $\zeta = z/h$, $(-1 \leq \zeta \leq 1)$. The thermo-elastic stresses in the plate are equal to [7.24]

$$\sigma_r^{(e)} = \sigma_\varphi^{(e)} = q(1 - 3\zeta^2), \tag{7.13}$$

where $q = \alpha E t(\tau)/6(1-\mu)$, $(0 \leq q \leq q_*)$.

For the sake of simplicity, the yield point will be treated in what follows as a temperature-independent quantity, unless stated otherwise. As in the previous example, the admissible collapse mechanism will be the one described by (7.1), (7.3). The plastic strain increments per cycle are

$$\Delta\epsilon_r = -\zeta h \Delta\kappa_r = 0, \quad \Delta\epsilon_\varphi = -\zeta h \Delta\kappa_\varphi = -\zeta h \frac{\Delta w_0}{R^2 \rho}. \tag{7.14}$$

---
[1] In what follows the axis $\zeta(z)$ is directed upwards whereas positive flexure is that of sagging.

7.2. Shakedown of a simply supported circular plate

Equation (2.88) assumes the form

$$\int_0^1 p_* \Delta w \rho \, d\rho = \int_0^1 M_0 \Delta \kappa_\varphi \rho \, d\rho - h \int_{\zeta_\partial} \sigma_\varphi^* \, d\zeta \int_0^1 \Delta \epsilon_\varphi \rho \, d\rho \qquad (7.15)$$

and the additional loading domain ζ_∂ obeys the condition

$$\sigma_{\varphi*}^{(e)} \Delta \epsilon_\varphi = \max_\tau (\sigma_\varphi^{(e)} \Delta \epsilon_\varphi) > 0. \qquad (7.16)$$

The loading program as assumed in (7.15), i.e. thermal cycling at $p = p_* = $ const., appears to be most disadvantageous as far as the shakedown of the plate is concerned. The reason for this is that, at stresses generated by the pressure (the hoop stresses can be found from (7.6)) and the strain increments (7.14), condition (7.16) is satisfied throughout the plate at one single value $p = p_*$, i.e. at one instant of time.

It can be readily seen in Fig. 7.4, plotted with the help of (7.13) and (7.14), that the additional loading region spreads over the layer $-\zeta_0 \leq \zeta \leq 0$ and $\zeta_0 \leq \zeta \leq 1$ ($\zeta_0 = \sqrt{3}/3$). The diagram of σ_φ^* is shown by shading.

On simple rearrangement of (7.15) we can write the incremental collapse condition in the form

$$\frac{p_*}{p_0} + \frac{5}{12} \frac{q_*}{q^0} = 1, \qquad (7.17)$$

where $q^0 = \sigma_s$ denotes the limiting value of q_* at $p = $ const. under the conditions of alternating plasticity.

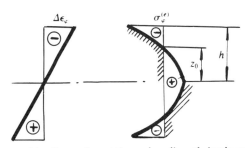

Fig. 7.4. Additional loading region at thermal cycling of simply supported plate.

7. Shakedown of plates

The alternating flow condition at arbitrary loading program can be derived from (2.90), (7.6) and (7.13). We get

$$\frac{p_* R^2}{16} \cdot \frac{10}{3} \cdot \frac{6}{(2h)^2} + 2q_* = 2\sigma_s$$

or, on rearrangement,

$$\frac{15}{16} \cdot \frac{p_*}{p_0} + \frac{q_*}{q^0} = 1. \tag{7.18}$$

The most unsafe points lie at the top and the bottom of the central normal to the middle surface. At constant pressure we get

$$q_*/q^0 = 1. \tag{7.19}$$

The shakedown diagram is shown in Fig. 7.5. At arbitrary loading program the shakedown domain is bounded by lines 1, 3, at $p = p_* =$ const., by lines 2, 3.

The obtained incremental collapse condition (7.17) can also be established by considering the limiting, statically admissible stress fields that correspond to a certain kinematically admissible collapse mechanism (additional loading method, see Sec. 5.5). Then one has to use the equilibrium equation

$$\frac{d(\rho M_r)}{d\rho} - M_\varphi = -\frac{pR^2}{2}\rho^2. \tag{7.20}$$

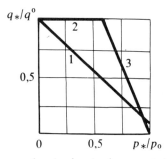

Fig. 7.5. Shakedown diagram for circular simply supported plate at thermal cycling.

7.2. Shakedown of a simply supported circular plate

Recalling the associated flow law and the collapse mechanism (7.3), we conclude that plasticity sets in when

$$M_\varphi > M_r > 0, \quad M_\varphi = M_0. \tag{7.21}$$

Hoop stress fields at characteristic instants of time inside the limiting cycle are given in Fig. 7.6. It can be seen that the moment M_φ will reach the yield point value M_0 if we do not take into account unloading occurring in some layers of the plate on the transition from one stage of loading to another, due to thermo-elastic stresses. Since the thermal stresses (7.13) do not depend on the radius, except for singularities at the contour, the presented field of hoop total stresses is at limiting cycle the same along each normal throughout the plate.

Since the pressure is constant at any stage, we have the equation, see Fig. 7.6,

$$M_\varphi = M_0 + h^2 \left[\int_{-\zeta_0}^{0} \sigma_\varphi^* \zeta \, d\zeta + \int_{\zeta_0}^{1} \sigma_\varphi^* \zeta \, d\zeta \right], \tag{7.22}$$

where σ_φ^* denote the thermo-elastic stresses (7.13) at $q = q_*$.

It is important to note that the second term on the right-hand side of (7.22) represents a generalized stress, namely the bending moment corresponding to the thermal stresses in accordance with the considered collapse mechanism, see Sec. 3.4. Calculations yield

$$M_\varphi = M_0 \left(1 - \frac{5}{12} \cdot \frac{q_*}{q^0} \right). \tag{7.23}$$

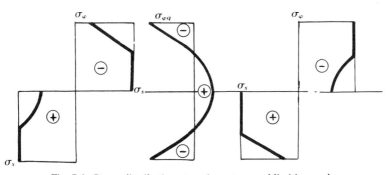

Fig. 7.6. Stress distribution at various stages of limiting cycle.

7. Shakedown of plates

Equation (7.20) enables us now to find the radial moment distribution. We get

$$M_r = M_\varphi - \frac{p_* R^2}{6}\rho^2 + \frac{C}{\rho}. \tag{7.24}$$

Due to the absence of any central cutout the constant C must vanish. Employing the boundary condition

$$(M_r)_{\rho=1} = 0, \tag{7.25}$$

as well as (7.23), we arrive at the known relationship (7.17).

7.3. Clamped circular plate

1. The problem of a clamped plate under conditions similar to those shown in Fig. 7.7 and considered in [7.11] may well serve as an illustration of the kinematical method. The temperature field (7.12) in an unrestrained plate generates the 'elastic' stresses (7.13) and it creates a spherical deflected surface with the radius

$$a = \frac{\alpha t_1(\tau)}{h}. \tag{7.26}$$

Clamping of the plate produces the support bending moments

$$M = Da(1+\mu) = \frac{\alpha t_1(\tau) D(1+\mu)}{h} = \frac{2\alpha E h^2 t_1(\tau)}{3(1-\mu)} = 4qh^2 \tag{7.27}$$

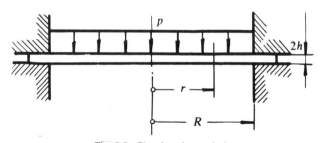

Fig. 7.7. Circular clamped plate.

7.3. Clamped circular plate

resisting flexure. Thus the temperature field (7.12) in a clamped plate generates the 'elastic' stresses

$$\sigma_r^{(e)} = \sigma_\varphi^{(e)} = q(1 - 6\zeta - 3\zeta^2), \tag{7.28}$$

where $0 \leq q \leq q_*$.

As in the previous example, let us employ the results of limit analysis [7.6], assuming the plastic deflections per cycle to be

$$\Delta w = \Delta w_0 \left(1 - \frac{\rho x}{1 + \ln x}\right), \quad \left(0 \leq \rho \leq \frac{1}{x}\right), \tag{7.29}$$

$$\Delta w = -\Delta w_0 \frac{\ln \rho}{1 + \ln x}, \quad \left(\frac{1}{x} \leq \rho \leq 1\right), \tag{7.30}$$

where $x = R/b$ and b designates an unknown radius separating the two regions of different flow regimes.

The curvature increments are, see (7.2),

$$\Delta \kappa_\varphi = \frac{\Delta w_0}{R^2} \cdot \frac{x}{\rho(1 + \ln x)}, \quad \Delta \kappa_r = 0, \quad \left(0 \leq \rho \leq \frac{1}{x}\right), \tag{7.31}$$

$$\Delta \kappa_\varphi = -\Delta \kappa_r = \frac{\Delta w_0}{R^2} \cdot \frac{1}{\rho^2(1 + \ln x)}, \quad \left(\frac{1}{x} \leq \rho \leq 1\right). \tag{7.32}$$

The plastic strain increments per cycle can be found, using expressions of the type (7.14).

Let us first consider the program of thermal cycling at $p = p_* = $ const. The question whether or not this program is most unfavourable in the given situation will be postponed for the moment, to be raised later on.

Using (7.28)–(7.32) and (7.14) we may write condition (2.81) in terms of additional (thermal) stresses in the form

$$\sigma_{ij}^* \Delta \epsilon_{ij0}'' = \sigma_\varphi^* \Delta \epsilon_\varphi > 0, \quad \left(0 \leq \rho \leq \frac{1}{x}\right), \tag{7.33}$$

$$\sigma_{ij}^* \Delta \epsilon_{ij0}'' = \sigma_\varphi \Delta \epsilon_\varphi + \sigma_r \Delta \epsilon_r = 0, \quad \left(\frac{1}{x} \leq \rho \leq 1\right). \tag{7.34}$$

7. Shakedown of plates

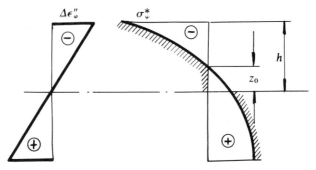

Fig. 7.8. Overload regions at thermal cycling of clamped plate.

The additional loading region in the central part of the plate is formed in the layers corresponding to the shaded branches of stress curves at $\sigma_\varphi^* \neq 0$, Fig. 7.8,

$$\zeta_0 = \frac{z_0}{h} = \frac{2\sqrt{3}-3}{3}.$$

In the outer part of the plate the additional loading region is completely absent because of $\sigma_\varphi = \sigma_r$ while $\Delta\epsilon_\varphi = -\Delta\epsilon_r$.

Equation (2.88) assumes the form

$$\int_0^1 p_* \Delta w \rho \, d\rho = \int_0^1 M_0 \Delta\kappa_\varphi \rho \, d\rho - h \int_{-1}^1 \sigma_\varphi^* \, d\xi \int_0^{1/x} \Delta\epsilon_\varphi \rho \, d\rho + \frac{M_0}{R^2} \left|\frac{d(\Delta w)}{d\rho}\right|_{\rho=1}. \qquad (7.35)$$

The integration range of the first integral on the right-hand side should be split-up according to (7.29) and (7.30). To within an accuracy of a multiplier the last right-hand side term represents the dissipation energy in the annular plastic hinge that forms on the built-in contour, cf. (2.87). In the contour hinge, where displacements suffer a jump, work done by thermo-elastic stresses is vanishing since the rotation angle increment takes place at that instant of time when the above stresses are absent.

After suitable computations and rounding off the condition for incremental collapse of the clamped plate becomes

$$\frac{p_*}{p_0} = \frac{1{,}07 x^2}{3x^2-1}\left(2 + \ln x - \frac{q_*}{q^0}\right), \qquad (7.36)$$

7.3. Clamped circular plate

where $p_0 = 11{,}26(M_0/R^2)$ (see [7.6]) and $q^0 = \frac{1}{4}\sigma_s$; for the mechanical interpretation of these quantities we refer to previous examples.

The parameter minimizing the shakedown load should be found correspondingly, employing the kinematical theorem. We get

$$3x^2 - 2\ln x - 5 = -2\frac{q_*}{q^0}, \qquad (7.37)$$

and thus (7.36) finally assumes the form

$$\frac{p_*}{p_0} = 0{,}535x^2. \qquad (7.38)$$

As the intensity of thermal cycling increases in the interval $0 \leq q_*/q^0 \leq 1$ in which alternating flow is impossible, the central part of the plate under the displacement increments field (7.29) is gradually spreading out to cover the whole plate at $q_*/q_0 = 1$, Fig. 7.9.

In this context the singularity is worth pointing out that at sufficiently large intensity of thermal cycling the loading program with pressure kept constant at its maximum, $p = p_*$, ceases to be the most unfavourable for the shakedown of the plate. This appears to be the case for $1/x > 0{,}827$ when there is formed the region $0{,}827 \leq \rho \leq 1/x$ in which the flow regime (7.31) is associated with the condition $M^{(e)}_{\varphi_p} < 0$ (Fig. 7.10, [7.23]). Hence in this case the maximum of the product $M^{(e)}_{\varphi_p}\Delta\kappa_\varphi$ corresponds to the absence of pressure, $p = 0$ where the index p denotes the bending moment due to the pressure.

In order to establish the incremental collapse condition at $1/x >$

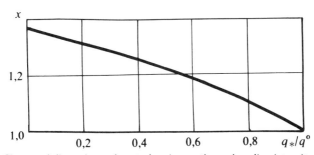

Fig. 7.9. Change of dimensions of central region as thermal cycling intensity increases.

7. Shakedown of plates

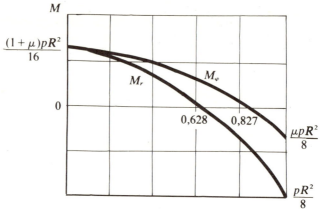

Fig. 7.10. Bending moment distribution in clamped plate under uniform pressure.

0,827, we have to rewrite the basic equation in the form

$$\int_0^1 M_0 \Delta\kappa_\varphi \rho \, d\rho - h \int_\zeta d\zeta \int \sigma_\varphi^* \Delta\epsilon_\varphi \rho \, d\rho + \frac{M_0}{R^2} \left| \frac{d(\Delta w)}{d\rho} \right|_{\rho=1} = 0, \qquad (7.39)$$

where σ_φ^* denotes the enclosing field of total stresses generated by both the loading and the temperature field. It is evident that the integration ranges must be split into three subranges according to the analytical expressions for $\Delta\kappa_\varphi$, $\Delta\epsilon_\varphi$ and σ_φ^*, namely: 0 to 0,827, 0,827 to $1/x$ and $1/x$ to 1. In the second subrange the moments arising from external load are equal to zero.

The above refers to the case in which $q^*/q^0 > 0,5$ and the value of $1/x$, derivable now not from (7.37), but from the other equation obtained on minimization of (7.39), differs relatively little from unity, $0.827 \leq 1/x \leq 1$. This is the actual reason for the shakedown condition to differ only negligibly from the condition (7.35). Calculations show that the ratio p_*/p_0 differs by about 3 per cent. This is a practically negligible error (compare lines 2 and 4 in Fig. 7.11). Thus, irrespective of the sequence of actions, we shall assume that the shakedown domain is bounded by the alternating flow condition instead of the incremental collapse condition. The shakedown domain at $p = p_* = $ const. is bounded by the lines 1, 2, at arbitrary program $0 \leq p \leq p_*$, $0 \leq q \leq q_*$ by line 3, Fig. 7.11.

2. Let us present the solution to the same plate assuming now that

7.3. Clamped circular plate

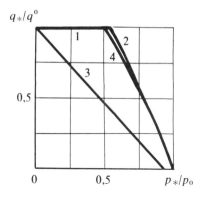

Fig. 7.11. Shakedown diagram for clamped plate at thermal cycling at a parabolic variation in temperature across the thickness.

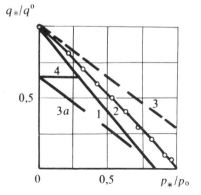

Fig. 7.12. Shakedown diagrams for clamped plate at thermal cycling obtained by using various methods at a linear variation in temperature across the thickness.

the temperature varies linearly across the thickness,

$$t(\tau, \zeta) = t_0(\tau) + \zeta t_1(\tau), \qquad (7.40)$$

where $0 \leq t_1(\tau) \leq t_*$, and not parabolically as in (7.12).

At such a temperature field in the clamped plate there is a full correspondence between the flexural mechanism and the linear distribution of thermo-elastic stresses across the thickness; the latter produce the bending moment [7.23]

$$M_{rr}^{(e)} = M_{\varphi\tau}^{(e)} = \frac{2\alpha E h^2 t_1(\tau)}{3(1-\mu)} = \frac{1}{4} q h^2, \quad (0 \leq q \leq q_*). \qquad (7.41)$$

A solution to this problem was obtained with the help of four methods, each employing generalized variables. These were

a) the statical method in the rigorous formulation [7.14], use being made of a special simplex method executed by a computer,

b) the statical method in the approximate formulation [7.25, 7.26], under the assumption that the residual moments along the radius were distributed according to (7.41), i.e. similarly to the thermal moments, to within a certain multiplier,

c) the kinematical method based on equation (7.35) with the second term on the right-hand side modified so as to be a function of internal forces and the corresponding generalized strains,

7. Shakedown of plates

d) the overload method.

For simplicity, in the applications of the last two methods it was assumed that $p(\tau) = p_* = \text{const.}$ in lieu of an arbitrary loading program because the discrepancies are known to be insignificant.

The obtained limiting cycle parameters are shown in Fig. 7.12. Line 1 corresponds to the approximate statical method and not surprisingly gives an estimation of allowable loading from below. The additional loading and kinematical methods yielded fully coinciding results, line 2; little circles indicate discrete numerical results of the rigorous statical method and they are very close to line 2, thus supplying the exact solution. The moment distributions at the basic stages of loading are shown in Fig. 7.13. Returning to Fig. 7.12, line 3 corresponds to the alternating flow condition $\Delta M = 2M_0$ expressed in terms of internal actions. This line is not critical but its location is valid only for a sandwich plate; in a solid plate the maximum stresses reach the yield point when the bending moment is one-and-a-half times smaller than M_0. Thus line 3 should be translated so as to occupy a new position 3a through the point $q_*/q^0 = \frac{2}{3}$, $p_*/p_0 = 0$. When the loading program incorporates thermal cycling at constant pressure, the alternating flow line 4 is relevant, starting from the above point in the horizontal direction. It is easy to conclude that in this case the drop in load-carrying capacity resulting from thermal cycling in the absence of alternating flow exceeds 60 per cent.

To illustrate the simplex method, as used in the lower bound approach, we have shown in Table 7.1 a general element of the simplex

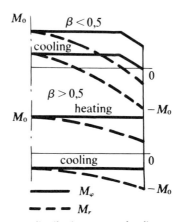

Fig. 7.13. Bending moment distributions at two loading states of limiting cycle.

7.3. Clamped circular plate

table for the interval $\sqrt{\frac{13}{33}} \leq \rho \leq \sqrt{\frac{13}{19}}$ ($\sqrt{\frac{13}{33}} \approx 0{,}628$, $\sqrt{\frac{13}{19}} \approx 0{,}827$) inside which $M^{(e)}_{\varphi p\tau} \geq 0$, $M^{(e)}_{r p\tau} \leq 0$, see Fig. 7.10. Let us recall that at given temperature field we have $M^{(e)}_{rq\tau} = M^{(e)}_{\varphi q\tau} > 0$ throughout the plate; here the subscripts p, q denote moments generated by pressure and temperature, respectively.

At given maximum temperature t_* the Melan theorem enables us to find the largest intensity of pressure p_* at which the time-independent, residual, bending moments M^0_r, M^0_φ, satisfying the equilibrium equation

$$\frac{d}{d\rho}(\rho M^0_r) - M^0_\varphi = 0, \tag{7.42}$$

belong to the domain bounded by the fictitious yield surface. Adopting the Tresca yield condition (2.12) in the form

$$\max(|M_r|, |M_\varphi|, |M_r - M_\varphi|) = M_0$$

and recalling (3.39), we arrive at the following description of the domain of admissible values for M^0_r, M^0_φ:

$$\max[\max_\tau(|M^0_r + M^{(e)}_{\varphi p\tau} + M^{(e)}_{rq\tau}|, |M^0_\varphi + M^{(e)}_{\varphi p\tau} + M^{(e)}_{\varphi q\tau}|,$$
$$|M^0_r - M^0_\varphi + M^{(e)}_{rp\tau} - M^{(e)}_{\varphi p\tau} + M^{(e)}_{rq\tau} - M^{(e)}_{\varphi q\tau}|)] \leq M_0. \tag{7.43}$$

Accounting for the signs of elastic bending moments shown earlier, we can express the moments in the form

$$-M_0 \leq M^0_\varphi \leq M_0 - \max_\tau M^{(e)}_{\varphi p\tau} - \max_\tau M^{(e)}_{\varphi q\tau},$$

$$-M_0 - \min_\tau M^{(e)}_{rp\tau} \leq M^0_r \leq M_0 - \max_\tau M^{(e)}_{rq\tau}, \tag{7.44}$$

$$-M_0 - \min_\tau(M^{(e)}_{rp\tau} - M^{(e)}_{\varphi p\tau}) \leq M^0_r - M^0_\varphi \leq M_0.$$

To simplify the system of constraints, let us solve the equation (7.42) for M^0_r. We get

$$\rho M^0_r = \int_0^\rho M^0_\varphi \, d\rho. \tag{7.45}$$

7. Shakedown of plates

Table 7.1
General element of the simplex table for the plate problem

...	$-x_{i-2}$	$-x_{i-1}$	$-x_i$	$-x_{i+1}$	$-x_{i+2}$	...	$-\bar{p}$	1
.	.					.		
	1						$1-\frac{19}{13}\rho_{i-2}^2$	$2-\bar{q}$
		1					$1-\frac{19}{13}\rho_{i-1}^2$	$2-\bar{q}$
			1				$1-\frac{19}{13}\rho_i^2$	$2-\bar{q}$
				1			$1-\frac{19}{13}\rho_{i+1}^2$	$2-\bar{q}$
					1		$1-\frac{19}{13}\rho_{i+2}^2$	$2-\bar{q}$
.						.		
	$\frac{1}{\rho_{i-1}}d^*_{i-2}$							$2-\bar{q}$
	$\frac{1}{\rho_{i-1}}d_{i-2}$	$\frac{1}{\rho_{i-1}}d^*_{i-1}$						$2-\bar{q}$
	$\frac{1}{\rho_i}d_{i-2}$	$\frac{1}{\rho_i}d_{i-1}$	$\frac{1}{\rho_i}d^*_i$					$2-\bar{q}$
	$\frac{1}{\rho_{i+1}}d_{i-2}$	$\frac{1}{\rho_{i+1}}d_{i-1}$	$\frac{1}{\rho_{i+2}}d_i$	$\frac{1}{\rho_{i+1}}d^*_{i+1}$				$2-\bar{q}$

7.3. Clamped circular plate

Table 7.1 (continued)

$\frac{1}{\rho_{i+2}}d_{i-2}$	$\frac{1}{\rho_{i+2}}d_{i-1}$	$\frac{1}{\rho_{i+2}}d_i$	$\frac{1}{\rho_{i+2}}d_{i+1}$	$\frac{1}{\rho_{i+2}}d_{i+2}$	·	·	$2-\bar{q}$
$\frac{-1}{\rho_{i-2}}d^*_{i-2}$					·	$\frac{33}{13}\rho^2_{i-2}-1$	·
$\frac{-1}{\rho_{i-1}}d_{i-2}$	$\frac{-1}{\rho_{i-1}}d^*_{i-1}$				·	$\frac{33}{13}\rho^2_{i-1}-1$	·
$\frac{-1}{\rho_i}d_{i-2}$	$\frac{-1}{\rho_i}d_{i-1}$	$\frac{-1}{\rho_i}d^*_i$			·	$\frac{33}{13}\rho^2_i-1$	·
$\frac{-1}{\rho_{i+1}}d_{i-2}$	$\frac{-1}{\rho_{i+1}}d_{i-1}$	$\frac{-1}{\rho_{i+1}}d_i$	$\frac{-1}{\rho_{i+1}}d^*_{i+1}$		·	$\frac{33}{13}\rho^2_{i+1}-1$	·
$\frac{-1}{\rho_{i+2}}d_{i-2}$	$\frac{-1}{\rho_{i+2}}d_{i-1}$	$\frac{-1}{\rho_{i+2}}d_i$	$\frac{-1}{\rho_{i+2}}d_{i+1}$	$\frac{-1}{\rho_{i+2}}d^*_{i+2}$	·	$\frac{33}{13}\rho^2_{i+2}-1$	·
·	·	·	·	·	·	·	·
$\frac{1}{\rho_{i-2}}d^*_{i-2}-1$					·	·	1
$\frac{1}{\rho_{i-1}}d_{i-2}$	$\frac{1}{\rho_{i-1}}d^*_{i-1}-1$				·	·	1
$\frac{1}{\rho_i}d_{i-2}$	$\frac{1}{\rho_i}d_{i-1}$	$\frac{1}{\rho_i}d^*_i-1$			·	·	1

7. Shakedown of plates

Table 7.1 (continued)

...	$-x_{i-2}$	$-x_{i-1}$	$-x_i$	$-x_{i+1}$	$-x_{i+2}$	...	$-\bar{p}$	1
	$\frac{1}{\rho_{i+1}}d_{i-2}$	$\frac{1}{\rho_{i+1}}d_{i-1}$	$\frac{1}{\rho_{i+1}}d_i$	$\frac{1}{\rho_{i+1}}d^*_{i+1}-1$				1
	$\frac{1}{\rho_{i+2}}d_{i-2}$	$\frac{1}{\rho_{i+2}}d_{i-1}$	$\frac{1}{\rho_{i+2}}d_i$	$\frac{1}{\rho_{i+2}}d_{i+1}$	$\frac{1}{\rho_{i+2}}d^*_{i+2}-1$			1
.	.	.	.	.	.	.	.	.
	$1-\frac{1}{\rho_{i-2}}d^*_{i-2}$						$\frac{14}{13}\rho^2_{i-2}$	1
		$1-\frac{1}{\rho_{i-1}}d^*_{i-1}$					$\frac{14}{13}\rho^2_{i-1}$	1
			$1-\frac{1}{\rho_i}d^*_i$				$\frac{14}{13}\rho^2_i$	1
				$1-\frac{1}{\rho_{i+1}}d^*_{i+1}$			$\frac{14}{13}\rho^2_{i+1}$	1
					$1-\frac{1}{\rho_{i+2}}d^*_{i+2}$		$\frac{14}{13}\rho^2_{i+2}$	1
.	.	.	.	.	.	.	.	.
z							-1	

7.3. Clamped circular plate

The integration constant is absent since M_r^0 at $\rho = 0$ must be finite. Let us rewrite now the right-hand side of (7.45) in the form of a finite sum, using nodal values of radius $\rho = \rho_i$, $i = 1, 2, \ldots$ and the constraints (7.44) imposed on those discrete points only. Instead of (7.45) we obtain

$$\rho_i M_{ri}^0 = \sum_{j=1}^{i} M_{\varphi j}^0 d_j, \tag{7.46}$$

where

$$d_j = \frac{1}{2}(\Delta\rho_{j-1} + \Delta\rho_j) \quad \text{for } j \neq 1 \text{ and } j \neq i,$$

$$d_1 = \frac{1}{2}\Delta\rho_1 \quad \text{for } j = 1,$$

$$d_i = \frac{1}{2}\Delta\rho_i \quad \text{for } j = i,$$

$$\Delta\rho_j = \rho_j - \rho_{j-1}.$$

The constraints become now

$$-x_i - \bar{p}\left(1 - \frac{19}{13}\rho_i^2\right) + 2 - \bar{q} \geq 0,$$

$$\frac{1}{\rho_i}\sum_{j=1}^{i} x_j d_j + 2 - \bar{q} \geq 0, \tag{7.47}$$

$$\frac{1}{\rho_i}\sum_{j=1}^{i} x_j d_j + \bar{p}\left(1 - \frac{33}{13}\rho_i^2\right) \geq 0,$$

$$x_i - \frac{1}{\rho_i}\sum_{j=1}^{i} x_j d_j + 1 \geq 0,$$

$$\frac{1}{\rho_i}\sum_{j=1}^{i} x_j d_j - x_i - \frac{14}{13}\bar{p}\rho_i^2 + 1 \geq 0,$$

where

$$x = 1 + \frac{M_\varphi^0}{M_0}, \quad \bar{p} = \frac{p_* R^2(1+\mu)}{16\sigma_s h^2}, \quad \bar{q} = \frac{\alpha E t_*}{3(1-\mu)\sigma_s} = \frac{q_*}{8\sigma_s}. \tag{7.48}$$

The above system is presented in Table 7.1 in which asterisk is used to

7. Shakedown of plates

denote d_j for $j = i$, and empty spaces indicate zeros. In computer aided calculations 12 nodes were used spaced non-uniformly in order to better represent the change of sign of bending moments as shown in Fig. 7.10. The dimension of the matrix of constraints was 13×61.

3. Let us return to the plate of Fig. 7.7 and discuss it by employing the kinematical theorem with the use of the simplex method. In accordance with the restated theorem (2.89) the problem of evaluating the limit pressure[1] p, in the sense of strain accumulation at given temperature parameter q can be stated as follows:

$$\min_{\Delta\epsilon_\varphi, \Delta\epsilon_r} p = ? \tag{7.49}$$

under the constraints

$$\int_0^1 p \Delta w \, d\rho \geq \int_0^1 (M^0_{r*} \Delta\kappa_r + M^0_{\varphi*} \Delta\kappa_\varphi) \, d\rho, \tag{7.50}$$

$$\frac{d}{dr}(\Delta w) = -R^2 \rho \Delta\kappa_\varphi, \quad \frac{d}{dr}(\rho \Delta\kappa_\varphi) = -\Delta\kappa_r, \quad \Delta w = 0 \text{ at } \rho = 1$$
$$\tag{7.51}$$

and under the normality condition

$$\Delta\kappa_{\alpha\beta} = \sum_\alpha \mu_\alpha \frac{\partial\phi(M^0_{\alpha\beta*})}{\partial M^0_{\alpha\beta*}}, \quad \mu_\alpha \geq 0. \tag{7.52}$$

Condition (7.52) follows from the associated flow rule (2.76) and it relates the bending moments on the fictitious yield surface M^0_{r*}, $M^0_{\varphi*}$ to the radial and circumferential curvature increments $\Delta\kappa_r$, $\Delta\kappa_\varphi$ per cycle.

As required by the methodology of Chapter 4, we adopt a numerical value for the multiplier to within an accuracy similar to that within which the relationships (7.50)–(7.52) determine the increments of deflections and curvatures, i.e. the collapse mechanism. To this end, let us write the equality

$$\int_0^1 \Delta w \, d\rho = 1. \tag{7.53}$$

---

[1] The pressure is assumed constant over the cycle.

7.3. Clamped circular plate

The curvature increments per cycle can be written as sums of terms corresponding to all the six possible flow regimes, Fig. 7.14:

$$\Delta\kappa_r = \Delta\kappa_r^+ - \Delta\kappa_r^- + \Delta\kappa_{r\varphi}^+ - \Delta\kappa_{r\varphi}^-,$$
$$\Delta\kappa_\varphi = \Delta\kappa_\varphi^+ - \Delta\kappa_\varphi^- - \Delta\kappa_{r\varphi}^+ + \Delta\kappa_{r\varphi}^-,$$
(7.54)

where

$$\Delta\kappa_r^+ \geq 0, \quad \Delta\kappa_r^- \geq 0, \quad \Delta\kappa_{r\varphi}^+ \geq 0,$$
$$\Delta\kappa_{r\varphi}^- \geq 0, \quad \Delta\kappa_\varphi^+ \geq 0, \quad \Delta\kappa_\varphi^- \geq 0.$$
(7.55)

The right-hand side of inequality (7.50) can be rearranged with the help of (7.54) and the internal forces can be shown on the fictitious inter-

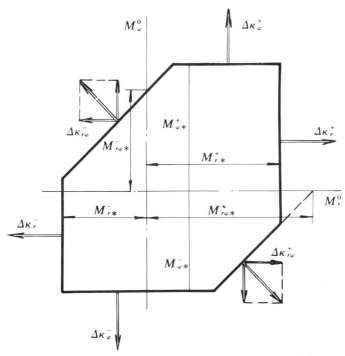

Fig. 7.14. Fictitious interaction surface as used in the calculations of clamped plate.

305

7. Shakedown of plates

action hexagon of Fig. 7.14. Recalling (7.53), we get

$$p \geq \int_0^R (M_r^+ * \Delta\kappa_r^+ + M_r^- * \Delta\kappa_r^- + M_\varphi^+ * \Delta\kappa_\varphi^+ + M_\varphi^- * \Delta\kappa_\varphi^- + \\ + M_{r\varphi}^+ * \Delta\kappa_{r\varphi}^+ + M_{r\varphi}^- * \Delta\kappa_{r\varphi}^-) \, d\rho. \tag{7.56}$$

It is possible to make further rearrangements aiming at eliminating the equations from the calculations to follow. Let us integrate the differential compatibility equations (7.51) making use of the boundary condition. The result is:

$$\rho \Delta\kappa_\varphi = -\int_0^\rho \Delta\kappa_r \, d\rho, \tag{7.57}$$

$$\Delta w = R^2 \int_0^1 d\rho \int_0^\rho \Delta\kappa_r \, d\rho. \tag{7.58}$$

Substituting the obtained function $\Delta w(r)$ into the equality (7.53), we have

$$R^3 \int_0^1 d\rho \int_0^\rho d\rho \int_0^\rho \Delta\kappa_r \, d\rho_2 = 1. \tag{7.59}$$

Let us replace the integrals appearing in (7.57) and (7.58) by finite sums and let us simultaneously assume that the function $\Delta\kappa_r(r)$ is piece-wise linear. Its components, see (7.54), can now be expressed as

$$\Delta\kappa_{ri}^+ = A_i^+ + B_i^+(\rho - \rho_i), \quad \Delta\kappa_{r\varphi i}^+ = C_i^+ + D_i^+(\rho - \rho_i),$$
$$\Delta\kappa_{ri}^- = A_i^- + B_i^-(\rho - \rho_i), \quad \Delta\kappa_{r\varphi i}^- = C_i^- + D_i^-(\rho - \rho_i), \tag{7.60}$$

where $i = 1, 2, \ldots, n$ is the number of the segment between the nodes $i, i+1$; the quantities A_i, C_i are non-negative on account of (7.55); and B_i, D_i are free variables.

Moreover, one of the two curvature components $\Delta\kappa_\varphi^+$, $\Delta\kappa_\varphi^-$ should be expressed in a similar fashion, for instance

$$\Delta\kappa_\varphi^- = E^- + D^-(\rho - \rho_i), \quad (E^- \geq 0) \tag{7.61}$$

because the other component can be determined from (7.57).

Substitution of (7.60), (7.61) into the fundamental inequality (7.56)

7.4. Circular plate with rigid boss

permits us now to arrive at the final form of the linear programming problem; here one of the components of the function $\Delta\kappa_r$ can be eliminated by means of (7.57). The constraints (7.55) are active on the ends of intervals only, i.e. at $r = r_i$.

The computation time needed for the shakedown analysis via linear programming based on the kinematical theorem is practically the same as the one necessary for the statical approach. However, the kinematical approach is found superior when the actual collapse mechanism is selected by using the minimization of the load factor, according to (7.49), with respect to certain parameters describing the considered class of mechanisms. A similar approach to the limit equilibrium problems was used by Rzhanitzin [7.27].

7.4. Circular plate with rigid boss

Consider the conditions for progressive bending of a circular simply supported plate consisting of two parts of different thickness, Fig. 7.15. The plate responds to a constant concentrated force P and a

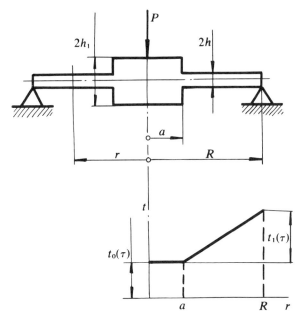

Fig. 7.15. Plate with central boss.

7. Shakedown of plates

variable temperature field

$$t(\tau, \rho) = t_0(\tau), \quad (0 \leq \rho \leq k),$$
$$t(\tau, \rho) = t_0(\tau) + t_1(\tau)\frac{\rho - k}{1 - k}, \quad (k \leq \rho \leq 1), \tag{7.62}$$

where $-t_* \leq t_1(\tau) \leq t_*$, $\rho = r/R$, $k = a/R$. Using the known solution to the constant thickness disc [7.28] and compatibility between the boss and the annular part, we may write the dimensionless thermo-elastic stress distribution in the form

$$\sigma_{rr}^{(e)} = q\varphi(\rho), \quad \sigma_{\varphi\tau}^{(e)} = q\psi(\rho), \tag{7.63}$$

where

$$q = \frac{\alpha E t_1(\tau)}{3(1 - k)}, \quad -q_* \leq q \leq q_*; \tag{7.64}$$

for $0 \leq \rho \leq k$

$$\varphi(\rho) = \psi(\rho) = \left(1 - \frac{3}{2}k + \frac{1}{2}k^3\right)\frac{\phi - \frac{1+\mu}{1-\mu}}{\phi + k^2}, \tag{7.65}$$

for $k \leq \rho \leq 1$

$$\varphi(\rho) = 1 - \rho - \left(\frac{1}{\rho^2} - 1\right)\frac{1}{2}k^2\frac{k(\phi + 3) - 2}{\phi + k^2}, \tag{7.66}$$

$$\psi(\rho) = 1 - 2\rho + \left(1 + \frac{1}{\rho^2}\right)\frac{1}{2}k^2\frac{k(\phi + 3) - 2}{\phi + k^2} \tag{7.67}$$

and

$$\phi = \left(\frac{1+\mu}{1-\mu} + \frac{h}{h_1}\right)\left(1 - \frac{h}{h_1}\right)^{-1}. \tag{7.68}$$

In the derivation of the above formulae the change of thickness at $r = a$ was assumed to be very rapid but continuous. This assumption will be retained in what follows. The thermal stresses stay clearly constant across the thicknesses.

7.4. Circular plate with rigid boss

According to the criterion (2.90) and the relationships (7.63) and (7.64), the alternating flow condition assumes the form

$$\max[q_*|\varphi(\rho)|, q_*|\psi(\rho)|, q_*|\varphi(\rho) - \psi(\rho)|] = 1. \tag{7.69}$$

In particular, for $\mu = \frac{1}{3}$, $k = \frac{1}{4}$ we obtain

$$q_* = q^0 = 1{,}015 \tag{7.70}$$

when $h/h_1 = 1$, and

$$q_*/q^0 = 0{,}965 \tag{7.71}$$

when $h/h_1 = 0$. These cases are represented in the shakedown diagram of Fig. 7.16 by the lines 1, 2 respectively.

Let us proceed to the incremental bending conditions. Our analysis will be based on the statical shakedown theorem and it will use Pontragin's maximum principle, see Sec. 4.4.

On account of (7.63)–(7.68) the generalized thermal stresses, for convenience referred to $M_0 = \sigma_s h_1^2$, can be expressed in the form

$$m_{rq} = \pm q_* \delta \varphi(\rho), \quad m_{\varphi q} = \pm q_* \delta \psi(\rho),$$
$$m_{r\varphi q} = \pm q_* \delta[\varphi(\rho) - \psi(\rho)], \tag{7.72}$$

where $\delta = 1$ for $k \le \rho \le 1$, $\delta = (h_1/h)^2$ for $0 \le \rho \le k$.

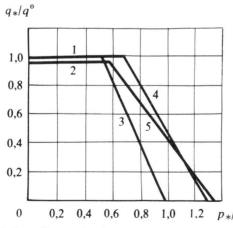

Fig. 7.16. Shakedown diagram for plate with central boss, various stiffness ratios.

7. Shakedown of plates

We seek a distribution of the time-independent moments m_r^0, m_φ^0 considered as sums of residual and 'elastic' bending moments, the latter exerted by the constant force P associated with the maximum of the load factor

$$p = \frac{P}{2\pi M_0}$$

at a given value of the temperature field factor q_*. The distribution should obey the equilibrium condition, both internally,

$$\frac{\mathrm{d}}{\mathrm{d}\rho}(\rho m_r^0) - m_\varphi^0 + p = 0, \tag{7.73}$$

and externally, $m_r^0(1) = 0$, as well as the inequalities insisted upon by the statical theorem. This means, using (7.72), we get

$$|m_r^0| \leq \delta[1 - q_*|\varphi(\rho)|], \quad |m_\varphi^0| \leq \delta[1 - q_*|\psi(\rho)|], \tag{7.74}$$

$$|m_r^0 - m_\varphi^0| \leq \delta[1 - q_*|\varphi(\rho) - \psi(\rho)|]. \tag{7.75}$$

For simplicity, let us confine ourselves to (7.74). This should lead to a certain overestimation of the load factor sought for since the hexagon (7.74), (7.75) is inscribed into the rectangle (7.74).

Let us introduce new variables

$$x \equiv \frac{m_r^0}{\delta[1 - q_*|\varphi(\rho)|]}, \quad u \equiv \frac{m_\varphi^0}{\delta[1 - q_*|\psi(\rho)|]} \tag{7.76}$$

to ensure the independence from the current radius ρ of both the domain of admissible variations in the phase coordinate and the control domain. The function x is taken to be the phase coordinate whereas the function u which possibly may have discontinuities of the first kind, plays the part of the control. On account of the constraints system (7.74) and the condition (7.69) securing the non-negativity of the right-hand sides of the inequalities (7.74), the equilibrium equation (7.73) assumes the form

$$\frac{\mathrm{d}x}{\mathrm{d}\rho} = \frac{1}{\rho\delta[1 - q_*|\varphi(\rho)|]}\left\{u\delta(1 - q_*|\psi(\rho)|) - p - x\frac{\mathrm{d}}{\mathrm{d}\rho}[\rho\delta(1 - q_*|\varphi(\rho)|)]\right\}, \quad |x| \leq 1, |u| \leq 1. \tag{7.77}$$

7.4. Circular plate with rigid boss

Expressing the cost function as the integral

$$I = \int_c^1 \frac{p}{1-c}\,d\rho,$$

where $(c, 1)$ is the longest interval inside which $|x| \leq 1$, we construct the Hamiltonian

$$H = \psi_0 \frac{p}{1-c} + \frac{\psi_1}{\rho\delta(1-q_*|\varphi(\rho)|)}\left\{u\delta(1-q_*|\psi(\rho)|) - \right.$$
$$\left. - p - x\frac{d}{d\rho}[\rho\delta(1-q_*|\varphi(\rho)|)]\right\}. \tag{7.78}$$

Using the maximum principle, we see that this Hamiltonian implies the equality

$$u = -\operatorname{sign} \psi_1. \tag{7.79}$$

The combined system

$$\frac{d\psi_0}{d\rho} = 0, \quad \frac{d\psi_1}{d\rho} = -\frac{\partial H}{\partial x} = \psi_1 \frac{\dfrac{d}{d\rho}[\rho\delta(1-q_*|\varphi(\rho)|)]}{\rho\delta(1-q_*|\varphi(\rho)|)} \tag{7.80}$$

yields

$$\psi_0 = \text{const.}, \quad \psi_1 = C_1\rho\delta(1-q_*|\varphi(\rho)|), \tag{7.81}$$

where C_1 is an integration constant and the signs of neither ψ_0 nor ψ_1 depend on ρ. Bearing in mind that ψ_0 is non-negative and the solution to the combined system should be non-trivial, we may employ the additional condition for the parameter problem [7.29], namely

$$\int_{\rho_i}^{\rho_{i+1}}\left\{\frac{\psi_0}{1-c} - \frac{\psi_1}{\rho\delta(1-q_*|\varphi(\rho)|)}\right\}d\rho = 0, \tag{7.82}$$

and conclude that $\operatorname{sign} \psi_1 = -1$. The expression (7.79) immediately implies that $u = 1$. Using this result in the equilibrium equation and remembering the outer boundary condition, we get

$$\rho\delta(1-q_*|\varphi(\rho)|)x = p(1-\rho) - \int_\rho^1 \delta(1-q_*|\psi(\rho)|)\,d\rho. \tag{7.83}$$

7. Shakedown of plates

It can be readily shown that the function $x(\rho)$ has only two local extrema; at the centre of the plate ($\rho = 0$) and where the change of thickness ($\rho = k$) occurs as approached from the annular part with the thickness $2h$. Consequently we put $c = 0$ or $c = k$ into the obtained expression. If $x(k) \leq x(0)$, the relationship between the shakedown values of p and q_* has the form

$$p = 1 + k(\delta - 1) - q_* \left[\int_k^1 |\psi(\rho)| \, d\rho + \left(\frac{h_1}{h}\right)^2 \left(1 - \frac{3}{2} + \frac{1}{2}k^3\right) \frac{\phi - \frac{1+\mu}{1-\mu}}{\phi + k^2} \right]. \tag{7.84}$$

If $x(k) > x(0)$, we have

$$p = \frac{1}{1-k} \left\{ 1 - q_* \left[\int_k^1 |\psi(\rho)| \, d\rho + k |\varphi(k)| \right] \right\}. \tag{7.85}$$

The ranges of validity of (7.84) and (7.85) can readily be established on substituting (7.84) into (7.83) at $\rho = k$. This yields

$$x(k) = \frac{(1-k)\left(\frac{h_1}{h}\right)^2 \int_0^k (1 - q_*|\psi(\rho)|) \, d\rho - k \int_k^1 (1 - q_*|\psi(\rho)|) \, d\rho}{k(1 - q_*|\varphi(k)|)}. \tag{7.86}$$

Now we can obtain these values of the ratio h_1/h, as functions of q_*, for which $x(k) < 1$. They supply the range of validity of (7.84).

Lines 3, 4, 5 in Fig. 7.16 correspond to the incremental bending condition specified for $k = \frac{1}{4}$, $\mu = \frac{1}{3}$. The straight line 3 is constructed for $h_1/h = 1$ and is totally determined by (7.84). The straight line 5 is plotted for $h_1/h \to \infty$, following (7.85). Line 4 corresponds to $h_1/h = 1{,}5$ and it is slightly kinked at $q = 0{,}4$; the ($q \leq 0{,}4$)-branch is deduced from (7.84), the other one from (7.85).

The above solution is felt to be a good illustration of transition from the global collapse mechanism to the partial one, as depending on the geometry of structure and intensity of thermal cycling.

7.5. Rectangular plates and plates of arbitrary shapes. Upper bounds on the limiting cycle parameters

From the point of view of incremental collapse the exact solutions to the shakedown situations in rectangular plates are rather hard to obtain. Particularly difficult situations arise when statical, either rigorous or approximate methods are to be employed because the partial differential equations have to be analysed under various complex boundary conditions. This also applies to the rigorous kinematical method, however, its approximate variant is found to be more easily applicable as it is in the case of the limit equilibrium situations. In this approximate approach certain a priori assumptions as to the possible collapse mechanisms turn out to be useful, especially if engineering intuition is coupled with some available experimental evidence. Extensive use is made here of the mechanisms comprising concentrated curvatures adjoining to yield lines [7.5, 7.9]. In other words, it is assumed that the incremental collapse of a plate is realised due to rotations in plastic hinges termed also yield lines.

As in other situations, the kinematical approach is very convenient and often makes it possible, after a correct choice of the collapse mechanism, to get results which may be surprisingly close to the exact ones. It suffices to assume the mechanism to within an accuracy of certain parameters to be found later from the load factor minimization.

1. To begin with, we shall consider a rectangular plate. This is a common element in various structures, especially in civil engineering. Let such a plate be clamped along the contour, Fig. 7.17, and subject to a constant uniformly distributed pressure p and cyclic fluctuations in temperature which is a function of current time τ and the dimensionless distance $\zeta = z/h$ from the middle surface.

Two types of temperature field will be considered,[1] namely

(a) $t(\tau, \zeta) = t_0(\tau) + \zeta t_1(\tau)$,

(b) $t(\tau, \zeta) = t_0(\tau) + \frac{1}{2}(1 + \zeta)^2 t_1(\tau)$,

(7.87)

where

$$0 \leq t_1(\tau) \leq t_*, \quad -1 \leq \zeta \leq 1. \tag{7.88}$$

[1] The following solutions pertaining to rectangular and square plates were arrived at by A.R. Belyakov.

7. Shakedown of plates

The associated thermal stresses calculated for a perfectly elastic plate are [7.24]

$$\sigma^{(e)}_{x\tau} = \sigma^{(e)}_{y\tau} = -\frac{\alpha E}{1-\mu}\left[t(\tau,\zeta) - \frac{1}{2}\int_{-1}^{1} t(\tau,\zeta)\,d\zeta\right], \tag{7.89}$$

where the x and y are arbitrary orthogonal directions.

Let us assume the pattern of yield lines forming a collapse mechanism in the shape of an inverted hipped roof, Fig. 7.17. The axes of relative rotations are labelled from 1 to 9. Each yield line has its local coordinate system n, l: the n-axis is directed normal to the particular yield line and the l-axis along it.

The condition for incremental collapse (2.88) assumes the form

$$\int_S p\Delta w\,dS = \sum_{k=1}^{9}\int_{l_k} h^2\,dl \int_{-1}^{1}\sigma^0_{n*}\Delta\theta_k\zeta\,d\zeta, \tag{7.90}$$

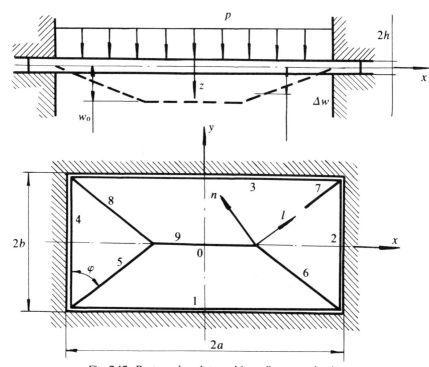

Fig. 7.17. Rectangular plate and its collapse mechanism.

7.5. Rectangular plates and plates of arbitrary shapes

where $S = 4ab$ is the area of the plan; $\Delta\theta_k$ stands for the rotation increment per cycle; σ_{n*}^0 denotes the normal stress belonging to the fictitious yield surface with the normal n and associated via (2.76) with the strain compatible with the displacement increment per cycle $\zeta\Delta\theta_k$; l_k designates the length of the yield hinge; k is the consecutive number of yield line as indicated in Fig. 7.17.

As in similar problems of limit equilibrium [7.5], the deformations in the middle plane of the plate are assumed negligibly small; hence the corresponding work of the stresses σ_{n*}^0 has been ignored in equation (7.90).

We shall assume that the Tresca material is described by (2.12). Suitable shaded fictitious yield polygons are shown in Fig. 7.18 in accordance with (7.89) and the equalities $\sigma_{x\tau}^{(e)} = \sigma_{y\tau}^{(e)} = \sigma_{n\tau}^{(e)} = \sigma_{l\tau}^{(e)}$ which correspond to points where $\sigma_{n\tau}^{(e)} < 0$.

The fictitious yield surface stresses entering the integrand of (7.90) can be found from the equalities

$$\sigma_{n*}^0 = \min(\sigma_s - \sigma_{n\tau}^{(e)}) \quad \text{for } \zeta\Delta\theta_k > 0,$$
$$\sigma_{n*}^0 = \max(-\sigma_s - \sigma_{n\tau}^{(e)}) \quad \text{for } \zeta\Delta\theta_k < 0. \tag{7.91}$$

At linear distribution of temperature (7.87a) formula (7.89) yields

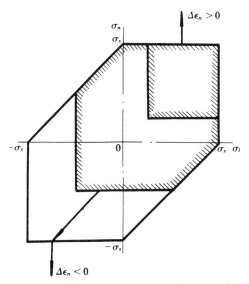

Fig. 7.18. Fictitious yield curves at stresses varying according to (7.89).

7. Shakedown of plates

sign $\sigma_{n\tau}^{(e)} = -\operatorname{sign} \zeta$. Noting that $\Delta\theta_k < 0$ for $k = 1, 2, 3, 4$ and $\Delta\theta_k > 0$ for the rest of the yield lines, we obtain from (7.91) that

$$|\sigma_{n*}^0| = \sigma_s - \frac{\alpha E t_*}{1-\mu}|\zeta|, \quad (k = 1 \div 4),$$
$$|\sigma_{n*}^0| = \sigma_s, \quad (k = 5 \div 9). \tag{7.92}$$

Hence the plastic flow appears to be isochronous within each of the yield lines; however it sets in at different times in different yield lines: in the contour at $t(\tau) = t_*$, and in the interior at $t(\tau) = 0$.

Substituting the stresses (7.92) into the criterion (7.90), we get the incremental collapse condition in the form

$$p\int_S \Delta w \, dS = M_0(1-q_*)\sum_{k=1}^{4} l_k\Delta\theta_k + M_0\sum_{k=5}^{9} l_k\Delta\theta_k, \quad \left(q_* = \frac{2\alpha E t_*}{3(1-\mu)\sigma_s}\right). \tag{7.93}$$

The required yield line lengths l_k, rotation increments $\Delta\theta_k$ and deflection increments Δw can be conveniently expressed in terms of the sides a, b, the angle φ and the maximum deflection increment Δw_0, see Fig. 7.17.

On simple rearrangements equation (7.93) takes the form

$$p = \frac{12M_0}{b^2} \cdot \frac{\lambda + \epsilon^{-1}}{3\lambda - \epsilon}\left(1 - \frac{1}{2}q_*\right), \tag{7.94}$$

where $\lambda = a/b$, $\epsilon = \operatorname{tg} \varphi$, $M_0 = \sigma_s h^2$. The condition $\min p = p_*$ is satisfied with respect to the parameter $\epsilon = \operatorname{tg} \varphi$ governing the yield lines pattern when

$$\epsilon_* = \sqrt{3 + \lambda^{-2}} - \lambda^{-1}. \tag{7.95}$$

In the specific case of a square plate ($\lambda = a/b = 1$) we have

$$\epsilon_* = \operatorname{tg} \varphi_* = 1, \quad \varphi_* = 45°.$$

Using (7.95) we can rewrite the incremental collapse condition (7.94) in the form

$$\frac{p_*}{p_0} + \frac{2}{3}\frac{q_*}{q^0} = 1, \tag{7.96}$$

7.5. Rectangular plates and plates of arbitrary shapes

where

$$p_0 = \frac{12M_0}{b^2} \cdot \frac{\lambda + \epsilon_*^{-1}}{3\lambda - \epsilon_*}, \quad q^0 = \frac{4}{3} \qquad (7.97)$$

are the known limit equilibrium and alternating plasticity quantities. In the considered case of $p = $ const. we have for the condition of alternating flow

$$\frac{q_*}{q^0} = 1. \qquad (7.98)$$

The obtained results are shown in Fig. 7.19 in the form of the shakedown diagram bounded by the lines 1 and 1'.

At a temperature field of the type (7.87b) the distribution (according to (2.81)) of the determining thermo-elastic stresses turns out to be non-isochronous not only for different yield lines but also for their points lying on the same normal to the middle surface of the plate. Similar computations give the equations

$$\frac{p_*}{p_0} + 0{,}501\frac{q_*}{q^0} = 1, \quad \frac{q_*}{q^0} = 1 \qquad (7.99)$$

represented in Fig. 7.19 by lines 2, 2'.

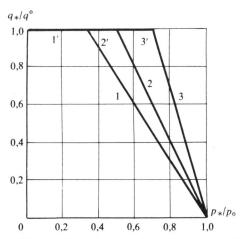

Fig. 7.19. Shakedown diagrams for rectangular plate at temperature fields (7.87), (7.88) and for square plate.

7. Shakedown of plates

2. For simplicity we shall deal with a square plate simply supported along all the edges, Fig. 7.20, and subject to constant uniform pressure and thermal cycling of type (7.87).

The thermo-elastic problem alone was solved in [7.23] by means of a series of trigonometric and hyperbolic functions:

$$\sigma_{xT}^{(e)} = -\alpha E t_1(\tau)\zeta\varphi(x, y),$$

$$\sigma_{yT}^{(e)} = -\alpha E t_1(\tau)\zeta[1 - \varphi(x, y)], \qquad (7.100)$$

$$\tau_{xyT}^{(e)} = -\alpha E t_1(\tau)\zeta\psi(x, y),$$

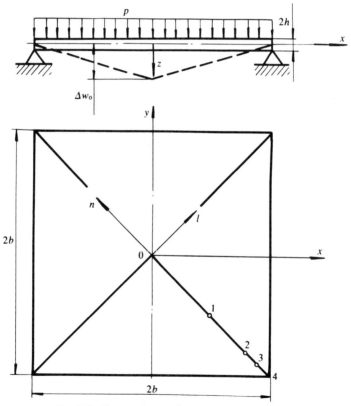

Fig. 7.20. Square plate and its collapse mechanism.

7.5. Rectangular plates and plates of arbitrary shapes

where

$$\varphi(x, y) = \frac{4}{\pi} \sum_{k=1,3,5,\ldots}^{\infty} \left(k \operatorname{ch} \frac{k\pi}{2}\right)^{-1} \sin \frac{k\pi}{2b}(b + x) \operatorname{ch} \frac{k\pi y}{2b},$$

$$\psi(x, y) = \frac{4}{\pi} \sum_{k=1,3,5,\ldots}^{\infty} \left(k \operatorname{ch} \frac{k\pi}{2}\right)^{-1} \cos \frac{k\pi}{2b}(b + x) \operatorname{sh} \frac{k\pi y}{2b}. \qquad (7.101)$$

These stresses are generated under the linear law of temperature variations by hinge supports, which resist distortions of the middle surface along the contour.

A detailed analysis shows that the normal stresses described by (7.100), (7.101) attain their largest value on the contour, clearly in sections perpendicular to the sides, and remain constant at all the points of the contour. The twisting action results in the shearing stresses that are largest at the points situated on the diagonals in sections parallel to the edge planes.

Let us assume that the deflection increments per cycle form the pyramid shown in Fig. 7.20; the yield lines simply coincide with the diagonals of the plate. Similarly as before, each line has its own local coordinate system: The n-axis is directed normal to the given yield line and the l-axis is tangent to it, Fig. 7.20. Condition (2.88) takes the form

$$\int_S p\Delta w \, dS = 4h^2 \int_0^{b\sqrt{2}} dl \int_{-1}^{1} \sigma_{n*}^0 \Delta \theta \zeta \, d\zeta. \qquad (7.102)$$

The thermo-elastic stresses acting in the planes with the normals n, l, which have to be found for determining the fictitious yield surface stresses are given by the formulae

$$\sigma_{l\tau}^{(e)} = -\alpha E \zeta t_1(\tau) \left[\frac{1}{2} + \psi\left(\frac{l}{\sqrt{2}}, \frac{l}{\sqrt{2}}\right)\right],$$

$$\sigma_{n\tau}^{(e)} = -\alpha E \zeta t_1(\tau) \left[\frac{1}{2} - \psi\left(\frac{l}{\sqrt{2}}, \frac{l}{\sqrt{2}}\right)\right], \qquad (7.103)$$

$$\tau_{nl\tau}^{(e)} = 0.$$

The fictitious yield polygons for the diagonal points 0, 1, 2, 3, 4, indicated in Fig. 7.20, are constructed in Fig. 7.21. Nearly all of them are hexagons except for the confined regions close to the corners of the

7. Shakedown of plates

plate where parallelograms appear. The formulae (7.100), (7.101) cease to be precise near the corners. This is why we choose to employ formulae (7.91) in order to determine the limiting values of the time-independent stresses. Thus the right-hand side of (7.102) can be rewritten as

$$\int_0^{b\sqrt{2}} \Delta\theta \min_\tau (M_0 - M_{n\tau}^{(e)}) \, dl, \qquad (7.104)$$

where $M_{n\tau}^{(e)}$ is the bending moment produced by the stresses (7.103); its distribution along the l-axis is shown in Fig. 7.22, and $M_0 = \sigma_s h^2$ is the unit yield moment.

It is clear that the integrand in (7.104) attains its minimum at $t(\tau) = t_*$ for $0 \leq l \leq l_0$ and at $t_1(\tau) = 0$ for $l_0 \leq l \leq b\sqrt{2}$.

The adopted collapse mechanism leads directly to the deflection increments

$$\Delta w = \Delta w_0 \left(1 - \frac{l}{b\sqrt{2}}\right) \qquad (7.105)$$

and the rotation increments in the yield lines

$$\Delta\theta = \frac{w_0 \sqrt{2}}{b}. \qquad (7.106)$$

Using (7.103), (7.105), (7.106), we may write the incremental collapse condition (7.102) in the form

$$\frac{p_*}{p_0} + 0{,}295 \frac{q_*}{q^0} = 1, \qquad (7.107)$$

where $p_0 = 6M_0/b^2$, $q^0 = 1{,}90$, $\mu = 0{,}3$.

Alternating plasticity (2.90) can occur at $p = p_* = \text{const.}$ if

$$\frac{q_*}{q^0} = 1. \qquad (7.108)$$

Critical in the latter sense are the points with the coordinates $z = \pm h$, lying in the middles of edges; the corners are not relevant since, as we pointed out earlier, the solution is not precise enough in the corner zones. The shakedown diagram is shown in Fig. 7.19, lines 3, 3'.

7.5. Rectangular plates and plates of arbitrary shapes

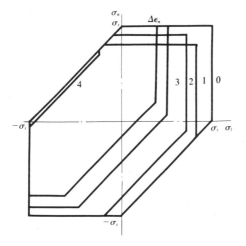

Fig. 7.21. Fictitious yield polygons for certain points of the square plate.

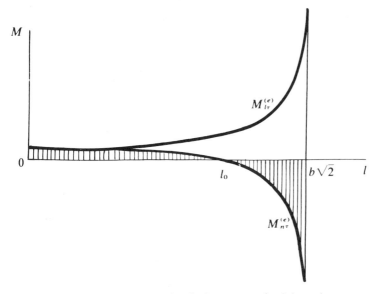

Fig. 7.22. Bending moment distribution as a result of thermal stresses.

7. Shakedown of plates

3. Consider the incremental collapse condition for a clamped plate with arbitrary contour subject to a constant concentrated force P, Fig. 7.23, and thermal cycling of the type (7.87). Uniform heating generates no stresses since the support allows for free in-plane motion of the plate. In the absence of lateral load the temperature field (7.87) causes no flexure of the clamped plate; the thermal stresses produce moments according to (7.41), see Sec. 7.3.

The mechanism of incremental collapse will be assumed to coincide with that of instantaneous collapse [7.5]. Hence a conical surface forms under the directed downward force P. Since the thermal stresses in the bottom layers are compressive, their moment will cause additional loading in the circular hinge ($M_\varphi^* \Delta \psi > 0$) and unloading in the radial hinges. When the force acts upwards or the heat source is 'upside down', the situation in yield lines is reversed; however, the final solution clearly does not depend on the adopted directions of actions.

Using (2.88), we get

$$P\Delta w_0 + 2\pi R M_\varphi \Delta \psi = 4\pi R M_0 \Delta \psi. \tag{7.109}$$

The right-hand side describes the energy increment dissipated per cycle in both the circumferential and radial yield lines.

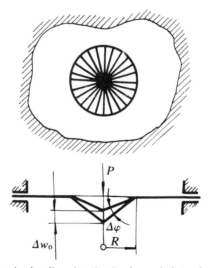

Fig. 7.23. Progressive bending situation in clamped plate of arbitrary contour.

Realising that $\Delta\psi = \Delta w_0/R$, we obtain the incremental collapse condition in the final form

$$\frac{P}{P_0} + \frac{2}{3} \cdot \frac{t_1^*}{t^0} = 1, \qquad (7.110)$$

where $t^0 = 2\sigma_s(1-\mu)/\alpha E$.

Similarly as in the limit equilibrium situation, the radius R within which the whole process takes place remains indefinite. Indeed, equation (7.110) describes merely the incipient collapse since geometry changes that follow as the cycling loading proceeds are not taken into account.

From the equation (7.110) it follows that at $t_1^*/t^0 = 1$, i.e. prior to alternating plastic flow, the load-carrying capacity of the plate drops to one third as a result of thermal cycling.

7.6. Shakedown of perforated plates

Perforated plates are used in many branches of technology. Not infrequently they have to respond to simultaneous action of a temperature field and repeated mechanical loading. A disturbance in the stress field resulting from a regular mesh of cutouts leads to an increase of the stresses and their amplitudes under cycling loading. It is clear that the influence of perforation on the condition for alternating plasticity is entirely determined by the level of stress concentrations.

Less obvious are the effects of multiple stress concentrations on the incremental collapse behaviour. In the limit equilibrium analysis of perfectly plastic perforated structures usually only the change in net cross-sections is accounted for.

In practical design vast applications are being made of approximate elastic methods to analyse perforated plates and shells based on certain equivalent models of continuum with respect to either strength or stiffness [7.30]. The analysis which will follow enables us to assess the validity of such an approach at variable repeated loading.

As in [7.19], consider the shakedown conditions of a rectangular plate with a regular square mesh of equal circular holes, Fig. 7.24. The overall dimensions of the plate are large enough as compared with the hole-to-hole distances,

$$l_1, l_2 \gg 2(t-r). \qquad (7.111)$$

7. Shakedown of plates

1. Let us first confine ourselves to a two-parameter mechanical loading of the type, Fig. 7.24,

$$p(\tau) = p = \text{const.}, \quad -m_* \leq m(\tau) \leq m_*, \tag{7.112}$$

where p, $m(\tau)$ stand, respectively, for the constant intensity of uniform stretching and the applied symmetrically cycling bending moment.

Let us employ the approximate approach and adopt the kinematically admissible collapse mechanism which assumes that in the net sections parallel to the y-axis the displacement increments per cycle in the x-direction are equal to

$$\Delta u(z, y) = \Delta u^0 > 0, \tag{7.113}$$

and they are accompanied by vanishing plastic strain increments $\Delta \epsilon_y$ (plane strain conditions).

Then the fundamental equation (2.88) of the kinematical theorem

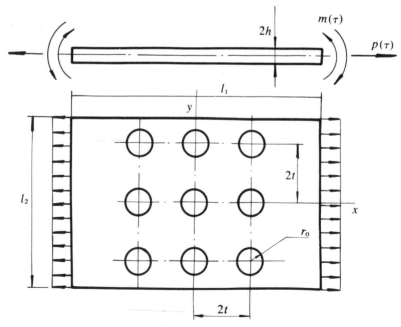

Fig. 7.24. Perforated plate.

7.6. Shakedown of perforated plates

can be written in the form

$$p\Delta u^0 t = \int_\tau^t \min_\tau [(\sigma_s - \sigma_{x\tau}^{(e)})\Delta u^0]\,dz, \qquad (7.114)$$

where σ_s is the yield point, and $\sigma_{x\tau}^{(e)}$ stand for the elastic stresses generated by the applied moment $m(\tau)$.

The right-hand side of the equation represents work done by the stresses corresponding to the fictitious yield surface on the displacement jumps at net sections. Both the symmetry and the constraint (7.111) have been employed.

For small values of the cutout coefficient $\lambda = r/t$, i.e. for sufficiently large spacing of holes, the stresses $\sigma_{x\tau}^{(e)}$ can be found with the help of a periodic extension of a known solution to the cylindrical flexure of a plate with a single hole. Savin [7.31] provides the following formulae for bending and twisting moments:

$$\begin{aligned}
M_x &= m\left[1 - \frac{5+3\nu}{2(3+\nu)} \cdot \frac{r^2(x^2-y^2)}{(x^2+y^2)^2} + \frac{1-\nu}{3+\nu} \cdot \frac{r^2(x^4+y^4-6x^2y^2)}{(x^2+y^2)^3} \right. \\
&\quad \left. - \frac{3(1-\nu)}{2(3+\nu)} \cdot \frac{r^4(x^4+y^4-6x^2y^2)}{(x^2+y^2)^4}\right], \\
M_y &= m\left[\frac{1-\nu}{2(3+\nu)} \cdot \frac{r^2(x^2-y^2)}{(x^2+y^2)^2} - \frac{1-\nu}{3+\nu} \cdot \frac{r^2(x^4+y^4-6x^2y^2)}{(x^2+y^2)^3} + \right. \qquad (7.115)\\
&\quad \left. + \frac{3(1-\nu)}{2(3+\nu)} \cdot \frac{r^4(x^4+y^4-6x^2y^2)}{(x^2+y^2)^4}\right], \\
M_{xy} &= -m\left[\frac{r^2xy}{(x^2+y^2)^2} - \frac{4(1-\nu)}{3+\nu} \cdot \frac{r^2xy(x^2-y^2)}{(x^2+y^2)^3} + \right.\\
&\quad \left. + \frac{6(1-\nu)}{3+\nu} \cdot \frac{r^4xy(x^2-y^2)}{(x^2+y^2)^4}\right].
\end{aligned}$$

Here the x, y-axes are taken parallel to the symmetry axes and originating from the centre of the hole.

The stresses $\sigma_{x\tau}^{(e)}$ in the net sections are distributed accordingly,

$$\sigma_{x\tau}^{(e)} = \frac{3zm(\tau)}{2h^3}\left[1 + \frac{7+\nu}{2(3+\nu)} \cdot \left(\frac{r}{y}\right)^2 - \frac{3(1-\nu)}{2(3+\nu)} \cdot \left(\frac{r}{y}\right)^4\right]. \qquad (7.116)$$

Quantitative analysis shows that the above formula remains good

7. Shakedown of plates

enough for $\lambda \leq 0.2$; for greater cutout coefficients the interaction between neighbouring holes results in substantial errors exceeding 5 per cent. In this case the elastic stresses can be found by using the method put forward by Grigoluk and Filshtinsky [7.30]. The bending moment distributions for various λ's are given in Table 7.2.

However, regardless of which stress distribution is used, (7.116) or a more accurate one, under the prescribed loading program, the integrand in (7.114) attains its minimum simultaneously at all points for which sign z = const. The simultaneity does not apply to all points of a normal to the middle surface; for $z > 0$, $m(\tau) = m_*$, for $z < 0$, $m(\tau) = -m_*$. Thus plastic deformation takes places simultaneously at all points lying on one half of the critical normal to the middle surface and therefore the right-hand side of (7.114) can be rewritten in the form

$$\int_r^t dy \int_{-h}^h \min_\tau [(\sigma_s - \sigma_{x\tau}^{(e)}) \Delta u^0] \, dz = \int_{-h}^h \min_\tau \left\{ \left[\int_r^t (\sigma_s - \sigma_{x\tau}^{(e)}) \, dy \right] \Delta u^0 \right\} dz. \tag{7.117}$$

Table 7.2

	$\lambda \setminus y/t$	0,2	0,4	0,6	0,8	0,9	1,0
$\dfrac{M_x}{m}$	0,2	1,86	1,31	1,17	1,12	1,11	1,11
	0,4		2,11	1,68	1,52	1,50	1,49
	0,6			2,81	2,46	2,38	2,35
	0,8				5,16	4,95	4,89
	0,9					10,16	9,92
$\dfrac{M_y}{m}$	0,2	0	−0,07	−0,04	−0,03	−0,02	−0,02
	0,4		0,01	−0,08	−0,08	−0,09	−0,09
	0,6			−0,002	−0,17	−0,17	−0,19
	0,8				−0,001	−0,01	−0,28
	0,9					0,02	−0,35

7.6. Shakedown of perforated plates

This means that the knowledge of the detailed stress distribution along the axis is immaterial; it is its resultant that counts. Thus the analysis can be based on averaged stresses in the net section by accounting for the cutout coefficient and disregarding the concentrations:

$$\sigma_{x\tau}^0 = \frac{3zm(\tau)}{2h^3} \cdot \frac{1}{1-\lambda}. \tag{7.118}$$

Substituting the maximum with respect to time magnitudes of these stresses

$$\max_\tau \sigma_{x\tau}^0 = \sigma_{x\tau}^{0*} = \frac{3|z|m_*}{2h^3} \cdot \frac{1}{1-\lambda} \tag{7.119}$$

into (7.114), we finally arrive at the incremental collapse condition

$$\frac{p}{p_0} + \frac{3}{4} \cdot \frac{m_*}{m_0} = 1, \tag{7.120}$$

where

$$p_0 = 2\sigma_s h(1-\lambda), \quad m_0 = \sigma_s h^2(1-\lambda) \tag{7.121}$$

are the ultimate loads at pure stretching and pure bending, respectively. Equation (7.120) is the same as the corresponding equation for the beam, see Sec. 5.6.

The incremental condition is represented in Fig. 7.25 by a solid line. Dashed lines show that the conditions for alternating flow occur in the planes $|z| = h$ at the circumference of cutouts with different cutout coefficients λ. The limiting parameters were computed according to the criterion (2.90) which states that the range of elastic stresses should not exceed the double yield point; this corresponds to the degeneration of the fictitious yield polygon into a line, or a point. The stresses of Table 7.2 were used as reflecting the concentrations.

2. Consider another loading program for the plate of Fig. 7.24. Let the actions p and m be applied in turns and let them vary inside the following limits per cycle:

$$0 \leqslant p(\tau) \leqslant p_* \quad \text{at } m(\tau) \equiv 0,$$
$$0 \leqslant m(\tau) \leqslant m_* \quad \text{at } p(\tau) \equiv 0. \tag{7.122}$$

7. Shakedown of plates

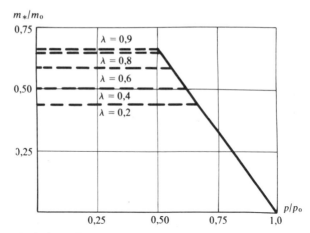

Fig. 7.25. Shakedown diagram for perforated plate, $p = \text{const.}$, $-m_* \leq m(\tau) \leq m_*$.

Since no constant load is present, the incremental collapse condition follows from the inequality

$$\int_r^t dy \int_{-h}^h \min_\tau [(\sigma_s - \sigma_{x\tau}^{(e)})\Delta u^0] \, dz \leq 0, \qquad (7.123)$$

where $\sigma_{x\tau}^{(e)}$ should now be understood as generated by each of the actions (7.122) separately.

For simplicity, assume the cutout coefficient λ to be small enough, so that the stress distribution established in [7.31] can be accepted. It was derived by periodic superposition of the solution obtained for the plate with an opening. The direct stresses in the net section at stretching $p(\tau)$ are

$$\sigma_{x\tau}^{(e)} = \frac{p(\tau)}{2h}\left[1 + 0.5\left(\frac{r}{y}\right)^2 + 1.5\left(\frac{r}{y}\right)^4\right], \quad r \leq y \leq t. \qquad (7.124)$$

The bending stresses result from the formula (7.116) which at $\mu = 0.3$ yields

$$\sigma_{x\tau}^{(e)} = \frac{3m(\tau)z}{2h^3}\left[1 + 1.106\left(\frac{r}{y}\right)^2 - 0.318\left(\frac{r}{y}\right)^4\right], \quad r \leq y \leq t. \qquad (7.125)$$

7.6. Shakedown of perforated plates

The distributions of direct and bending stresses in the net section close to the hole are dissimilar: The concentration coefficients are 3 and 1,8, respectively. The stresses at $z = h$ and $p_*/p_0 = 3/2 \ (m_*/m_0)$, cf. (7.121), are shown in Fig. 7.26. The situation arises in which, at given loading program, the enveloping stresses corresponding to the minimum of the integrand in (7.123) and indicated by shading are non-isochronous not only across the thickness of the plate, as in the previous program, but also along the net section. The situation is somewhat analogous to that considered in Sec. 7.1.

To sum up, the presence of stress concentrations in the adopted program exerts a distinct influence on the incremental collapse conditions. In a plate of considerable width and provided with a single hole this influence is rather negligible; in the perforated plate it can be quite significant. This is visualized in the shakedown diagram shown in Fig. 7.27 where line 2 is constructed by disregarding concentrations, i.e. for smeared out stresses. It should be remembered that, at purely mechanical variable repeated loads, the differences between the load factors leading to the incremental and the instantaneous collapse are rarely greater than those shown here.

3. The in-plane non-isochronous enveloping stresses can be also encountered in perforated plates at non-stationary temperature fields. Let us consider a periodical flow of coolant through the holes of a

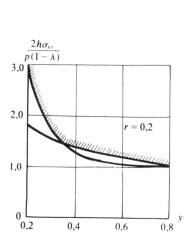

Fig. 7.26. Stress distributions near the cutout of a plate at stretching and bending.

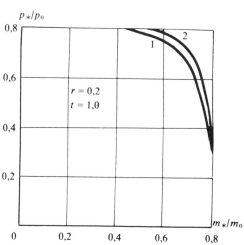

Fig. 7.27. Shakedown diagram for perforated plate at successive stretching and bending.

329

7. Shakedown of plates

perforated plate stretched by $p(\tau) = \eta = \text{const.}$, $m(\tau) \equiv 0$ as shown in Fig. 7.24. Let the thermal cycling be axi-symmetric near each hole and given by

$$-t_* \leq t(\tau) \leq 0 \quad \text{for } r \leq \rho \leq 2r,$$
$$t(\tau) \equiv 0 \quad \text{for } 2r < \rho \leq t, \tag{7.126}$$

where ρ is the distance from the centre of a hole. For sufficiently small cutout parameters λ resulting in practically no interaction between neighbouring holes, the thermal stresses in the net sections parallel to the y-axis can be found as suitable periodical extensions of the solution to the thermo-elasticity problem of the plate with a single hole [7.23]. We obtain

$$\sigma_{x\tau}^{(e)} = -\frac{1}{2}\alpha E t(\tau)\left[1 + \left(\frac{r}{y}\right)^2\right] \quad \text{for } r \leq y \leq 2r,$$
$$\sigma_{x\tau}^{(e)} = \frac{3}{2}\alpha E t(\tau) \cdot \left(\frac{r}{y}\right)^2 \quad \text{for } 2r < y \leq t. \tag{7.127}$$

Given (7.126), the integrand in the inequality (7.114) attains its minimum at the stresses

$$\sigma_{x\tau}^* = \frac{1}{2}\alpha E t_*\left[1 + \left(\frac{r}{y}\right)^2\right] \quad \text{for } r \leq y \leq 2r,$$
$$\sigma_{x\tau}^* = 0 \quad \text{for } 2r < y \leq t. \tag{7.128}$$

The obtained stress distribution is non-isochronous, hence the load-carrying capacity of the perforated plate drops as a result of thermal cycling. The limiting condition for incremental stretching of the plate follows from equation (7.114) and it becomes now

$$\frac{p}{p_0} + \frac{3}{2} \cdot \frac{\lambda}{1-\lambda} \cdot \frac{t_*}{t^0} = 1, \tag{7.129}$$

where $t^0 = 2\sigma_s/\alpha E$ denotes the level of temperature t_* at the beginning of alternating flow around the cutout.

The limiting cycle parameters are shown diagrammatically in Fig. 7.28. Lines 1 and 2 indicate the incremental collapse for $\lambda = 0.1$ and

7.6. Shakedown of perforated plates

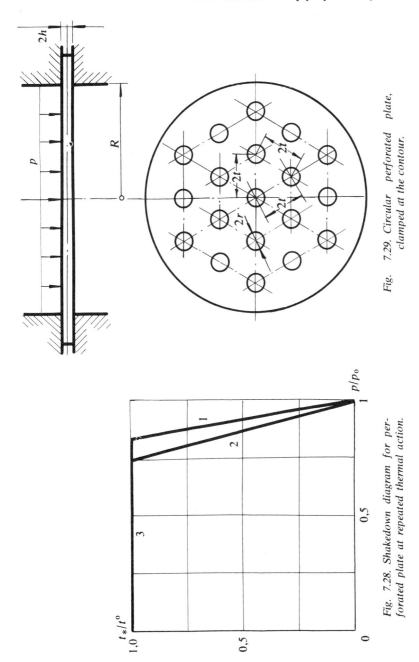

Fig. 7.29. Circular perforated plate, clamped at the contour.

Fig. 7.28. Shakedown diagram for perforated plate at repeated thermal action.

7. Shakedown of plates

$\lambda = 0{,}15$, respectively; line 3 warns against the alternating flow. The greater the cutout coefficient λ, the more pronounced is the decrease in the load-carrying capacity.

4. The presence of a mesh of holes can under certain types of cyclic loading generate specific mechanisms of partial incremental collapse not encountered in solid plates. Consider, for example, a circular clamped plate perforated according to a regular triangular pattern, Fig. 7.29, and subject to a time-independent pressure p and thermal cycling of the type

$$t(t) = t_0(\tau) + \zeta t_1(\tau), \quad \zeta = \frac{z}{h}, \qquad (7.130)$$

$$-1 \leqslant \zeta \leqslant 1, \quad -t_* \leqslant t_1(\tau) \leqslant t_*.$$

Taking into account the in-plane simultaneity of plastic flow under the temperature field (7.130), we can accept that the averaged elastic stresses in the net sections have the form

$$\sigma_{rt}^{(e)} = \frac{1}{\psi}\sigma_{rt}, \quad \sigma_{\varphi\tau}^{(e)} = \frac{1}{\psi}\sigma_{\varphi t}, \qquad (7.131)$$

where ψ designates the reduction coefficient determined by using one of the methods discussed in [7.30, 7.32, 7.33]; and σ_{rt}, $\sigma_{\varphi t}$ stand for the thermo-elastic stresses in the solid plate [7.23] given by the formulae

$$\sigma_{rt} = \sigma_{\varphi t} = \frac{\alpha E t_1(\tau)}{1-\mu}\zeta = \zeta q(\tau),$$

$$-q_* \leqslant q(\tau) \leqslant q_*, \quad q_* = \frac{\alpha E t_*}{1-\mu}. \qquad (7.132)$$

Consider two possible mechanisms. The first one is described by (7.29), (7.30); it forms in a solid plate under analogous conditions. The other one forms due to the concentration of plastic strains along certain lines traced on the middle surface similarly as in the rectangular plate studied in Sec. 7.5. It is here reasonable to expect the yield lines to form in the weakest sections of the triangular mesh; thus the mechanism will have the form of a hexagonal pyramid as shown in Fig. 7.30.

The incremental collapse condition for the first mechanism can be formulated exactly in the same manner as it was done for the clamped

7.6. Shakedown of perforated plates

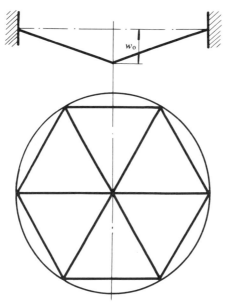

Fig. 7.30. Possible collapse mechanism of circular perforated plate.

solid plate, Sec. 7.3; the only differences consist in the coefficient $\psi < 1$ appearing in (7.131) and in the necessity of integration over a multi-connected region when seeking the expression for work done by fictitious yield surface stresses. The integrals can be approximately determined by applying suitably defined cutout coefficients γ_1 and γ_2. The first one is equal to the ratio of the perforated area of the plate to the unperforated one, see Fig. 7.29,

$$\gamma_1 = 1 - \frac{\pi\sqrt{3}}{6}\lambda^2, \quad \lambda = \frac{r}{t}. \tag{7.133}$$

The other one reflects the influence of holes on the lengths of the peripheral yield lines,

$$\gamma_2 = 1 - \lambda. \tag{7.134}$$

Finally, the incremental collapse condition takes the form

$$\frac{p}{p_0} = \frac{1{,}07}{3x^2 - 1}\left[\gamma_1(1 + \ln x) + \gamma_2 - \frac{2}{3}q_*\frac{\gamma_1 + \gamma_2}{\psi\sigma_s}\right], \tag{7.135}$$

333

7. Shakedown of plates

where $p_0 = 11{,}26(M_0/R)$ is the load-carrying capacity of a solid plate at limit equilibrium. The radius x can be found on insisting that the load parameter be at its minimum (as in Sec. 7.3); this leads to the equation

$$3x^2 - 2\ln x - 3 - 2\delta = -\frac{4}{3} q_* \frac{1+\delta}{\psi\sigma_s}, \quad \delta = \frac{\gamma_2}{\gamma_1}. \tag{7.136}$$

Using this in (7.135), we get

$$\frac{p}{p_0} = 0{,}535 x^2 \gamma_1, \quad (0 \leq q_* \leq A),$$

$$\frac{p}{p_0} = 0{,}535(\gamma_1 + \gamma_2)\left(1 - \frac{2}{3}\frac{q_*}{q^0}\right), \quad (A \leq q_* \leq 1), \tag{7.137}$$

where

$$A = \frac{2}{3} \cdot \frac{\delta}{1+\delta} q^0, \tag{7.138}$$

$$q^0 = \psi\sigma_s, \quad \left(\frac{q_*}{q^0} = 1\right), \tag{7.139}$$

q^0 being determined from the alternating flow condition (2.90).

The shakedown diagrams for various perforations λ are shown by solid lines in Fig. 7.31, they were constructed by using (7.137)–(7.139); the necessary reduction coefficients ψ were taken from [7.30], and the criterion for equivalence was that of equal stiffnesses.

When the mechanism shown in Fig. 7.30 is adopted, the limiting condition resulting from the kinematical theorem (2.88) takes the form

$$p \int_S \Delta w \, dS = 6 \int_{l_0} M_1^0 \Delta\theta_1 \, dl + 6 \int_{l_0} M_2^0 \Delta\theta_2 \, dl, \tag{7.140}$$

where

$$\Delta w = \Delta w_0 \left(1 - \frac{r}{R}\right), \quad \Delta\theta_1 = \Delta\theta_2 = \frac{2\Delta w_0}{R\sqrt{3}},$$

$$M_1^0 = M_2^0 = M_0 - \frac{2}{3\psi} q_* h^2, \quad l_0 = R(1-\gamma_2), \quad \int_S \Delta w \, dS = \frac{R^2\sqrt{3}}{2} \Delta w_0.$$

7.6. Shakedown of perforated plates

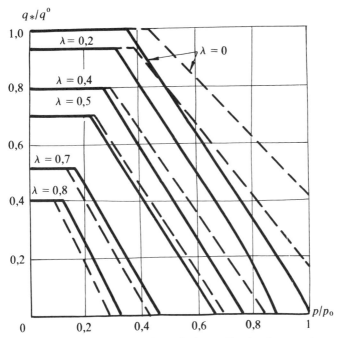

Fig. 7.31. Shakedown diagrams for perforated plate collapsing in accordance with two possible mechanisms.

Since the cycle is assumed to be symmetric, $-q_* \leq q(\tau) \leq q_*$, the additional loading occurs successively in all the yield hinges: at one stage in all the radial ones, at another in all the peripheral ones. Since $M_1^0 = M_2^0$, equation (7.140) can be rearranged to become

$$\frac{p}{p_0} = 1{,}42\gamma_2 \left(1 - \frac{1}{8} \cdot \frac{q_*}{q^0}\right). \tag{7.141}$$

Dashed lines in Fig. 7.31 depict the situation. For smaller degrees of perforation ($\lambda < 0{,}5$) the first mechanism provides better upper bounds on the limiting cycle parameters for the incremental collapse conditions. When the cutout coefficients are larger, then the partial collapse mechanism consisting of concentrated radial plastic hinges unencountered in the solid circular plate becomes critical.

Let us note that, unlike in the analysis of elastic perforated plates, the incremental collapse analysis based on an equivalent model of solid plate required the introduction of more than one reduction coefficient.

7. Shakedown of plates

References

[7.1] Gvozdev, A.A., *Load-carrying capacity of structures based on limit equilibrium*, in Russian, Strojizdat, Moscow, 1949.
[7.2] Sokolovsky, W.W., *Theory of plasticity*, in Russian, Gostekhizdat, Moscow, 1950.
[7.3] Ilyushin, A.A., *Plasticity*, in Russian, Gostekhizdat, Moscow, 1948.
[7.4] Feinberg, S.M., Limit stress principle, in Russian, *Prikladnaya Mat. Mekh.*, *12*, Nr 1, 1948.
[7.5] Rzhanitzin, A.R., *Structural analysis taking account of plastic properties of materials*, in Russian, Gosstrojizdat, Moscow, 1954.
[7.6] Hopkins, H.C., Prager, W., The load carrying capacities of circular plates. *J. Mech. Phys. Solids*, *2*, 1953.
[7.7] Rabotnov, Yu.N., Approximate technical theory of elastic-plastic shells, in Russian, *Prikladnaya Mat. Mekh.*, *15*, Nr 1-2, 1951.
[7.8] Hodge, Ph.G., *Limit analysis of rotationally symmetric plates and shells*. Prentice-Hall, 1963.
[7.9] Shapiro, G.S., Behaviour of plates and shells beyond elastic limit, in Russian, in *Proceed. II All-Union Congr. on Theor. and Appl. Mech. Solid Mech.* Nauka, Moscow 1966.
[7.10] Gokhfeld, D.A., Some problems of shakedown in plates and shells, in Russian, in *Shells and Plates Theory*, Proceed. VI All-Union Conf., (Baku, 1966), Nauka, Moscow, 1966.
[7.11] Gokhfeld, D.A., On Koiter's theorem in shakedown of non-uniformly heated elastic-plastic bodies, in Russian, *Prikl. Mekh.*, *3*, Nr 8, 1967.
[7.12] Gokhfeld, D.A., Shakedown theorems and methods for elastic-plastic bodies, in Russian, in *Thermal stresses in structural members*, V. 7, Naukova Dumka, Kiev, 1967.
[7.13] Belyakov, A.R., Cherniavsky, O.F., Shakedown conditions for circular plates under multi-parameter loading, in Russian, in *Thermal stresses in structural members*, V. 14, Naukova Dumka, Kiev, 1974.
[7.14] Gokhfeld, D.A., Cherniavsky, O.F., Linear programming in some two-dimensional limit equilibrium and shakedown situations in statical formulation, in Russian, in *Thermal stresses in structural members*, V. 7, Naukova Dumka, Kiev, 1967.
[7.15] Gokhfeld, D.A., Cherniavsky, O.F., Linear programming in shakedown problems in kinematical formulation, in Russian, in *Thermal stresses in structural members*, V. 9, Naukova Dumka, Kiev, 1970.
[7.16] Cherniavsky, O.F., Limit analysis and shakedown problems using the maximum Pontragin principle, in Russian, *Izv. AN SSSR, Mekh. Tvard. Tela*, Nr 4, 1970.
[7.17] König, J.A., Shakedown theory of plates, *Arch. mech. sto.*, *21*, Nr 5, 1969.
[7.18] Sawczuk, A., Janas, M., König, J.A., *Plastic analysis of structures* (in Polish), Ossolineum, Wrocław, 1972.
[7.19] Gokhfeld, D.A., Kononov, V.M., Cherniavsky, O.F., Shakedown of perforated plates, in Russian, in *Shells and plates theory*, Proceed. X All-Union Conf., Kutaysi, *1*, Mecnierba, Tbilisi, 1975.
[7.20] Belytschko, T., Plane stress shakedown analysis by finite elements. *Int. J. Mech. Sci.*, *14*, 1972.
[7.21] Muspratt, M.A., Stochastic shakedown analysis of concrete slabs. *Internat. Journ. for Numerical Methods in Engineering 5*, 1973, 303–309.

References

[7.22] Ikrin, W.A., Approximate assessment of stress and strain states at shakedown of circular and annular plates, in Russian, in *Strength problems in machinery*, Proceed. Cheliabinsk Polyt. Inst., Nr 151, Cheliabinsk, 1974.

[7.23] Timoshenko, S., Voinovsky-Krieger, S., *Theory of plates and shells*. McGraw-Hill, New York, 1959.

[7.24] Parkus, H., *Instationäre Wärmespannungen*, Springer, Wien, 1959.

[7.25] Prager, W., Structural analysis beyond elastic limit at thermal cycling, in Russian, *Mekhanika*, 3(43), 1958.

[7.26] Rosenblum, W.J., On shakedown theory of elastic-plastic bodies, in Russian, *Izv. AN SSSR, OTN*, Nr 6, 1958.

[7.27] Rzhanitzin, A.R., Limit equilibrium analysis of shells, by linear programming, in Russian, in *Proceed. VI All-Union Conf. on Shells and Plates*, (Baku, 1966), Nauka, Moscow, 1966.

[7.28] Boley, B.A., Weiner, J.H., *Theory of themal stresses*. Wiley, New York, 1960.

[7.29] Butkovsky, A.G., *Optimum control theory for systems with distributed parameters*, in Russian, Nauka, Moscow, 1965.

[7.30] Grigoluk, E.P., Filshtinsky, L.A., *Perforated plates and shells*, Nauka, Moscow, 1970.

[7.31] Savin, G.A., *Stress concentration near holes*, in Russian, Gostekhizdat, Moscow, 1951.

[7.32] Gogolev, A.Ya., Analysis of tubular heat exchangers at limit load, in Russian, *Energomashinostrojenie*, Nr 4, 1963.

[7.33] Dolinsky, W.M., Kovalevsky, B.S., Flexural rigidity of tubular grids, in Russian, *Chemical and Oil Machinery*, Nr 9, 1970.

8

Shakedown of shells at mechanical and thermal cycling

The shakedown problem in a shell, one of the widely used structural elements, is of particular importance.

The first non-trivial shakedown problem, namely the incremental collapse analysis of a shell was solved in [8.1] with the help of the reformulated Koiter theorem (see Chap. 2); the conditions for accumulation of strains were established for a hollow thin cylinder under constant ring load and thermal cycling. This approach was further developed by Sawczuk in [8.2, 8.3] to solve certain load situations in terms of generalized variables introduced in an analogous manner to that employed in the limit analysis theory.

At roughly the same time an independent study was made [8.4] on the behaviour of thin tubes subjected to cyclic internal pressure and temperature. Such tubes are encountered in the fuel elements of nuclear reactors. Both the incremental collapse and the alternating flow analyses were made with no reference to the fundamental shakedown theorems; instead a method was employed resembling the overload method considered in Chap. 5 and assuming a piece-wise linear stress state across the thickness of the shell. Practically the same results were later arrived at in England [8.5].

Further achievements in the kinematical and statical methods in tackling shakedown situations can be found in [8.6–8.9] and others. The authors claimed that the determination of incremental collapse conditions was the principal problem in shakedown theory; those effects were most probable in the conditions of cyclic non-isothermal loading. We should mention here a series of publications [8.10–8.13 and others] devoted to the shakedown of pressure vessels of various shapes. Statical methods based on the Melan theorem were used in both the approximate formulation, when the residual stress field is given, and the exact one when the optimum solution via linear programming is looked for. The types of unsafe state,

8 Shakedown of shells

defining conditions for shakedown, i.e. both the alternating flow and the incremental collapse, were not analysed; however, the considered situations such as stress concentrations at connections of shells as well as one-parameter loading indicate that alternating plastic flow was critical to shakedown. This means that the solutions to many problems could have been arrived at in a simpler manner, see Chap. 2.

In the present chapter a number of shakedown situations will be considered for shells of various shapes under mechanical and thermal actions. As usual, our attention will be mainly focused on the incremental collapse and the conditions conducive to this type of unsafe behaviour. And also, on those examples, we shall demonstrate and elucidate the basic methods of limit analysis.

8.1. Cylindrical shell under ring load and thermal cycling

1. Load-carrying capacity of such a shell at single loading was found by A.A. Ilyushin [8.14] with the use of the Huber–Mises yield condition. In order to establish the conditions for incremental collapse let us start by adopting the simplest possible approximation to the yield condition put forward by Drucker [8.15], namely the so-called square yield condition, Fig. 8.1

$$\max(|n_\varphi|, |m_x|) = 1, \tag{8.1}$$

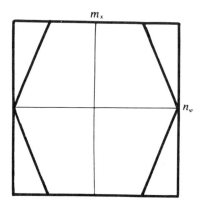

Fig. 8.1. Yield condition for cylindrical shell in the absence of axial load.

8.1. Cylindrical shell under ring load and thermal cycling

where

$$n_\varphi = N_\varphi/N_0, \quad m_x = M_x/M_0, \quad N_0 = 2\sigma_s h, \quad M_0 = \sigma_s h^2,$$

N_φ is the circumferential direct stress resultant (hoop normal force), and M_x is the meridional bending moment.

The shell on Fig. 8.2 is assumed to be long enough for the end boundary conditions to be ineffective. The temperature will cycle according to [8.16]

$$t(\tau, \zeta) = t_0(\tau) + \zeta t_1(\tau), \quad 0 \leq t_1(\tau) \leq t_* \tag{8.2}$$

and heat is assumed to be supplied from outside. The thermo-elastic stresses are

$$\sigma_{\varphi\tau}^{(e)} = \sigma_{x\tau}^{(e)} = -\frac{\alpha E t_1(\tau)}{1-\mu}\zeta, \quad \zeta = \frac{\bar{z}}{h}. \tag{8.3}$$

The postulated collapse mechanism, shown in Fig. 8.2, is described by

$$\Delta w = \Delta w_0(1-\xi), \quad \xi = \frac{X}{L_0} \quad (0 \leq \xi \leq 1). \tag{8.4}$$

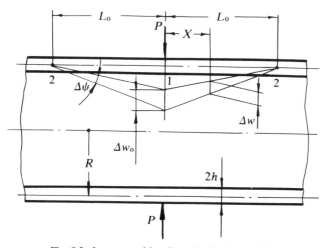

Fig. 8.2. Incremental bending of cylindrical shell.

341

8 Shakedown of shells

It follows immediately that

$$\Delta\psi = \pm\frac{\Delta w_0}{L_0}, \quad \Delta\epsilon''_\varphi = -\frac{\Delta w}{R} = -\frac{\Delta w_0}{R}(1-\xi). \tag{8.5}$$

We shall assume now, and check later, that the most unfavourable loading program occurs at $P = P_* = \text{const}$. A domain in which the thermal stresses lead to additional loading can be found by looking at strain increments (8.5) and the thermal stress distribution (8.3). With the help of Fig. 8.3 we conclude that the additional loading domain occupies the outer portion of the shell, $0 \le \xi \le 1$, together with the middle ring yield hinge 1 whose strain diagram is shown in Fig. 8.3 as $\Delta\epsilon_{x1} = \Delta\epsilon_{x1}(\psi)$. When the temperature field in the shell becomes uniform, the plastic strain rates become non zero in the inner part of the shell, $-1 \le \xi \le 0$, and in the extreme ring yield hinges 2 (unloading region). When one of the directions is reversed, either that of the ring load P or of the heat flux, the overloading and the unloading regions will be interchanged.

Equation (2.88) takes the form

$$P_*\Delta w_0 = 2L_0 h \int_0^1 d\xi \int_{-1}^1 \min_\tau[(\sigma_s - \sigma^{(e)}_{\varphi\tau})\Delta\epsilon''_\varphi]\,d\zeta +$$
$$+ 4 \min_\tau[(M_0 - M^{(e)}_{x\tau})\Delta\psi], \tag{8.6}$$

where

$$M^{(e)}_{x\tau} = \frac{2\alpha E h^2 t_1(\tau)}{3(1-\mu)}.$$

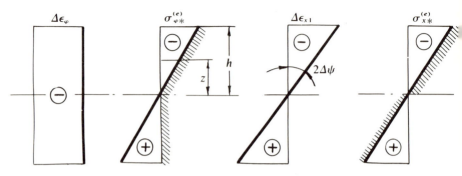

Fig. 8.3. Additional loading domains in a shell at thermal cycling.

342

8.1. Cylindrical shell under ring load and thermal cycling

The right-hand side of (8.6) will be at minimum for $\sigma_{\varphi\tau}^{(e)} = \sigma_{\varphi\tau}^*$, $M_{x\tau}^{(e)} = M_{x\tau}^*$ where asterisks denote values at $t_1(\tau) = t_*$.

After necessary rearrangements and using the above conditions for additional loading region together with conditions $\sigma_{\varphi\tau}^{(e)} = 0$, $M_{x\tau}^{(e)} = 0$ for the unloading region we get

$$P_* = \frac{2\sigma_s h}{R}\left(L_0 + \frac{2Rh}{L_0}\right) - \beta \frac{\sigma_s h}{R}\left(L_0 + \frac{8}{3}\cdot\frac{Rh}{L_0}\right), \tag{8.7}$$

where $\beta = (t_*/t_1^0)$ and, as usual, $t_1^0 = (2\sigma_s(1-\mu))/(\alpha E)$. The above solution depends on the spacing of yield hinges L_0. Since we are using the upper bound approach based on the kinematical theorem, the best estimate will result from the minimization of (8.7). This yields

$$L_0^2 = \frac{4Rh\left(1 - \frac{2}{3}\beta\right)}{2 - \beta}. \tag{8.8}$$

At $\beta = 0$ the expressions (8.7), (8.8) supply the known limit load solution [8.17]

$$P_0 = 4\sqrt{2}\,\frac{M_0}{\sqrt{Rh}}. \tag{8.9}$$

From (8.8) it follows that the increase in the intensity of thermal cycling is accompanied by the decrease in the spacing L_0 between the yield hinges. On account of (8.8), (8.9) the incremental collapse condition can be conveniently written as

$$\frac{P_*}{P_0} = \sqrt{\left(1 - \frac{1}{2}\beta\right)\left(1 - \frac{2}{3}\beta\right)}. \tag{8.10}$$

For $0 \leq \beta \leq 1$, equation (8.10) is nearly linear as can be seen in Fig. 8.4, line 3.

It is easy to show that the 'elastic' stresses from the applied load P [8.18] do positive work at all the points of the tube where the displacements differ from zero in accordance with the adopted mechanism. This means that $P = \text{const.}$ creates indeed the most unfavourable incremental collapse conditions.

To the alternating flow conditions there correspond two lines in Fig. 8.4: line 1 with the equation

8 Shakedown of shells

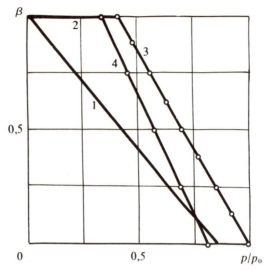

Fig. 8.4. Shakedown diagram for cylindrical shell.

$$1{,}17\frac{P_*}{P_0} + \beta = 1 \qquad (8.11)$$

for arbitrary loading programs and line 2 for thermal cycling at $P = $ const. Thus the entire shakedown domain is bounded by the line 1 in the first case and by lines 2 and 3 in the second. Thermal cycling at $P = $ const. is thus found to reduce the load-carrying capacity of the tube to almost 60 per cent (at $\beta \leq 1$, i.e. when the alternating plasticity is still absent).

Fig. 8.5 shows the stress distributions in the cylindrical shell at two distinct stages of the limiting cycle (8.10): at the end of heating (a) and of cooling (b). These fields differ by the thermo-elastic stresses (8.3) and they are seen to be statically admissible at any instant of time. Thus, within the system of assumptions made and under the yield condition (8.1), the solution (8.10) can be considered to be exact.

It remains to determine the minimum length of the tube for which the above solution is valid. Provided the end effects on the thermal stresses are ignored, the necessary analysis can follow that made by Eason and Shield for the limit equilibrium situation of the same shell, see [8.17].

8.1. Cylindrical shell under ring load and thermal cycling

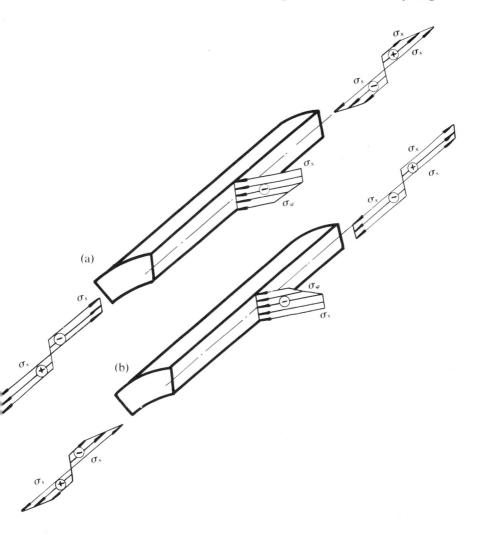

Fig. 8.5. Stress distribution in cylindrical shell during the limiting cycle.

2. Reconsider the same shell on the basis of the statical theorem via linear programming. The suitable analysis consists in constructing the fictitious interaction curves based on the Tresca yield condition (2.12), their piece-wise linear approximations, making the problem discrete, and finally, computing the limiting cycle parameters by means of the simplex method.

8 Shakedown of shells

For convenience, let us first change the limits of variation of the function $t_1(\tau)$ determining the temperature intervals across the thickness, as given by the temperature field (8.2). The average level of temperature will be insignificant since the yield point is kept temperature-independent. Let

$$-\frac{1}{2}t_* \leq t_1(\tau) \leq \frac{1}{2}t_*. \tag{8.12}$$

The symmetrically cycling thermal stresses are now expressed by

$$\sigma^{(e)}_{\varphi\tau} = \sigma^{(e)}_{x\tau} = -\zeta q(\tau)\sigma_s, \quad -q_* \leq q(\tau) \leq q_*, \tag{8.13}$$

where

$$q_* = \frac{\alpha E t_*}{2(1-\mu)\sigma_s}, \quad -1 \leq \zeta \leq 1.$$

The fictitious yield polygons in the presence of the Tresca criterion (2.12), time-dependent stresses (8.13) and at $P = $ const. are described by

$$\max(|\sigma^0_{x*}| + |\zeta|q_*\sigma_s, |\sigma^0_{\varphi*}| + |\zeta|q_*\sigma_s, |\sigma^0_{x*}| - |\sigma^0_{\varphi*}|) - \sigma_s = 0. \tag{8.14}$$

In the $\sigma^0_{x*}, \sigma^0_{\varphi*}$-plane we obtain hexagons if $q_*|\zeta| < \frac{1}{2}$ and squares if $q_*|\zeta| \geq \frac{1}{2}$, see Fig. 8.6. For $q_* = 1$ and $|\zeta| = 1$ the domain of admissible values of time-independent stresses degenerates into a point and the alternating flow commences.

We construct now the fictitious interaction curves. According to the Kirchhoff–Love hypotheses the increments of meridional and circumferential plastic strains per cycle $\Delta\epsilon''_x, \Delta\epsilon''_\varphi$ are related to the increments of elongation and curvature by

$$\Delta\epsilon''_x = \Delta e_x + \zeta h \Delta\kappa_x, \quad \Delta\epsilon''_\varphi = \Delta e_\varphi. \tag{8.15}$$

The expression (3.7) assumes now the form

$$M^0_{x*}\Delta\kappa_x + N^0_{x*}\Delta e_x + N^0_{\varphi*}\Delta e_\varphi = \int_{-h}^{h} [\sigma^0_{x*}(\Delta e_x + \zeta h \Delta\kappa_x) + \sigma^0_{\varphi*}\Delta e_\varphi] \, d\zeta.$$

(8.16)

8.1. Cylindrical shell under ring load and thermal cycling

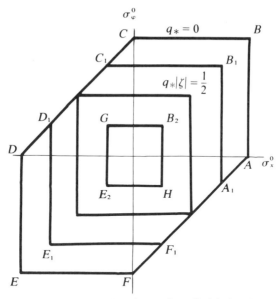

Fig. 8.6. Fictitious yield polygons for cylindrical shell.

It follows from the global equilibrium of the shell shown in Fig. 8.2 that the meridional normal force N_x^0 is absent. This fact enables us to find a relationship between Δe_x and $\Delta \kappa_x$. It will suffice to consider the case $\Delta e_\varphi > 0$, $\Delta \kappa_x > 0$, since the fictitious interaction curves have to be symmetric with respect to both the axes $M_x^0 = 0$ and $N_\varphi^0 = 0$.

In accordance with the strain increment distribution across the thickness, Fig. 8.7, and the relationships (2.76) reflecting the associated flow rule, there can be present in general three different flow regimes in the cross-section of the shell, provided the fictitious interaction curve is a hexagon, $q_*|\zeta| < \tfrac{1}{2}$. These correspond to
a) $\eta_1 \leq \zeta \leq 1$ when $\Delta\epsilon_\varphi'' > 0$, $\Delta\epsilon_x'' > 0$, the flow régime is represented by the point B_1, Fig. 8.6, and we have

$$\sigma_{x*}^0 = \sigma_{\varphi*}^0 = (1 - q_*|\zeta|)\sigma_s; \tag{8.17}$$

b) $\eta_2 \leq \zeta \leq \eta_1$ when $\Delta\epsilon_\varphi'' > 0$, $\Delta\epsilon_x'' < 0$ and $\Delta\epsilon_\varphi'' > |\Delta\epsilon_x''|$, the flow régime is represented by the point C_1, and we have

$$\sigma_{\varphi*}^0 = (1 - q_*|\zeta|)\sigma_s, \quad \sigma_{x*}^0 = -q_*|\zeta|\sigma_s; \tag{8.18}$$

8 Shakedown of shells

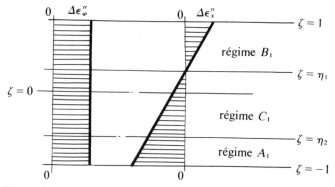

Fig. 8.7. Strain increments to construct the fictitious interaction curves; flow régimes indicated.

c) $\eta_2 \geq \zeta \geq -1$ when $\Delta\epsilon_\varphi'' > 0$, $\Delta\epsilon_x'' < 0$ but $\Delta\epsilon_\varphi'' < |\Delta\epsilon_x''|$, the flow régime is represented by the point D_1, and we have

$$\sigma_{\varphi *}^0 = q_*|\zeta|\sigma_s, \quad \sigma_{x *}^0 = -(1 - q_*|\zeta|)\sigma_s. \tag{8.19}$$

Substituting (8.17)–(8.19) into the right-hand side of (8.16) and equalising the multipliers of the generalized strains appearing after integration in the left- and the right-hand sides, we get

$$n_{x *}^0 = \frac{1}{2}(-\eta_1 - \eta_2 - q_* \eta_2^2), \tag{8.20}$$

$$m_{x *}^0 = 1 - \frac{\eta_1^2 + \eta_2^2}{2} - \frac{1}{3} q_*(1 + \eta_2^3), \tag{8.21}$$

$$n_{\varphi *}^0 = \frac{1}{2}(1 - \eta_2) - \frac{1}{4} q_* \eta_2^2, \tag{8.22}$$

where

$$n_{x *}^0 = N_{x *}^0 / N_0, \quad m_{x *}^0 = M_{x *}^0 / M_0, \quad n_{\varphi *}^0 = N_{\varphi *}^0 / N_0,$$
$$N_0 = 2\sigma_s h, \quad M_0 = \sigma_s h^2. \tag{8.23}$$

By making (8.20) vanish we obtain

$$\eta_1 = -\eta_2 - q_* \eta_2^2. \tag{8.24}$$

8.1. Cylindrical shell under ring load and thermal cycling

It is worth mentioning that the integration of (8.16) has been for

$$0 \geq \eta_2 \geq -1, \tag{8.25}$$

since it is easy to show that otherwise the condition $N_{x*}^0 = 0$ is violated.
When the parameter q_* is large enough, the points of the shell for which $q_*|\zeta| \geq \frac{1}{2}$, i.e.

$$|\zeta| \geq a, \quad \text{where} \quad a = \tfrac{1}{2} q_*^{-1}$$

are associated with the square domain B_2GE_2H, Fig. 8.6, of the admissible values of constant stress components; the points closer to the median surface, $|\zeta| < a$, continue to be associated with the hexagonal domain $A_1B_1C_1D_1E_1F_1$.

Fig. 8.8 shows the strain distribution and the flow regimes in the particular case in which

$$\eta_1 < a, \quad \eta_2 > -a. \tag{8.27}$$

Similar rearrangements show that (8.21), (8.24), (8.25) remain unchanged whereas (8.22) is replaced by

$$\eta_{\varphi*}^0 = 1 - \frac{1}{2}(\eta_2 + a) - \frac{1}{4} q_*(1 + \eta_2^2 - a^2). \tag{8.28}$$

The further analysis of the other relationships between η_1, η_2, a which do not coincide with (8.27) is virtually analogous. The domains of admissible

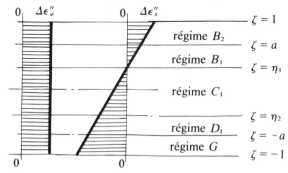

Fig. 8.8. Strain increments to construct the fictitious interaction curves; greater values of q_*; flow régimes indicated.

8 Shakedown of shells

values of constant stress resultants for particular values of the thermal parameter q_* are shown in Fig. 8.9 by solid lines. The obtained boundaries are curvilinear in spite of the piece-wise linearity of the fictitious yield curves. At $q_* = 0$, the parametric equations (8.21), (8.22), (8.24) lead to the results that coincide with the known ultimate relationship for the cylindrical shell [8.14, 8.17].

To employ now the linear programming procedures, we have to linearize the obtained curves. In the limit equilibrium situations ($q_* = 0$) the two widely used approximations of interaction curves are the square and the hexagonal conditions [8.17]. Let us first confine ourselves to the simplest rectangular approximation shown in Fig. 8.9 by dashed lines. The ultimate values of the dimensionless hoop force and the meridional bending moment are mutually independent and form the so-called one-moment limited-interaction condition

$$|m_{x*}^0| = 1 - \frac{2}{3} q_*, \tag{8.29}$$

$$|n_{\varphi*}^0| = 1 - \frac{1}{2} q_*. \tag{8.30}$$

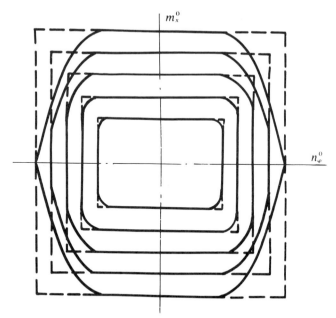

Fig. 8.9. Fictitious interaction curves for cylindrical shell at various values of temperature field parameter.

8.1. Cylindrical shell under ring load and thermal cycling

We can rewrite the equilibrium condition of the shell shown in Fig. 8.2 in the form [8.18]

$$m_x^0 = \int_0^x dl \int_0^l n_\varphi^0 \, d\zeta - px + m_x^0(0), \tag{8.31}$$

where

$$x = \frac{X\sqrt{2}}{\sqrt{Rh}}, \quad p = \frac{P\sqrt{Rh}}{M_0\sqrt{2}}. \tag{8.32}$$

Given the numerical value of the temperature field parameter, the ultimate load parameter can be found from the condition

$$\max p = ?. \tag{8.33}$$

The problem (8.29)–(8.33) will now be discretized by assuming the function $n_\varphi^0(x)$ to be continuous and linear within each of the equal intervals and by imposing the constraints (8.29), (8.30) only at the junctions of those intervals. The longitudinal moment m_x^0 will be eliminated from the computations with the help of (8.31). We obtain the linear programming problem

$$\max p = ?,$$

$$|n_{\varphi k}^0| \leq 1 - \frac{1}{2} q_*,$$

$$\left| (\Delta x)^2 \left(\frac{2k-3}{4} n_{\varphi 1}^0 + \frac{1}{4} n_{\varphi k}^0 + \sum_{i=2}^{k-1} \frac{k-i+1}{2} n_{\varphi i}^0 \right) - \right. \tag{8.34}$$

$$\left. - p\Delta x(k-1) + m_x^0(0) \right| \leq 1 - \frac{2}{3} q_* \quad \text{for } k = 2, 3, \ldots, s,$$

$$|m_x^0(0)| \leq 1 - \frac{2}{3} q_*.$$

The numbering of the cross-sections is shown in Fig. 8.2. A solution to the problem (8.34) for shells of various lengths was obtained by using a computer. The simplex method was programmed with variable integration step Δx, the parameter q_* was given numerically. It turned out that for $x > 3$ the shakedown load parameter p was practically independent of the shell length. Moreover, the parameter p was hardly

8 Shakedown of shells

dependent on the integration step when $\Delta x < 1$; it remained practically constant when $\Delta x < 0{,}25$ (at given q_*).

The relationship between the limiting values of p and the thermal cycling intensity is shown in Fig. 8.4 by little circles which happen to lie on the line 3. The hoop forces were constant along the shell, $n_x^0 = 1 - \frac{1}{2}q_*$ for $0 \leq x \leq l$. Longitudinal bending moments reached their ultimate magnitudes in two-cross-sections: at $x = 0$ and $x = l$, where the coordinate $l \leq 3$ is depending on the magnitude of q_*.

Thus the incremental collapse solution obtained earlier on the basis of the approximate kinematical method coincides with the exact solution, line 3; in fact, both solutions stemmed from the same approximation of the interaction surface. Fig. 8.9 shows that the rigorous solution obtained according to the square approximation of the yield condition is an upper bound on the limiting cycle parameters.

It is interesting to solve the problem for a piece-wise linear fictitious interaction curve inscribed in the curve (8.21), (8.22) and to obtain the lower bound estimation. Let us adopt the inscribed hexagon

$$|m_x^0| \leq 1 - \tfrac{2}{3}q_*,$$

$$|n_\varphi^0| + \frac{1}{2}|m_x^0| = 1 - \frac{1}{2}q_*. \tag{8.35}$$

The bounds of the domains of admissible constant stress resultants determined by (8.35) are shown in Fig. 8.10 by thick solid lines for various magnitudes of q_* starting from $q_* = 0$. The bounds corresponding to (8.21), (8.22) are drawn in thin solid lines. Figs. 8.9 and 8.10 show that the degree of approximation to the exact relationships (8.21), (8.22) varies with the increase in the thermal parameter q_*.

Exactly in the same fashion as before we can introduce the discrete model for reducing the shakedown problem under the conditions (8.35) to a linear programming problem. The results are shown in Fig. 8.4 by line 4 obtained for $\Delta x = 1$. Comparison of line 4 with line 3 indicates that the differences between the results corresponding to the rectangular approximation (8.29), (8.30) and the hexagonal approximation (8.35) do not exceed 17 per cent at $q_* = 0$, and absolute differences decrease as the parameter q_* increases.

The relationship between the limiting value of the load parameter p and the integration step Δx at $q_* = 0$ is illustrated in Fig. 8.11 where the reciprocals of Δx are marked on the horizontal axis. The dashed line corresponds to the value 1,73 of the ultimate load parameter which is the exact analytical result to be considered in the next section. The solid line

8.1. Cylindrical shell under ring load and thermal cycling

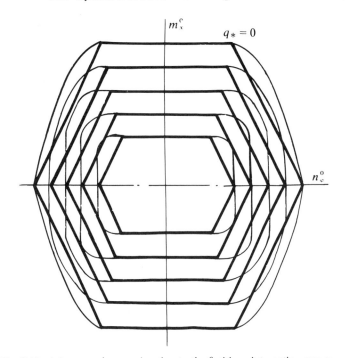

Fig. 8.10. A hexagonal approximation to the fictitious interaction curves.

shows the discrete model solution which is found to be fully acceptable for $\Delta x < 1$ ($\Delta X < \frac{1}{2}\sqrt{2Rh}$).

The hoop force distribution under the condition (8.35) is shown in Fig. 8.12 at $q_* = 0$ which represents limit equilibrium of the shell, for various integration steps Δx. The forces obtained for $\Delta x = 1$ and $\Delta x = 0{,}5$ differ considerably at $0 \leq x \leq 1$, becoming close enough to each other at $1 < x < 3$. Thus it follows from Fig. 8.11 that the differences between respective values of ultimate loads are relatively small. At shell cross-sections that are sufficiently far away from the plane of the ring load P, $x > 3$, the n_φ^0-distribution is unstable and it does not converge to any definite solution. However it can be shown that those portions of the shell remain elastic in spite of the plastic collapse of the structure. Thus the hoop force distribution on them has nothing to do with the magnitude of yield point load. Moreover, the ensuing stress distribution at collapse does not have to be unique.

The thick diagram in Fig. 8.12 shows the hoop force field at limit equilibrium, found under the square condition, $n_{\varphi*}^0 = -1$ for $0 < x < 2$.

353

8 Shakedown of shells

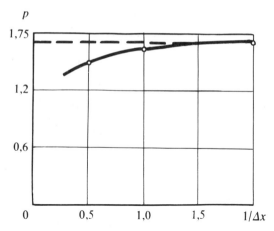

Fig. 8.11. Influence of integration step on the magnitude of limit load.

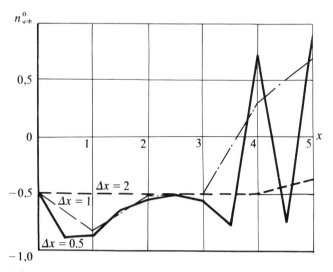

Fig. 8.12. Influence of integration step on the distribution of circumferential forces at limit equilibrium.

8.2. Limit analysis of shells by the Pontragin principle

8.2. Limit analysis of cylindrical elastic-plastic shells by the Pontragin maximum principle

1. Consider a cylindrical shell subjected at one end to a time-independent ring force and a ring bending moment, Fig. 8.13. Assuming the conditions of either instantaneous or incremental collapse we shall find the relationships between the ultimate loads Q_1 and M_1, as depending on the applied variable actions and the shell dimensions such as length, radius and thickness.

Using the dimensionless variables [8.1, 8.7], we can write the differential equilibrium equations for a hollow cylinder of a constant thickness $2h$ in the form [8.18]

$$dm_x^0/dx = r_x^0, \quad dr_x^0/dx = -n_\varphi^0, \tag{8.36}$$

where $r_x^0 = Q_x^0\sqrt{Rh}/(M_0\sqrt{2})$ is the dimensionless shear force, and n_φ^0, m_φ^0 stand for the time-independent dimensionless hoop force and the longitudinal bending moment, respectively. For the considered shell the equations (8.36) should be solved under the boundary conditions

$$r_x^0(0) = 0, \quad m_x^0(0) = 0,$$
$$r_x^0(l) = r_1^0, \quad m_x^0(l) = m_1^0, \quad (l = L\sqrt{2}/\sqrt{Rh}). \tag{8.37}$$

To find the domain of admissible time-independent load factors r_1^0, m_1^0, let us determine how the max r_1^0 and $-\max(-r_1^0) = \min r_1^0$ depend on the

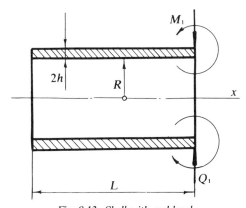

Fig. 8.13. Shell with end load.

355

8 Shakedown of shells

given value m_1^0 or alternatively, how the max m_1^0 and min m_1^0 depend on r_1^0.

Since the yield point σ_s is assumed to be constant, the cycle of varying elastic stresses can be reduced to become symmetric for all the points of the shell. Thus the fictitious interaction surfaces for the constant stress resultants are also symmetric and their construction can be conveniently simplified. Let us adopt a somewhat simpler, rectangular approximation to those surfaces given by the equations

$$|n_{\varphi*}^0| = \varphi_1(x, q_*), \quad |m_{x*}^0| = \varphi_2(x, q_*), \tag{8.38}$$

where φ_1 and φ_2 denote the non-negative-valued functions of the coordinate x and of the variable external action factors conventionally denoted as q_*.

The two differential equilibrium equations (8.36) in the three unknown functions can be looked upon as equations of a control process. As shown in [8.18], the functions m_x^0, r_x^0 are continuous whereas the function n_φ^0 tolerates discontinuities of the first kind. Thus the latter will play the part of the control while m_x^0 and r_x^0 will be the phase coordinates. In accordance with the restated statical formulation of the problem we should find an optimum control which makes the phase coordinate $r_x^0 = r_1^0$ attain its maximum (minimum) at the end of the trajectory, $x = 1$, for the prescribed magnitude of m_1^0.

To ensure that the right-hand sides of (8.38) are independent of x, let us adopt the new variables

$$u = \frac{n_\varphi^0}{\varphi_1(x, q_*)}, \quad y = \frac{m_x^0}{\varphi_2(x, q_*)}. \tag{8.39}$$

Remembering that the functions φ_1, φ_2 are both non-negative-valued, we see that the constraints on the internal forces resulting from the accepted approximation of the fictitious interaction surface (8.38) take the form

$$|u| \leq 1, \quad |y| \leq 1. \tag{8.40}$$

Substitution of (8.39) into (8.36) leads to the equations

$$\frac{dy}{dx} = \frac{1}{\varphi_2}\left(r_x^0 - y\frac{d\varphi_2}{dx}\right), \quad \frac{dr_x^0}{dx} = -u\varphi_1. \tag{8.41}$$

8.2. Limit analysis of shells by the Pontragin principle

The maximization of r_1^0 in the presence of the constraints (8.40), (8.41) for the open region of the admissible phase coordinates can be made with the help of the theorem [8.19] discussed in Chapter 4. However it is more convenient to employ the specially modified theorem [8.20] considered in Sec. 4.1 for seeking the extremum of a phase coordinate at the end of a trajectory (see (4.14)). Following this procedure, we obtain the relevant Hamiltonian and the associated system of equations in the form

$$H = \frac{\psi_1}{\varphi_1}\left(r_x^0 - y\frac{d\varphi_1}{dx}\right) - \psi_2 u \varphi_2, \tag{8.42}$$

$$\frac{d\psi_1}{dx} = -\frac{\partial H}{\partial y} = \frac{\psi_1}{\varphi_1} \cdot \frac{d\varphi_1}{dx}, \quad \frac{d\psi_2}{dx} = -\frac{\partial H}{\partial r_x^0} = -\frac{\psi_1}{\varphi_1}, \quad \psi_2(l) = -1. \tag{8.43}$$

At the maximum of r_1^0, the control u makes the function H reach its minimum and vice versa. Bearing in mind the constraints (8.40) and the non-negativity of φ_2, we conclude that

$$u(x) = \operatorname{sign} \psi_2. \tag{8.44}$$

The solution

$$\psi_1 = C\varphi_1, \quad \psi_2 = C(l - x) - 1 \tag{8.45}$$

to the system (8.43), where C is an unknown constant, shows that the function ψ_2 has no more than two intervals inside which its sign remains constant. Therefore (8.44) can be rewritten in the form

$$u(x) = 1 \text{ for } x = a, \quad u(x) = -1 \text{ for } a \leqslant x \leqslant l, \tag{8.46}$$

where a is the coordinate of the unknown control switching-point, i.e. of the section at which the control $u(x)$ changes sign.

Let us substitute (8.46) into the system (8.41) and solve the latter for r_x^0 and y. Using (8.39), we can determine the integration constants from the boundary conditions (8.37) at $x = 0$; we can find a from the equality $m_x^0(l) = m_1^0$, insisting that the functions $m_x^0(x)$, $r_x^0(x)$ be both continuous.

As an illustration let us consider a somewhat simpler situation in which $\varphi_1(x, q_*) = \varphi_2(x, q_*) = 1$, i.e. the limit analysis problem of the shell

8 Shakedown of shells

of a prescribed length l. Employing the above obtained results, we get

$$a = l\left(1 - \sqrt{\frac{1}{2} + \frac{m_1^0}{l^2}}\right), \quad \max r_1^0 = l\left(2\sqrt{\frac{1}{2} + \frac{m_1^0}{l^2}} - 1\right). \tag{8.47}$$

Insisting on $\min(r_1^0)$, we get

$$a = l\left(1 - \sqrt{\frac{1}{2} - \frac{m_1^0}{l^2}}\right), \quad \min r_1^0 = -l\left(1 - 2\sqrt{\frac{1}{2} - \frac{m_1^0}{l^2}}\right). \tag{8.48}$$

The formulae (8.47) and (8.48) show that for $(m_1^0/l^2) \geq 0{,}5$ the value a is not contained inside the interval $(0,1)$. This means that the hoop force does not change sign and remains constant throughout the shell,

$$|n_\varphi^0| = 1. \tag{8.49}$$

The plastic collapse mechanism of a shell of suitable dimensions consists in the hoop squeezing (or stretching) along the whole length. When $(m_1^0/l^2) < 0{,}5$, the collapse is accompanied by rotation of generators around the ring $x = a$ and the hoop force n_φ has opposite signs on the two sides of the section $x = a$.

Solutions (8.47), (8.48) are valid only when $(m_x^0) < 1$ inside the interval $(0, l)$, i.e. when no yield hinge circles are formed in the shell. Otherwise some additional conditions should be satisfied [8.19] so that we can find the extremum of r_1^0 in the presence of sections (or finite annular zones) at which $|m_x^0(x)| = 1$.

However, on assuming that inside the length $(0, l)$ no more than one hinge circle can form, i.e. that the equality $|m_x^0| = 1$ is met at no more than one section (at $x = x_*$ where $0 \leq x_* \leq l$, $r_x^0(x_*) - 0$), the problem can be reduced to the one considered above by simply measuring the coordinate x from the plane of the hinge circle, instead of the shell end, and adopting $l - x_* = l_*$ as an unknown quantity (clearly, $m_x(0)$ does not vanish). The additional condition $H(l_*) = 0$ [8.20] provides l_*. The results are:

$$\max r_1^0 = l\sqrt{2\left(\frac{1}{l^2} + \frac{m_1^0}{l^2}\right)} \quad \text{for} \quad \frac{1}{l} \leq 1 - \sqrt{\frac{1}{2} + \frac{m_1^0}{l^2}} \leq 2, \tag{8.50}$$

$$\min r_1^0 = -l\sqrt{2\left(\frac{1}{l^2} - \frac{m_1^0}{l^2}\right)} \quad \text{for} \quad \frac{1}{l} \leq 1 - \sqrt{\frac{1}{2} - \frac{m_1^0}{l^2}} \leq 2. \tag{8.51}$$

8.2. Limit analysis of shells by the Pontragin principle

The domain of safe load parameters r_1^0, m_1^0 is shown in Fig. 8.14. The lines 1 correspond to $0 \le l \le \sqrt{2}$ and they are related to the equalities (8.47), (8.48). When $\sqrt{2} \le l \le 2$, we have for specific values of l additional boundaries, following from the condition $|m_1^0| \le 1$ and shown as dashed lines 2. In neither of the above intervals, i.e. nowhere inside $0 \le l \le 2$ any plastic hinges are formed.

When $2 \le l \le 4$, the boundary of the safe domain of r_1^0/l and m_1^0/l^2 is determined, as before by (8.47), (8.48) and by the condition $|m_1^0| \le 1$ (lines 1 and 2). However the additional conditions (8.50), (8.51) are now active, shown in Fig. 8.14 by dash-dotted lines 3. Plastic collapse can occur here either with, or without hinge circles. For instance, when $m_1^0/l^2 = C$ is prescribed, the hinge circle will be present provided the vertical straight line determined by C intersects the corresponding line 3 within the safe domain, Fig. 8.14. When this vertical line is found to intersect the line 2, no hinge circle forms. Thus, for $2 \le l \le 4$ the presence or absence of the hinge circle at collapse of the tube loaded with constant moment M_1 is conditioned by the direction of the end force Q_1.

Finally, when $l > 4$ the yield hinge circle is definitely expected to form and the safe actions domain is bounded by (8.50), (8.51) and $|m_1^0| \le 1$.

The diagram of Fig. 8.14 can readily be linearized so that it can be employed, in particular, in the computations of complex tubular structures with the help of linear programming; here a direct application of the maximum principle seems to be rather difficult.

2. Attempts to use the hexagonal approximation to the yield condition (8.35), see thick lines in Fig. 8.10, lead even at $q_* = 0$ to considerable difficulties. To get rid of the relationship between the constraints imposed on the control and the phase coordinates, let us introduce the two control parameters

$$u_1 = |n_\varphi^0| + \frac{1}{2}|m_x^0| \quad \text{and} \quad u_2 = \operatorname{sign} n_\varphi^0 = \frac{n_\varphi^0}{|n_\varphi^0|} \tag{8.52}$$

that uniquely determine n_φ^0 (at known m_x^0). We get

$$n_\varphi^0 = u_1 u_2 - \frac{1}{2}|m_x^0|u_2. \tag{8.53}$$

We replace the conditions (8.35) by

$$|m_x^0| \le 1, \quad 0 \le u_1 \le 1, \quad u_2 = 1 \quad \text{or} \quad u_2 = -1. \tag{8.54}$$

8 Shakedown of shells

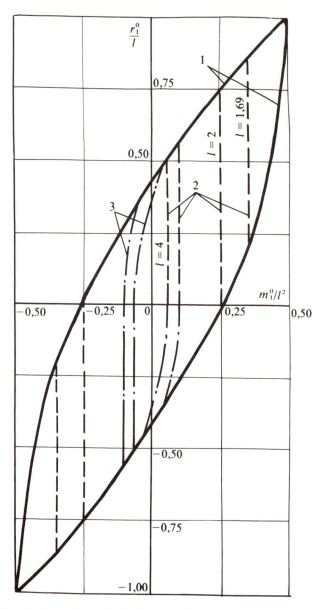

Fig. 8.14. Domain of safe load parameters for the shell of Fig. 8.13.

8.2. Limit analysis of shells by the Pontragin principle

Then the system of equations (8.36) takes the form

$$\frac{dm_x^0}{dx} = r_x^0, \quad \frac{dr_x^0}{dx} = -u_1 u_2 + \frac{1}{2}|m_x^0|u_2. \tag{8.55}$$

The Hamiltonian and the associated system are

$$H = \psi_1 r_x^0 + \psi_2 \left(\frac{1}{2}|m_x^0| - u_1\right) u_2, \tag{8.56}$$

$$\frac{d\psi_1}{dx} = -\frac{\partial H}{\partial m_x^0} = -\frac{1}{2}\psi_2 u_2 \text{ sign } m_x^0,$$

$$\frac{d\psi_2}{dx} = -\frac{\partial H}{\partial r_x^0} = -\psi_1, \quad \psi_2(l) = -1. \tag{8.57}$$

Insisting on max r_1^0 shows that the Hamiltonian is at minimum when

$$u_1 = 1, \quad u_2 = \text{sign } \psi_2. \tag{8.58}$$

Combining (8.55)–(8.58), we obtain the following system of conditions for the sought for max $r_1^0 = r_x^0(l)$ at $|m_x^0| < 1$:

$$\frac{dm_x^0}{dx} = r_x^0, \quad r_x^0(0) = 0,$$

$$\frac{dr_x^0}{dx} = \left(\frac{1}{2}|m_x^0| - 1\right) \text{sign } \psi_2, \quad m_x^0(0) = 0,$$

$$\frac{d\psi_1}{dx} = -\frac{1}{2}|\psi_2| \text{ sign } m_x^0, \quad m_x^0(l) = m_1^0, \tag{8.59}$$

$$\frac{d\psi_2}{dx} = -\psi_1, \quad \psi_2(l) = -1.$$

Finding the solution is difficult since the above system is linear only on those intervals of coordinates inside which the functions ψ_2 and m_x^0 do not change sign; those intervals are not known beforehand and must be established in the process of solving the system (8.59).

A rather complicated analysis of the cases in which $m_x^0(x)$ changes sign inside the interval $(0, l)$ either once, or twice, or more, shows that the solution to (8.59) is unique for any fixed pair of values l and m_1^0. An

8 Shakedown of shells

example of dependence of the parameter $r_1^0 = r_x(l)$ on the shell length is shown in Fig. 8.15 by line 1, valid for $m_1^0 = 0$. The same is shown for the square condition (8.1) by line 2.

It is worth mentioning that in the case of non-linear yield surfaces the methods of optimum control theory, i.e. the maximum principle, enable us to reduce the limit analysis problem to finding the solution to a certain system of non-linear differential equations [8.19]. However, the difficulties such as the use of the hexagonal yield condition (8.35), lead to cumbersome computations, so that once again the linear programming tools prove to be more suitable for handling the problems.

3. The above solutions to the limit analysis problems, based on various approximations of the interaction surface, were given for shells subject to time-independent actions and therefore instantaneous plastic collapse was the only unsafe state possible. It is clear that the relationships (8.39)–(8.41) together with the continuity condition of the function $m_x^0(x)$ make it possible to solve similar problems of limit equilibrium in more complex situations; for instance, taking account of the temperature-dependent yield point or admitting the shakedown situation of the cylindrical shell as shown in Fig. 8.2.

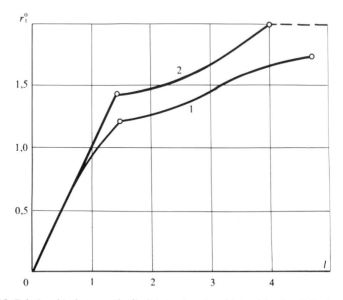

Fig. 8.15. Relationship between the limit intensity of end lateral load and the length of the shell at $m_1^0 = 0$.

8.2. Limit analysis of shells by the Pontragin principle

Let the ring load P be constant in time and the temperature vary according to (8.2). A shell of this kind was analysed in Sec. 8.1, but there its length L was sufficiently large to be considered infinite. Here a shell with finite length will be dealt with. For simplicity, the fictitious yield surface (8.29), (8.30) is assumed to apply along the whole shell length and the boundary conditions are assumed to have no effect on the thermo-elastic stresses. The above simplifications can be relaxed, but at considerable cost as regards brevity of the involved formulae.

Due to symmetry, only one half of the shell has to be analysed, as shown in Fig. 8.16. Combining (8.29), (8.30) and (8.38), we get:

$$\varphi_1(x, q_*) = 1 - \frac{1}{2} q_*, \quad \varphi_2(x, q_*) = 1 - \frac{2}{3} q_*. \tag{8.60}$$

Unlike in the condition (8.37), for the shell of Fig. 8.16 the bending moment at $x = 1$ can take on any value, provided it obeys the constraints (8.29). The optimum control problem which arises in seeking max r_1^0 of such a type is termed a problem with the free right end of the trajectory. According to [8.20], the associated system (8.43) has to include the supplementary boundary condition (4.16) which here takes the form

$$\psi_1(l) = \mu_1 \operatorname{sign} y(l) = \mu_1 \operatorname{sign} m_x^0(l), \tag{8.61}$$

where μ_1 denotes a non-negative multiplier, non-vanishing only for $|y(l)| = 1$, i.e. where the hinge circles form. The system (8.43) together with the condition (8.61) leads to the equality

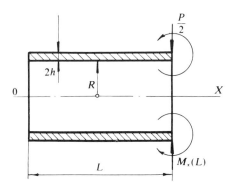

Fig. 8.16. Cylindrical shell of finite length.

8 Shakedown of shells

$$u(x) = \text{sign}[\mu_1(l-x)\text{ sign } y(l) - 1]. \tag{8.62}$$

Using (8.39), we deduce from (8.62) that

$$n_\varphi^0(x) = 1 - \frac{1}{2}q_* \quad \text{for} \quad 0 \leq x \leq a,$$
$$n_\varphi^0(x) = -\left(1 - \frac{1}{2}q_*\right) \quad \text{for} \quad a \leq x \leq l, \tag{8.63}$$

where

$$a = l - \frac{1}{\mu_1 \text{ sign } m_x^0(l)}. \tag{8.64}$$

If a lies outside the interval $(0, l)$ or $\mu_1 = 0$, i.e. no hinge circle forms in the section $x = l$, the second condition (8.63) is valid along the whole shell, i.e. we have an occurrence of 'no-shift control'.

Integrating the system (8.36) with the help of (8.63), using the continuity conditions for m_x^0, r_x^0 and the given boundary conditions, we get

$$p = \left(1 - \frac{1}{2}q_*\right)l \quad \text{for} \quad l \leq \sqrt{\frac{2(1-\frac{2}{3}q_*)}{1-\frac{1}{2}q_*}}. \tag{8.65}$$

For

$$\sqrt{\frac{2(1-\frac{2}{3}q_*)}{1-\frac{1}{2}q_*}} \leq l \leq 4\sqrt{\frac{1-\frac{2}{3}q_*}{1-\frac{1}{2}q_*}}$$

we obtain

$$p = \left(1 - \frac{1}{2}q_*\right)\left(2\sqrt{\frac{1}{2}l^2 + \frac{1-\frac{2}{3}q_*}{1-\frac{1}{2}q_*}} - l\right),$$
$$a = l - \sqrt{\frac{1}{2}l^2 + \frac{1-\frac{2}{3}q_*}{1-\frac{1}{2}q_*}}. \tag{8.66}$$

For still larger l the magnitude of the parameter p is given by

$$p = 2\sqrt{\left(1-\frac{1}{2}q_*\right)\left(1-\frac{2}{3}q_*\right)}, \tag{8.67}$$

8.2. Limit analysis of shells by the Pontragin principle

i.e. it remains the same as for

$$l = 4\sqrt{\frac{1-\frac{2}{3}q_*}{1-\frac{1}{2}q_*}}.$$

The last result was arrived at in Sec. 8.1, where long enough shells were considered by means of the approximate kinematical method as well as numerically with the use of linear programming.

The shakedown diagram in the dimensionless coordinates $p/p_0 = P/P_0$, $\beta = q_*/q_0 = t_*/t_0$ is shown in Fig. 8.17. Account has been taken the facts that, according to (8.67), for the long shell the limit equilibrium load factor is equal to $p_0 = 2$ and the limiting value of the temperature field factor with respect to alternating flow is equal to $q^0 = 1$ according to (8.13) and (2.90). This permits us to compare the present diagram with that of Fig. 8.4, constructed for the long shell in the previous section.

As usual, the horizontal straight line $\beta = 1$ corresponds to the alternating plastic flow at $P = \text{const}$. For $l \leq 2{,}67$ the incremental collapse condition is expressed by (8.66) and it can be satisfied without

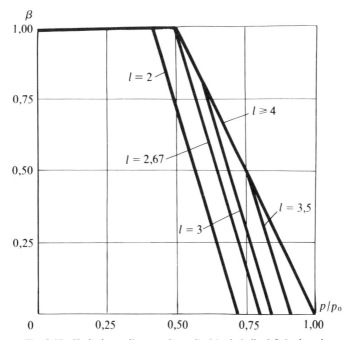

Fig. 8.17. Shakedown diagram for cylindrical shell of finite length.

8 Shakedown of shells

formation of any hinge circles. For $2{,}67 < l < 4$ the limiting cycle boundary consists of two segments: at low intensity of thermal cycling we have, as before, the equations (8.66); at higher intensities the formula (8.67) associated with the formation of a yield hinge circle becomes valid. Finally, for $l > 4$ the hinge circle forms at any thermal cycling intensity.

It is worth mentioning that no extra difficulties arise when one has to take account of distributed loads such as internal or external pressure or centrifugal forces.

8.3. Shakedown of cylindrical shells under moving mechanical load

As mentioned earlier, the differences between the ultimate values of purely mechanical variable loads determined under the instantaneous and the progressive plastic collapse conditions are found to be often insignificant. However, it should be expected that those differences at moving mechanical load can be substantial because the enveloping stress distributions are in this case more uniform than in the case of fixed load and at the same time non-isochronous. Such situations are encountered in the design of bridges, railway tracks and cranes, see for instance [8.21–8.23], as well as in the plastic working processes such as rolling or drawing and in contact problems dealing with roller and ball bearings.[1]

Consider a long cylindrical shell, Fig. 8.18, subject to two equal ring loads which move slowly and repeatedly from one end of the shell to the other while preserving constant spacing $2l$.

For small spacing l the above problem is clearly reduced to repeated passages of the ring load P (in Secs. 8.1 and 8.2 a fixed load P was considered); and for sufficiently large spacing l, to repeated passages of the ring load $0{,}5P$.

Let us find now the upper bound on the incremental collapse load factor, using the approximate kinematical method. In the considered case of infinitely long shell the elastic stress distribution is quasi-stationary with respect to the coordinate system moving together with the ring load. The suitable incremental collapse mechanism should thus consist in uniform, but non-simultaneous, plastic compression in the direction of the hoop. The instantaneous distribution of the increments

---

[1] Certain contact problems of the shakedown theory in the presence of moving load will be considered in Chapter 10.

8.3. Shells under moving mechanical load

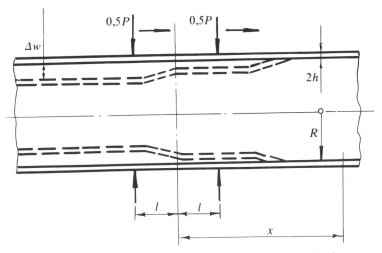

Fig. 8.18. Cylindrical shell under repeated action of moving load.

of residual displacements is roughly shown in Fig. 8.18; the passage of load along the tube generates a moving wave of plastic deformation. The hoop plastic strain increments per cycle depend on the radial displacement increments of the median surface, namely

$$\Delta\epsilon_\varphi'' = -\frac{\Delta w}{R}, \quad (\Delta\epsilon_\varphi'' < 0). \tag{8.68}$$

The longitudinal displacement increments and the corresponding longitudinal plastic strain increments $\Delta\epsilon_x''$ are assumed to vanish throughout the tube. Thus condition (2.88) for the commencement of incremental collapse takes the form

$$\int_{-h}^{h} \sigma_{\varphi *}^0 \Delta\epsilon_\varphi'' \, dz = 0. \tag{8.69}$$

Integration along the shell is here not involved due to the already stated fact that neither the determining hoop stresses $\sigma_{\varphi T}^*$ nor, in consequence, the fictitious yield surface hoop stresses $\sigma_{\varphi *}^0$ depend on the axial coordinate. Let us find the latter assuming the Tresca–Saint Venant material of the shell. If, according to the flow rule in the form (2.76), the strain vector $(\Delta\epsilon_\varphi'' < 0, \Delta\epsilon_x'' = 0)$ is associated with a smooth portion of the fictitious yield surface, then

8 Shakedown of shells

$$\sigma^0_{\varphi *} = \max_{\tau}(-\sigma_s - \sigma^{(e)}_{\varphi\tau}), \tag{8.70}$$

where $\sigma^{(e)}_{\varphi\tau}$ denotes the instantaneous distribution of hoop elastic stresses across the thickness, generated by the two ring forces $0,5P$ and varying with the axial coordinate x of the cross-section. For a long tube we have that $-\infty < x < \infty$. The right-hand side of (8.70) should be maximized with respect to x which plays here the part of current 'time', since on the passage of load, at each fixed cross-section of the tube the stresses are successively generated corresponding to their overall instantaneous distribution along the tube, $\sigma^{(e)}_{\varphi\tau} = \sigma^{(e)}_\varphi(x)$.

If the vector $(\Delta\epsilon''_\varphi < 0, \Delta\epsilon''_x = 0)$ is associated with a corner of the fictitious yield surface, the expression (8.70), as was shown in Chapter 2, yields the upper bound on the hoop stresses belonging to the fictitious yield surface.

Substituting (8.68) and (8.70) into the equality (8.69), we obtain

$$\int_{-h}^{h} \min_{x}\left\{[-\sigma_s - \sigma^{(e)}_\varphi(x)]\left(-\frac{\Delta w}{R}\right)\right\} dz = 0;$$

hence

$$\int_{-h}^{h} [\sigma_s + \min_{x} \sigma^{(e)}_\varphi(x)] \, dz = \int_{-h}^{h} [\sigma_s + \sigma^*_\varphi] \, dz = 0. \tag{8.71}$$

The total elastic stresses generated by the two ring loads equal to $0.5P$ can be readily found by applying the superposition principle to the solution [8.18] obtained for a tube subject to a single ring load. After simple rearrangements the relevant formulae take the form

$$\frac{\sigma^{(e)}_\varphi}{\sigma_s} = -p\left[\varphi_1(\beta x) - \frac{3\mu}{\sqrt{3(1-\mu^2)}} \zeta\psi_1(\beta x)\right], \tag{8.72}$$

$$\frac{\sigma^{(e)}_x}{\sigma_s} = \frac{3}{\sqrt{3(1-\mu^2)}} p\zeta\psi_1(\beta x), \tag{8.73}$$

where

$$p = \frac{PRh\beta}{4M_0}, \quad \beta^4 = \frac{3(1-\mu^2)}{4R^2 h^2}, \tag{8.74}$$

8.3. Shells under moving mechanical load

$$\varphi_1(\beta x) = \frac{1}{2}\varphi[\beta(x+l)] + \frac{1}{2}\varphi(\beta|x-l|),$$
$$\psi_1(\beta x) = \frac{1}{2}\psi[\beta(x+l)] + \frac{1}{2}\psi(\beta|x-l|),$$
(8.75)

$$\varphi(\beta x) = e^{-\beta x}(\cos\beta x + \sin\beta x),$$
$$\psi(\beta x) = e^{-\beta x}(\cos\beta x - \sin\beta x).$$
(8.76)

The functions $\varphi_1(\beta x)$ and $\psi_1(\beta x)$ are plotted in Fig. 8.19 for $\beta l = 0{,}3$ and $\beta l = 0$ by solid and dashed lines, respectively.

The enveloping distribution of variable elastic stresses σ_φ^* entering (8.71) follows from the formula (8.72):

$$\sigma_\varphi^* = -p\max_{\beta x}\left[\varphi_1(\beta x) - \frac{3\mu}{\sqrt{3(1-\mu^2)}}\zeta\psi_1(\beta x)\right].$$
(8.77)

The magnitude βx stands for the distance between the considered cross-section and the ring loads, Fig. 8.18, and varies within the interval $-\infty \leq \beta x \leq \infty$. In the moving coordinate system the enveloping distribution of hoop stresses in the fixed cross-section corresponds to the same βx, but to different instants of time τ.

The enveloping distribution of the stresses σ_φ^* across the tube thickness is shown for $\beta l = 0{,}3$ by solid line, Fig. 8.20. The straight segment ABC corresponds to $\beta x = 0$. For a part of the thickness, $-1 \leq \zeta \leq 0{,}5$, the hoop elastic stresses reach their minimum when the cross-section is equidistant from both ring loads. For the opposite quarter of the thickness, $0{,}5 \leq \zeta \leq 1$, the stresses $\sigma_\varphi^{(e)}$ are at minimum for $\beta x = 0{,}6$; this is shown by the curve CD in Fig. 8.20.

On substituting (8.77) into (8.71) we get after some computations the upper bound on the incremental collapse load factor. For $\beta l = 0{,}3$ we obtain $p_* = 1{,}06$.

The alternating flow condition follows from the criterion (2.90) and takes here the form

$$\max_\tau(\sigma_{x\tau}^{(e)} - \sigma_{\varphi\tau}^{(e)}) - \min_\tau(\sigma_{x\tau}^{(e)} - \sigma_{\varphi\tau}^{(e)}) = 2\sigma_s$$
(8.78)

yielding the appropriate load factor $p_0 = 1{,}19$, whereas the critical points lie on the inner surface of the tube. Thus, for $\beta l = 0{,}3$, the load-carrying capacity of the tube is limited by the incremental collapse.

Let us quote here also the magnitudes of ultimate load at instantaneous plastic collapse. Those are

8 Shakedown of shells

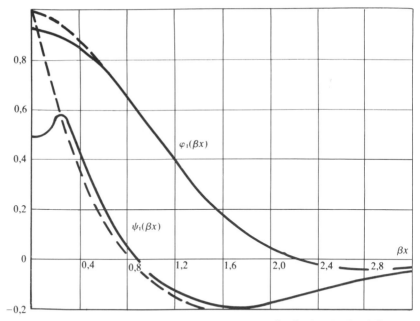

Fig. 8.19. Functions to determine the instantaneous stress distributions in the shell of Fig. 8.18.

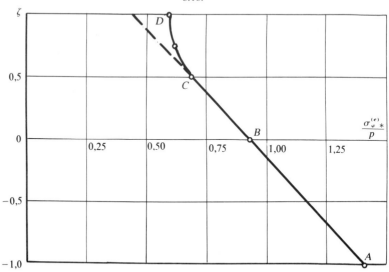

Fig. 8.20. Determination of enclosing quasi-stationary stress distribution in the shell at repeated load passage.

8.3. Shells under moving mechanical load

$$p_0 = \beta l + \sqrt[4]{3(1-\mu^2)} \quad \text{for} \quad \beta l \leq \sqrt[4]{3(1-\mu^2)}, \tag{8.79}$$

$$p_0 = 2\sqrt[4]{3(1-\mu^2)} \quad \text{for} \quad \beta l \geq \sqrt[4]{(3(1-\mu^2)}. \tag{8.80}$$

For $\mu = 0{,}3$ we obtain $\sqrt[4]{3(1-\mu^2)} \approx 1{,}285$. The corresponding mechanisms are shown in Fig. 8.21a, b. For $\beta l = 0{,}3$ it follows from formula (8.79) that the dimensionless ultimate load amounts to $p_0 = 1{,}58$. The difference in comparison with the limiting cycle load is very substantial, and this means that it is the moving load which should be employed to deform the tubes plastically. In other words, the moving ring load necessary for the process would be 1,5 times lower than the ring load at rest.

Let us now investigate the influence of the profile of the ring load on the limiting conditions for the alternating flow as well as for the incremental and instantaneous plastic collapse. Compare first the above obtained results with those corresponding to a single ring force P, i.e. to $l = 0$, and then with those corresponding to a band load of intensity P/l, Fig. 8.22, both for $\beta l > 3$. In each case we are considering the repeated passages of constant load.

The functions determining the instantaneous distribution of elastic stresses for $l = 0$ are shown in Fig. 8.19; those branches which do not coincide with the curves for $\beta l = 0{,}3$ are dashed. The function $\varphi_1(\beta x)$ generating the average hoop stresses at each cross-section is seen to change negligibly when βl drops from 0,3 to zero; its largest increase is from 0,93 to 1,0. At the same time, the function $\varphi_1(\beta x)$ generating the

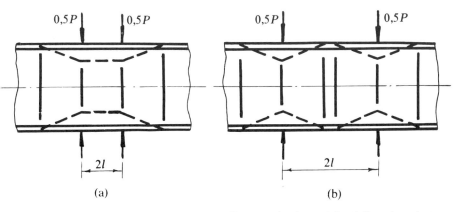

Fig. 8.21. Admissible instantaneous plastic collapse mechanisms of the shell as dependent on the spacing between two ring loads.

8 Shakedown of shells

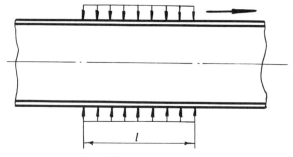

Fig. 8.22. Moving band load.

stresses at bending is seen to increase appreciably; its maximum grows from 0,58 to 1,0, whereas its minimum remains practically the same, namely −0,2. The danger of incremental collapse is indicated, according to (8.71), by the level of the average, across the thickness, of the hoop stresses in the 'determining' cross-section of the tube. The danger of alternating plastic flow is indicated, in accordance with (8.78), by the largest variation of local stresses per cycle. Thus we have good grounds to expect that under constant resultant load the decrease in βl will exert strong influence on the condition for alternating flow.

Indeed, calculations show that for $\beta l = 0$ the limiting cycle load parameter is $p_* = 0,95$ with respect to the incremental collapse and $p_0 = 0,82$ with respect to the alternating flow; instantaneous collapse occurs at $p_0 \simeq 1,28$. Thus, the load-carrying capacity of the tube is now limited by the alternating flow, unlike for $\beta l = 0,3$ when the incremental collapse was critical.

In the case of moving band load, Fig. 8.22, the non-uniformity of instantaneous distribution of 'elastic' stresses along the coordinate x is decreasing and therefore alternating flow ceases to be critical. As the parameter βl increases, the difference between the incremental and the instantaneous collapse load factors tends to shrink. For $\beta l \geqslant 3$ it does not exceed 5 per cent.

8.4. Incremental collapse of a conical shell

Following [8.9], we shall consider a general procedure to approximately establish the conditions for incremental collapse of a rotationally symmetric conical shell with variable thickness, Fig. 8.23, acted upon by both mechanical and thermal agencies. Let the elastic meridional and

8.4. Incremental collapse of a conical shell

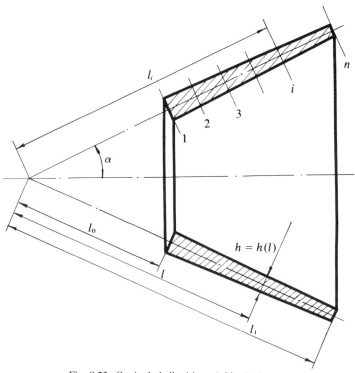

Fig. 8.23. Conical shell with variable thickness.

hoop stresses $\sigma^{(e)}_{l\tau}, \sigma^{(e)}_{\varphi\tau}$ generated by the variable external actions be determined to within an accuracy of the parameters $q_i(\tau)$, $i = 1, 2, \ldots, n$:

$$\sigma^{(e)}_{l\tau} = \sum_{i=1}^{n} q_i(\tau)\sigma^{(e)}_{li}(l, z), \quad \sigma^{(e)}_{\varphi\tau} = \sum_{i=1}^{n} q_i(\tau)\sigma^{(e)}_{\varphi i}(l, z), \tag{8.81}$$

where $\sigma^{(e)}_{li}(l, z)$, $\sigma^{(e)}_{\varphi i}(l, z)$ designate the distributions at $q_i(\tau) = 1$ of elastic stresses from appropriate i-th actions. The coordinates l, z are measured along the meridian of the median surface and along its normal, respectively.

The temperature-dependent yield point σ_{st}, varying in general inside a cycle, is expressed in the form

$$\sigma_{st} = \sigma_s - \Delta\sigma_{st}, \quad t = t(\tau), \tag{8.82}$$

373

8 Shakedown of shells

where the σ_s denote the yield point at the lowest temperature of the cycle, and $\Delta\sigma_{st}$ are the yield limit variations at each point of the shell reflecting the program of temperature variations. The latter will be formally treated as certain fictitious variable stresses to be added to the elastic stresses generated by variable external loads.

Let us employ the statical shakedown theorem, adopting generalized variables. Our analysis will begin with the construction, as in the approximate method in Chapter 3, of the fictitious interaction surface in the space of generalized stresses.

Let the actual yield surface for the conical shell be prescribed. The direct stress resultants and the bending moments belonging to this surface will be conveniently referred to either as $N_0 = 2\sigma_s h$ or $M_0 = \sigma_s h^2$, and denoted by n_l, n_φ, m_l, m_φ. The generalized stresses generated by variable agencies and allowing for the changes in the yield point will be designated by $n_{l\tau}^*$, $n_{\varphi\tau}^*$, $m_{l\tau}^*$, $m_{\varphi\tau}^*$. They are derivable from the equality of type (3.39)

$$\int_{-h}^{h} \max_\tau \{[\sigma_{l\tau}^{(e)} + \Delta\sigma_{st}](\Delta e_l + z\Delta\kappa_l) + [\sigma_{\varphi\tau}^{(e)} + \Delta\sigma_{st}](\Delta e_\varphi + z\Delta\kappa_\varphi)\}\,dz =$$
$$= N_0(n_{l\tau}^*\Delta e_l + n_{\varphi\tau}^*\Delta e_\varphi) + M_0(m_{l\tau}^*\Delta\kappa_l + m_{\varphi\tau}^*\Delta\kappa_\varphi), \qquad (8.83)$$

where Δe_l, Δe_φ, $\Delta\kappa_l$, $\Delta\kappa_\varphi$ are the increments of plastic elongations and curvatures per cycle duly associated with n_l, n_φ, m_l, m_φ by means of the flow rule.

The domain of admissible values of the time-independent generalized stresses n_l^0, n_φ^0, m_l^0, m_φ^0 is bounded by the fictitious interaction surface described by the equations (3.37) containing the terms:

$$n_{l*}^0 = n_l - n_{l\tau}^*, \quad n_{\varphi*}^0 = n_\varphi - n_{\varphi\tau}^*, \quad m_{l*}^0 = m_l - m_{l\tau}^*, \quad m_{\varphi*}^0 = m_\varphi - m_{\varphi\tau}^*. \qquad (8.84)$$

In order to find the ultimate values of the time-independent internal actions n_{l*}^0, $n_{\varphi*}^0$, m_{l*}^0, $m_{\varphi*}^0$, it is necessary to associate the ratios of components of the generalized strain increment vector (Δe_l, Δe_φ, $\Delta\kappa_l$, $\Delta\kappa_\varphi$) with each given point of the actual interaction surface, and substitute them into equation (8.83). The latter will then provide the internal actions of the variable external load which should be in turn substituted into the equalities (8.84).

In Figs. 8.24–8.27 the construction of the domain of the admissible constant agency for the Tresca material (2.12) is schematically shown in

8.4. Incremental collapse of a conical shell

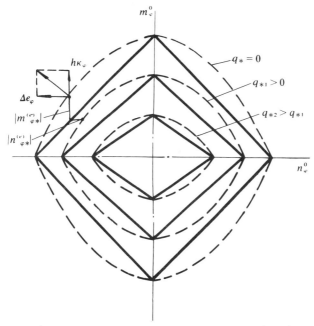

Fig. 8.24. Intersections of the fictitious yield surfaces by the plane $n_l^0 = m_l^0 = 0$ for various intensities of variable actions q_*.

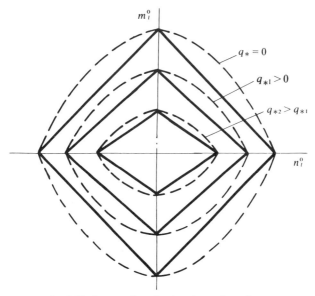

Fig. 8.25. Intersections by the plane $n_\varphi^0 = m_\varphi^0 = 0$.

8 Shakedown of shells

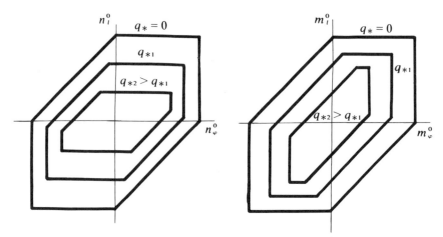

Fig. 8.26. Intersections by the plane $m_l^0 = m_\varphi^0 = 0$.

Fig. 8.27. Intersections by the plane $n_l^0 = n_\varphi^0 = 0$.

the case of a symmetric cycle of one-parameter external action $-q_* \leq q(\tau) < q_*$ at constant yield stress (8.81). The following intersections of the fictitious yield surface are shown in the figures: by the plane $n_l^0 = m_l^0 = 0$ (Fig. 8.24, dashed lines), by the plane $n_\varphi^0 = m_\varphi^0 = 0$ (Fig. 8.25, dashed lines), $m_l^0 = m_\varphi^0 = 0$ (Fig. 8.26), $n_l^0 = n_\varphi^0 = 0$ (Fig. 8.27). It is clear that in the situation considered the points of the fictitious surface corresponding to either pure stretching, $\Delta\kappa_l = \Delta\kappa_\varphi = 0$, or pure bending, $\Delta e_l = \Delta e_\varphi = 0$, remain on suitable coordinate axes, see Figs. 8.26, 8.27.

The actual interaction surface of the shell can be approximated in the known manner by an inscribed piece-wise linear surface with coinciding points of pure stretching and pure bending [8.24]:

$$\max[(|n_\varphi| + |m_\varphi|), (|n_l| + |m_l|), (|n_\varphi - n_l| + |m_\varphi - m_l|)] = 1. \tag{8.85}$$

We shall use a similar approximation to describe the fictitious interaction surface. Insisting on common points of pure stretching and pure bending, we can write

$$\max\left[\frac{|n_{\varphi*}^0|}{1-|n_{l\tau}^*|} + \frac{|m_{\varphi*}^0|}{1-|m_{l\tau}^*|}, \frac{|n_{l*}^0|}{1-|n_{l\tau}^*|} + \frac{|m_{l*}^0|}{1-|m_{l\tau}^*|}, \right. \\ \left. \frac{|n_{l*}^0 - n_{\varphi*}^0|}{1-|(n_{l\tau} - n_{\varphi\tau})^*|} + \frac{|m_{l*}^0 - m_{\varphi*}^0|}{1-|(m_{l\tau} - m_{\varphi\tau})^*|}\right] \equiv 1, \tag{8.86}$$

8.4. Incremental collapse of a conical shell

where $n_{l\tau}^*$, $n_{\varphi\tau}^*$, $m_{l\tau}^*$, $m_{\varphi\tau}^*$, $(n_{l\tau} - n_{\varphi\tau})^*$, $(m_{l\tau} - m_{\varphi\tau})^*$ denote the variable generalized stresses calculated in accordance with (8.83) at the following ratios of generalized strains, respectively:

$$\begin{aligned}
&\Delta e_l \neq 0, &&\Delta e_\varphi = \Delta\kappa_l = \Delta\kappa_\varphi = 0, \\
&\Delta e_\varphi \neq 0, &&\Delta e_l = \Delta\kappa_l = \Delta\kappa_\varphi = 0, \\
&\Delta\kappa_l \neq 0, &&\Delta e_l = \Delta e_\varphi = \Delta\kappa_\varphi = 0, \\
&\Delta\kappa_\varphi \neq 0, &&\Delta e_l = \Delta e_\varphi = \Delta\kappa_l = 0, \\
&\Delta e_l = -\Delta e_\varphi, &&\Delta\kappa_l = \Delta\kappa_\varphi = 0, \\
&\Delta\kappa_l = -\Delta\kappa_\varphi, &&\Delta e_l = \Delta e_\varphi = 0.
\end{aligned} \qquad (8.87)$$

Let us, for instance, establish the magnitudes entering (8.84) for a shell of constant thickness, $h(l) = h = \text{const.}$, subject to repeated temperature, varying linearly only across the thickness. The thermo-elastic stresses far enough from the edges of the shell are [8.18]:

$$\sigma_{\varphi\tau}^{(e)} = \sigma_{l\tau}^{(e)} = q(\tau)\zeta\sigma_s, \quad \zeta = z/h,$$

$$q(\tau) = \frac{\alpha E t_1(\tau)}{2(1-\mu)\sigma_s}, \quad -q_* \leq q(\tau) \leq q_*. \qquad (8.88)$$

The equality (8.83) can be used in the cases of pure stretching and pure bending. In the presence of (8.87) and (8.88) and for $\sigma_{st} = \sigma_s = \text{const.}$ we get:

$$n_{l\tau}^*\Delta e_l + n_{\varphi\tau}^*\Delta e_\varphi = \frac{h}{N_0}\int_{-1}^{1} \max_\tau [q(\tau)\zeta\sigma_s(\Delta e_l + \Delta e_\varphi)]\,d\zeta =$$

$$= \frac{1}{2}q_*(\Delta e_l + \Delta e_\varphi),$$

$$m_{l\tau}^*\Delta\kappa_l + m_{\varphi\tau}^*\Delta\kappa_\varphi = \frac{h^2}{M_0}\int_{-1}^{1} \max_\tau [q(\tau)\zeta^2\sigma_s(\Delta\kappa_l + \Delta\kappa_\varphi)]\,d\zeta = \qquad (8.89)$$

$$= \frac{2}{3}q_*(\Delta\kappa_l + \Delta\kappa_\varphi)$$

and therefore

$$n_{l\tau}^* = n_{\varphi\tau}^* = \frac{1}{2}q_*, \quad m_{l\tau}^* = m_{\varphi\tau}^* = \frac{2}{3}q_*, \quad (n_{l\tau} - n_{\varphi\tau})_*^* = (m_{l\tau} - m_{\varphi\tau})^* = 0. \qquad (8.90)$$

8 Shakedown of shells

In accordance with the statical shakedown theorem stated in terms of generalized variables (Sec. 3.6), the time-independent internal actions belong to the domain bounded by the fictitious yield surface and at the same time satisfy the equilibrium conditions. The latter, for an open conical shell of Fig. 8.23, have the form [8.18]:

$$n_l^0 x = \int_1^x n_\varphi^0 u \, d\zeta - \int_1^x q_l^0 u \xi \, d\xi + n_{l1}^0, \qquad (8.91)$$

$$m_l^0 x = m_{l1}^0 + \int_1^x m_\varphi^0 u^2 \, d\zeta + k(x-1)n_{l1}^0 + \int_1^x d\xi \int_1^\xi k(n_\varphi^0 - q_l^0 \eta) \, d\eta, \qquad (8.92)$$

where $u = h(\xi)/h(x)$, $x = l/l_0$, $k = 2l_0 \operatorname{ctg} \alpha / h$,

$$n_{l1}^0 = (n_l^0)_{x=1}, \quad m_{l1}^0 = (m_l^0)_{x=1},$$

q_l^0 is the meridional component of load distributed on the median surface of the shell and referred to N_0; and $p = p(x)$ denotes the axial load parameter in the cross-section l. In general, the equations (8.91), (8.92) should be supplemented by the stress boundary conditions.

Rewriting the constraints on the generalized stresses resulting from the conditions (8.86), (8.91) in discrete form similarly as was done in Sec. 7.3 for the circular plate, and supplementing the obtained system by the optimality criterion, i.e. the maximization of the unknown load parameter, we arrive at a linear programming problem that can be solved on a computer with the help of the standard simplex-method program.

To give a numerical example, we shall analyse a conical shell of constant thickness rotating with constant angular velocity; the dimensions are: $h = 2,5$ mm, $l_0 = 292$ mm, $l_1 = 380$ mm, $\alpha = 20°$. The edges are supported in such a way that the meridians cannot rotate, i.e. no stress constraints are imposed on the bending moment m_l. In addition the axial forces are restricted by the limiting values of the friction forces in support devices. Let us accept for example $n_{l1} \leq 0,5$. The thermal stresses are determined by (8.88).

Substituting (8.91) and (8.92) into the inequalities resulting from the description of the fictitious yield surface (8.86) and applying the boundary conditions formulated above, we obtain the discrete form of the incremental collapse problem

$$\max p = ?, \qquad (8.93)$$

8.4. Incremental collapse of a conical shell

$$\frac{1}{1-\frac{1}{2}q_*}|n_{\varphi i}^0| + \frac{1}{1-\frac{2}{3}q_*}|m_{\varphi i}^0| \leq 1, \tag{8.94}$$

$$\frac{1}{1-\frac{1}{2}q_*}\left|n_{l1}^0 - \frac{1}{3}p(x_i^3-1) + \frac{1}{2}\Delta x I_i\right| +$$

$$+ \frac{1}{1-\frac{2}{3}q_*}\left|m_{l1}^0 + \frac{1}{2}\Delta x \sum_{k=1}^{i-1}[(m_\varphi^0)_k + (m_\varphi^0)_{k+1}] + k(x_i-1)n_{l1}^0 +\right.$$

$$\left. + \frac{1}{4}\Delta x^2 k \sum_{k=1}^{i-1}[I_k + I_{k+1}] - \frac{1}{3}pk\left(\frac{1}{4}x_i^4 - x_i + \frac{3}{4}\right)\right| \leq x_i, \tag{8.95}$$

$$\frac{1}{1-\frac{1}{2}q_*}\left|n_{l1}^0 - \frac{1}{3}p(x_i^3-1) + \frac{1}{2}\Delta x I_i - n_{\varphi i}^0 x_i\right| + \frac{1}{1-\frac{2}{3}q_*} \times$$

$$\times \left|m_{l1}^0 + \frac{1}{2}\Delta x \sum_{k=1}^{i-1}[(m_\varphi^0)_k + (m_\varphi^0)_{k+1}] + k(x_i-1)n_{l1}^0 +\right.$$

$$\left. + \frac{1}{4}\Delta x^2 k \sum_{k=1}^{i-1}[I_k + I_{k+1}] - \frac{1}{3}pk\left(\frac{1}{4}x_i^4 - x_i + \frac{3}{4}\right) - m_{\varphi i}^0\right| \leq x_i,$$

$$i = 1, 2, 3, \ldots, n, \tag{8.96}$$

where

$$I_i = \sum_{k=1}^{i-1}[(n_\varphi^0)_k + (n_\varphi^0)_{k+1}],$$

$$p = \gamma\omega^2(l_0 \sin\alpha)/g\sigma_s, \quad \Delta x = x_i - x_{i-1} = \text{const}. \tag{8.97}$$

The numbering of cross-sections is indicated in Fig. 8.23. The ultimate values of the centrifugal force factor p can be found for numerically given values of q_* by solving the linear programming problem. The results of the incremental collapse analysis are shown in Fig. 8.28 as line 1 bounding the safe domain. Line 2 corresponds to the value of q_* whose excess leads in the Tresca shell to alternating plastic flow. This value, $q_* = 1$, can be easily found from condition (2.90) which states that alternating plasticity occurs when the range of variation of the equivalent stresses per cycle exceeds double the yield point. As seen in the shakedown diagram, a drop in the load-carrying capacity of the shell due to thermal cycling can amount to about 40 per cent without even causing the alternating flow.

The distribution of internal forces along the shell at the limit load ($q_* = 0$) is shown in Fig. 8.29. The hoop actions n_φ^0, m_φ^0 were obtained directly from the linear programming solution; the meridional bending

8 Shakedown of shells

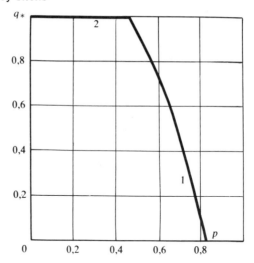

Fig. 8.28. Shakedown diagram for rotating conical shell at thermal cycling.

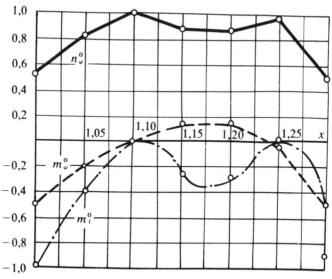

Fig. 8.29. Internal actions distribution along the meridian of conical shell at limit equilibrium.

8.4. Incremental collapse of a conical shell

moment m_l^0 and normal force n_l^0 were calculated afterwards with the help of (8.91), (8.92). The latter is too small to be shown diagrammatically; it varies monotonically from 0,022 at the left-hand edge of the shell to 0,091 at the right-hand one.

The flow régimes at various cross-sections of the shell at the yield point state are listed in Table 8.1. As follows from the associated flow law, the corresponding collapse mechanism is rather complex; at the ends of the shell there form yield hinge circles, and in between, the régimes take place simultaneously, each corresponding to a different yield hyperplane. In consequence, the internal actions belong to the corners of the approximated interaction surface. All the plastic generalized strains, i.e. the meridional and hoop elongations and curvatures are non-vanishing and therefore the normals to the median surface rotate with respect to points other than those lying on this surface. Let us remember that the associated flow law enables us to determine the directions of the strain increments without knowing quantitatively the ratios between them.

The flow régimes at the limiting cycle, when the value of q_* is approaching unity, are shown in Table 8.2. The hoop bending moments are now absent throughout the shell, the hoop normal forces are constant along the generator, $n_\varphi^0 = 0,5$. The sum of these constant forces and the variable forces clearly amounts to its limiting value $n_\varphi^0 + n_{\varphi*}^{(e)} = 1$.

Table 8.1
Flow régimes in conical shell at limit load, $q_* = 0$.

x	Flow régimes at yield point state		$(q_* = 0)$
1,0	$-n_\varphi^0 + m_\varphi^0 + 1 = 0,$	$n_l^0 - n_\varphi^0 + m_l^0 - m_\varphi^0 + 1 = 0,$	$-n_l^0 + m_l^0 + 1 = 0$
1,05	$-n_\varphi^0 + m_\varphi^0 + 1 = 0,$	$n_l^0 - n_\varphi^0 + m_l^0 - m_\varphi^0 + 1 = 0$	
1,10	$-n_\varphi^0 + m_\varphi^0 + 1 = 0,$	$n_l^0 - n_\varphi^0 + m_l^0 - m_\varphi^0 + 1 = 0,$	$-n_\varphi^0 - m_\varphi^0 + 1 = 0$
1,15	$-n_\varphi^0 - m_\varphi^0 + 1 = 0,$	$n_l^0 - n_\varphi^0 - m_l^0 + m_\varphi^0 + 1 = 0$	
1,20	$-n_\varphi^0 - m_\varphi^0 + 1 = 0,$	$n_l^0 - n_\varphi^0 - m_l^0 + m_\varphi^0 + 1 = 0$	
1,25	$-n_\varphi^0 + m_\varphi^0 + 1 = 0,$	$n_l^0 - n_\varphi^0 + m_l^0 - m_\varphi^0 + 1 = 0$	
1,30	$-n_\varphi^0 + m_\varphi^0 + 1 = 0,$	$n_l^0 - n_\varphi^0 + m_l^0 - m_\varphi^0 + 1 = 0,$	$n_l^0 + m_l^0 + 1 = 0$

8 Shakedown of shells

Table 8.2
Flow régimes in conical shell at incremental collapse, $q_* \to 1$.

x	Flow régimes at limiting cycle		$(q_* \to 1)$
1,0	$-2n_\varphi^0 - 3m_\varphi^0 + 1 = 0,$	$-2n_\varphi^0 + 3m_\varphi^0 + 1 = 0,$	$-2n_l^0 + 3m_l^0 + 1 = 0$
1,05	$-2n_\varphi^0 - 3m_\varphi^0 + 1 = 0,$	$-2n_\varphi^0 + 3m_\varphi^0 + 1 = 0$	
1,10	$-2n_\varphi^0 - 3m_\varphi^0 + 1 = 0,$	$-2n_\varphi^0 + 3m_\varphi^0 + 1 = 0$	
1,15	$-2n_\varphi^0 - 3m_\varphi^0 + 1 = 0,$	$-2n_\varphi^0 + 3m_\varphi^0 + 1 = 0,$	$-2n_l^0 - 3m_l^0 + 1 = 0$
1,20	$-2n_\varphi^0 - 3m_\varphi^0 + 1 = 0,$	$-2n_\varphi^0 + 3m_\varphi^0 + 1 = 0$	
1,25	$-2n_\varphi^0 - 3m_\varphi^0 + 1 = 0,$	$-2n_\varphi^0 + 3m_\varphi^0 + 1 = 0$	
1,30	$-2n_\varphi^0 - 3m_\varphi^0 + 1 = 0,$	$-2n_\varphi^0 + 3m_\varphi^0 + 1 = 0,$	$-2n_l^0 + 3m_l^0 + 1 = 0$

The meridional normal forces are small; their dimensionless value does not exceed 0,045. From the associated flow law it follows that the hoop elongation increments Δe_φ are present in the whole shell; the yield hinge circles form at both ends of the shell and in the midsection; the rotations of the normals take place about points lying close to the median surface.

The qualitative changes in the collapse mechanism due to increasing intensity of thermal cycling are related to the fact that, among all the constraints on n_l^0, n_φ^0, m_l^0, m_φ^0, see (8.86), those having the form

$$\frac{|n_{l*}^0 - n_{\varphi*}^0|}{1 - |(n_{l\tau} - n_{\varphi\tau})^*|} + \frac{|m_{l*}^0 - m_{\varphi*}^0|}{1 - |(m_{l\tau} - m_{\varphi\tau})^*|} \leq 1$$

do not depend at the adopted temperature field on the thermal parameter, see the relationships (8.90); thus they cease to be definite as the parameter q_* increases.

Finally, let us focus our attention on two singular features of the application of the simplex method to the shakedown problems. In the authors' opinion they are rather significant in the practical computations, although not obvious. The first peculiarity consists in the fact that the constraints matrix, as a rule, has a large number of zero coefficients; the system (8.94)–(8.97) produces about 50 per cent of zeros. The same conclusion can be drawn by inspecting the solutions for cylindrical

8.5. Shakedown of spherical shells

shells and circular plates obtained in Secs. 8.1 and 7.3, respectively. On the other hand, in the column of the free terms the same magnitudes appear repeatedly which results in the appearance of zero-free terms in the course of computations. When a computer is used, there can appear some 'numerical zeros' as a result of round-off errors. The ratio between two such small values can be comparable to the ratio of non-zero free terms; thus the 'numerical zero' may become a pivotal element. Although reshaping the simplex table with respect to such elements leads to good final results, it can substantially enlarge the required computing time. A suitable analysis of the above calculations for the conical shell as well as the cylindrical shell of Sec. 8.1 has shown that the consumption of extra computation time involved in the operations with negligibly small pivotal elements can reach 30–40 per cent of the total time. Only a reasonable a priori limit on the accuracy of computations can prevent a waste of computer time.

The second peculiarity is also connected with computing time; it depends to a large degree on how close to the optimum solution the starting solution really is. The selection of the latter is usually made with the help of a standard algorithm of the simplex method. However, it would be of interest to attempt adopting the starting solution as close as possible to the optimum one by using, instead of the formal mathematical rules, sound engineering judgement and experience gained from other similar situations. True enough, such an approach within the framework of the statical method is rather difficult. However, it often appears simple and straightforward when the kinematical method is employed with an intuitive a priori selection of the starting solution in the form of a certain kinematically admissible field of displacement increments. Thus the use of the kinematical approach can widen the class of complex shakedown situations to be solved numerically.

8.5. Shakedown of spherical shells

1. Let a rotationally symmetric spherical cap of constant thickness $2h$, Fig. 8.30, be clamped along its edge and subject to internal pressure p and repeated temperature

$$t(\tau, \zeta) = t_0(\tau) + \zeta t_1(\tau), \quad \zeta = \frac{z}{h}, \quad 0 \leq t_1(\tau) \leq t_*, \tag{8.98}$$

8 Shakedown of shells

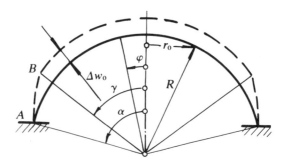

Fig. 8.30. Spherical cap.

where $2t_l(\tau)$ is the temperature interval between the inner and the outer surface of the shell measured along the normal. The axis z is directed towards the centre of the sphere, $-h \le z \le h$.

Let us establish the conditions for progressive accumulation of strains and for alternating plastic flow. The yield point σ_s will, for simplicity, be kept constant independently of the temperature. First, the pressure will remain constant inside the thermal cycle. Both the upper and lower bound on the limiting cycle parameters can be obtained from the conditions of incremental collapse.

The upper bound will be estimated by using the approximate kinematical method. Condition (2.88) takes the form

$$R \int_0^\alpha p\Delta w(\varphi) \sin\varphi \, d\varphi =$$

$$= \int_0^\alpha d\varphi \int_{-1}^1 (\sigma_{\theta*}^0 \Delta\epsilon_\theta'' + \sigma_{\varphi*}^0 \Delta\epsilon_\varphi'') \, d\zeta + \sum_\mu \sin\varphi_\mu \int_{-1}^1 \sigma_{\theta*}^0 \Delta\theta_\mu' \, d\zeta, \quad (8.99)$$

where $\Delta w(\varphi)$ is the radial displacement increment per cycle; $\Delta\theta_\mu'$ stand for the jumps in meridional displacement increment at the cross-sections $\mu = 1, 2, \ldots$ whose locations are determined by the angles φ_μ; $\Delta\epsilon_\theta''$, $\Delta\epsilon_\varphi''$ are the kinematically admissible increments of the meridional and hoop plastic strains; and $\sigma_{\theta*}^0$, $\sigma_{\varphi*}^0$ are the corresponding meridional and hoop stresses belonging to the fictitious yield surface.

The incremental collapse mechanism is assumed to be, see dashed line in Fig. 8.30,

$$\Delta w(\varphi) = -\Delta w_0 = \text{const.} \quad \text{for } 0 \le \varphi \le \gamma,$$

$$\Delta w(\varphi) = -\Delta w_0 \left(1 - \frac{\varphi - \gamma}{\alpha - \gamma}\right) \quad \text{for } \gamma \le \varphi \le \alpha, \quad (8.100)$$

8.5. Shakedown of spherical shells

$$\Delta\theta \equiv 0 \qquad \text{for } 0 \leq \varphi \leq \alpha,$$

where $\Delta\theta$ is the meridional displacement increment per cycle.
Using the compatibility conditions for spherical shells [8.18]

$$R\Delta\epsilon_\theta = \frac{d}{d\varphi}(\Delta\theta) - \Delta w - \zeta\frac{d}{d\varphi}\left[\Delta\theta + \frac{d}{d\varphi}(\Delta w)\right],$$

$$R\Delta\epsilon_\varphi = \Delta\theta \operatorname{ctg}\varphi - \Delta w - \zeta\left[\Delta\theta + \frac{d}{d\varphi}(\Delta w)\right]\operatorname{ctg}\varphi, \qquad (8.101)$$

we can find the plastic strain increments corresponding to the displacement field (8.100); these are

$$\Delta\epsilon_\theta'' = \Delta\epsilon_\varphi'' = \Delta w_0/R \quad \text{for } 0 \leq \varphi \leq \gamma,$$

$$\Delta\epsilon_\theta'' = (\Delta w_0/R)\cdot\frac{\alpha-\varphi}{\alpha-\gamma}, \quad \Delta\epsilon_\varphi'' = \frac{\Delta w_0}{R(\alpha-\gamma)}\left[\alpha-\varphi-\frac{\zeta h}{R}\operatorname{ctg}\varphi\right] \qquad (8.102)$$

$$\text{for } \varphi > \gamma.$$

In sections AA and BB of the cap, Fig. 8.30, there occur jumps in the meridional displacement increments,

$$|\Delta\theta_1'| = |\Delta\theta_2'| = |\zeta h\Delta w_0/R(\alpha-\gamma)|. \qquad (8.103)$$

Let us now discuss the stresses $\sigma_{\theta*}^0, \sigma_{\varphi*}^0$ on the fictitious yield surface. Under the adopted assumptions, e.g. the temperature-independence of the yield point, only the range of thermo-elastic stresses, and consequently, only the interval of temperature variations across the thickness is significant. Thus the cycle (8.98) can be replaced by the symmetric one, $-\frac{1}{2}t_* \leq t_1(\tau) \leq \frac{1}{2}t_*$, and the fictitious yield surface can be constructed in a more convenient manner. The thermo-elastic stresses in a clamped cap are:

$$\sigma_{\theta\tau}^{(e)} = \sigma_{\varphi\tau}^{(e)} = \zeta q(\tau)\sigma_s, \quad -q_* \leq q(\tau) \leq q_*, \quad q_* = \frac{\alpha E t_*}{2(1-\mu)\sigma_s}. \qquad (8.104)$$

In accordance with (2.56) and at variable stresses (8.104), for the Tresca material, (2.12), the fictitious yield surface is described by

$$\max(|\sigma_{\varphi*}^0| + q_*\sigma_s|\zeta|, |\sigma_{\theta*}^0| + q_*\sigma_s|\zeta|, |\sigma_{\varphi*}^0 - \sigma_{\theta*}^0|) - \sigma_s = 0. \qquad (8.105)$$

8 Shakedown of shells

Applying the associated flow rule in the form (2.76) and using (8.102), (8.105), we get

for $0 \leq \varphi \leq \gamma$: $\Delta\epsilon_\theta'' = \Delta\epsilon_\varphi'' > 0$, $\sigma_{\theta*}^0 = \sigma_{\varphi*}^0 = \sigma_s(1 - q_*|\zeta|)$,

for $\gamma \leq \varphi \leq \alpha$: $\Delta\epsilon_\theta'' > 0$, $\sigma_{\theta*}^0 = \sigma_s(1 - q_*|\zeta|)$,

for $\alpha - \varphi > \zeta h \operatorname{ctg}\varphi/R$: $\Delta\epsilon_\varphi'' > 0$, $\sigma_{\varphi*}^0 = \sigma_s(1 - q_*|\zeta|)$, (8.106)

for $\alpha - \varphi < \zeta h \operatorname{ctg}\varphi/R$: $\Delta\epsilon_\varphi'' < 0$, $\sigma_{\varphi*}^0 = -\sigma_s(1 - q_*|\zeta|)$.

Substituting the stresses (8.106) as well as the strain and displacement increments (8.100), (8.102), (8.103) into equation (8.99), followed by the minimization of p with respect to γ leads to the upper bound on the limiting cycle parameter p_* ($p^+ \geq p_*$). We get

$$\frac{p^+ R}{4\sigma_s h} = 1 - \frac{1}{2}q_* + \frac{1}{4}\frac{h}{R}\left(1 - \frac{2}{3}q_*\right)\frac{2\sin\alpha}{\alpha - \sin\alpha}. \tag{8.107}$$

In computing the above result, the magnitude $|\alpha - \varphi - \zeta h \operatorname{ctg}\varphi/R|$, entering into one of the integrands of (8.99), was replaced by $\alpha - \varphi + |\zeta h \operatorname{ctg}\varphi/R|$. This considerably simplified the calculations at the cost of a certain overestimation of the limiting cycle parameter.

We shall now estimate the lower bound by using the approximate statical method. To begin with, we shall construct the fictitious interaction surface in the space of generalized stresses. As in the previous section, we shall use the simplified procedure, see Sec. 3.4. Let the actual interaction surface be approximated by

$$\max(|N_\theta| - N_0, |N_\varphi| - N_0, |N_\theta - N_\varphi| - N_0, |M_\theta| - M_0,$$
$$|M_\varphi| - M_0, |M_\theta - M_\varphi| - M_0) = 0, \tag{8.108}$$

where N_θ, N_φ, M_θ, M_φ stand for hoop and meridional normal forces and bending moments, respectively [8.24].

Applying the equalities (3.37)–(3.40) and accounting for (8.104), we can describe the fictitious surface as follows:

$$\max\left[|N_{\theta*}^0| - N_0\left(1 - \frac{1}{2}q_*\right), |N_{\varphi*}^0| - N_0\left(1 - \frac{1}{2}q_*\right),\right.$$
$$|N_{\theta*}^0 - N_{\varphi*}^0| - N_0, |M_{\theta*}^0| - M_0\left(1 - \frac{2}{3}q_*\right),$$
$$\left.|M_{\varphi*}^0| - M_0\left(1 - \frac{2}{3}q_*\right), |M_{\theta*}^0 - M_{\varphi*}^0| - M_0\right] = 0. \tag{8.109}$$

8.5. Shakedown of spherical shells

The piece-wise linear surface (8.109) and the exact surface have common points on all the coordinate axes. However, neither the limited-interaction surface (8.108) nor, consequently, (8.109) allow for mutual effects of bending and stretching. Thus, under certain circumstances, the ultimate internal forces determined by (8.109) can be not only smaller, but also larger, than those associated with the exact solution. The corresponding analysis of numerical errors is presented in [8.24].

According to the Melan theorem restated in terms of generalized stresses (Sec. 3.6), the time-independent internal forces in the shell are in equilibrium. The appropriate equations are

$$\frac{d}{d\varphi}(N_\theta^0 \sin \varphi) - N_\varphi^0 \cos \varphi = Q_\theta^0 \sin \varphi,$$

$$(N_\varphi^0 + N_\theta^0 - pR)\sin \varphi = -\frac{d}{d\varphi}(Q_\theta^0 \sin \varphi), \qquad (8.110)$$

$$\frac{d}{d\varphi}(M_\theta^0 \sin \varphi) - M_\varphi^0 \cos \varphi = RQ_\theta^0 \sin \varphi,$$

and they are valid for the domain bounded by the fictitious yield surface (8.109)

$$|N_\theta^0| \le N_0\left(1 - \frac{1}{2}q_*\right), \quad |N_\varphi^0| \le N_0\left(1 - \frac{1}{2}q_*\right), \quad |N_\theta^0 - N_\varphi^0| \le N_0,$$
$$|M_\theta^0| \le M_0\left(1 - \frac{2}{3}q_*\right), \quad |M_\varphi^0| \le M_0\left(1 - \frac{2}{3}q_*\right), \quad |M_\theta^0 - M_\varphi^0| \le M_0. \qquad (8.111)$$

Let us proceed by assuming that the distributions of generalized stresses obey (8.110), (8.111). Next we find the load factors p, q_*. In the collapse mechanism (8.102) the curvature increments per cycle have been assumed to vanish for $0 \le \varphi \le \gamma$. The associated flow rule immediately provides that

$$M_\theta = M_\varphi = Q_\theta = 0, \quad N_\varphi = N_\theta = \text{const.} \qquad (8.112)$$

Let us assume that the distribution (8.112) spreads all over the cap, $0 \le \varphi \le \alpha$.

The internal actions (8.112) clearly satisfy both the equilibrium condition and the constraints (8.111) on the bending moments. In addition, (8.112) yields the equality

8 Shakedown of shells

$$\frac{1}{2}pR = N_\varphi = N_\theta. \tag{8.113}$$

Maximizing p in the presence of the constraints (8.111) on the normal forces, we obtain the following lower bound on the limiting cycle parameter p_*, where $p^- \leq p_*$:

$$\frac{p^- R}{4\sigma_s h} = 1 - \frac{1}{2}q_*. \tag{8.114}$$

An analysis shows that for a thin cap the obtained lower bound on p_* differs but little from the upper bound (8.107). Both are shown in Fig. 8.31 by dashed and solid lines, respectively, for the cap with the ratio $h/R = 0{,}025$. It can readily be seen that for a closed spherical shell, $\alpha = 180°$, the solution (8.114) is exact. For thick as well as shallow caps, $\sin \alpha \ll 1$, the difference between the two estimates is significant. The lower bound can be improved here by a better approximation of the flow regimes over the segment AB, Fig. 8.30. To this end, let us consider a cap subtending a small angle α and assume the statically admissible field of time-independent internal actions in analogy with the circular plate whose limit equilibrium was studied in [8.24]:

$$M_\varphi = \text{const.}, \quad Q_\theta = \frac{1}{2}pr_0 = \frac{1}{2}pR\varphi. \tag{8.115}$$

If we substitute those quantities into (8.110), and then make suitable rearrangements, using the inequalities (8.111), we get the following lower bound on the limiting cycle parameter:

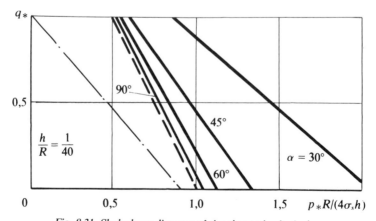

Fig. 8.31. Shakedown diagram of the clamped spherical cap.

$$\frac{p^- R}{4\sigma_s h} = \frac{1}{2}\frac{h}{R}\frac{1-\tfrac{2}{3}q_*}{1-\alpha\,\mathrm{ctg}\,\alpha}. \tag{8.116}$$

In particular, for the cap with $h/R = 0{,}025$ and $\alpha < 10°$ this relationship provides a better lower bound than (8.114).

2. Let us analyse a simply-supported spherical cap ($\beta > 3$) under constant internal pressure and thermal cycling of the same type as before. The thermo-elastic stresses at $\mu = 0{,}3$ can now be found after [8.25]:

$$\sigma_{\theta\tau}^{(e)} = \zeta q(\tau)(1 - e^{-\beta}\cos\beta)\sigma_s, \quad \beta = 1{,}285(\alpha - \varphi)\sqrt{\frac{R}{h}},$$
$$\sigma_{\varphi\tau}^{(e)} = -q(\tau)[-2{,}3 e^{-\beta}\sin\beta + (1 - 0{,}3 e^{-\beta}\cos\beta)\zeta]\sigma_s, \tag{8.117}$$
$$-q_* \leq q(\tau) \leq q_*.$$

The fictitious yield surface for the Tresca shell is described by

$$\max\{|\sigma_{\theta*}^0| + q_*(1 - e^{-\beta}\cos\beta)\zeta\sigma_s, |\sigma_{\varphi*}^0| + q_*|-2{,}3 e^{-\beta}\sin\beta +$$
$$+ (1 - 0{,}3 e^{-\beta}\cos\beta)\zeta|\sigma_s, |\sigma_{\theta*}^0 - \sigma_{\varphi*}^0| +$$
$$+ q_*|-2{,}3 e^{-\beta}\sin\beta + 0{,}7 e^{-\beta}\cos\beta\zeta|\sigma_s\} = \sigma_s. \tag{8.118}$$

The above equations practically coincide with (8.105) for points lying far enough from the simply-supported edge, $\beta > 3$.

The normal forces belonging to the fictitious interaction surface at vanishing curvature increments are

$$N_{\varphi*}^0 = 2\sigma_s h\left\{1 - \frac{1}{2}q_*[2{,}3 e^{-\beta}\sin\beta + (1 - 0{,}3 e^{-\beta}\cos\beta)(1-\delta^2)]\right\},$$
$$N_{\theta*}^0 = 2\sigma_s h\left[1 - \frac{1}{2}q_*(1 - e^{-\beta}\cos\beta)\right], \tag{8.119}$$

where

$$\delta = \frac{2{,}3 e^{-\beta}\sin\beta}{1 - 0{,}3 e^{-\beta}\cos\beta}.$$

For the purpose of obtaining a lower bound on the limiting cycle parameters, let us adopt the time-independent action field in the form

8 Shakedown of shells

(8.112). Both the internal equilibrium conditions (8.110) and suitable boundary conditions are duly met. Insisting on the maximum of the pressure factor in the presence of (8.119), we arrive at the lower bound on p_*:

$$\frac{p^- R}{4\sigma_s h} = 1 - \frac{1}{2} q_*(1 - e^{-\beta_0} \cdot \cos \beta_0), \quad \beta_0 = 1{,}285\alpha \sqrt{\frac{R}{h}}, \qquad (8.120)$$

where $p^- \leq p_*$.

An upper bound, $p^+ \geq p_*$, will be found in a similar fashion as for the built-in cap. Accepting the collapse mechanism (8.102), we get for $\alpha \geq 30°$, $\beta_0 > 3$ the solution

$$\frac{p^+ R}{4\sigma_s h} = 1 - \frac{1}{2} q_* + \frac{1}{4} \frac{h}{R} \left(1 - \frac{2}{3} q_*\right) \frac{\sin \alpha}{\alpha - \sin \alpha}. \qquad (8.121)$$

The obtained results are plotted in Fig. 8.32; the dashed line corresponds to the lower bound (8.120) for $30° \leq \alpha \leq 90°$, solid lines correspond to the upper bound (8.121) for various angles α.

According to the criterion (2.90), alternating plastic flow can take place at $q_* = 1$; for the simply-supported cap this equality holds when $\alpha \geq 30°$, $\beta_0 \geq 3$, for the built-in one the equality is true for all angles α.

In the cases in which the pressure p varies inside the cycle, the fictitious yield surface should be constructed accordingly. A detailed analysis shows that for an arbitrary loading program neither the upper

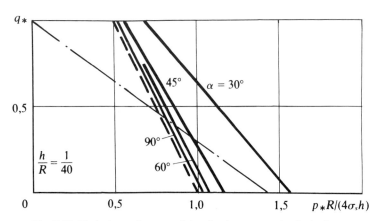

Fig. 8.32. Shakedown diagram of the simply supported spherical cap.

8.6. Combustion chamber of liquid fuel rocket engine

nor the lower bound on the incremental collapse load factors differ appreciably from the above ones obtained for p = const. However the alternating flow conditions change more dramatically when the program becomes arbitrary; the conditions are depicted in Figs. 8.31 and 8.32 by dot-dashed lines.

8.6. Combustion chamber of liquid fuel rocket engine

As one more example of application of the shakedown theory let us analyse a combustion chamber of the liquid fuel rocket engine. The chamber constitutes a double-shell structure, Fig. 8.33a, consisting of inner and outer parts joined by suitably spaced and sufficiently stiff connectors. It is the inner shell that has to withstand the most severe conditions of a gas stream of temperature 3000 to 3500°C. It is cooled by means of a liquid fuel component flowing in-between the shells in the direction from the collector to the engine head. The outer shell, well protected against excessive heat, serves as a basic load-carrying member of the structure.

For the purpose of qualitative analysis of the ultimate behaviour of the combustion chamber at repeated starts we shall employ the computational procedure [8.26] in which the shell is idealized by a sandwich ring whose faces are not bridged by any direct forces, Fig. 8.33b. If axial forces are absent the only equilibrium equation takes the form

$$\sigma'h' + \sigma''h'' = pR, \qquad (8.122)$$

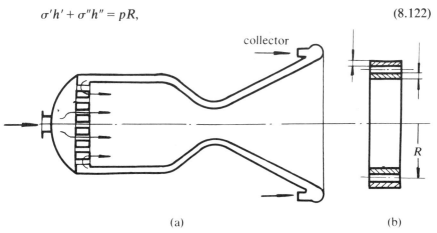

Fig. 8.33. Combustion chamber of liquid fuel rocket engine.

8 Shakedown of shells

where p is the gas pressure at a given cross-section of the chamber.

Assuming perfectly rigid connectors, the compatibility of the two shells is expressed by

$$\epsilon' = \epsilon''. \tag{8.123}$$

The symbol $\epsilon = \epsilon_e + \epsilon_p + \epsilon_t$ denotes the sum of elastic, plastic and thermal longitudinal strains, respectively, the first two being generated by the pressure only. Quantities referring to the inner ring are marked prime, a double prime marks quantities referring to the outer ring.

The above equations actually coincide with (1.1) and (1.2), or with their rearranged counterparts (1.15) and (1.16), see Fig. 1.2. This is why we can use here the method worked out in Chapter 1; another variant of the graphical representation of the behaviour of a combustion chamber is offered in [8.27].

The domain of admissible changes in the hoop direct stress resultants is shown in Fig. 8.34 as a solid rectangle. Dashed sides visualize a certain drop in the yield forces in the inner shell as a result of elevated temperature; the maximum temperature is here accounted for. Corner S corresponds to the limit load of the structure at room temperature; its projection on the p-axis determines the ultimate pressure. Point S' is

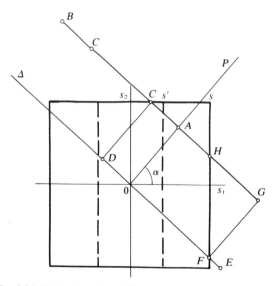

Fig. 8.34. *Behaviour of a shell at repeated gun-like actions.*

8.6. Combustion chamber of liquid fuel rocket engine

also associated with the state of yield point load, but in the presence of decreased yield hoop force due to heating. The ratios between the yield hoop forces of particular shells and between respective stiffnesses are here assumed to be very close to unity; in computations they should be properly adopted in accordance with the actual parameters of the chamber considered.

Consider two possible starting régimes of a rocket engine. The so-called gun-like, or blast, start consists in very rapid mounting of pressure in the chamber followed by relatively mild approach to the maximum temperature. After a working period of the engine, the liquid fuel component is cut off followed by abrupt drop in pressure accompanied by gradual cooling. The behaviour of the combustion chamber under such conditions is shown in Fig. 8.34.[1] The segment OA corresponds to the almost instantaneous application of the internal pressure, AB concerns the thermal deformation on heating, BC is related to the plastic hoop elongation of the outer shell, CD relates to the drop in pressure, DE concerns the thermal deformation on cooling and, finally, EF corresponds to the plastic hoop elongation of the inner shell. The latter can be in fact still larger, as it commences to mount before complete cooling takes place, i.e. when the yield point is lower. Point F indicates the end of the first cycle which leads to a system of residual forces: tension in the inner shell and compression in the outer.

Beginning with the second cycle, the process becomes completely stabilized. The stress profile now is the following: FG corresponds to the increase in pressure, GH concerns to the associated plastic deformation of the inner shell, HK relates to the thermal deformation on heating, KC concerns the plastic hoop elongation of the outer shell, and the segments CD, DE and EF follow the stress profile already passed during the first cycle. As results from (1.8) and (1.9), each cycle is accompanied by equal increments of the plastic deformation in both shells. Since those increments per cycle appear to be kinematically admissible, every cycle to follow will be similar provided the strain-hardening is absent.

The second characteristic starting régime takes place when the start is preceded by heating; the corresponding stress profile is shown in Fig. 8.35. In the first cycle we have the following stages: the segment OA corresponds to a certain initial pressure, AB relates to thermal defor-

---

[1] Let us note that the actual loads and deformations are related to the segments of the diagram that are parallel to the axes s_1 and s_2 by means of the formulae (1.7), (1.8), with due allowance for the scale.

8 Shakedown of shells

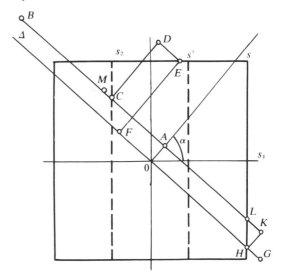

Fig. 8.35. *Behaviour of the preheated shell at repeated actions.*

mation on heating, BC concerns the plastic hoop contraction of the inner shell, CD relates to the maximum increase in pressure, DE concerns the associated plastic hoop elongation of the outer shell, EF is related to the pressure drop in the chamber, FG to the thermal deformation on cooling and, finally, GH corresponds to the plastic hoop elongation of the inner shell. Each following cycle will trace the stress profile $HKLMCDEFGH$. The increment of one-sign deformations per cycle, of elongations in particular, appears to be significantly smaller than that generated at the same magnitudes of relevant parameters but under the conditions of the blast starting (cf. segments KC in Figs. 8.34 and 8.35). However, apart from the deformations tending to become elongations, the inner shell also suffers a deformation of the opposite sign, i.e. tends to be contracted circumferentially, see segment MC.

Thus, although at preheated starts the behaviour of the combustion chamber appears to be closer to the shakedown conditions than at the blast starts, in the former case it is necessary to take precautions against thermal fatigue as a possible result of alternating plastic deformation of the inner shell. It is also obvious that certain changes in the particular parameters of the considered structure can lead to possibilities of other unsafe strain states in the constitutent shells.

In the above analysis the intercasing of pressure has not been taken into account. It can readily be shown that the resulting self-equilibrating

8.7. Fast nuclear reactor fuel element

hoop forces in a sandwich shell exert an influence merely in the first cycle.

The geometrical interpretation makes it also possible to evaluate the creep effects present at those time intervals when the inner shell is heated to very high temperature. In the case of blast start the relevant state is characterized by the point C, Fig. 8.34. A creep of the inner casing under hoop stretching leads to the translation of this point in the positive direction of the Δ-axis. However, such a shift is clearly impossible since the force in the outer casing cannot exceed its ultimate value. This implies that the irreversible creep deformation of the inner casing will be immediately compensated by equal short-term plastic deformation of the outer casing. As was mentioned in Chapter 1, the increase in overall deformability of the shell due to creep may slow down, as the number of cycles increases, owing to the strain-hardening of the material.

In the case of preheated start of the engine the creep effects will be present in the stress states C and E. In the former state the internal forces will be somewhat redistributed and the point C will translate towards the point A, as a result of additional compression of the inner casing. Further increase in pressure will lead to a decrease in the short-term deformation DE of the outer casing. Then the situation will develop as in the case of blast start. However let us draw attention to the fact that the creep effects at the stress states C and E are of opposite directions and therefore the possibility of thermal fatigue of the inner casing increases.

The adopted idealization of the considered combustion chamber does not reflect all the features of its working conditions. In particular, it neglects the variations of the temperature field along the shells and in time, also the loss of stability of the inner casing (its progressive buckling) in-between the neighbouring connectors and so on. Neither was the strain-hardening taken into consideration (the results on two-parameter systems obtained in Sec. 1.3 could be helpful here) nor the interaction between the short-term and the long-term deformation states.

8.7. Fast nuclear reactor fuel element

The behaviour of thin tubes subjected to internal pressure and cyclic temperature field that varies across the thickness was studied in [8.4]. The problem emerged in connection with the analysis of shells of fuel

8 Shakedown of shells

elements encountered in nuclear reactors and subjected to repeated on and off actions. In discussing this topic we shall assume that the internal pressure of gaseous breakdown products is constant and the temperature varies linearly across the thickness when heat is transferred from within the tube. Although the tube is closed, the longitudinal stresses are ignored and so are the end-effects. On these premises we find the hoop stress at a generic point to be

$$\sigma = \sigma_p - \zeta\sigma_t, \quad \zeta = z/h, \tag{8.124}$$

where

$$\sigma_p = \frac{pR}{2h}, \quad \sigma_t = \frac{1}{2}\alpha E \Delta t, \quad \Delta t = \frac{\Delta t'}{1-\mu} \tag{8.125}$$

and $\Delta t'$ is an actual interval of temperatures across the wall of the tube. μ is Poisson's ratio and the denominator $1-\mu$ reflects a change in the hoop thermal stress due to the actual plane stress state at non-uniform heating of the shell. All further computations are based on the fictitious temperature interval described in (8.125).

The analysis is based on some developments of the method put forward in [8.28] and basically similar to the additional loading method presented in Chapter 3. At the first stage of the analysis the temperature-dependence of the yield point, the strain-hardening of the material and the creep phenomena are all ignored. The full shakedown diagram for such a situation is shown in Fig. 8.36 in the dimensionless plane of $\bar{\sigma}_p = \sigma_p/\sigma_s$ and $\bar{\sigma}_t = \sigma_t/\sigma_s$. The following subdomains can be distinguished: A corresponding to shakedown, B to alternating plastic flow, C to one-sign deformation, mounting with each cycle, D to the combination of both types of cyclic plasticity and, finally, E which is the domain beyond the line 5 corresponding to the conditions for instantaneous plastic collapse, i.e. to the exhaustion of the load-bearing capacity.

The shakedown region A is, in turn, split up into three subregions: A' corresponding to purely elastic behaviour at the initial stage of loading, A'' to shakedown after plastic deformations have taken place in the initial semi-cycle at the outer sheet of the tube, and A''' corresponding to shakedown after plastic deformations of both outer and inner sheets.

The diagram shown differs basically from the full shakedown

8.7. Fast nuclear reactor fuel element

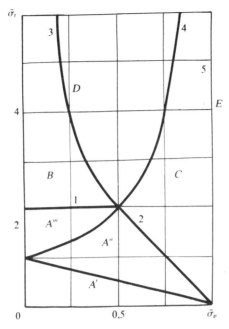

Fig. 8.36. Full shakedown diagram for the shell of nuclear reactor fuel element.

diagram constructed in Sec. 5.6 for a beam under variable repeated loading. In the case of purely mechanical actions each of the constituent loads alone was able to give rise to instantaneous plastic collapse; the influence of its variability was traceable only at alternating plasticity situations. Unlike this, the diagram of Fig. 8.36 shows again how pronounced a drop in the load-carrying capacity can be caused by thermal cycling (a steady temperature field would not have exerted any influence in this case unless the mechanical properties had changed with temperature).

Let us discuss the method of establishing the boundaries of the shakedown domain. It is based on the assumption that at the end of each semi-cycle the hoop stress distribution remains continuous and piecewise linear. (The process of stabilization in which, after plastic deformations at the initial loading cycles, a suitable distribution of self-equilibrating stresses takes place is considered in detail in [8.5]). Thus we encounter two zones across the thickness of the considered tube: At one zone the steady cycle stresses reach the yield point, $|\sigma| = \sigma_s = $ const. at suitable instants of time, whereas at the other, which underwent

8 Shakedown of shells

unloading before, the stress gradient remains constant, since it is determined by the thermo-elastic stress distribution. The resulting normal force can be found from the equilibrium of the shell subject to internal pressure; in other words, the area of the actual stress diagram and the area of the averaged stress diagram should both be equal at any instant of time.

The development of one-sign plastic strains at thermal cycling is visualized in Fig. 8.37; the stress fields at the end of the cooling semi-cycle and at the end of the heating semi-cycle are denoted by (a) and (b), respectively. The segment of 'overlapping' plastic regions is found to have length

$$a = h\left[1 - 2\sqrt{\frac{\sigma_s - \sigma_p}{\sigma_t}}\right]. \tag{8.126}$$

Putting $a = 0$, we get the limiting cycle relationship

$$\sigma_p + \frac{1}{4}\sigma_t = \sigma_s \tag{8.127}$$

which is represented by line 2 in the shakedown diagram.

If the absolute magnitude of the largest compressive stresses at a certain finite zone reach the yield point, the one-sign strains will combine with the local alternating flow. A suitable limiting condition can be found with the help of Fig. 8.37; we get

$$(h - a)\frac{\sigma_t}{h} = 2\sigma_s.$$

Eliminating a by means of (8.126), we obtain

$$(\sigma_s - \sigma_p)\sigma_t = \sigma_s^2 \tag{8.128}$$

which is represented by line 4 in the shakedown diagram.

Let us now evaluate the magnitude of the residual strain increment per cycle for the subdomain C, Fig. 8.36. Since the tube is uniformly loaded throughout and, consequently, no flexure takes place, the basic condition for deformations is that the total strains, comprising thermal, elastic and plastic components at any instant of time, do not depend on the coordinate z across the thickness of the tube. Residual strain fields

8.7. Fast nuclear reactor fuel element

per two successive semi-cycles constructed under this condition are shown in Fig. 8.37c, d. The determination of the residual strains per the heating semi-cycle is illustrated in Fig. 8.38 which shows: a) the thermal strains increasing from the periphery to the centre of the tube (uniform heating is of no significance here), b) the elastic strains corresponding, according to Fig. 8.37, to the stress variations per semi-cycle and, finally, c) the plastic strains as a component of the total strain. The latter are clearly found by requiring that the sum of thermal, elastic and plastic strains is constant across the thickness, see dashed line in Fig. 8.38a. It follows that the residual strain increments per cycle are equal to the difference between the free thermal strains at the points where the coordinates z differ by the length $2a$. Therefore, using (8.125), (8.126), we get

$$\Delta \epsilon_c = \alpha \Delta t \frac{a}{h} = \frac{2\sigma_s}{E} \left[1 - 2 \sqrt{\frac{\sigma_s - \sigma_p}{\sigma_t}} \right]. \tag{8.129}$$

Accordingly, the change in the tube diameter $D = 2R$, as the number of thermal cycles N increases, is found to be

$$\Delta D = 2RN\Delta\epsilon_c = \frac{4\sigma_s RN}{E} \left[1 - 2 \sqrt{\frac{\sigma_s - \sigma_p}{\sigma_t}} \right]. \tag{8.130}$$

Any increase in the diameter of the shell of the fuel element must be kept reasonably low as not to hinder proper cooling; this increase could diminish the cross-sections through which the coolant is flowing. The ductility of the material is here recognized as another limiting factor since radiation is known to suppress plastic properties, especially at elevated temperature [8.4]. Formula (8.130) is thus very useful for making appropriate checks.

Another unsafe state of the considered tube can be caused by cycles of alternating plastic flow in circumferential direction. The corresponding boundary line of the shakedown diagram (Fig. 8.36, line 1) reflects the known condition (2.90),

$$\sigma_t = 2\sigma_s. \tag{8.131}$$

In accordance with the adopted linear temperature field across the thickness, and consequently linear distribution of stress, the alternating plastic flow commences at both surfaces of the tube at the same time.

8 Shakedown of shells

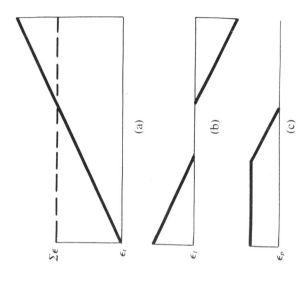

Fig. 8.38. Residual strains per semi-cycle during heating.

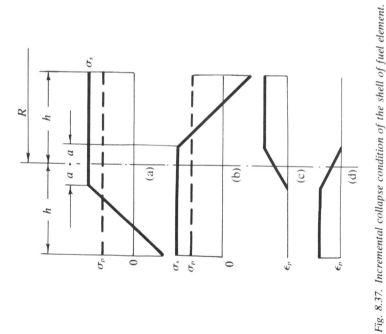

Fig. 8.37. Incremental collapse condition of the shell of fuel element.

8.7. Fast nuclear reactor fuel element

Lines 1 and 2 in Fig. 8.39a show the total stress fields at the beginning and at the end of each semi-cycle, respectively. Now we can establish the condition under which the plastic zones commence to overlap and the alternating flow is accompanied by strain accumulation. Similarly as before, we obtain the equation of line 3,

$$\sigma_t \sigma_p = \sigma_s^2 \tag{8.132}$$

which separates the subdomains B and D, see Fig. 8.36.

The total plastic strains per cycle, corresponding to the subdomain B are equal to zero; yielding at either semi-cycle has opposite direction whereas the one-sign strain accumulation is still absent, see Fig. 8.39b. By constructing diagrams similar to those of Fig. 8.38, we can now find the range of plastic strains per semi-cycle generated at the surfaces of the tube,

$$\Delta \epsilon_{hc} = \alpha \Delta t \frac{b}{2h} = \frac{1}{E}(\sigma_t - 2\sigma_s). \tag{8.133}$$

The service life of the tube should be estimated also from the point of view of thermal fatigue. Making use, for instance, of the Coffin formula [8.29], we can evaluate the number of cycles prior to collapse,

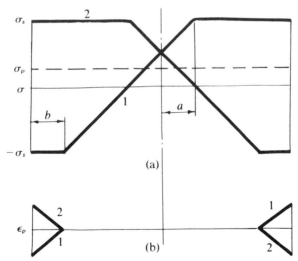

Fig. 8.39. Alternating flow and range of plastic strains per semi-cycle.

8 Shakedown of shells

$$N = \left(\frac{\delta}{2\Delta\epsilon_{hc}}\right)^2 = \frac{E^2\delta^2}{4(\sigma_t - 2\sigma_s)^2}, \qquad (8.134)$$

where δ denotes the residual strain in the statical tension test of a specimen made of the material involved.

The above assessment is by no means exact since some factors have not been allowed for, such as the influence of hold-time periods when the temperature is kept constant, the interaction between the cyclic short-time and the long-term deformation processes as well as the resulting defects in the considered shell.

The approximate approach to the design of the parameters of a shell of the nuclear reactor fuel element should use the two analyses corresponding to the two possible types of cyclic plastic deformation, each yielding conditions for unserviceability of the shell; the residual strain accumulation leading to unacceptable increase in the diameter, formula (8.130), and alternating flow resulting in the accumulation of damage and thermal fatigue, formula (8.134).

In the second part of [8.4] an attempt was made to reject some of the previously adopted assumptions. The possibility was investigated of the appearance of a temperature-dependent yield point which remains constant across the thickness at each loading stage. Attention was also focused on the effects of strain-hardening of the material. In particular, assuming linear strain-hardening in the conditions of subdomain C, the stress and strain states in the semi-cycles to follow were described. The results show a gradual slowing down of an increase in the irreversible strains in an asymptotic approach to the shakedown situation, see Chapter 1. The total residual strain prior to shakedown was found to be

$$\epsilon_0 = \frac{\sigma_s}{E} + \frac{1}{E_T}\left(\sigma_p - \sigma_s + \frac{1}{4}\sigma_t\frac{E - E_T}{E}\right), \qquad (8.135)$$

where E_T is the strain-hardening modulus.

It is interesting to compare the above result with the experimental evidence [8.30] on tubular specimens under thermal cycling in the presence of constant internal pressure. Test measurements showed an increase in the diameter of the tubes, i.e. an incremental collapse, accompanied by steady decrease in the strains accumulating per cycle. As soon as the strain increments per cycle faded away, the internal pressure increased to reach the next level; this again resulted in larger strain increments followed as before by their drop, and so on.

8.7. Fast nuclear reactor fuel element

The obtained measurements are presented diagrammatically in Fig. 8.40, where ΔR stands for a periodic increase in diameter, and N denotes the number of cycles measured each time from the transition to a new level of internal pressure. The little circles, dots, crosses and triangles correspond to the internal pressures of 400, 460, 480 and 520 kG/cm², respectively. The basic specimen parameters were: $a = 9$ mm, $b = 10{,}25$ mm, $2h = 1{,}25$ mm, $\sigma_s = 25$ kG/mm², $E = 2 \cdot 10^4$ kG/mm², $E_T = 4 \cdot 10^2$ kG/mm². Mechanical properties were measured at the ambient temperature of 20°C. Stresses generated by the internal pressure $p_{max} = 520$ kG/cm² and the maximum thermal stresses were: $\sigma_p = 40$ kG/cm², $\sigma_t = 80$ kG/cm². Using the approximate formula (8.135), we get that the strain prior to shakedown amounts to $\epsilon_0 = 0{,}088$ which corresponds to the diameter increment $\Delta R \cong 1{,}8$ mm. The increment measured after 570 thermal cycles, the last 270 having been performed at the internal pressure $p_{max} = 520$ kG/cm², was about 2,5 mm. Considering some inaccuracies in the starting data as well as the approximate character of the adopted assumptions, the obtained result seems to be satisfactory. A somewhat better assessment of the expected deformation can be arrived at by allowing for some variations in hoop stresses which

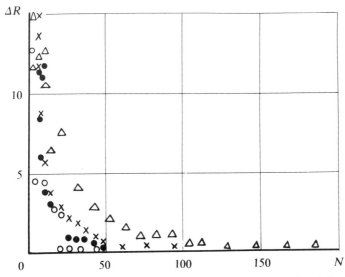

Fig. 8.40. Experimental data on the strain accumulation per cycle in the shell at thermal cycling and different values of constant pressure p (p = 400, 460, 480, 520 kG/cm²).

8 Shakedown of shells

result from an increase of the diameter, and the accompanying thinning of the tube wall, and also by taking account of the temperature-dependent yield point.

In [8.4] it was also suggested to allow, in a similar manner, for the strain-hardening in the alternating flow conditions of the subdomain B. Such results would seem to be rather unreliable, as they would be based on the isotropic strain-hardening model in which the Bauschinger effect is ignored. Over the last years the material properties under cyclic loading have been investigated in more detail; see Chapter 1 and relevant references.

Creep is accounted for on the assumption that the steady period of the reactor performance is long enough for the thermal stresses in the fuel element tubes to relax completely. Such a simplification does clearly lead to the shrinking of the shakedown domain. When the respective limiting condition is violated, increase in the deformation of the tube over a cycle is caused partly by creep in steady thermal conditions, and partly by the short-term plastic strains at cooling. Understandingly enough, the above assumptions lead to very rough estimates on the real behaviour.

The results presented in [8.4] are incorporated in the American Code on pressure vessels design [8.31]. The problem considered above was solved in [8.32] for the conditions of creep. At the stage of long-term loading the isochronic creep curve was approximated by the ideal elastic-plastic response. Computations were made employing the methods explained here.

It is worth mentioning that the conditions of elastic adaptation obtained in this section could be established with the help of the approximate kinematical method explained in Chapter 5. This could be done equally well with or without taking into account the creep behaviour.

References

[8.1] Gokhfeld, D.A., Some problems in shakedown theory of plates and shells, in Russian, in *Proc. VI. All-Union Conf. on Theory of Shells and Plates*, (Baku 1966), Nauka, Moscow, 1966.

[8.2] Sawczuk, A., On incremental collapse of shells under cyclic loading. *IUTAM Symp. (Copenhagen 1967). Theory of thin shells.* Springer, Berlin, 1969.

[8.3] Sawczuk, A., Evaluation of upper bounds to shakedown loads for shells. *J. Mech. Phys. Solids*, 17, 1969, 291–301.

References

[8.4] Bree, J., Elastic-plastic behaviour of thin tubes subjected to internal pressure and intermittent high-heat fluxes with application to fast-nuclear reactor fuel elements. *J. Strain Anal.*, Nr 3, 1967, 226–238.

[8.5] Beer, F.J., Plastic growth of pressurized shell through interaction of steady pressure and cyclic thermal stress. *Therm. Stress and Therm. Fatigue.* London, 1971, 340–347.

[8.6] Gokhfeld, D.A., Cherniavsky, O.F., On incremental collapse conditions in axially symmetric shakedown problems using linear programming, in Russian, *AN SSSR, Mechanica Tverd. Tela.*, Nr 3, 1970.

[8.7] Gokhfeld, D.A., Cherniavsky, O.F., Methods solving problems in the shakedown theory of continus. *Proc. Int. Symp. on Found. of Plasticity, Warsaw, 1972.* Noordhoff, Leiden, 1973.

[8.8] Cherniavsky, O.F., Cherniaev, E.F., Shakedown of cylindrical shells at repeated thermal and mechanical actions, in Russian, in *Thermal stresses in structural members*, 14, Naukova Dumka, Kiev, 1974.

[8.9] Cherniavsky, O.F., Load-carrying capacity of perfectly plastic conical shell at thermal cycling, in Russian, in *Thermal stresses in structural members*, Nr 10, Naukova Dumka, Kiev, 1970.

[8.10] Leckie, F.A., Shakedown pressures for flush cylinder-sphere shell intersection. *J. Mech. Eng. Sci*, Vol. 7, Nr 1, 1965, 367–371.

[8.11] Leckie, F.A., Penny, R.K., Shakedown loads for radial nozzles in spherical pressure vessels. *Int. of Solids Struct.*, 3, Nr 5, 1967, 743–755.

[8.12] Fox, L.D., Kraus, H., Penny, R.K., Shakedown of pressure vessels with ellipsoidal heads. *Pap. ASME*, N PVP-34, 1971.

[8.13] Findlay, G., Moffat, D., Stanley, P., Torispherical drumheads: a limit analysis and shakedown investigation. *J. Strain Analysis*, 6, Nr 3, 1971, 147.

[8.14] Iliushin, A.A., *Plasticity*, in Russian, Gostekhizdat, Moscow, 1948.

[8.15] Drucker, D.C., Limit analysis of cylindrical shells under axially-symmetric loading. *Proc. 1st Midwestern Conf. Solid Mech.*, Urbana, 1953, 158–163.

[8.16] Melan, E., Parkus, H., *Wärmespannungen infolge stationärer Temperaturfelder.* Springer, Wien, 1953.

[8.17] Hodge, Ph.G., *Plastic analysis of structures.* McGraw-Hill, New York, 1959.

[8.18] Timoshenko, S., Woinowsky-Krieger, S., *Theory of plates and shells.* McGraw-Hill, New York, 1959.

[8.19] Pontryagin, L.C., *Mathematical theory of optimum control*, in Russian, Nauka, Moscow, 1965.

[8.20] Rozonoer, L.I., The Pontragin maximum principle in the theory of optimum systems, in Russian, *Automatika i Telemekhanika*, 20, Nr 10, 11, 12, 1959.

[8.21] Eyre, D.G., Galambos, T.V., Shakedown of grids. *J. Struct. Div. Proc. ASCE*, 99, Nr 10, 1973, 2049–2060.

[8.22] Franciosi, V., Augusti, G., Sparacio, R., Collapse of arches under repeated loading. *J. Struct. Div. Proc. ASCE*, 90, Febr. 1964, 165–201.

[8.23] Chernov, N.L., On gradual development of plastic deformation in steel beams under moving load, in Russian, *Izv. vish. uchebn. zaved., Stroitelstvo i Architektura*, Nr 1, 1973.

[8.24] Hodge, Ph.G., *Limit analysis of rotationally symmetric plates and shells.* Prentice-Hall, 1963.

8 Shakedown of shells

[8.25] *Strength, Stability, Vibrations,* in Russian, handbook ed. by I.A. Birger and Ya.G. Panovko, Mashinostrojenie, Moscow, 1968.
[8.26] Feodosiev, V.I., *Strength of thermally stresses connections in liquid fuel rocket engines,* in Russian, Oborongiz, Moscow, 1963.
[8.27] Zarubin, W.S., Study on the loading history of non-uniformly heated cylindrical shell, in Russian, in *Thermal stresses in structural members, 4,* Naukova Dumka, Kiev, 1964.
[8.28] Miller, D.R., Thermal stress ratchet mechanism in pressure vessels, *Trans. ASME,* ser. D., *81,* Nr 2, 1959.
[8.29] Coffin, L.E., Jr., A study of the effects of cycling thermal stresses on a ductile material. *Trans. ASME,* 76, Nr 6, 1954.
[8.30] Weicman, R.I., On plastic strength of tubes under internal pressure and thermal cycling, in Russian, in *Problems of mechanical stability,* Mashinostrojenie, Moscow, 1964.
[8.31] ASME Boiler and pressure vessels Code, Section VIII, Div. I, 1971.
[8.32] O'Donnel, W.I., Porowski, I., Upper bounds for accumulated strains due to creep ratcheting. *Trans. ASME, 196,* Nr 3, 1974, 150–154.

9

Strain accumulation at thermal cycling under negligible mechanical loading

The previous chapters were chiefly concerned with a decrease in the mechanical load-carrying capacity of structures in the presence of thermal cycling. Incremental collapse under thermal actions was shown to occur at loads lower than those associated with the instantaneous collapse situations.

However, strain accumulation at thermal cycling alone has also been found to take place. In other words, incremental collapse is also possible when the mechanical applied loads are small enough to be neglected. Many examples of such a behaviour can be quoted in the metallurgy of both ferrous and nonferrous metals, such as deformations of containers, crystallizers, casting moulds and dies, carriages of firing machines and so on. This phenomenon can also be frequently encountered in other industries in which high-temperature working processes are unavoidable. In some cases the strain accumulation can be so extensive as to lead to unacceptable, from the point of view of serviceability, changes in geometry of the structure in question. This problem has received much attention in nuclear energy industry [9.1 and others]. Strain accumulation in the absence of applied loads can also be observed in parts of gas turbine engines (combustion chambers [9.2], jet rings, blades), in ultrasonic aircraft (fuselages, brakes) and even in radio-electronics (valve electrodes).

Since the term 'incremental collapse' in the absence of any mechanical loads becomes rather misleading (there is no direct transition to the instantaneous collapse situation), we shall use in what follows the term 'progressive (or incremental) thermal cycling deformation' or, following the English terminology connected with this subject, simply 'ratchetting', to indicate the simple analogy between the strain ac-

9 Strain accumulation at thermal cycling

cumulation and the ratchet-wheel that transforms oscillatory motion of a pawl into rotary motion of a toothed wheel.

In the beginning ratchetting was studied only experimentally. An extensive survey of investigations made by various authors can be found in the monograph [9.3] by Davidenkov and Likhatchev. At that time importance was attached mainly to the effects of the crystallographic structure and physical-mechanical properties of metals and alloys. Tests were performed on cylindrical specimens made chiefly of pure metals in which ratchetting effects were well pronounced; to a much lesser degree on structural alloys. Such aspects of ratchetting as influence of structural shapes and dimensions on the performance of structures under thermal cycling were hardly touched upon. In this chapter we shall analyse in more detail those aspects of strain accumulation.

Particularly the problem of warping, or buckling, of thin-walled structures is of great practical importance, especially when these are caused by thermal cycling and variable loading. This is duly reflected in the survey by Bolotin and Grigoluk [9.37] where a definite relation is hinted at between the notions of shakedown and stability, although a detailed study of this connection is thought to be premature. Much attention is focused on surveying all available up-to-date results, both theoretical and experimental, no matter how limited they may seem to be.

9.1. Strain accumulation conditions based on the kinematical theorem of shakedown. Influence of the type of temperature field

In the absence of mechanical loading or, more precisely, under its negligible influence, the basic inequality of the kinematical theorem (2.86) for the one-sign strain accumulation assumes the form

$$\int \sigma_{ij*}^0 \Delta \epsilon_{ij0}'' \, dv = \int \min_{\tau}[(\sigma_{ij} - \sigma_{ij\tau}^{(e)})\Delta \epsilon_{ij0}''] \, dv < 0. \tag{9.1}$$

It is assumed here that at any point of the body only one flow régime is realized during each loading cycle. For steady mechanical properties of the material and, in particular, for constant yield point we specify that

$$\int [\sigma_{ij}\Delta \epsilon_{ij0}'' - \max_{\tau}(\sigma_{ij\tau}^{(e)}\Delta \epsilon_{ij0}'')] \, dv = \int (\sigma_{ij} - \sigma_{ij\tau}^*)\Delta \epsilon_{ij0}'' \, dv < 0. \tag{9.2}$$

9.1. Strain accumulation conditions

In the expressions (9.1), (9.2) the inequalities are strict since conditions are sought under which the one-sign strains accumulate.

In order to illustrate the effects of different temperature fields on the ratchetting let us consider two contrasting cases:

a) simple temperature field, i.e. varying in proportion to a common multiplier,

b) temperature field which remains quasi-stationary with respect to the coordinate system moving along with the heat source.

In the first case, with constant mechanical properties, we have a one-parameter thermo-elastic stress field, namely

$$\sigma_{ij\tau}^{(e)} = a\sigma_{ij}^0, \quad a = a(\tau), \quad a_0 \leq a \leq 0, \tag{9.3}$$

where σ_{ij}^0 is a reference field of thermo-elastic stresses corresponding to a certain 'unit' value of the parameter.

Stresses corresponding to the extremum value of the parameter a with isochronous distribution

$$\sigma_{ij0}^{(e)} = a_0 \sigma_{ij}^0, \tag{9.4}$$

satisfy the condition

$$\int \sigma_{ij0}^{(e)} \Delta \epsilon_{ij0}'' \, dv = 0, \tag{9.5}$$

due to the virtual work principle (2.15). A kinematically admissible field of plastic strain increments is here denoted by $\Delta \epsilon_{ij0}''$.

If the considered cycle is the limiting one, in the sense of strain accumulation, and $\Delta \epsilon_{ij0}''$ describes the actual mechanism of deformation, then the determining (in accordance with the condition (2.81)) thermo-elastic stresses $\sigma_{ij\tau}^*$ in the additional loading region, where $\sigma_{ij\tau}^* \Delta \epsilon_{ij0}'' > 0$, reach their extremum magnitudes, $\sigma_{ij\tau}^* = \sigma_{ij0}^{(e)}$ just as $\sigma_{ij\tau}^* = 0$ in the unloading region where $\sigma_{ij0}^{(e)} \Delta \epsilon_{ij0}'' \leq 0$.

In this situation, using (9.5), we can write

$$\int \sigma_{ij\tau}^* \Delta \epsilon_{ij0}'' \, dv = \int_{v_1} \sigma_{ij0}^{(e)} \Delta \epsilon_{ij0}'' \, dv = -\int_{v_2} \sigma_{ij0}^{(e)} \Delta \epsilon_{ij0}'' \, dv$$

$$= \frac{1}{2} \int |\sigma_{ij0}^{(e)} \Delta \epsilon_{ij0}''| \, dv, \tag{9.6}$$

9 Strain accumulation at thermal cycling

where v_1 and v_2 are the volumes of the additional loading and unloading regions, respectively; the total volume of the body is thus $v = v_1 + v_2$.

Let us assume now that the thermal cycling intensity does not exceed that corresponding to the commencement of alternating plastic flow. Then the criterion (2.90) for one-parameter loading provides

$$\frac{1}{2}\sigma_{ij0}^{(e)} = \sigma_{ij}^{(a)}, \tag{9.7}$$

where $\sigma_{ij}^{(a)}$ is an admissible stress state.

Thus we have

$$\frac{1}{2}|\sigma_{ij0}^{(e)}\Delta\epsilon_{ij0}''| = |\sigma_{ij}^{(a)}\Delta\epsilon_{ij0}''| \leq \sigma_{ij}\Delta\epsilon_{ij0}'', \tag{9.8}$$

where σ_{ij} are the yield surface stresses associated, according to the flow rule (2.76), with the plastic strain increments $\Delta\epsilon_{ij0}''$. These stresses correspond in the mechanism of progressive deformation to the direction of the vector $\Delta\epsilon_{ij0}''$.

Making use of (9.6) and (9.8), we get

$$\int \sigma_{ijr}^{*}\Delta\epsilon_{ij0}''\,dv \leq \int \sigma_{ij}\Delta\epsilon_{ij0}''\,dv, \tag{9.9}$$

and we conclude that the condition (9.2) cannot be satisfied. This means that in the considered situation of repeated actions of a one-parameter temperature field the shakedown domain is completely determined by the alternating flow condition; the one-sign strains can basically accumulate only in the presence of alternating yielding.

A similar reasoning was already given in Chapter 1 in which simple bar systems made of isotropic material were analysed. It was shown there that the situation changes, once the yield point ceases to be temperature-independent; this dependency gives rise to a privileged direction of deformation, since $\frac{1}{2}\sigma_{ij0}^{(e)}$ can exceed the yield point at the maximum temperature inside the cycle, cf. (9.7).

In general, an unsteady temperature field does not necessarily lead to an isochronous distribution of the extremal thermo-elastic stresses and the equation (9.5) no longer applies. The corresponding expansion of the additional loading region and a more uniform distribution of the enclosing stresses obeying (2.81) both result in the diminishing of the quantity at the

9.1. Strain accumulation conditions

left-hand side of inequality (9.2), thus making one-sign deformation more likely to occur.

Thus, just as in the case of thermal one-parameter cycling, if the ultimate state of a structure is derivable from the alternating flow condition then the more irregular the temperature distribution in the cycle gets, the more important becomes the strain accumulation condition in the form of either (9.1) or (9.2). The temperature fields which are kept quasi-stationary with respect to the coordinate system moving along with the heat source have the extremum properties in the discussed sense; they can develop in bodies of large dimensions such as plates or hollow cylinders which are homogeneous, both physically and geometrically. Nearly quasi-stationary temperature fields are encountered in certain metal treatment processes such as welding and casting in which the heat source moves along the object [9.4, 9.5].

The thermo-elastic stresses generated by the quasi-stationary temperature field and referred to the moving coordinate system are time-independent and their envelopes (2.81) are independent of the coordinate in the direction of the motion. This is why the expressions in inequality (9.2) can be integrated only over the two remaining dimensions, i.e. width and thickness. Thus in the case of a tube with axi-symmetric quasi-stationary temperature field, the progressive deformation (ratchetting) condition assumes the form

$$\int_{-h}^{h} (\sigma_{\alpha\beta} - \sigma^*_{\alpha\beta\tau})\Delta\epsilon''_{\alpha\beta 0}\,dz < 0, \quad \alpha, \beta = 1, 2. \tag{9.10}$$

A structure can easily be selected in which, at a quasi-stationary temperature field, the determining stresses are equal at all points at which the strain increments per cycle do not vanish in the actual ratchetting mechanism. Assuming such a situation also for an admissible mechanism, we obtain the sufficient condition for the strain accumulation in its simplest form

$$(\sigma_{\alpha\beta} - \sigma^*_{\alpha\beta\tau})\Delta\epsilon''_{\alpha\beta 0} < 0. \tag{9.11}$$

In this case ratchetting will take place if the maximum thermo-elastic (fictitious) stresses are a little larger than the yield point.

The bar system considered in Sec. 1.3 is a good example of the discussed situation. As was shown, (9.11), the one sign strains commence to accumulate at considerably lower maximum cycle temperature

9 Strain accumulation at thermal cycling

than necessary for the alternating plasticity and in the first case the sufficient condition is specified by the single yield point whereas in the second by double yield point, i.e. the sum of tension and compression yield point stresses. It is proper to mention that in the presence of repeated actions of quasi-stationary thermal fields the condition of alternating flow should be based on the instantaneous distribution of the thermo-elastic stresses, as indeed the stresses from a moving heat source vary at each point of the body according to such a distribution. Thus a situation can by no means be ruled out in which both the limiting conditions, i.e. ratchetting and alternating flow, become operative at the same time. Here the temperature dependence of the yield point plays the decisive part as to the character of the cyclic deformation process; the consequences usually are that strain accumulation, mounting with each consecutive cycle, can take place at such values of the temperature field parameters at which the alternating flow is still absent.

A number of examples will follow, including real structural situations, in which ratchetting is encountered. Some results will also be discussed of specific experiments enabling the studied effects to be investigated under extremum conditions.

9.2. Repeated action of moving heat source on a bar system

Conditions for thermal ratchetting can be conveniently discussed by considering some bar systems. One of the simplest models is shown in Fig. 9.1 consisting of a bundle of equal parallel bars connecting two rigid plates that can translate without rotation. The bars are one by one heated so as to reach a certain temperature t_*; heating a particular bar is assumed to last so long that the remaining ones have enough time to cool down to the initial temperature. The situation is fully analogous to that considered in Sec. 1.3. However, it is the kinematical theorem that will now be employed.

According to condition (2.81) and to the two assumed kinematically admissible ratchet mechanisms (tension or compression of bars with equal strain increments per cycle) we find the envelope of the stress diagram shown in Fig. 9.2. Equilibrium implies that

$$\sigma_k^- = (n-1)\sigma_k^+, \tag{9.12}$$

where k is the number of a particular bar, n denotes the number of all

9.2. Repeated action of moving heat source on a bar system

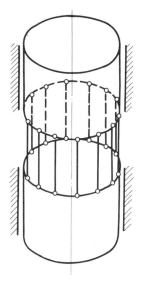

Fig. 9.1. Model for strain accumulation due to moving heat sources.

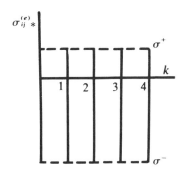

Fig. 9.2. Distribution of the determining thermal stresses in the system of Fig. 9.1.

bars, minus and plus superscripts imply compression and tension, respectively, while σ^- is the absolute value of the corresponding stress.

Since at the quasi-steady temperature field the determining stresses are equal in all the constituent bars, the subscript k will be from now on suppressed.

In accordance with the condition (9.11), the ratchet mechanism can form if the largest stresses exceed the yield point; since $\sigma^- > \sigma^+$ when $n > 2$, the actual mechanism will indeed lead to the shortening of the bars after each cycle, as a result of their compression. The ratchetting condition is therefore expressed by the inequality

$$\sigma^- > \sigma_s. \tag{9.13}$$

Simple calculations yield that

$$\sigma^- = \alpha E t_* \frac{n-1}{n}, \tag{9.14}$$

hence the maximum temperature of the limiting cycle is

$$t_0 = \frac{\sigma_s}{\alpha E} \cdot \frac{n}{n-1}. \tag{9.15}$$

9 Strain accumulation at thermal cycling

Let us now establish what will happen at $t_* > t_0$, i.e. beyond the shakedown limits. First let us assume that the system was initially in a virgin (unstressed) state. Heating of the first bar results in its plastic contraction

$$\epsilon_p = -\alpha t_* + \frac{\sigma_s}{E} \cdot \frac{n}{n-1}. \tag{9.16}$$

This will lead during the cooling of the bar to the generation of residual stresses in the system. The preheated bar will be stretched with the stress

$$\sigma^0 = -\sigma_s + \alpha E t_* \frac{n-1}{n}, \tag{9.17}$$

the remaining bars will be squeezed with smaller stresses.

It must be emphasized now that the above system of residual stresses is disadvantageous to the further development of a cycle. The presence of compression residual stresses implies that every next heated bar yields at a lower temperature than did the previous one. Moreover, the amount of strain at the same maximum temperature t_* will be larger.

Stress distributions at all the stages can be determined in an elementary manner once certain regularities are made explicit. First of all, it must be realized that the residual stresses in each consecutive bar after its cooling do not depend on any initial stresses having been present prior to its heating; this is because unloading takes place on cooling, starting from the yield point reached before. Next, since the system of residual stresses is clearly self-equilibrating, a stress change in one (heated) bar is immediately compensated by suitable uniform stress changes in the remaining bars.

Let us consider a specific four-bar system and let us assume that the maximum heating temperature is such as to generate thermo-elastic stresses, see formula (9.12), amounting to $\sigma^- = 1{,}6\sigma_s$. Thus the residual stresses in a heated bar will be equal, after its cooling, to

$$\sigma^0 = -\sigma_s + \sigma^- = 0{,}6\sigma_s. \tag{9.18}$$

After this stage each of the remaining three bars will be residually squeezed with the stress $0{,}2\sigma_s$. At the next stage, i.e. after heating and

9.2. Repeated action of moving heat source on a bar system

cooling of a consecutive bar, its residual stresses will be tensile and equal to those determined by (9.18). The stress change in this bar, amounting now to $0{,}8\sigma_s$, should be evenly distributed among the remaining bars. Proceeding in a similar fashion, we can easily obtain residual stress distributions at all the stages of the first cycle as well as for the following cycles, Fig. 9.3. The process will stabilize. Indeed, a certain quasi-steady residual stress state is generated in the system; remaining

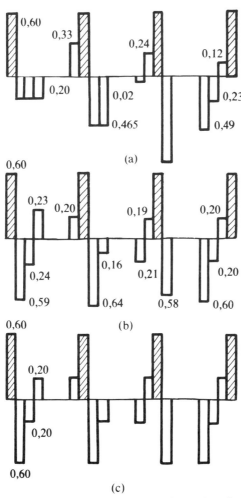

Fig. 9.3. Residual dimensionless stresses at cyclic motion of a heat source.

9 Strain accumulation at thermal cycling

constant, it translates after each stage by one step, i.e. from one bar to the next, according to the motion of heat source. The process of stabilization practically terminates during the second cycle. The asymptotic state (Fig. 9.3c) can be described by

$$\sigma_i^0 = -\sigma_k^0 \left[1 - \frac{2(i-k-1)}{n-1} \right], \quad (i > k),$$

$$\sigma_i^0 = \sigma_k^0 \left[1 + \frac{2(i-k)}{n-1} \right], \quad (i \leq k),$$

(9.19)

where σ_k^0 is the residual stress in a heated bar, according to (9.17).

The residual per cycle strain in each bar amounts to

$$\Delta\epsilon^0 = -\frac{2}{E}\left(\alpha E t_* - \frac{n}{n-1}\sigma_s\right).$$

(9.20)

Since the above deformation is clearly kinematically admissible, the system returns after each cycle to the previous residual stress state and the process is repeated anew. In the absence of strain-hardening the total strains are proportional to the number of cycles.

The formula (9.20) shows that the residual strains per cycle increase as the maximum heating temperature is raised. However the increase cannot be unbounded since the residual stresses, see (9.17), cannot exceed the yield point. Hence the temperature

$$t^0 = \frac{2\sigma_s}{\alpha E} \cdot \frac{n}{n-1} = 2t_0,$$

(9.21)

can be found at which the residual strains per cycle will attain their maximum

$$|\Delta\epsilon_{max}^0| = \frac{2\sigma_s}{E} \cdot \frac{n}{n-1}.$$

(9.22)

At $t_* > 2t_0$ the residual strain per cycle ceases to increase and the alternating flow becomes responsible for further irreversible deformation.

The occurrence of strain accumulation at thermal cycling is unrelated here to the temperature-dependency of the yield point. The

9.3. Ratchetting of a thin cylindrical shell

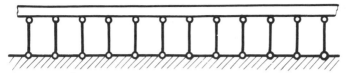

Fig. 9.4. Another variant of bar system as illustration of strain accumulation at repeated actions of moving heat source.

latter, if accounted for, would affect the quantitative results only; in particular, it would lower the specific temperatures t_0 and t^0. Creep is found to affect the process in a similar manner.

Other bar systems can also be employed in order to illustrate the plastic strain accumulation at repeated passages of a heat source. For instance, consider a beam of finite flexural stiffness as shown in Fig. 9.4, resting on elastic-perfectly plastic bars subjected to heating. On account of the known beam analogy, this model can approximate the behaviour of a shell. With its help some shell structures can be analysed subjected to repeated actions of different temperature fields such as thermal wave or thermal front. However, the cylindrical shell is found to lend itself to a direct treatment.

9.3. Ratchetting of a thin cylindrical shell under repeated action of quasi-stationary temperature field

1. Let us first consider the possibilities for strain accumulation in a long tube under repeated passages along its axis of a certain idealized quasi-steady temperature field, Fig. 9.5a, assumed to be axially symmetric with respect to the tube and constant across its thickness. The thermo-elastic stresses can be readily determined by insisting on continuity of the displacements and rotations in the cross-section at the beginning of the applied temperature field. Due to antisymmetry, Fig. 9.5b, the only internal actions that need be considered in this section are the shear forces Q_0. Known relationships [9.5] lead to

$$Q_0 = \alpha t \beta^3 D a, \tag{9.23}$$

where

$$\beta = \sqrt[4]{\frac{3(1-\mu^2)}{4a^2h^2}}, \quad D = \frac{2Eh^3}{3(1-\mu^2)}.$$

9 Strain accumulation at thermal cycling

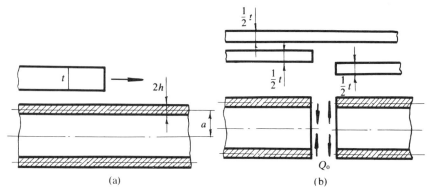

Fig. 9.5. Idealization of the temperature field in a tube.

The maximum hoop direct stresses at this section amount to

$$\sigma_{\varphi\,\text{max}} = \frac{Q_0 E}{2\beta^3 Da} = 0{,}5\alpha Et \qquad (9.24)$$

whereas, due to the specific load symmetry conditions, the bending stresses in the cross-section considered are absent. The maximum magnitude of the bending stresses

$$\sigma_{x\,\text{max}} \approx 0{,}29\alpha Et \qquad (9.25)$$

is attained in the cross-section of the tube at the distance $\beta x = 0{,}8$ from the front of the temperature field; in what follows those small bending stresses will be ignored.

The zone of maximum hoop stresses which is compressive to the left and tensile to the right of the front of temperature field shifts along the tube as the heating source moves ahead. The enveloping distributions of tensile and compressive dimensionless stresses are shown in Fig. 9.6 by dashed lines; both ordinates have the same absolute values. Thus the limiting cycle proves to be determined by the alternating flow condition.

However the one-sign strains cease to be indeterminate as soon as the yield point begins to be treated as a temperature-dependent quantity; the instantaneous distribution of the limiting cycle stresses is shown in Fig. 9.7. Since the temperature rise results in a drop of the yield point, the latter is lower to the left of the front of the temperature field than to

9.3. Ratchetting of a thin cylindrical shell

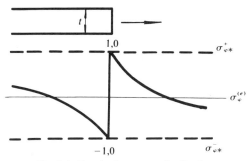

Fig. 9.6. Enveloping stress distribution.

the right. Therefore the condition

$$\sigma_{\varphi\tau}^* = \max_\tau |\sigma_{\varphi\tau}^{(e)}| > \sigma_{st} \qquad (9.26)$$

is associated with the tendency of progressive squeezing of the shell in the circumferential direction; the alternating flow due to (2.91) does not take place.

If, for instance, the yield point depends linearly on the temperature,

$$\sigma_{st} = \sigma_s(1 - nt), \qquad (9.27)$$

the limiting temperature t_0, an excess of which will immediately create one-sign cyclic strains, is found to be

$$t_0 = \frac{\sigma_s}{0.5\alpha E + n\sigma_s}. \qquad (9.28)$$

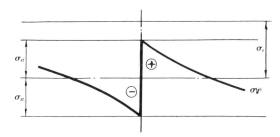

Fig. 9.7. Limiting cycle stress distribution for temperature-dependent yield point.

9 Strain accumulation at thermal cycling

For low-carbon steel with $\sigma_s = 24\,\text{kG/mm}^2$, $\alpha E = 0{,}25\,\text{kG/mm}^2\,°\text{C}$ and $n = 12{,}5 \times 10^{-4}\,1/°\text{C}$, we get $t_0 = 150°\text{C}$. The temperature at which the ratchetting can commence under repeated passages of temperature field is found to be rather low; in particular, it is considerably lower than the temperature at which the strains were accumulating in the heated tubes which were repeatedly submerged into a coolant [9.3, 9.6].

The condition for the occurrence of one-sign strains at the considered type of temperature field depends mainly on the temperature gradients along the tube, the gradients across the wall of a thin tube playing a minor, if any, part. This can be shown by applying the expression (2.81) to the actual ratchetting mechanism.

At suitable temperature gradients the passage of a quasi-steady temperature field causes a longitudinally uniform squeezing of the tube; the dimensions along the generator and the thickness should both satisfy the condition of preservation of volume of the material (e.g. for an infinitely long tube under plane strain the decrease in its diameter should be fully compensated by suitable increase in the thickness of its wall).

Residual stresses which would cause in an unclamped tube the appearance of axial or hoop direct stress resultants, are here impossible; the hoop forces are not self-equilibrating due to their uniform distribution along the tube whose end-effects are neglected. Thus only the bending stresses can be generated as a possible result of the presence of non-uniform plastic strains across the thickness; though in the considered idealized situation there is no reason at all for such an inhomogeneity. However, the bending stresses, even if present, are not capable of suppressing the strain accumulation. Hence it can be expected that the effects of each consecutive passage of the temperature field will not differ from the effects of the first passage as concerns the amount of accumulated residual strain. It will be shown later that these results of simple analysis are well corroborated by the available experimental evidence.

As a second example of strain accumulation exerted by repeated actions of moving heat source one can consider an infinite plate. Let us imagine two concentrated heat sources located symmetrically with respect to the middle surface of the plate and ensuring local rise of temperature, constant across the thickness. Such a situation is typical for welding. Substantial compressive stresses, generated as a result of excessive heating, are expected to lead to plastic squeezing of material inside a certain radius; even more so because the heating will be accompanied by a decrease in the yield point. If the periodic heat source is at rest, the appeearance of alternating flow is the main result of repeated heating and

9.3. Ratchetting of a thin cylindrical shell

cooling, the latter being accompanied by residual tensile stresses. If the heat source is moving along a certain path, a kinematically admissible cycle of plastic strain rates can be realized. The effects of each subsequent passage will be the same and they will consist in a gradual thickening of a bulge forming along the heat source path as a result of lateral compression, Fig. 9.8. When the heat spot travels along a straight path there are no residual forces in the planes parallel to the path because of the conflicting requirements of homogeneity and self-equilibrium.

Similar deformations can be expected in a tube under quasi-steady local heating whose path is the circle coinciding with the cross-section of the tube. Unlike in the plate, non-uniformity of heating across the thickness will not generate bending. Such a situation is encountered during butt welding two tubes together [9.7]. A qualitative analysis with the help of the shakedown theory of the bulging mechanism at repeated passages of the heat point (electrode) was made in [9.8].

2. A quasi-steady, with respect to a moving frame of reference, temperature field takes place also in the case of a long tube monotonously submerged in, or emerged from a liquid of differing temperature. Repeated action implies periodic emersions or submersions, each followed by the restoration of the initial temperature level; we shall for the time being assume that the temperature at the latter stage changes slowly and uniformly so as to generate no thermal stresses in the tube.

The actual temperature fields accompanying the submergence of the tube are in fact different from the idealized thermal front visualized in Fig. 9.5. They depend on the initial temperatures of the tube, the surrounding liquid and the gaseous media, on the direction and velocity of

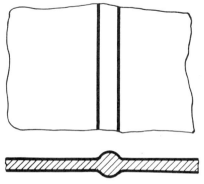

Fig. 9.8. Bulging of a plate subjected to repeated passages of concentrated heat source.

9 Strain accumulation at thermal cycling

submersion or emersion as well as on a number of thermo-physical and geometrical parameters. An analytical solution to a similar situation can be worked out on the basis of results shown in the monograph [9.9] for the case of filling a cylindrical cavity with liquid or emptying it.

To obtain an approximate solution to the problem of a hot tube submerged in a liquid, the initial temperature of the tube $t_s(x)$ is assumed to be equal to the temperature of surrounding gaseous medium, volumes of both gas and liquid are assumed infinite and the temperature varying along the axis of the tube only. Under such conditions the temperature distribution in the tube above and below the liquid level can be described in the case of quasi-steady régime of submergence with constant velocity by the formulae

$$t_1(x_1) = t_l + \frac{\Delta t}{1 + \frac{\bar{\alpha}_1}{\bar{\alpha}_2}} e^{-(\bar{\alpha}_1 x_1/h)}, \qquad (9.29)$$

$$t_2(x_2) = t_g - \frac{\Delta t}{1 + \frac{\bar{\alpha}_2}{\bar{\alpha}_1}} e^{-(\bar{\alpha}_2 x_2/h)}, \qquad (9.30)$$

where $t_1(x_1)$ and $t_2(x_2)$ stand for the temperatures of the tube (t_s) respectively below and above the liquid table (for the coordinates we refer to Fig. 9.9a), t_l and t_g denote the temperatures of liquid and gas as indicated by subscripts, $\Delta t = t_g - t_l$, $t_g > t_l$, is the temperature drop. The auxiliary coefficients are

$$\bar{\alpha}_1 = A + \sqrt{A^2 + \frac{2\alpha_1 h}{\lambda}}, \qquad (9.31)$$

$$\bar{\alpha}_2 = -A + \sqrt{A^2 + \frac{2\alpha_2 h}{\lambda}}, \qquad (9.32)$$

$$A = \frac{Vhc\rho}{2\lambda}, \qquad (9.33)$$

where α_1, α_2 denote the coefficients of heat emission from the liquid and from the gas to the tube, respectively, $2h$ is the thickness of the tube, c, ρ, λ are respectively the thermal capacity, density and thermal conductivity of the material, V stands for the velocity of submersion ($V < 0$) and emersion ($V > 0$).

The temperature field in the tube near the liquid level is determined by the relationship between $\bar{\alpha}_1$ and $\bar{\alpha}_2$. For $\bar{\alpha}_1 = \bar{\alpha}_2$ the distribution is

9.3. Ratchetting of a thin cylindrical shell

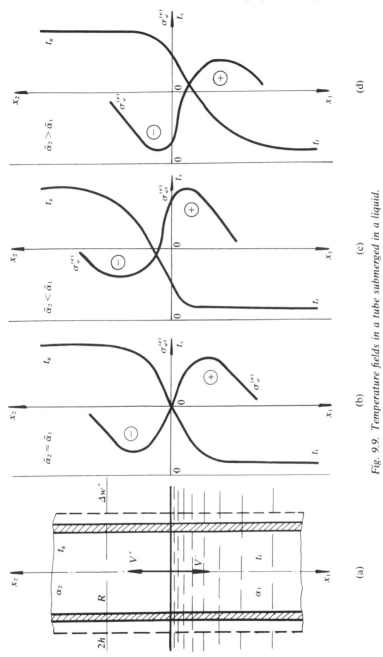

Fig. 9.9. Temperature fields in a tube submerged in a liquid.

9 Strain accumulation at thermal cycling

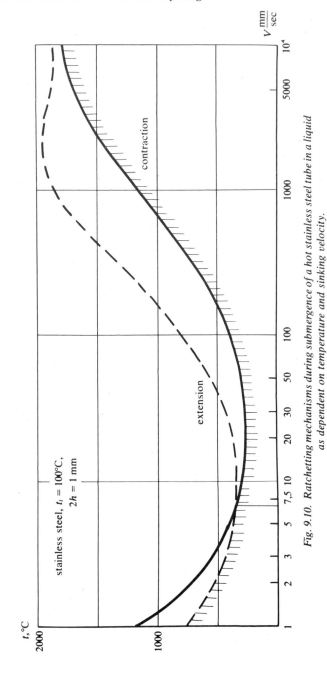

Fig. 9.10. Ratchetting mechanisms during submergence of a hot stainless steel tube in a liquid as dependent on temperature and sinking velocity.

9.4. Ratchetting of a crystallizer

antisymmetric with respect to the level of the liquid, Fig. 9.9b. A similar property holds for the thermo-elastic hoop stress distribution $\sigma_\varphi^{(e)}$; as mentioned before, the longitudinal bending stresses will be ignored as rather insignificant. The cycle of stress variations is symmetric at each point of the tube. As in the previously considered example, the condition for alternating plasticity is critical to the shakedown situation. However, as soon as the yield point is considered temperature-dependent, the condition for progressive strain accumulation becomes critical. Since the yield point is known to drop at elevated temperatures, the repeated submersions or emersions under suitable limiting conditions will lead to hoop squeezing of the tube.

During the passage of the tube from gas to liquid or vice versa when the coefficients $\bar{\alpha}_1$ and $\bar{\alpha}_2$ are not equal to one another the asymmetric stress cycles, in which the absolute values of maximum stresses in tension and compression differ, come to play a role. In this case strain accumulation is critical to shakedown although the yield point remains insensitive to temperature. The type of those strains (stretching or squeezing) depends on the ratio of moduli of the current thermo-elastic hoop stresses of opposite signs. The latter depends on the temperature profile in the vicinity of liquid level and is determined by the ratio of $\bar{\alpha}_1$ to $\bar{\alpha}_2$, see Fig. 9.9c, d.

When all the thermo-physical and geometrical parameters of the tube are constant, the decisive factor is the tube velocity and its direction. By varying the velocity we can generate by repeated passages under equal heat drop, either extension or contraction of the tube in the circumferential direction. In Fig. 9.10 we can see the relationships between the sinking velocity and the temperature of a stainless steel tube ($2h = 1$ mm) repeatedly submerged into liquid with the temperature $t_l = 100°C$. In this situation there can form either the extension or the contraction ratchetting mechanism. The mechanism which really occurs is clearly that corresponding to the lowest temperature. Under the adopted conditions progressive contraction of the tube is most likely to take place at almost any velocity. It is only for sufficiently small velocities that the hoop strains can change sign.

9.4. Ratchetting of a crystallizer of a semi-continuous tube casting machine

One of the mechanical devices whose serviceability is limited mainly by excessive progressive deformations is a crystallizer shown schematic-

9 Strain accumulation at thermal cycling

ally in Fig. 9.11a as part of a semi-continuous tube casting machine. Liquid cast-iron with temperature of about 1300°C is poured in-between the cylindrical sleeves 1 and 2 made of low-carbon steel. The outer shell, to the analysis of which this section is devoted, is cooled with water at temperature 15°C passing between its outer surface and the case 3. The working cycle is the following: casting of iron to reach a certain level, lifting of a ready tube with the liquid iron level kept almost constant and, after the required length of the tube has been reached, termination of casting followed by cooling of the whole device with continuous supply of water. One cycle lasts about 10 minutes.

The average, with respect to the thickness of the tube, temperature profiles in the shell 1 are shown in Fig. 9.11b as tested at four instants of time measured from the beginning of casting. The profiles resemble those of the travelling thermal wave. As the wave moves upwards, the longitudinal temperature gradients are found to increase as do the temperatures themselves; the radial temperature gradients across the thickness of the tube increase accordingly. During a certain period of time after the casting process has been commenced the thermal wave reaches the level at which the casting enters a stationary regime. It is at this instant of time that the temperature of the tube in contact with liquid metal reaches its maximum, as do both the longitudinal and radial temperature gradients. After the tube is pulled out of the device rapid cooling takes place.

Inspection shows that in the upper part of the outer shell 1 a waist (or choke) tends to form (Fig. 9.12), developing with each casting and leading to a substantial change in the required thickness of cast tubes. It should be here mentioned that the mechanical loads are rather insignificant and cannot be made responsible for this permanent deformation.

In order to establish the condition for the strain accumulation with each cycle, let us employ the kinematical shakedown theorem in the form (2.88). The known data about the working cycle are the experimentally measured temperature profiles (Fig. 9.11b) and the temperature-dependent yield point of the material. The limiting cycle is assumed to differ from the working cycle only by a proportional change of thermo-elastic stresses, the yield point distribution in the shell remaining constant in both.

Thus we have

$$\sigma_{x\tau}^{(e)} = n\sigma_{x\tau}^{(p)}, \quad \sigma_{\varphi\tau}^{(e)} = n\sigma_{\varphi\tau}^{(p)}, \quad (n > 0) \tag{9.34}$$

where $\sigma_{x\tau}^{(e)}, \sigma_{\varphi\tau}^{(e)}$ denote the longitudinal and hoop thermo-elastic stresses

9.4. Ratchetting of a crystallizer

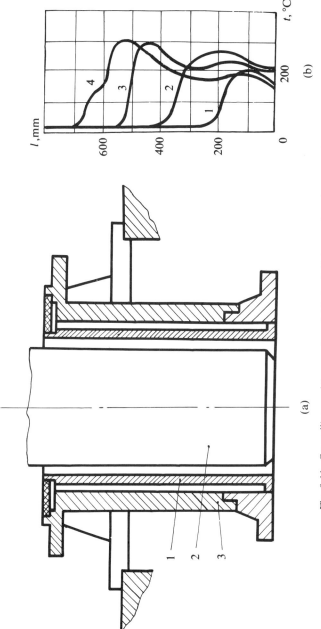

Fig. 9.11. Crystallizer and temperature fields in shell 1 at various instants of a cycle.

9 Strain accumulation at thermal cycling

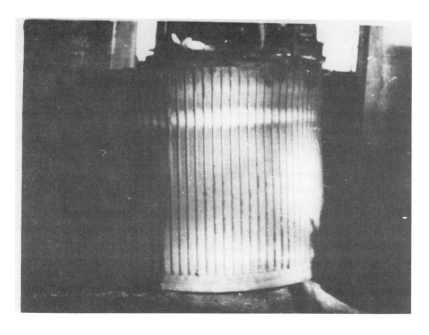

Fig. 9.12. Formation of a waist in the crystallizer shell.

9.4. Ratchetting of a crystallizer

in the limiting cycle and $\sigma_{x\tau}^{(p)}, \sigma_{\varphi\tau}^{(p)}$ are the stresses in the working cycle. At the accepted temperature-independence of the elastic moduli the equalities (9.34) correspond to proportional changes in the temperature gradients.

In the presence of (9.34) the relationship (2.88), applied to the thin hollow cylinder, assumes the form

$$\int_L dx \int_{-h}^{h} \min_{\tau}[(\sigma_x - n\sigma_{x\tau}^{(p)})\Delta\epsilon_x'' + (\sigma_\varphi - n\sigma_{\varphi\tau}^{(p)})\Delta\epsilon_\varphi''] \, dz +$$
$$+ \sum_\mu \int_{-h}^{h} \min_{\tau}[(\sigma_x - n\sigma_{x\tau}^{(p)})\Delta u_\mu] \, dz = 0. \tag{9.35}$$

To determine the magnitude of the shakedown safety factor n we assume that the kinematically admissible ratchetting mechanism consists of three plastic hinge circles between which the deflection increments $\Delta w(x)$ in each cycle change linearly and the longitudinal displacements of the middle surface are absent, Fig. 9.13. Put analytically, this amounts to

$$\Delta u(x) \equiv 0; \quad \Delta w(x) = x\Delta w_0/L_1 \quad \text{for } 0 \leq x \leq L_1, \tag{9.36}$$

$$\Delta w(x) = \Delta w_0[1 - (x - L_1)/L_2] \quad \text{for } L_1 \leq x \leq L_1 + L_2. \tag{9.37}$$

From the compatibility in the rotationally symmetric situation [9.11] it follows that the following strain increments are produced by (9.36), (9.37):

$$\Delta\epsilon_x'' = \frac{d(\Delta u)}{dx} - z\frac{d^2(\Delta w)}{dx^2} = 0 \quad \text{for } x \neq 0, x \neq L_1, x \neq L_1 + L_2,$$

$$\Delta\epsilon_\varphi'' = -\frac{\Delta w(x)}{R} = -\frac{\Delta w_0}{R} \cdot \frac{x}{L_1} \quad \text{for } 0 \leq x \leq L_1, \tag{9.38}$$

$$\Delta\epsilon_\varphi'' = -\frac{\Delta w(x)}{R} = -\frac{\Delta w_0}{R}\left(1 - \frac{x - L_1}{L_2}\right) \quad \text{for } L_1 \leq x \leq L_1 + L_2,$$

where z is the normal to the middle surface of the shell.

At the hinge circles 1, 2, 3 (Fig. 9.13) we encounter discontinuities in the longitudinal displacement increments

$$\Delta u_1 = -z\frac{\Delta w_0}{L_1}, \quad \Delta u_2 = -\Delta u_1 - \Delta u_3, \quad \Delta u_3 = -z\frac{\Delta w_0}{L_2}. \tag{9.39}$$

9 Strain accumulation at thermal cycling

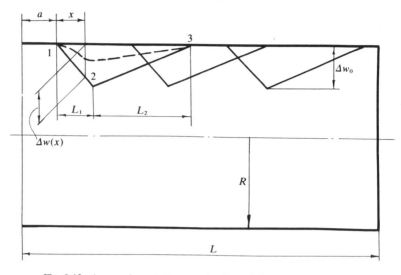

Fig. 9.13. Assumed ratchetting mechanism of the crystallizer shell.

It is further assumed that at each point of the deformed part of the tube made of the Tresca material (2.12) a unique flow régime takes place corresponding to some instantaneous stress distribution. As stated in Chapters 2 and 3, this assumption considerably simplifies the necessary calculations and it enables us to omit part of the construction of the fictitious yield surface at the cost of a certain overestimation of the limiting cycle parameters.

It follows from the above assumptions and the flow rule (2.76), that negative hoop strains (9.38) are associated with the stress $\sigma_\varphi = -\sigma_s$ whereas the hinge circle longitudinal strains, related to the displacement discontinuities (9.39) are associated with the stresses $\sigma_x = \sigma_s$ for $\Delta u_\mu > 0$ and with the stresses $\sigma_x = -\sigma_s$ for $\Delta u_\mu < 0$, $\mu = 1, 2, 3$.

Substituting now the strain and the displacement increments (9.38), (9.39) as well as the corresponding yield surface stresses into equation (9.35), we arrive at

$$\int_0^{L_1} \frac{x}{L_1} \, dx \int_{-h}^{h} \min_\tau(\sigma_s + n\sigma_{\varphi\tau}^{(p)}) \, dz + \int_{L_1}^{L_1+L_2} \left(1 - \frac{x - L_1}{L_2}\right) dx \times$$
$$\times \int_{-h}^{h} \min_\tau(\sigma_s + n\sigma_{\varphi\tau}^{(p)}) \, dz + R \int_0^{h} \left\{ \frac{1}{L_1} [\min_\tau(\sigma_s + n\sigma_{x\tau}^{(p)})_{x=0}] + \right.$$
$$\left. + \left(\frac{1}{L_1} + \frac{1}{L_2}\right) [\min_\tau(\sigma_s - n\sigma_{x\tau}^{(p)})_{x=L_1}] + \frac{1}{L_2} [\min_\tau(\sigma_s + n\sigma_{x\tau}^{(p)})_{x=L_2}] \right\} z \, dz +$$

9.4. Ratchetting of a crystallizer

$$+ R \int_0^h \left\{ \frac{1}{L_1} [\min_\tau(\sigma_s - n\sigma_{x\tau}^{(p)})_{x=0}] + \left(\frac{1}{L_1} + \frac{1}{L_2} \right) [\min_\tau(\sigma_s + n\sigma_{x\tau}^{(p)})_{x=L_1}] + \right.$$
$$\left. + \frac{1}{L_2} [\min_\tau(\sigma_s - n\sigma_{x\tau}^{(p)})_{x=L_2}] \right\} (-z) \, dz = 0. \tag{9.40}$$

Knowing the temperature profiles, Fig. 9.11b, we can determine the yield stresses σ_s together with the stresses $\sigma_{x\tau}^{(p)}, \sigma_{\varphi\tau}^{(p)}$ at various stages of the cycle. Thus equation (9.40) can be solved with respect to the factor n by using the trial and error procedure, since at different values of n the minima of the various integrands in (9.40) can be attained at a point of the tube at different instants of time. To solve (9.40), we should divide for each preassigned value of n the deformed part of the tube ($0 \leq x \leq L_1 + L_2$) into a number of sections and then we should evaluate for various values of the coordinate z the magnitudes

$$\min_\tau(\sigma_s + n\sigma_{\varphi\tau}^{(p)}) \tag{9.41}$$

at those sections. Similarly, we should evaluate the magnitudes

$$\min_\tau(\sigma_s \pm n\sigma_{x\tau}^{(p)}) \tag{9.42}$$

at the hinge circles $x = 0$, $x = L_1$, $x = L_2$. The corresponding constructions are shown in Fig. 9.14a, b for two different sections; both the thermo-elastic stresses and the values of the yield point have been calculated for actual, working temperature fields. Thin lines indicate the differences between the yield points and the relevant thermo-elastic stresses at a number of instants of time. The envelopes of the distributions are shown by shading; in Fig. 9.14a they correspond to the instants $\tau_1 = 109$ sec and $\tau_2 = 464$ sec, in Fig. 9.14c to 86 sec and 114 sec, all measured from the beginning of casting.

To obtain the best (lowest) upper bound on the progressive ratchetting safety factor n the parameters α, L_1, L_2 locating the plastic hinge circles were subject to variation. The magnitudes of n for a number of variants are shown in Table 9.1. The best n amounts to 0,65 which means that the actual safety margin as regards progressive ratchetting ($k \leq n = 0,65$) is substantially lower than unity and the one-sign strains are expected to accumulate in the working cycle in the region $0,1 \leq x/L \leq 0,3$, see Table 9.1. It should be stressed that both the shape and

9 Strain accumulation at thermal cycling

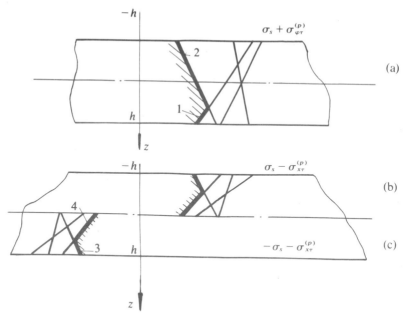

Fig. 9.14. Towards determination of minima in the expressions (9.41), (9.42).

the dimensions of this region, obtained analytically, differ little from those encountered in the service life of crystallizers, see dashed line in Fig. 9.13.

The shakedown analysis clearly fails to supply the magnitude of the maximum deflection increment in the working cycle. The calculations of deformations of a crystallizer tube are given in [9.13]; they are based on the deformation theory of plasticity and the technical theory of elastic-

Table 9.1
Results for the crystallizer shell.

N	1	2	3	4	5	6	7	8	9
a/L	0,1	0,1	0,1	0,2	0,3	0,4	0,0	0,1	0,1
L_1/L	0,1	0,2	0,3	0,1	0,1	0,1	0,1	0,1	0,1
L_2/L	0,3	0,2	0,1	0,3	0,3	0,3	0,3	0,1	0,6
n	0,651	0,677	0,877	0,930	1,826	3,23	0,705	0,707	0,895

9.4. Ratchetting of a crystallizer

plastic shells after Rabotnov [9.12]. Account was taken of the temperature-dependent yield point, whereas the strain-hardening and the creep effects were both neglected. The elastic-plastic deformation process under non-stationary thermal actions was investigated by means of a step-by-step application of an additional temperature field starting with a certain instant of time at which the strains were known to be still elastic. Even with computers, such a procedure is considerably time-consuming because a series of cycles must be calculated before the stresses stabilize. Luckily enough, for metal casting the residual stresses have been found to differ very little, starting from the second or the third cycle.

The computational results for deflections after the second and the third cycles are presented in Fig. 9.15a. The results of measurements taken on a crystallizer after five castings are shown in Fig. 9.15b. The curves match rather well, although a big scatter of test results is here unavoidable.

It is again stressed that the adopted idealized model of the process is by all means approximate. In particular, in actual service of crystallizers the external actions are rather mildly cyclic since the levels of liquid metal vary with different castings even in the steady casting régime. Deflection accumulation after a certain number of castings may

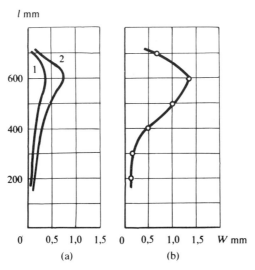

Fig. 9.15. Comparison of analytical results for progressive deformation of a crystallizer sleeve with experimental data.

9 Strain accumulation at thermal cycling

therefore be lower than computed at the cost of expansion of the deformed zone, i.e. its 'smearing out'.

On the other hand, no effects of gas cavities near the tube surface were accounted for in the analysis; such blowholes are known to increase the temperature gradients thus leading to a rapid increase of displacements.

9.5. Ratchetting of a conical shell-type structure (slag container)

The deformations of cups for transporting furnace slag are found to be similar to those of crystallizer sleeves considered in the previous section. The characteristic necking that forms above the supporting ring, i.e. at that part of the cup where gravity load is practically insignificant, is clearly seen in Fig. 9.16; the undeformed cup had straight generators. The periodic changes in the level of heat transferring medium such as cast iron or slag are a common feature of thermal actions in the crystallizer of a slag container. The generated temperature fields are no longer quasi-stationary in the moving frame of reference. However, in a certain sense they are similar, namely the zone of maximum temperatures and their gradients moves during the cycle along the generators while suffering certain small changes.

The time variations of the temperature field in the cup during slag filling

Fig. 9.16. Deformations of a slag container.

9.5. Ratchetting of a conical shell-type structure

and its transportation are visualized in Fig. 9.17. The temperatures were measured in centigrades in nine consecutive cycles. Solid lines correspond to the temperatures of the inner surface, dashed lines to those of the outer surface. Circles indicate readings of thermo-couples referred to the middle surface of the cup.

Along with temperature measurements the temperature field was examined with the help of the electro-thermal analogy [9.14] by means of a special discrete model device. Particular attention was devoted to readings relevant to the temperature gradients along the generators and across the thickness of the cup as the level of slag lifts to reach its maximum. The required precision of the data was achieved by taking a large number of discrete elements in the device modelling the thermal process.

The temperature fields[1] measured at various stages of the working cycle were subsequently used to compute the thermo-elastic stresses as well as the corresponding values of the yield point.

The determination of the thermo-elastic stresses in a conical shell of constant thickness (Fig. 9.18) in the absence of mechanical applied loads consists in solving the system of two differential equations [9.15]

$$l\frac{d^2 V}{dl^2} + \frac{dV}{dl} - \frac{V}{l} = E2h\theta \operatorname{ctg} \alpha + (1+\mu)l\frac{dN_t}{dl},$$

$$l\frac{d^2\theta}{dl^2} + \frac{d\theta}{dl} - \frac{\theta}{l} = -\frac{V}{D}\operatorname{ctg}\alpha + \frac{l}{D}\cdot\frac{dM_t}{dl},$$

(9.43)

where $V = R_2 Q$, R_2 denotes the current radius, Q is the shear force, θ stands for the rotation angle of the normal to the middle surface, whereas $D = E(2h)^3/12(1-\mu^2)$ specifies the flexural stiffness of the shell. Further,

$$N_t = \frac{E\alpha}{1-\mu}\int_{-h}^{h} t\,dz, \quad M_t = \frac{E\alpha}{1-\mu}\int_{-h}^{h} tz\,dz,$$

where $t = t(l, z)$ denotes the law of temperature distribution, and z is the coordinate along the outward normal to the middle surface of the shell. Both the elastic and the thermal properties of the material are kept temperature-independent.

---
[1] They should be considered as averaged ones as there was considerable scatter of temperature readings.

9 Strain accumulation at thermal cycling

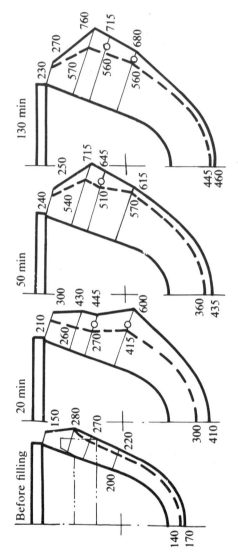

Fig. 9.17. Temperature variations in a container at slag pouring followed by overheating.

9.5. Ratchetting of a conical shell-type structure

The system (9.43) was solved by means of the difference factorization method [9.16] which made it possible to reduce the second-order difference equations to first order difference equations and then to programme the remaining calculations on a computer. From the boundary conditions at one end of the integration range certain auxiliary quantities were calculated at points $l_0, l_1, l_2, \ldots, l_m$ (straight course) and then using the boundary conditions at the other end, the values of the sought for functions were found (reverse course).

The boundary conditions for a conical shell with free edges are the following:

$$\text{for } l = l_0: V = 0, \quad M_l = -D\left(\frac{\mathrm{d}\theta}{\mathrm{d}l} + \mu \frac{\theta}{l}\right) + M_t = 0; \tag{9.44}$$

$$\text{for } l = l_m: V = 0, \quad M_l = 0, \tag{9.45}$$

where M_l is the meridional bending moment. The internal forces can be determined by using the calculated values of the functions V and θ with the help of the corresponding relations given in [9.15]:

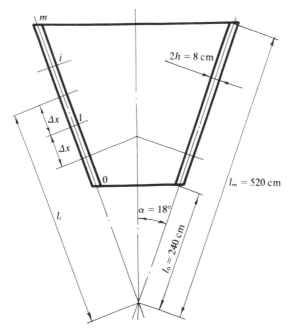

Fig. 9.18. Idealized cup of a slag container.

9 Strain accumulation at thermal cycling

$$M_l = -D\left(\frac{d\theta}{dl} + \mu\frac{\theta}{l}\right), \quad M_\theta = -D\left(\frac{\theta}{l} + \mu\frac{d\theta}{dl}\right),$$

$$N_l = -\frac{V}{l}, \quad N_\theta = -\frac{dV}{dl},$$

(9.46)

where M_l, M_θ, N_l, N_θ denote the meridional and hoop bending moments and membrane forces, respectively. The total meridional and hoop normal stresses are

$$\sigma_l = \frac{N_l}{2h} + \frac{12M_l}{(2h)^3}\cdot z, \quad \sigma_\theta = \frac{N_\theta}{2h} + \frac{12M_\theta}{(2h)^3}\cdot z.$$

(9.47)

To determine the thermo-elastic stresses, the temperature fields were measured for many instants of time. Two computer-aided results are shown in Fig. 9.19. Also the temperature distributions along the inner and outer surface are given. Due to their relatively small values, the meridional membrane force N_l and the shearing force were both neglected. The material constants were $E = 2{,}1\times 10^6$ kG/cm², $\mu = 0{,}3$, $\alpha = 11{,}5\times 10^{-6}$ 1/°C, corresponding to the grade of cast steel used [9.17].

The computed thermo-elastic stresses exceeded considerably the yield point shown in Table 9.2 for the corresponding temperatures.

As we know, the sufficient condition for alternating plastic flow is that the variation of the fictitious elastic stresses at any point of the shell must exceed the double yield point; for non-isothermal loading it must exceed the sum of yield points at respective temperatures. Let us find how much the thermo-elastic stresses should increase, the yield point being kept constant, so that we obtain the limiting cycle in the sense of alternating flow. The problem consists in finding for all the material volumes the minimum value of the multiplier n at which one of the following inequalities holds good:

$$\min_\tau[\sigma_{st} - n_l\sigma_l] - \max_\tau[-\sigma_{st} - n_l\sigma_l] \leq 0,$$

(9.48)

$$\min_\tau[\sigma_{st} - n_\theta\sigma_\theta] - \max_\tau[-\sigma_{st} - n_\theta\sigma_\theta] \leq 0,$$

(9.49)

$$\min_\tau[\sigma_{st} - n_{l\theta}(\sigma_l - \sigma_\theta)] - \max_\tau[-\sigma_{st} - n_{l\theta}(\sigma_l - \sigma_\theta)] \leq 0.$$

(9.50)

9.5. Ratchetting of a conical shell-type structure

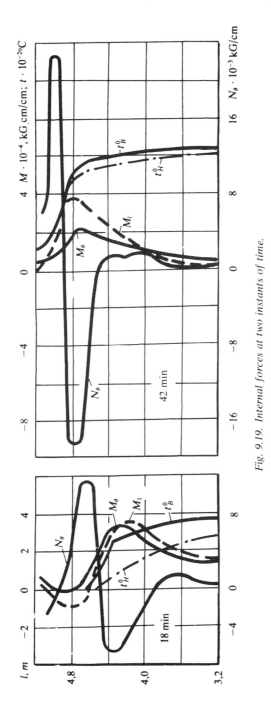

Fig. 9.19. Internal forces at two instants of time.

9 Strain accumulation at thermal cycling

Table 9.2.
The yield point for the cup as dependent on temperature.

t, °C	20	200	300	400	450	500	550	600
σ_s, kG^2/cm^2	2100	1700	1600	1600	1400	1300	1200	800

Thus

$$n = \min(n_l, n_\theta, n_{l\theta}). \tag{9.51}$$

All the quantities entering (9.48)–(9.50) are functions of current time, consistent with that at which the temperature fields develop. The computational results are shown in Fig. 9.20; curve 1 gives n_θ for $z = +h$, curve 2 shows n_θ for $z = -h$, curve 3 shows n_l for $z = +h$, curve 4 shows n_l for $z = -h$. Being much larger, the multiplier $n_{l\theta}$ is not shown. The smallest n is $n_{\theta\,\min} = 0{,}6$ and it corresponds to the point with the coordinates $l = 420$ cm, $z = -h$. In the regions $300 \leq l \leq 400$ cm, $460 \leq l \leq 480$ cm the values of n_l (for $z = \pm h$) and n_θ (for $z = -h$) are less than one. Thus in those regions we can expect the meridional and hoop cracks to form during service of the slag container.

Indeed, in a number of cups the cracks did really form after a few hundreds of cycles: meridional ones in the region $380 \leq l \leq 450$ cm and hoop ones in the regions $370 \leq l \leq 390$ cm and $420 \leq l \leq 440$ cm. Rather poor coincidence of the calculated and observed regions can partly be explained by the fact that in the observations the level of slag differed from pouring to pouring by as much as 20 cm and sometimes even more. Let us also remember that the shakedown analyses permit us to determine only the beginning of failure. This has to be kept in mind when comparing the corresponding results with observations of fracture under actual operating conditions. The coordinate l was measured as shown in Fig. 9.18. The above regions correspond to the sections below and above the supporting ring, see Fig. 9.16, i.e. to the highest level of slag.

To investigate ratchetting, let us employ the approximate kinematical approach assuming the ratchetting mechanism as shown in Fig. 9.21. Analytically, we have

$$\Delta w = -a\frac{l - l_1}{l_2 - l_1}, \quad \Delta u = 0 \quad \text{for } l_1 \leq l \leq l_2, \tag{9.52}$$

9.5. Ratchetting of a conical shell-type structure

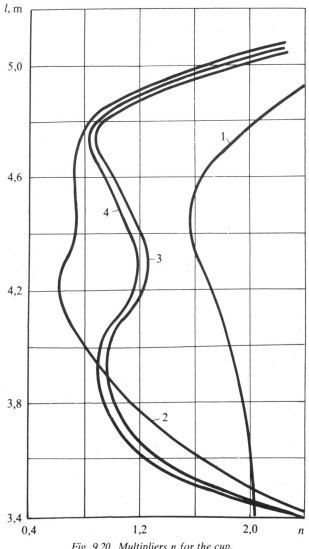

Fig. 9.20. Multipliers n for the cup.

9 Strain accumulation at thermal cycling

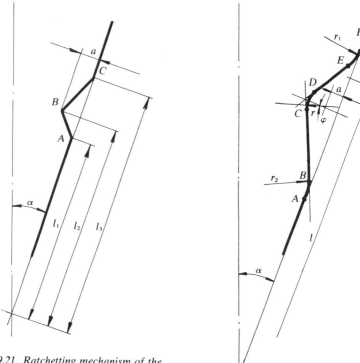

Fig. 9.21. Ratchetting mechanism of the slag container.

Fig. 9.22. Refined ratchetting mechanism.

$$\Delta w = -a \frac{l_3 - l}{l_3 - l_2}, \quad \Delta u = 0 \quad \text{for } l_2 \leq l \leq l_3, \tag{9.53}$$

where Δw is the deflection increment along the normal to the middle surface, and Δu is the displacement increment along the generator of the cup. The corresponding strain increments are [9.15]

$$\Delta \epsilon_l = \frac{d(\Delta u)}{dl} - z \frac{d^2(\Delta w)}{dl^2}, \quad \Delta \epsilon_\theta = \frac{\Delta u + \Delta w \operatorname{ctg} \alpha}{l} - \frac{1}{l} z \frac{d(\Delta w)}{dl}. \tag{9.54}$$

On account of (9.52), (9.53) we get

$$\Delta \epsilon_l = 0, \quad \Delta \epsilon_\theta = -\frac{a}{l} \cdot \frac{l - l_1}{l_2 - l_1} \operatorname{ctg} \alpha + z \frac{a}{l(l_2 - l_1)} \quad \text{for } l_1 \leq l \leq l_2, \tag{9.55}$$

$$\Delta \epsilon_l = 0, \quad \Delta \epsilon_\theta = -\frac{a}{l} \cdot \frac{l_3 - l}{l_3 - l_2} \operatorname{ctg} \alpha - z \frac{a}{l(l_3 - l_2)} \quad \text{for } l_2 \leq l \leq l_3. \tag{9.56}$$

9.5. Ratchetting of a conical shell-type structure

Condition (2.88) for the progressive strain accumulation takes the form

$$\int_{l_1}^{l_3} 2\pi l \sin\alpha \, dl \int_{-h}^{h} (\sigma^0_{l*}\Delta\epsilon_l + \sigma^0_{\theta*}\Delta\epsilon_\theta) \, dz + 2\pi \sin\alpha \times$$

$$\times \sum_{i=1}^{3} l_i \int_{-h}^{h} (\sigma^0_{l*}\Delta u')_i \, dz = 0. \qquad (9.57)$$

The meridional and hoop stresses σ^0_{l*}, $\sigma^0_{\theta*}$ belonging to the fictitious yield hexagon will be computed on the assumption that in the limiting cycle the plastic strain rates at each point of the cup are different from zero for only one value of the thermo-elastic stresses. Thus the following stresses correspond to the yielding regimes (9.55), (9.56):

$$\sigma^0_{\theta*} = \min_{\tau}(\sigma_{st} - n\sigma_{\theta\tau}) \quad \text{for } \Delta\epsilon_\theta > 0,$$

$$\sigma^0_{\theta*} = \max_{\tau}(-\sigma_{st} - n\sigma_{\theta\tau}) \quad \text{for } \Delta\epsilon_\theta < 0, \qquad (9.58)$$

and, in the yield hinge circles ($l = l_1, l_2, l_3$)

$$\sigma^0_{l*} = \min_{\tau}(\sigma_{st} - n\sigma_{l\tau}) \quad \text{for } \Delta\epsilon_l > 0,$$

$$\sigma^0_{l*} = \max_{\tau}(-\sigma_{st} - n\sigma_{l\tau}) \quad \text{for } \Delta\epsilon_l < 0. \qquad (9.59)$$

Substituting (9.58), (9.59) into (9.57) and recalling (9.52), (9.53), (9.55), (9.56), we arrive at

$$\frac{1}{l_2 - l_1}\int_{l_1}^{l_2} dl \int_{-h}^{h} \min_{\tau}\{(\sigma_s\delta_1 - n\sigma_{\theta\tau})[z - (l - l_1)\operatorname{ctg}\alpha]\}\, dz +$$

$$+ \frac{1}{l_3 - l_2}\int_{l_2}^{l_3} dl \int_{-h}^{h} \min_{\tau}\{(\sigma_s\delta_2 - n\sigma_{\theta\tau})[-z - (l_3 - l)\operatorname{ctg}\alpha]\}\, dz +$$

$$+ \frac{l_1}{l_2 - l_1}\int_{-h}^{h} \min_{\tau}[(\sigma_s\delta_3 - n\sigma_{l\tau})z]_{l=l_1}\, dz +$$

$$+ l_2\left(\frac{1}{l_2 - l_1} + \frac{1}{l_3 - l_2}\right)\int_{-h}^{h} \min_{\tau}[(-\sigma_s\delta_3 - n\sigma_{l\tau})(-z)]_{l=l_2}\, dz +$$

$$+ \frac{l_3}{l_3 - l_2}\int_{-h}^{h} \min_{\tau}[(\sigma_s\delta_3 - n\sigma_{l\tau})z]_{l=l_3}\, dz = 0, \qquad (9.60)$$

443

9 Strain accumulation at thermal cycling

where

$$\delta_1 = \text{sign}[z - (l - l_1)\operatorname{ctg}\alpha], \quad \delta_2 = \text{sign}[-z - (l_3 - l)\operatorname{ctg}\alpha],$$

$$\delta_3 = \text{sign } z, \quad \left(\text{sign } k = \frac{k}{|k|}\right).$$

Equation (9.60) should be solved with respect to the unknown n by the method of trial and error, similarly as in the case of the crystallizer sleeve where, at different values of n, the integrands also attained their minima at the same point of the shell at different instants of time. To enhance the upper bound, a number of ratchetting mechanisms was considered whose parameters are collected in Table 9.3. Some computational results are shown in the first column of Table 9.4, among them the best estimate corresponding to the mechanism with the parameters shown in the eleventh row of Table 9.3.

Table 9.3
Parameters of considered admissible collapse mechanisms of the cup.

N	l_1, cm	l_2, cm	l_3, cm
1	344	392	440
2	360	392	424
3	376	392	408
4	376	424	472
5	392	424	456
6	408	424	440
7	408	456	504
8	424	456	488
9	440	456	472
10	440	472	504
11	456	472	488

9.5. Ratchetting of a conical shell-type structure

Table 9.4
Progressive collapse safety factors for the cup.

N	n formula (9.60)	n formula (9.57)
4	1,415	–
7	1,54	–
10	1,63	1,24
11	1,33	0,93

The upper bound results can be further improved by assuming a somewhat refined class of the ratchetting mechanisms shown in Fig. 9.22. In these mechanisms the deflection increments over the segments BC and DE are assumed linearly distributed along the generators, cf. (9.55), (9.56), whereas the connecting segments AB, CD and EF are circular arcs. In other words, concentrated hinge circles are smeared out to form plastic bands of finite width. In order to determine the multiplier n, we accept expression (9.57) as sufficiently general for that purpose. We then insert $\Delta u' = 0$ and establish the strain increments with the help of the relationships (9.54) and Fig. 9.22. In particular, for the segment BC we have

$$\Delta w = -a - r \cos \varphi. \tag{9.61}$$

This, on substituting into (9.54) and taking $\Delta u = 0$, yields

$$\Delta \epsilon_l = -\frac{z}{r \cos^3 \varphi}, \quad \Delta \epsilon_\theta = \frac{1}{l}[-(a + r \cos \varphi) \operatorname{ctg} \alpha - z \operatorname{tg} \varphi]. \tag{9.62}$$

From (9.62) for $-h \le z \le 0$, it follows that

$$\Delta \epsilon_l > 0, \tag{9.63}$$

$$\Delta \epsilon_\theta < 0 \quad \text{for } -z \operatorname{tg} \varphi < (a + r \cos \varphi) \operatorname{ctg} \alpha, \tag{9.64}$$

$$\Delta \epsilon_\theta > 0 \quad \text{for } -z \operatorname{tg} \varphi > (a + r \cos \varphi) \operatorname{ctg} \alpha. \tag{9.65}$$

9 Strain accumulation at thermal cycling

Moreover,

$$|\Delta\epsilon_\theta| > \Delta\epsilon_l \quad \text{for } (a + r\cos\varphi)\operatorname{ctg}\alpha > -\frac{z}{\cos\varphi}\left(\frac{l}{r\cos^2\varphi} + \sin\varphi\right). \tag{9.66}$$

The second inequality in (9.66) is satisfied when

$$r(a+r) > (5 \div 10)h^2.$$

The selection of the fictitious yield curve stresses $\sigma_{l*}^0, \sigma_{\theta*}^0$ associated with (9.63), (9.64) and (9.65), (9.66), according to the flow law, is shown in Fig. 9.23. It is worth emphasizing that an approximate description of stresses belonging to the fictitious yield curve (9.58), (9.59) was not employed when the latter was not degenerate at the given point of the shell and at the given value of n. In the degenerate case simply the stresses (9.58), (9.59) were substituted into (9.57).

Since this part of the cup for which the fictitious yield curve was degenerate was relatively small, the applied approach is expected to yield rather insignificant errors in the determination of the progressive ratchetting condition.[1]

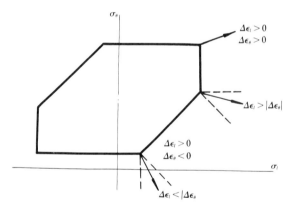

Fig. 9.23. Fictitious yield hexagon for one of the cup points.

---

[1] However, it must be mentioned that the approach is approximate and does not exactly fit the framework of the shakedown theory, because it involves an attempt to establish conditions for strain accumulation in the situation in which the shakedown domain is bounded by the alternating flow criterion. This problem will be discussed in more detail in Chapter 11.

9.6. Experimental investigation of deformations

The computational results for two ratchetting mechanisms, described by the parameters shown in rows 10 and 11 of Table 9.3 and with the hinge circles replaced by smooth bands, are presented in Table 9.4. The best upper bound on the multiplier n turns out to be equal to 0,93 which means that the ratchetting of the cup should be encountered already in the working cycle. The progressive deformation region was calculated to be contained within $456 \leq l \leq 498$ cm. In fact, it was spreading over $350 \leq l \leq 480$ cm which can be explained, as in the case of the crystallizer, by the mildly cyclic character of the real actions and by different levels of slag in different pourings.

When comparing the results for the crystallizer sleeve and the slag cup it must be realized that the number of pourings in the service life is for the latter by one order of magnitude larger than for the former. This is why, irrespective of the higher safety factor (0,93 and 0,65), the strain accumulation per life of the cup appears to be much more pronounced.

In neither case were any creep effects taken into account; they could have changed the ratchetting conditions quantitatively by lowering the values of n. Characteristically enough, in both cases the periods of stationary temperature fields are rather short as compared with the periods of transient processes.

In some parts of the structures creep is capable of developing after short-term plastic strains of opposite sign have occurred; this is known to produce a considerable increase in the average rates of creep strains. However, the properties of particular materials in such conditions still have to be carefully investigated [9.18].

9.6. Experimental investigation of deformations under variable repeated action of quasi-stationary (in the moving coordinate system) temperature fields

Tests on thermal ratchetting aimed at checking the theoretical results were made with the use of specially designed benches. Thin-walled tubular specimens were tested with outer diameter of 30 to 50 mm and wall thickness of 1,2 to 6 mm. The tubes were heated by means of a high-frequency current supplied by special generators. The advantage of such treatment consists in an easy generation of rapid and intensive local heating; its drawback is that only metals and alloys with good magnetic properties can be used. At heating interspersed with water

9 Strain accumulation at thermal cycling

cooling temperature fields with considerable gradients were generated, sufficient for the maximum fictitious thermo-elastic stresses to exceed the yield point of the material of the specimens.

Consider a device shown in Fig. 9.24 used to generate the rotationally symmetric temperature field moving downwards. The particular parts are: 1 an electric motor to slowly rotate the specimen 4 in order to secure its uniform circumferential heating, 2 a coupler, 3 and 7 two holders. The inductor 5 generates a circle of intensive heating. Due to rapid cooling of the adjacent zone of the tube, either by water in basin 6 or by an annular sprayer, there occur large temperature gradients along the generators. Centering support 8 with a very soft coil spring does not practically prevent the specimen from axial thermal expansion. The motion of the tube with respect to both the heating and cooling sources is realized, either manually or motor-driven, by lead-screw 9 and slide 10 connected to frame 11.

The temperature field in the tubular specimen at a generic instant of time is shown in Fig. 9.25. It can be looked upon as quasi-stationary provided it is referred to the coordinate system whose origin lies in the plane of the inductor and the tube is long enough. One thermocouple fixed in the middle of the tube was used to measure the temperatures registered by means of an oscillograph.

Each cycle consisted of the downward passage of the tube with a suitable velocity enabling it to reach the prescribed maximum temperature and of a rapid (idle) upward passage to occupy the original position.

According to the previous analysis, at sufficiently large temperature gradients between the inductor and the coolant level the ensuing thermal wave should result in a plastic hoop squeezing of the most heated band of the tube. As mentioned before, depending on the temperature field profile, i.e. on the passage velocity and direction, the bulging of the tube is also possible with each passage as a result of subsequent plastic hoop elongations of cooler parts of the specimen.

The tests turned out to corroborate the theoretical results obtained in Sec. 9.3. A side view of the specimen (a) and its axial section (b) are both shown in Fig. 9.26. Together with plastic hoop squeezing there were encountered changes in the spacing of marks and in the thickness of the wall. According to the assumed incompressibility, the sum of three mutually perpendicular strains should vanish. Indeed, either of the two elongations was equal to roughly a half of the contraction.

Since the strains after the passage of the temperature field are

9.6. Experimental investigation of deformations

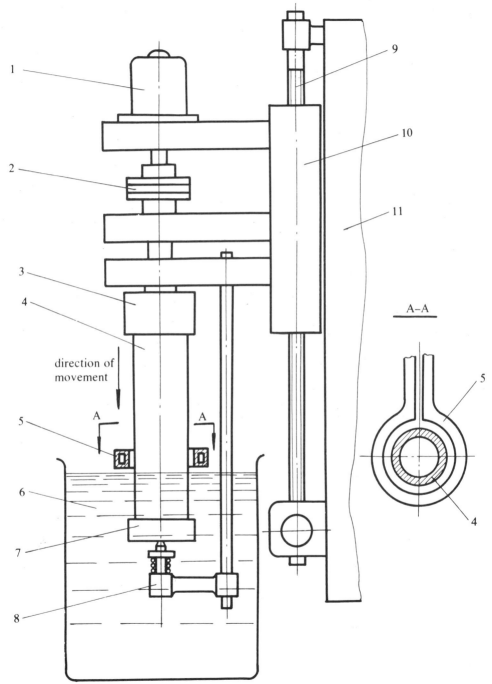

Fig. 9.24. Device for creating quasi-stationary temperature field in a shell specimen.

9 Strain accumulation at thermal cycling

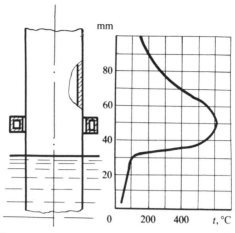

Fig. 9.25. Instantaneous distribution of symmetric temperature in the specimen.

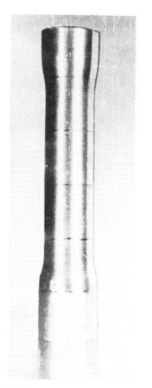 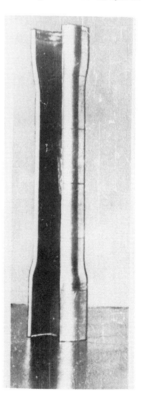

Fig. 9.26. View of the specimen and its longitudinal section after repeated actions of axi-symmetric quasi-stationary temperature field.

9.6. Experimental investigation of deformations

uniform, except at the ends of the specimen, residual membrane stresses cannot occur, due to equilibrium requirements. Experimental evidence has indeed shown that strain increments per cycle are constant as long as the strain-hardening is absent.

The development of deformations is shown in Fig. 9.27, as dependent on the number of passages and the maximum temperatures at the bore of the tube made of low-carbon steel; the outer diameter was 38 mm, and the wall thickness 2 mm. With the actions applied, very substantial strains reaching some tenth part of a percent per cycle (passage) were taking place at moderate heating, unlike in the experiments reported in [9.3] in which relatively higher temperatures were accompanied by smaller strains.

The amount of strain accumulated per cycle depends not only on the maximum temperature but clearly also on its maximum gradient. It must be remembered that the computations of Sec. 9.3 showed the existence of strain accumulation at the idealized temperature profile when the temperature jump exceeded 150°C. The latter depends on the heating and cooling conditions as well as on the thermal conductivity of

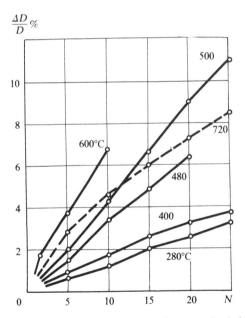

Fig. 9.27. Development of deformation (decrease in diameter of tubular specimen) as the number of cycles N increases; various maximum temperatures indicated.

9 Strain accumulation at thermal cycling

the material. In tests depicted in Fig. 9.27 the clearance between the inductor and the water level and also the water temperature were kept constant.

Influence of wall thickness on the strains per cycle was also investigated. Increase of wall thickness was found to result in decreasing strains, under the same heating conditions; dashed curve in Fig. 9.27 is plotted for a 6 mm thick tube (as compared with 2 mm in the remaining tests).

At departure from axial symmetry due to heating conditions and/or variable thickness, a certain amount of flexure was observed, together with a diminished bore. In some cases, particularly for very thin tubes, the cross-section distorted similarly as in the case of external pressure leading to the loss of stability of the shell.

In Fig. 9.28 five curves are plotted to show the development of the ratchetting strains at $t_{max} = 600°C$ for several alloys. Tubular specimens were used with diameter 38 mm and 2 mm thick, made of various metal alloys. Fig. 9.29 shows the relationships between the increments of deformation $\Delta D/D$ and the maximum temperatures t_{max} for the three grades of steel described in the caption of Fig. 9.28. The curves are plotted for average increments per the first 10–15 cycles of the process; Figs. 9.27 and 9.28 suggest that the relationships are nearly linear. The tested tubes were 2 to 2,5 mm thick and had the outer diameter $D = 50$ mm. Below a certain limiting temperature, depending on the particular material, no strain

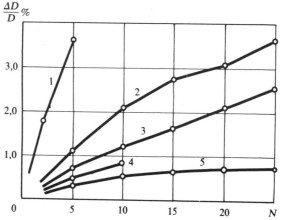

Fig. 9.28. Growth of deformation at repeated quasi-stationary temperature field ($t_{max} = 600°C$): 1 – low-carbon steel, 2 – alloy steel of pearlitic kind, 3 – cast iron, 4 – high-carbon steel, 5 – stainless steel.

9.6. Experimental investigation of deformations

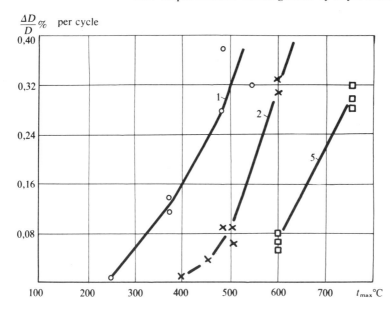

Fig. 9.29. Growth of deformation per cycle against the maximum heating temperature; steel grades 1, 2, 5 as in Fig. 9.28.

accumulation was detected. For instance, these temperatures were 240°C for low-carbon steel (1) and 360°C for alloy steel (2). For stainless steel (5) no exact measurements could be taken due to its weak magnetic properties.

To estimate the limiting temperature of the cycle analytically, the thermo-elastic stresses corresponding to the real temperature field as shown in Fig. 9.25 will be assumed to vary proportionally to the maximum temperature inside the cycle. Analysis of temperature fields with various values of t_{max} observed in the experiments has shown that the above assumption is reasonable, at least in a definite range of temperatures.

The ratchetting condition (2.88), specified for the repeated actions of a quasi-stationary relatively to the moving frame of reference temperature field on an axi-symmetric cylindrical shell, takes the form

$$\int_{-h}^{h} (\sigma_{x*}^0 \Delta\epsilon_{x0} + \sigma_{\varphi*}^0 \Delta\epsilon_{\varphi 0})\, dz < 0. \qquad (9.67)$$

It goes without saying that integration with respect to the axial coordinate is not done due to the quasi-steady stress state; only the middle sections of the tube are considered.

453

9 Strain accumulation at thermal cycling

Assuming the Tresca material, see (2.12), we have that the interior of the fictitious yield surface, Fig. 9.30, is bounded by the inequalities

$$\max_{\tau}(-\sigma_{st} - \sigma_{\varphi\tau}^{(e)}) \leq \sigma_{\varphi*}^0 \leq \min_{\tau}(\sigma_{st} - \sigma_{x\tau}^{(e)}),$$

$$\max_{\tau}(-\sigma_{st} - \sigma_{x\tau}^{(e)}) \leq \sigma_{x*}^0 \leq \min_{\tau}(\sigma_{st} - (\sigma_{\varphi\tau}^{(e)} - \sigma_{\varphi\tau}^{(e)})], \quad (9.68)$$

$$\max_{\tau}[-\sigma_{st} - (\sigma_{x\tau}^{(e)} - \sigma_{\varphi\tau}^{(e)})] \leq \sigma_{x*}^0 - \sigma_{\varphi*}^0 \leq \min_{\tau}[\sigma_{st} - (\sigma_{x\tau}^{(e)} - \sigma_{\varphi\tau}^{(e)})],$$

where the subscripts x, φ denote longitudinal and hoop stresses, respectively.

One and the same flow régime in the cycle is assumed to set in at all the points of the normal to the middle surface of the tube; the possible régimes are associated with the sides AB, EF, FA in Fig. 9.30. At $\mu \geq 0$ they are

$$\Delta\epsilon_{x0} = \mu, \quad \Delta\epsilon_{\varphi 0} = 0, \quad \Delta\epsilon_{z0} = -\mu, \quad \sigma_{x*}^0 = \min_{\tau}(\sigma_s - \sigma_{x\tau}^{(e)}), \quad (9.69)$$

$$\Delta\epsilon_{x0} = \mu, \quad \Delta\epsilon_{\varphi 0} = -\mu, \quad \Delta\epsilon_{z0} = 0, \quad \sigma_{x*}^0 - \sigma_{\varphi*}^0 = \min_{\tau}[\sigma_s - (\sigma_{x\tau}^{(e)} - \sigma_{\varphi\tau}^{(e)})], \quad (9.70)$$

$$\Delta\epsilon_{x0} = 0, \quad \Delta\epsilon_{\varphi 0} = -\mu, \quad \Delta\epsilon_{z0} = \mu, \quad \sigma_{\varphi*}^0 = \max_{\tau}(-\sigma_s - \sigma_{\varphi\tau}^{(e)}), \quad (9.71)$$

or any combination of these.

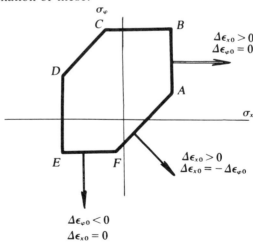

Fig. 9.30. Fictitious yield hexagon for the tubular specimen.

9.6. Experimental investigation of deformations

The ratchetting mechanism consists in a uniform hoop squeezing of the tube accompanied by a simultaneous increase in its length, régimes (9.70), or by its unchanged length, régime (9.71). It can be assumed also, (9.69), that increase of the length is not accompanied by a change in the diameter of the tube. Other possibilities of collapse mechanisms have been checked to lead to worse upper bounds on the limiting temperature of the cycle.

As an example, consider a tubular low-carbon steel specimen ratchetting according to (9.70). The temperature-dependent yield point for this grade of steel [9.19] is given in Table 9.5; the calculated thermoelastic stresses, corresponding to $t_{max} = 605°C$ are shown in Table 9.6. On account of (9.70), condition (9.67) takes the form

$$2h \int_{-1}^{1} \min_{\tau}[\sigma_{st} - n(\sigma_{x\tau}^{(e)} - \sigma_{\varphi\tau}^{(e)})]\,dz = 0, \qquad (9.72)$$

Table 9.5
The yield point stress for low-carbon steel as dependent on temperature [9.19].

t, °C	20	200	300	400	500	600
σ_s, kg/cm²	2300	2400	1600	1250	1000	700

Table 9.6
Instantaneous distribution of thermo-elastic stresses in the tubular specimen at steady temperature field.

X, cm	t, °C	$\sigma_{\varphi\tau}^{(e)}$, kG/cm²	$\sigma_{x\tau}^{(e)}$, kG/cm²
3,6	90	340	$-1500z$
4,0	140	667	$-1810z$
4,4	235	$541 - 52z$	$-1415z$
4,8	350	$409 - 70z$	$-82z$
5,2	475	$156 - 17z$	$2360z$
5,6	560	$-63 + 170z$	$4810z$
6,0	605	$-1248 + 587z$	$7210z$
5,38	582	$-3545 + 1200z$	$7620z$
6,42	578	$-3545 + 8475z$	$7580z$
6,8	420	$1767 + 6910z$	$4468z$
7,2	290	$746 + 5250z$	$2423z$
7,6	205	$197 + 3880z$	$2603z$
8,0	145	$29 + 2760z$	$1745z$

9 Strain accumulation at thermal cycling

the inequality sign being duly suppressed; where $n = t_0/t_{max}$, and $t_{max} = 605°C$ is the maximal temperature of the operating cycle for which thermo-elastic stresses were calculated (see Table 9.6).

Expecting the limiting maximum temperature t_0 to lie inside the interval $200°C \leq t_0 \leq 300°C$, we can assume, according to Table 9.5, that the yield point depends linearly on temperature. Having the results collected in Table 9.6, we can specify condition (9.72); thus getting

$$\int_0^1 [2400 - 8(578n - 200) - (3545 + 6420z)n] \, dz +$$
$$+ \int_{-1}^0 [2400 - 8(582n - 200) - (3545 - 895z)n] \, dz = 0. \quad (9.73)$$

From equation (9.73) we get $n = 0.4$, i.e. $t_0 = 240°C$. Practically the same result follows from the flow régime (9.71), whereas the régime (9.69) leads to a much higher t_0, i.e. to a worse upper estimate.

Similar analyses show that for the tubes made of alloy steel of pearlitic type and of stainless steel the limiting temperatures are 400 and 430°C, respectively, the same ratchetting mechanisms having been adopted. Fair coincidence of analytical and test results is found for low-carbon steel and alloy steel, see curves 1, 2 in Fig. 9.29; for stainless steel, curve 5, the coincidence is poorer as the experimental data at $t_{max} = 600°C$ are absent.

Let us finally remark that the above calculations were based on the approximate determination of the limiting stresses; in particular, no account was taken of the fact that on the surfaces of the tube the amplitude of 'elastic' stresses might exceed the double yield point. Thus the strain accumulation was assumed to take place under the conditions of alternating plastic flow.

9.7. Incremental buckling at thermal fluctuations. Control and interpretation of experiments, the present-day concepts

Generation of thermal stresses in thin-walled structures is often accompanied by buckling which is either reversible or irreversible. In the absence of mechanical load the deflections increase according to the increase in temperature. At constant temperature the duration of the process is rather insignificant, since along with the mounting deformations, the thermal stresses relax.

9.7. Incremental buckling at thermal fluctuations

However, at repeated thermal actions, structures are observed to undergo substantial warping which moreover accumulates with each cycle. The discussed effect can be encountered in many structures subject to non-steady temperature fields of high intensity, especially in metallurgy. Two photos are given here for illustration: general flexure of a caisson of a melting shaft furnace, Fig. 9.31 and warping of its cross-section, Fig. 9.32. Formation of corrugations of the casing of a rotary furnace is shown in Fig. 9.33.

Suitable examples can also be found in other branches of technology in which high-temperature processes are relevant. The working wheel of a radial-axial flow gas turbine of a compressor is shown in Fig. 9.34; attention is drawn to progressive warping of a thin web of the wheel. A development of deflections in-between the blades was also

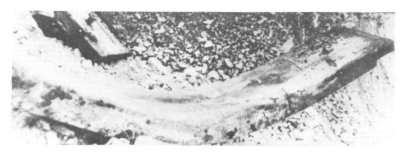

Fig. 9.31. Flexure of a caisson.

Fig. 9.32. Cross-sectional warping of a caisson.

9 Strain accumulation at thermal cycling

Fig. 9.33. Corrugations of a rotary furnace casing.

Fig. 9.34. Corrugations of a gas turbine working wheel.

9.7. Incremental buckling at thermal fluctuations

encountered in experiments as shown in Fig. 9.35, where the numbers indicate spacings around the periphery. Buckling was present at only some spacings; its absence was due to the locally thicker, though within the dimensional tolerances, rim of the wheel. Deflections did not form abruptly, and usually not earlier than after the second cycle; then they began to grow [9.20].

Special devices were designed [9.21] to investigate progressive thermal buckling of initially flat circular plates with a rigid central boss. One of the plates after repeated heating by means of an annular inductor fed by high-frequency current is shown in Fig. 9.36 (thickness 2 mm, outer diameter 100 mm, boss diameter 50 mm). The first stage of heating, just after the generator was switched on, consisted in a rapid rise in temperature in an annular peripheral zone of the plate. Then the zone was gradually growing towards the central boss and was accompanied by continuing rise in maximum temperature. Unforced cooling began at the moment of switch-off, with no means provided to force it. To obtain uniform heating along the concentric circles, the plate was slowly rotated around its horizontal axis. In the first couple of cycles the plate remained plane and then some residual deflections started to show; their increments were largest after 8 to 10 cycles and then started to fade away and disappear. Tests with variable thickness and diameters of the plate showed a substantial influence of the initial geometry on both the form of the deflected surface, i.e. the number of half-waves around the periphery and on the intensity of deflections, the maximum temperature having been kept equal.

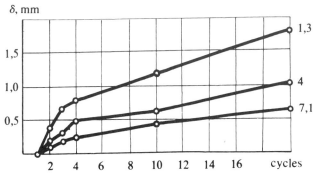

Fig. 9.35. *Development of deflections of turbine working wheel against the number of working cycles.*

9 Strain accumulation at thermal cycling

Fig. 9.36. Deplanation of a circular plate with rigid boss as a result of thermal cycling.

Progressive buckling of hollow cylinders was described in Sec. 9.6, as caused by repeated passages of a quasi-steady temperature field of the type of thermal wave. At moderate deflections the rotational symmetry was preserved. However, on further increase in the number of cycles, ovalization of the cross-section was accompanied by longitudinal folds, see Fig. 9.37. These effects were observed even earlier in the thinner specimens, especially when some initial imperfections were present such as non-uniform wall thickness or eccentricity of the annular inductor resulting in some departure from the axial symmetry of the temperature field.

As with single loading, qualitatively new forms of equilibrium (collapse mechanisms) are encountered in progressive buckling at sufficiently large parameters of the external actions. The ensuing deformations lead to more pronounced effects on the behaviour of structures over the following cycles than it is the case with ordinary incremental collapse.

9.7. Incremental buckling at thermal fluctuations

Fig. 9.37. Distortion of a tube after repeated passages of rotationally symmetric temperature field.

Repeated actions of the extremum, in the sense of ratchetting, temperature fields under conditions of non-negligible geometrical changes can cause deformations that mount with every cycle. This was distinctly observed in tests on tubular thin-walled specimens under moving concentrated heat source. Excessive local heating gives rise to buckling which develops along the path of the 'hot point', forming a corrugation [9.22, 9.23]. On prolonged heating the deformations are basically kinematically admissible. When the annular corrugation is closed, the only possible residual stresses are those of the bending type; stresses resulting from axial forces would not be self-equilibrated. Thus the subsequent passages of the hot point along the same path result

9 Strain accumulation at thermal cycling

mainly in an increasing amplitude of the corrugation. The already accumulated strains are also relevant to the process.

The test bench is shown in Fig. 9.38, after [9.22]; the test results are visualized in Fig. 9.39. Temperature fields measured at different arbitrary instants are plotted in Fig. 9.40. Circular corrugations were formed in specimens made of various steels. For example, measurements for the tubes made of pearlitic low alloy steel of diameter 38 mm are given in Fig. 9.41. Curves 1, 2 correspond to the wall thickness 1 mm and the

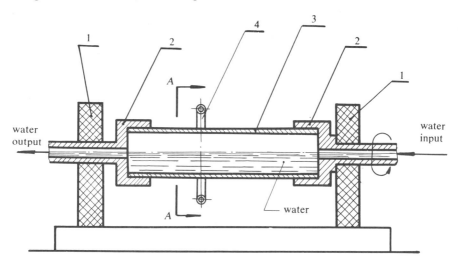

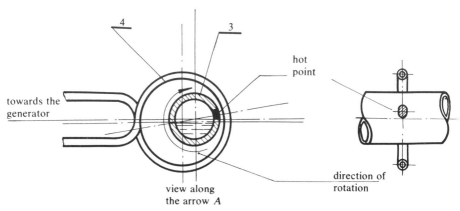

Fig. 9.38. Bench for creating a quasi-steady local temperature field in a tubular specimen.

9.7. Incremental buckling at thermal fluctuations

temperatures of 750 and 650°C, respectively; curves 3, 4, 5 to the thickness 2 mm and temperature of 750, 650 and 600°C, respectively.

Depending on the path of the hot spot other than circular corrugations were obtained. In particular, thread-type corrugations are shown in Fig. 9.39c, formed by means of moving the specimen in an axial direction and rotating it at the same time.

More and more attention has over the last decade been paid to buckling of thin-walled structures at elevated temperatures. In what follows we shall touch only upon those problems in which the mechanical loading is negligible. In other words, we shall deal with buckling

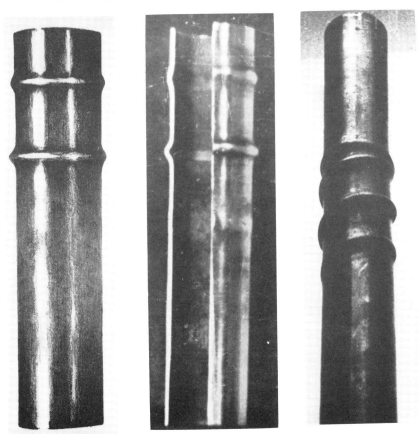

Fig. 9.39. Corrugations of a tubular specimen at repeated actions of quasi-steady local temperature field.

9 Strain accumulation at thermal cycling

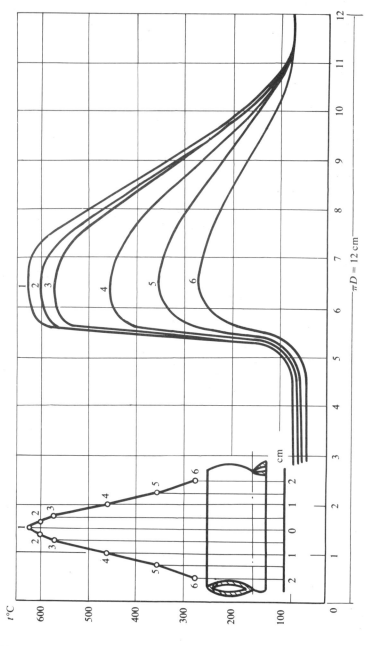

Fig. 9.40. Instantaneous temperature distributions in a specimen tested on the bench shown in Fig. 9.38.

9.7. Incremental buckling at thermal fluctuations

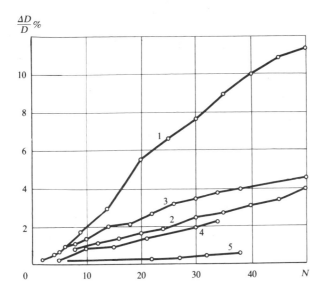

Fig. 9.41. *Relative elevation of a corrugation against the number of passages for various temperatures and wall thicknesses.*

under thermal forces generated as a result of non-uniform heating and under conditions in which free thermal strains are confined by some external constraints imposed on the structural member considered. Papers devoted to thermal buckling are rather scarce, we should mention [9.24–9.32].

In tackling the buckling problems at variable repeated load there has been employed, along with the classical notions of stability theory, also the deformational approach. It consists in assuming a state of initial imperfections; for instance, initial deflections of a beam or a plate. The first approach was used for instance in [9.33], where the loss of stability of a shell was assumed to take place after a number of loading cycles as a result of strain accumulation and the increasing geometry changes. The critical load was found with the help of the Shanley concept of prolonged loading. The second approach was employed in [9.34] where the strains were studied accumulating due to mechanical load in a plate with initial curvature; the member was subject to alternating loads acting in the plane occupied initially by an ideal member. The problem of progressive flexure of a beam at thermal cycling was studied independently in [9.35] by using a similar approach.

9 Strain accumulation at thermal cycling

In exploring the buckling effects under repeated loading we can, as in other problems of progressive strain accumulation, employ limit analysis and investigate the mechanism of inelastic deformations of geometrically non-linear systems. It should, however, be emphasized that the transition from one of the approaches to the other is much less clear than in the case of geometrically linear situations; this is due to the fact that the shakedown conditions may correspond to substantial changes in the geometry; these should have been accounted for in the limit analysis.

It is here proper to mention an analogy between the problems of limit analysis and of stability, both under single and repeated loading, see [9.36, 9.37].

Creep effects such as stress relaxation in the progressive buckling problems were investigated in [9.38]. The singular behaviour of the material at cyclic inelastic straining prior to buckling was discussed in [9.33] under the conditions of plane stress; in [9.39] an axial thrust caused by cycling stress relaxation was considered.

In the sections to follow we shall investigate on simple models the behaviour of structures that change their geometry at thermal cycling. First the deformations will be studied by means of a step-by-step procedure, followed by the shakedown analysis; singularities of the two procedures will be duly indicated.

9.8. Incremental buckling at thermal cycling as demonstrated on a bar system

We shall consider, after [9.35], the behaviour of an initially deflected pin-supported bar at thermal cycling. To make the qualitative description of the problem as straightforward as possible, let us make a number of simplifying assumptions; some of them are similar to those adopted in the analysis of instability in the presence of creep effects [9.12, 9.40, 9.41].

The bar of rectangular cross-section is assumed to consist of three parts: two symmetrically arranged outer parts 1 undergo a purely elastic deformation whereas the middle, relatively short element 2 behaves in a rigid-perfectly plastic fashion, Fig. 9.42. These assumptions reflect the fact that, when the strain-hardening is either weak or absent, plastic strains concentrate in the central zone of the bar.

The initial deflection of the element 1 is assumed to follow a lobe of a sine curve,

9.8. Incremental buckling on a bar system

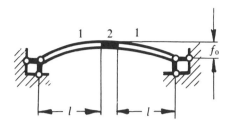

Fig. 9.42. Progressive buckling at thermal cycling.

$$v_0 = f_0 \sin \pi x/2l. \tag{9.74}$$

It is next assumed that further deflections of the part 1 resulting from slow heating and cooling, (while the temperature of the supporting devices is kept constant) develop according to the equation

$$v = \varphi x + f \sin \pi x/2l, \tag{9.75}$$

where φ denotes an angle of rotation caused by the plastic curvature of the element 2.

We consider the approximate differential description for the deflected axis

$$EI(v'' - v_0'') = M(x), \tag{9.76}$$

and we insist that it should be satisfied only at the point $x = l$ and its symmetric counterpart (method of collocation). We thus obtain that

$$f = \frac{f_0 + N\varphi l/N_{eu}}{1 - N/N_{eu}}, \quad \text{where } N_{eu} = \frac{\pi^2 EI}{(2l)^2}. \tag{9.77}$$

The other equation will be derived from the fact that the spacing of the supports remains unchanged; using axial elongation and curvature we obtain

$$\alpha t l - \frac{Nl}{EF} - \Delta l_p - \frac{1}{2} \int_0^l (v'^2 - v_0'^2) \, dx = 0,$$

where Δl_p is the plastic elongation of the element 2, the initial length of the bar being $s = 2l$.

9 *Strain accumulation at thermal cycling*

Using (9.74) and (9.75), we get

$$\alpha t l = \frac{Nl}{EF} + \Delta l_p + \frac{1}{2}\varphi^2 l + \varphi f + \frac{\pi^2}{16l}(f^2 - f_0^2). \tag{9.78}$$

The rigid-perfectly plastic insert 2 cannot deform unless there develops in it the limit state, i.e. unless the known interaction equation

$$\frac{M}{M_0} + \left(\frac{N}{N_0}\right)^2 = 1 \tag{9.79}$$

is satisfied [9.42], where $M = N(\varphi l + f)$ is the bending moment at the midspan. The ultimate bending moment and axial force are

$$M_0 = \frac{1}{4}\sigma_s b h^2, \quad N_0 = \sigma_s b h, \tag{9.80}$$

where b, h stand for the width and the depth of the bar, respectively.

The ratio of elongation rate to flexure rate is determined by the associated flow law; in this particular case the above ratio can be equally well derived from the law of preservation of plane sections, that is

$$\dot{\varphi} = \frac{(\Delta \dot{l}_p)}{\zeta}, \tag{9.81}$$

where ζ is the distance between the centre of gravity of the cross-section and the current location of neutral axis amounting to

$$\zeta = (\varphi l + f)\left[\sqrt{1 + \frac{h^2}{4(\varphi l + f)^2}} - 1\right]. \tag{9.82}$$

The assumption that the insert 2 can deform only under condition (9.79) has an approximate character since it stems from the adopted assumption of rigid-perfectly plastic behaviour and therefore it fails to reflect the possibility of non-isochronous plastic flow within the bar cross-section.

Let us now replace in (9.81) the infinitesimally small increments by small finite ones and assume that the coordinate ζ remains constant in the respective temperature intervals. Then we can write

9.8. Incremental buckling on a bar system

$$\Delta\varphi = \frac{\Delta(\Delta l_p)}{\zeta}. \tag{9.83}$$

Thus we have arrived at a simultaneous system of equations in the basic unknowns N, f, Δl_p, φ, the argument being $\theta = \alpha t$. For the sake of convenience let us introduce the following dimensionless quantities:

$$y = f/l, \quad z = \frac{\pi^2 k^2}{3\epsilon_s} \cdot \frac{N}{N_{eu}}, \quad \eta = \zeta/l, \quad \delta_p = \Delta l_p/l,$$
$$y_0 = f_0/l, \quad k = h/4l, \quad \epsilon_s = \sigma_s/E. \tag{9.84}$$

Then the system of equations takes the form

$$\theta = \epsilon_s z + \delta_p + 0{,}5\varphi^2 + \varphi y + \frac{\pi^2}{16}(y^2 - y_0^2), \tag{9.85}$$

$$\frac{3\epsilon_s}{\pi^2 k^2} z = \frac{y - y_0}{y + \varphi}, \tag{9.86}$$

$$z\left[z + \frac{1}{k}(y + \varphi)\right] = 1, \tag{9.87}$$

$$\eta\Delta\varphi = \Delta\delta_p, \tag{9.88}$$

$$\eta = 2k|z|, \tag{9.89}$$

and now the changes in the dimensionless internal forces and deformations can be studied as a result of yielding of element 2 at temperature fluctuations. The computations begin with the initial semi-cycle (first heating). Initially, the deformations are elastic and only the two equations (9.85) and (9.86) survive at that stage, while $\varphi = \delta_p = 0$. The obtained functions $y = y(\theta)$ and $z = z(\theta)$ yield a valid solution as long as the left-hand side of (9.87) stays less than one. Afterwards, when it becomes equal to one, all the constituent equations must be solved simultaneously, because a further increase in temperature gives rise to plastic flow in the insert 2.

At the beginning of cooling (first semi-cycle) plastic flow ceases to develop and the computations must start anew to solve the equations (9.85) and (9.86) in which δ_p and φ preserve the magnitudes reached at the end of heating. At sufficiently high maximum temperature of heating the plastic flow is resumed at the end of cooling and its sign is in general

9 Strain accumulation at thermal cycling

opposite to that accompanying the flow at heating. However, due to non-linearity, the relationships between the elongations and rotations will be different at heating and at cooling; a certain amount of deformation will thus be accumulated per cycle.

The loading and unloading paths are shown in Fig. 9.43 against the background of the interaction curve plotted in the dimensionless N/N_0, M/M_0-plane. By inspecting the inclination of the outward normal vector, establishing the ratio of the angular to the linear deformation, we can clearly see that the plastic flow at heating, arc 1–2, is dominated by the rotation rate while at cooling, arc 3–4, the elongation rate plays a more important part.

The computer-aided results for various input data are shown in the figures to follow:

In Fig. 9.44 the dimensionless axial thrust z is shown against the temperature for $\theta_{max} = 12{,}7 \times 10^{-3}$. The periods of heating and cooling can be readily distinguished, each comprising an elastic and an elastic-plastic stage.

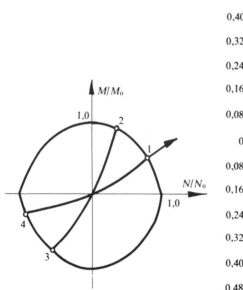
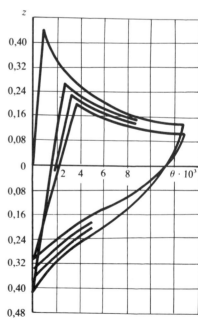

Fig. 9.43. Interaction curve for a bar under simultaneous compression and bending; strain accumulation mechanism shown.

Fig. 9.44. Variation of axial force in a bar under compression and bending at thermal cycling.

9.8. Incremental buckling on a bar system

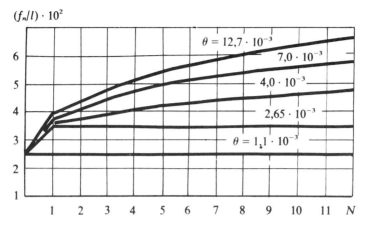

Fig. 9.45. Mounting of deflections at thermal cycling.

Fig. 9.45 visualizes the changes in the maximum residual deflection as the number of cycles increases. The curves are plotted for various maximum temperatures θ_{max}; the values of the parameters were

$$k = 0{,}02, \quad \epsilon_s = 12 \cdot 10^{-4}, \quad y_0 = 0{,}025.$$

Plastic deformations accompany the first cycle when $\theta_{max} > 1100 \cdot 10^{-6}$. The deflections begin to mount when $\theta_{max} > 2650 \cdot 10^{-6}$; at lower maximum temperature the bar shakes down just after the first cycle

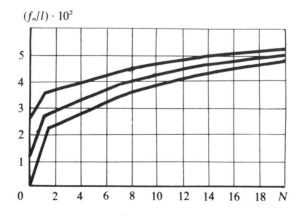

Fig. 9.46. Influence of initial curvature on the deflection of a bar.

471

9 Strain accumulation at thermal cycling

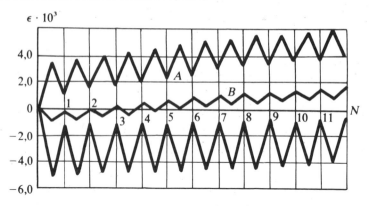

Fig. 9.47. Cyclic strains in different fibres of a bar at thermal cycling.

is terminated. With increasing θ_{max} both the value of the residual deflection and its gradient after the first cycle are found to increase. As the number of cycles increases, the deflection gradient gradually diminishes.

The influence of the initial curvature on the deflection accumulation is shown in Fig. 9.46; it decreases with the increase of the thermal cycle number.

Variations in plastic strains of the element 2 at repeated heating are shown in Fig. 9.47; the curves A, B and C correspond, respectively, to the convex surface, to the centre of gravity and to the concave surface of the bar. One-sign strains, especially excessive in outer fibres, are seen to accumulate in the conditions of alternating flow.

9.9. Conditions for progressive warping in the light of shakedown theory

Let us discuss the other, directly based on the shakedown theory, approach to the analysis of progressive buckling at thermal cycling. The material will be as usual assumed to be elastic-perfectly plastic and cyclically stable.

Let a structure, having a certain prescribed initial state characterized by its geometry, internal forces and material properties, be subject to cyclic actions of mechanical and thermal nature such that the changes in its geometry due to irreversible strains taken per each particular cycle are small enough to be ignored when determining stresses and strains in the subsequent cycle. Nevertheless, the strain

9.9. Conditions for progressive warping

accumulation per a series of cycles can be considerable. Thus we are prepared to introduce the notion of quasi-steady cyclic states, the first of which begins after the stabilization stage (consisting of initial loading cycles) and lasts over a number of cycles ($i = k+1, k+2, \ldots, m$). This number is limited by the adopted condition as to the allowable deflections, or associated stress changes. The stabilization process itself is usually ignored, i.e. we can assume $k = 0$. We further assume that, if after m cycles the deflections exceed those tolerable, a new quasi-steady state sets in, without any transition phase, corresponding to the appropriately changed geometry and again lasting a limited number of cycles etc.

Thus, a continuous evolution of the stress cycle as a result of irreversible strain accumulation is replaced by a step-wise process. Hence the following formulation of the shakedown problem becomes possible: Establish, for a given pattern of external actions, the dependence of the limiting cycle parameters on the step-wise varying 'initial' geometry of the structure. In this case the distribution of displacement increments on transition to the next stage is determined by suitable distribution of plastic strain rates, i.e. the collapse mechanism at the end of the previous state. A 'step' at which the geometry changes to reach the subsequent stage is to be properly selected.

The philosophy behind the above approach is similar to that employed in the plastic analysis theory; it allows for the effects of geometrical non-linearities on the load-carrying capacity at monotonically increasing loads [9.43]. Its application to cycling loads is governed by the necessity of taking into account the geometry changes in the very course of the cycle when analysing purely elastic behaviour of the structure in order to establish the limiting cycle parameters at the adopted quasi-steady state. In the case of a very flexible structure those changes may affect the values of the limiting cycle parameters to a large extent; the stiffer the structure the less important will those changes be.

If the geometrical non-linearity in the elastic régime is ignored and only the irreversible strains are taken into account, accumulating on transition to a new quasi-steady cyclic state, the conditions for shakedown at each stage of the geometry changes can be established with the help of the standard versions of the statical and the kinematical theorems, see Chapter 2.

If the geometry changes inside the cycle are substantial, the equilibrium equations must be appropriately modified. To determine the limiting cycle parameters it is necessary to allow for the effects of

9 Strain accumulation at thermal cycling

geometry changes not only on the elastic stresses caused by external actions but also on the residual stresses corresponding to the invariable (at shakedown) plastic strains. A certain analogy exists here with the effects of temperature-dependent elastic moduli, though some extra complications are encountered because we cannot apply the rule of superposition to stresses generated by different agencies.

Both in the first (simpler) and in the second case the obtained relationship between the limiting cycle parameters and the 'initial' geometry of the structure enables us to assess the possibility of shakedown at the given external action factors as well as to roughly evaluate the expected deformations which will accumulate prior to shakedown, whereas the number of cycles evidently remains unknown. Moreover, this relationship makes it possible, on account of the differences between the given load factors and the varying parameters of the limiting cycle, to assess the rate at which the strains accumulate as well as its variation as the cycles continue.

In the most general situation in which the problem of determining the elastic stresses is geometrically non-linear, the fundamental shakedown theorems can be formulated as follows:

a) Statical theorem (on safe actions). A structure with prescribed initial geometry will behave purely elastically after a number of non-steady cycles if for $\tau = 0$ there can be found a distribution of self-equilibrating stresses $\overset{0}{\rho}_{ij}$ at which the stresses $\tilde{\sigma}_{ij\tau}^{(e)}$ constitute a safe stress state at a generic point of the structure over the entire cycle. The latter stresses are calculated in an elastic manner from given external actions taking into account the initial state characterized by the stresses $\overset{0}{\rho}_{ij}$ and the geometrical non-linearity.

b) Kinematical theorem (on unsafe actions). A structure will never shake down, i.e. the cyclic plastic strains will continue to accumulate, if under the external actions varying in time in accordance with the prescribed program, a kinematically admissible cycle of plastic strain rates $\dot{\epsilon}_{ij0}''$ can be selected, obeying the known (see Chapter 2) relationships

$$\Delta\epsilon_{ij0}'' = \frac{1}{2}(\Delta u_{i0,j} + \Delta_{j0,i}), \quad \Delta\epsilon_{ij0}'' = \int_0^T \dot{\epsilon}_{ij0}\, d\tau, \quad \Delta u_{i0} = \int_0^T \dot{u}_{i0}\, d\tau, \quad (9.90)$$

for which, at any admissible self-equilibrating stress state $\overset{0}{\rho}_{ij}$ corresponding to an arbitrarily chosen 'initial' instant ($\tau = 0$) inside the cycle, the following condition is satisfied:

$$\int_0^T d\tau \left[\int (\sigma_{ij} - \tilde{\sigma}_{ij\tau}^{(e)}) \dot{\epsilon}_{ij0}'' \, dv + \sum_\mu \int_{S_\mu} (\sigma_{ij} - \tilde{\sigma}_{ij\tau}^{(e)}) n_j \Delta u_{i0\mu} \, dS \right] \leq 0. \quad (9.91)$$

The peculiarity of the above formulation as compared with those of Chapter 2 is that the stresses $\tilde{\sigma}_{ij\tau}^{(e)}$ include the changes (as referred to ρ_{ij}^0) in the self-equilibrating stresses in the course of a cycle; in other words, they simultaneously comprise the stresses from external actions with due allowance for initial state and geometry changes.

Rearrangements similar to those made in Chapter 2, as regards the classical formulation of the Koiter theorem, enable us to rewrite the inequality (9.91) in a form which is more convenient when seeking the approximate estimates:

$$\int \min_\tau [(\sigma_{ij} - \tilde{\sigma}_{ij\tau}^{(e)}) \Delta \epsilon_{ij0}''] \, dv + \sum_\mu \int_{S_\mu} \min_\tau [(\sigma_{ij} - \tilde{\sigma}_{ij\tau}^{(e)}) n_j \Delta u_{i0\mu}] \, dS \leq 0. \quad (9.92)$$

The formulation (9.91) of the kinematical theorem is not the only one possible. First of all, it ignores the duality requirements posed by mathematical programming, as regards the previous formulation of the statical theorem. The reason for this is that the determination of the limiting cycle parameters is a minimax problem: the minimization with respect to plastic strain rates $\dot{\epsilon}_{ij0}''$ is to be accompanied by the maximization with respect to the initial field of self-equilibrating stresses ρ_{ij}^0. However in some applications the above formulation is found to yield acceptable approximations to the actual solutions.

Maier [9.44] suggested that in the considered situations the changes in the geometry of structure should be allowed for by adding a 'geometrical term' to the equilibrium equations. The term is constructed by multiplying the stiffness matrix, corresponding to a certain initial configuration, by the displacement vector generated by an additional time-dependent load. Regretfully, no specific examples of the proposed approach were given, neither were the limitations of its applicability discussed.

9.10. Examples of progressive warping analysis

Example 1. We begin by demonstrating the above described approach to the analysis of progressive buckling on the example of a bar with

9 Strain accumulation at thermal cycling

constant cross-section, and pin-supported at both ends, Fig. 9.48a. The bar is subject to slow uniform heating up to the temperature t_* followed by complete cooling down to zero (the thermal conditions are the same as in the bar of Sec. 9.8, Fig. 9.42).

Initially, i.e. before the first thermal cycle starts, the axis of the bar is assumed to be straight. Let us find the temperature at which the progressive buckling commences. To this end, let the bar be subjected to a small lateral load P, Fig. 9.48b, which produces a slight departure from the initial equilibrium of the straight axis. Let the force be applied prior to each heating and released at the end of each cycle, i.e. just after the termination of the cooling of the bar. The following situations are here conceivable: a) the bar returns elastically, after each cycle, to its initial straight form, b) an irreversible deformation accompanies each cycle, though, if the deflections are small enough as compared with those caused by the force P (i.e. the initial state remains unchanged on transition from one cycle to the next), the strain accumulation gradually disappears and the bar shakes down, c) under the same conditions as in the previous situation a steady state sets in with constant increment of irreversible deflection per each cycle.

The shakedown theorems enable us, as was pointed out in Sec. 9.9, to establish a relationship between the limiting (steady) cycle parameters and the 'initial' geometry of the bar. For simplicity, let us assume that in the limiting cycle the bending stresses generated by the axial force due to deflections caused by the lateral force P are negligibly small as compared with the bending stresses generated by the lateral force itself. This assumption is by no means typical in the stability problems and it will be suppressed later on, in a more general approach. The stress problem has

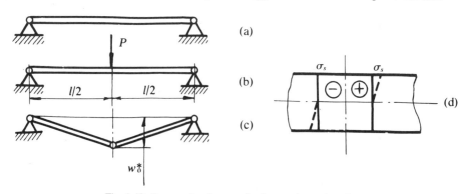

Fig. 9.48. *Progressive flexure of a bar at thermal cycling.*

9.10. Examples of progressive warping analysis

thus become a geometrically linear one, and the conditions for the commencement of progressive buckling can be found from the standard, but restated, theorems of Melan and Koiter, see Chapter 2.

Let us employ the kinematic theorem by assuming the collapse mechanism shown in Fig. 9.48c. The basic relationship (2.88) readily leads to the conclusion that, as the force is tending to zero, the progressive collapse condition is

$$\alpha E t_* = 2\sigma_s. \tag{9.93}$$

In the limiting cycle the plastic strain rates are non-vanishing in the upper part of the cross-section at maximum heating (contraction) and in the lower part at full cooling (elongation). The sign of flexure depends on the presence of an arbitrarily small, but finite, lateral force P. In the case of perfectly straight bar the condition for commencement of progressive buckling at thermal cycling coincides with the condition for alternating flow; the latter spreads all over the cross-section of the bar.

In Fig. 9.48d there is visualized instability as a result of alternating flow in a zone large enough for the formation of a kinematically admissible mechanism. It is interesting to note that condition (9.93) remains unaltered when the lateral force P is applied to another point, i.e. when the plastic hinge is formed in a different section. This is a case of a neutral equilibrium, in other words, of a peculiar bifurcation of equilibrium forms. A more general formulation of this example, allowing to consider geometrical changes when computing the elastic stresses, will be presented in the next example. Summarizing, let us stress that condition (9.93) could have been also obtained by choosing some other reason for buckling, instead of the mechanical lateral load, a small curvature or small difference between the yield points at compression and tension.

Example 2. Making use of the approximate kinematical method, we shall now consider the progressive buckling situation in the geometrically non-linear formulation for the beam shown in Fig. 9.49a. Let the beam axis be initially deflected according to the formulae

$$w_0(x) = w_0^* \frac{2x}{l} \quad \text{for } 0 \leq x \leq \frac{1}{2}l,$$

$$w_0(x) = w_0^* \frac{2(l-x)}{l} \quad \text{for } \frac{1}{2}l \leq x \leq l. \tag{9.94}$$

9 Strain accumulation at thermal cycling

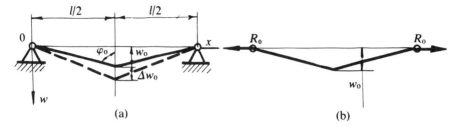

Fig. 9.49. Progressive flexure of an initially deflected beam.

The uniform temperature varies periodically in the interval

$$0 \leq t(\tau) \leq t_*, \qquad (9.95)$$

whereas at the beginning of the cycle, $t(\tau) = 0$, a residual tensile thrust R_0 is assumed to be present, Fig. 9.49b.

For the purpose of determining the additional deflections $w(x)$ we shall adopt the approximate equation of the deflected axis of a beam. This clearly imposes definite restrictions on the initial and additional lateral displacements, and the accuracy of the results will increase with increasing rigidity of the beam.

By integrating the suitable equation for simultaneous elastic bending and compression of a bar whose ends are prevented from spreading apart, we obtain the following relationships for the axial force at heating and the associated additional bending:

$$w(x) = w_0 \left(\frac{\sin(2x/l)u}{u \cos u} - \frac{2x}{l} \right) \left(1 + \frac{\sigma_0}{au^2} \right), \qquad (9.96)$$

$$au^2 + \epsilon_s \left(\frac{w_0}{l} \right)^2 \left[\operatorname{tg}^2 u + \left(\frac{\operatorname{tg} u}{u} - 1 \right) \left(\frac{4}{1 + \sigma_0/au^2} - 3 \right) \right] + \sigma_0 = q(\tau), \qquad (9.97)$$

where

$$u = pl/2, \quad p^2 = (\Delta R - R_0)/EI, \quad \sigma_0 = R_0/\sigma_s F,$$
$$q(\tau) = \alpha E t(\tau)/\sigma_s, \quad a = 4EI/\sigma_s Fl^2, \quad \epsilon_s = \sigma_s/E, \quad 0 \leq q(\tau) \leq q_*. \qquad (9.98)$$

Both the Young modulus E and the yield point σ_s are here for simplicity kept temperature-independent and insensitive to the duration of loading (i.e. creep is disregarded); I, F are, respectively, the cross-sectional

9.10. Examples of progressive warping analysis

moment of inertia and area, and ΔR stands for the increment of thrust associated with the change in temperature from zero to t_*.

The membrane and the bending stresses at the section $x = l/2$, conveniently referred to the yield point, can be obtained from (9.96)–(9.98); we get

$$\sigma_N = -au^2, \quad \sigma_M = \zeta ac\frac{w_0}{l}u^2\left[1-\left(\frac{\operatorname{tg} u}{u}-1\right)\left(1+\frac{\sigma_0}{au^2}\right)\right], \qquad (9.99)$$

where $\zeta = z/h$ is the dimensionless distance from the centroid (positive upwards), and h is the depth of the beam, $c = Flh/I$.

At $t(\tau) = 0$ the stresses at midspan are

$$\sigma_{N0} = \sigma_0, \quad \sigma_{M0} = \zeta c \sigma_0 \frac{w_0}{l}. \qquad (9.100)$$

The fictitious elastic stress distributions at the midspan, computed from (9.99), (9.100) are plotted in Fig. 9.50 for three different ratios w_0/l. The remaining data were: rectangular 5 mm × 10 mm cross-section, $2l = 120$ mm, $E/\sigma_s = 1000$, $\sigma_0 = 1$, $q = 0$ and 2.

Let us adopt the progressive buckling mechanism in the form shown by a dashed line in Fig. 9.49a. The plastic strains, generated by the deflection increment Δw_0, are assumed to be concentrated at the midsection of the beam; here discontinuity in axial displacements Δu_{01} takes place causing the angle φ_0 to grow together with the length of the beam axis. When the cross-section is symmetric, the jump Δu_{01} is described by the function of lateral deflection and the coordinate ζ given by

$$\Delta u_{01} = 4\Delta w(-\zeta h/2l + w_0/l). \qquad (9.101)$$

It is readily seen that

$$\Delta u_{01} < 0 \quad \text{for } \zeta > \zeta_0, \quad \text{and} \quad \Delta u_{01} \geq 0 \quad \text{for } \zeta \leq \zeta_0,$$

where

$$\zeta_0 = (w_0/l)/(2l/h). \qquad (9.102)$$

Let the deflections be such that the maximum tensile stresses are reached at all points of the midsection $x = l/2$ at the instant of complete

9 Strain accumulation at thermal cycling

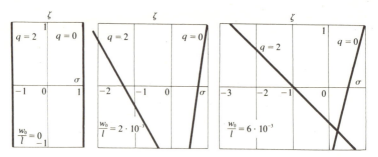

Fig. 9.50. 'Elastic' stress distribution at the midspan of a beam at cooling and maximum heating for various ratios w_0/l.

cooling, $t(\tau) = 0$, while the maximum compressive stresses are reached at maximum temperature. It is seen in Fig. 9.50 that, for instance at $q = 2$, this condition is fulfilled if $w_0/l < 6 \cdot 10^{-3}$. Thus, at those points of the cross-section where in the limiting cycle the contraction increments differ from zero (i.e. at $\zeta > \zeta_0$) the minimum of the integrand in the inequality (9.92) is attained for the stresses (9.99); at those points where elongations appear (i.e. at $\zeta < \zeta_0$) this minimum is attained for the stresses (9.100). Inequality (9.92) becomes

$$\int_{\zeta_0}^{1} \left\{ -1 + au^2 + \zeta ac \frac{w_0}{l} u^2 [1 + (\mathrm{tg}\, u/u - 1)(1 + \sigma_0/au^2)] \right\}$$
$$\times \left[-\zeta + \left(\frac{w_0}{l}\right)\frac{l}{h} \right] d\zeta + \int_{-1}^{\zeta_0} \left[1 - \sigma_0 - \zeta c \sigma_0 \frac{w_0}{l} \right] \left[-\zeta + \left(\frac{w_0}{l}\right)\frac{l}{h} \right] d\zeta < 0.$$
(9.103)

The critical magnitude of q is to be derived by solving simultaneously inequality (9.103) and equation (9.97). The results for the same data as before are plotted in Fig. 9.51. Line 1 corresponds to the progressive buckling condition in the case when the maximum residual stresses (9.100) reach the yield point,

$$\sigma_0 = \frac{1}{1 + \left(\dfrac{w_0}{l}\right) c}.$$
(9.104)

Line 2 fits the case $\sigma_0 = 0$. According to the kinematical theorem, as applied to non-linear systems, the solution represented by line 1 is closer

9.10. Examples of progressive warping analysis

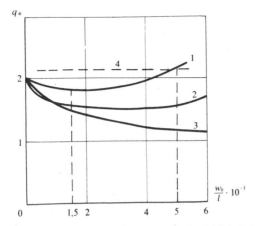

Fig. 9.51. Temperature of the limiting cycle against the initial deflection (as regards progressive buckling).

to the parameters of the actual limiting cycle obtained by maximization with respect to initial stresses.

When $w_0 = 0$, the progressive buckling condition (9.103) coincides, as expected, with the condition (9.93) based on the geometrically linear theory; an ideally shaped bar can buckle progressively only when its entire cross-section is subject to alternating plastic flow. Of a similar nature is clearly the case of geometrically ideal shells and plates; when the alternating flow develops in a region large enough for the deformation mechanism to form, new forms of equilibrium should be expected to occur.

The stress analysis of a beam with initial imperfections such as an initial deflection $w_0 \neq 0$, see Fig. 9.50, shows that progressive buckling can take place when the alternating flow spreads not over the whole section but only over its certain part (the upper one, $\zeta > 0$).

The alternating flow conditions in the beam are depicted by line 3 in Fig. 9.51; the analysis is analogous to that pertaining to progressive collapse; to begin with, the collapse mechanism, $\Delta w = 0$ at $\dot{\epsilon}''_{ij0} \neq 0$, and the initial stresses are assumed, next follow the calculations of 'elastic' stresses and the limiting cycle parameter, using (9.91). From Fig. 9.51 it is seen that alternating plasticity becomes operative at a lower thermal factor q_* than progressive buckling. This is why the analysis of conditions for the latter should here be treated as merely a qualitative one.

Returning to the curve 1 in Fig. 9.51, attention should be drawn to the occurrence of a minimum. Indeed, the plastic strain increments per

9 Strain accumulation at thermal cycling

cycle are expected to increase as the difference between the actual temperature and the limiting cycle temperature increases; if the former corresponds to, say, the dashed line 4, the deflection increments per cycle should initially increase to reach the maximum at $w_0/l = 0{,}0015$ and afterwards decrease to become zero at $w_0/l = 0{,}0050$.

Similar computations made for beams with various flexural rigidities show that on increasing rigidity, the function $q_* = q_*(w_0/l)$ becomes gradually monotone, i.e. its minimum shifts to the left of the q_*-axis. On the other hand, when the flexural resistance of the beam decreases, the minimum of the curve 1, Fig. 9.51, is more pronounced ($w_0 = 0$ corresponds to $q_* = 2$ as before) and the elastic curvature at heating is found to mount. For a sufficiently large initial deflection w_0 the serviceability of the beam is limited by the elastic deflections rather than by the accumulation of the inelastic ones.

Also the progressive buckling of a pin-supported bar at thermal cycling was investigated in [9.38]. The discussed effect was explained by creep, leading to the complete stress relaxation at each semi-cycle of heating. Indeed, creep is undoubtedly found to be one of the relevant factors that substantially influence both the conditions for progressive buckling and its further development. This influence can be evaluated within the framework of the presented formulations of the theorem in a way similar to that used in the problems of linear shakedown theory, Sec. 6.4.

Example 3. Consider a flat circular plate, Fig. 9.52a, subjected to cyclic heating from its periphery to its centre. The non-steady temperature field will be idealized as follows: The temperature of the rim will be assumed to increase to the prescribed level t_* and then the concentric thermal front will be supposed to move at $t_* = $ const. towards the centre of the plate, Fig. 9.52b; after heating the entire plate to the temperature t_* uniform cooling will take place.

The thermo-elastic stresses are

$$\sigma_{rr}^{(e)} = \frac{\alpha E t_*}{2}\left(\frac{\gamma}{\rho}\right)^2 (1-\rho^2), \quad \sigma_{\varphi\tau}^{(e)} = -\frac{\alpha E t_*}{2}\left(\frac{\gamma}{\rho}\right)^2 (1+\rho^2) \quad \text{at } 1 \leq \rho \leq \gamma, \tag{9.105}$$

$$\sigma_{rr}^{(e)} = \sigma_{\varphi\tau}^{(e)} = \frac{\alpha E t_*}{2}(1-\gamma^2) \quad \text{at } \gamma > \rho \geq 0, \tag{9.106}$$

where $\rho = r/R$, $\gamma = c/R$.

9.10. Examples of progressive warping analysis

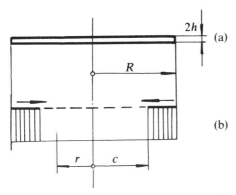

Fig. 9.52. Circular plate under the action of concentric thermal front.

The instantaneous distribution of the hoop stresses is plotted in Fig. 9.53 with dashed line. Enclosing distributions, associated with various ratios of the hoop to the radial strain components, are expressed by

$$\max_\tau \sigma_{\varphi\tau}^{(e)} = \frac{\alpha E t_*}{2}(1-\rho^2), \quad \min_\tau \sigma_{\varphi\tau}^{(e)} = -\frac{\alpha E t_*}{2}(1+\rho^2), \tag{9.107}$$

$$\max_\tau \sigma_{r\tau}^{(e)} = \frac{\alpha E t_*}{2}(1-\rho^2), \quad \min_\tau \sigma_{r\tau}^{(e)} = 0, \tag{9.108}$$

$$\max_\tau(\sigma_{r\tau}^{(e)} - \sigma_{\varphi\tau}^{(e)}) = \alpha E t_*, \quad \min_\tau(\sigma_{r\tau}^{(e)} - \sigma_{\varphi\tau}^{(e)}) = 0. \tag{9.109}$$

The enveloping hoop stresses are plotted in Fig. 9.53 with solid lines.

It is clear that at $\alpha E t_* = 2\sigma_s$ the fictitious yield curves degenerate

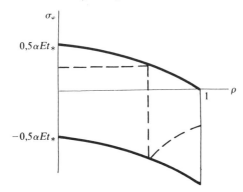

Fig. 9.53. Thermo-elastic instantaneous and enclosing hoop stresses in a plate.

9 Strain accumulation at thermal cycling

into a straight line simultaneously for all the points of the plate. As mentioned before, in such circumstances the kinematical theorem states that arbitrary ratchetting mechanisms can form, including buckling.

The considered above computational algorithm fits very well the situation described in Sec. 9.7. In practice, however, a definite buckling mechanism will form, depending on the thickness-to-radius ratio rather than on random factors. Two circumstances are here apparent. Firstly, the plate ceases to remain plane at temperatures far from those causing progressive buckling (the stresses taken into account fail to support this form of equilibrium). Secondly, the relationships of the type shown in Fig. 9.51 no longer apply to various 'acquired' forms of the plate. This is why the peculiarities of the postcritical behaviour of structures lead in any particular case to the formation of a unique mechanism of progressive buckling.

References

[9.1] ASME Boiler and Pressure Vessel Code, sec. III, Nuclear Vessels, 1963.
[9.2] Kistyan, L.K., Neyman, A.M., Serdelevitch, G.E., Combustion chambers of locomotive gas-turbine engines, in Russian, Mashinostrojenie, Moscow, 1965.
[9.3] Davidenkov, N.N., Likhatchev, V.A., Irreversible deformation of metals at thermal cycling, in Russian, Mashgiz, Moscow, 1962.
[9.4] Rosenthal, D., The theory of moving sources of heat and its application to metal treatment. Trans. ASME, 68, Nr 8, 1946.
[9.5] Parkus, H., Instationäre Wärmespannungen. Springer, Wien, 1959.
[9.6] Ermakov, P.I., Studies of irreversible deformations of cylinder at thermal cycling, in Russian, in 'Thermal stresses in structural members', 4, Naukova Dumka, Kiev, 1964.
[9.7] Grinenko, W.I., Belkin, A.S., Astafurova, N.I., Butt welding of non-rotating tubes from the steel HI9N9T by means of selfpressing, in Russian, Svarotchnoye Proizvodstvo, Nr 10, 1963.
[9.8] Gokhfeld, D.A., On mechanism of mounting strains at repeated passages of moving heat source, in Russian, Svarotchnoye Proizvodstvo, Nr 4, 1965.
[9.9] Zarubin, W.S., Temperature Fields in Aircraft, in Russian, Mashinostrojenie, Moscow, 1966.
[9.10] Gokhfeld, D.A., Plagov, I.M., Cherniavsky, O.F., Strain accumulation in a shell at thermal cycling and negligible mechanical load, in Russian, in 'Theory of shells and plates', Proc. IX All-Union Conf., (Leningrad, 1973), Sudostrojenie, Leningrad, 1975.
[9.11] Timoshenko, S.P., Woinowsky-Krieger, Theory of plates and shells. McGraw-Hill, New York, 1959.
[9.12] Rabotnov, Ju.N., Creep Problems in Structural Members. North-Holland, Amsterdam, 1969.

References

[9.13] Gokhfeld, D.A., Laptevsky, A.G., Plagov, I.M., On ratchetting at thermal cycling, in Russian, in 'Thermal strength of materials and structural members', Naukova Dumka, Kiev, 1967.

[9.14] Tetelbaum, I.M., Electrical Modelling, in Russian, Fizmatgiz, 1959.

[9.15] Kovalenko, A.D., Grigorenko, Ya.M., Ilin, L.A. Theory of thin conical shells, in Russian, AN SSSR, Kiev, 1963.

[9.16] Dymarksy, Ya.S. et al., Programmers handbook, in Russian, Vol. 1, Leningrad, 1963.

[9.17] Mikhajlov-Mikheev, P.B., Handbook on Metallic Materials used in Turbine and Engine Industry, in Russian, Mashgiz, Moscow, 1961.

[9.18] Gokhfeld, D.A., Sadakov, O.S., A mathematical model of medium for analysing the inelastic deforming of structures subjected to repeated actions of load and temperature. Trans. 3rd Intern. Conf. on Struct. Mech. in Reactor Techn., 5, L 5/7, London, 1975.

[9.19] Besukhov, N.I. et al., Strength, Stability and Vibrations at Elevated Temperatures, in Russian, Mashinostrojenie, Moscow, 1965.

[9.20] Kononov, K.M., Strength of axial-flow gas turbine at repeated starts, in Russian, in 'Thermal strength of materials and structural members', Naukova Dumka, Kiev, 1965.

[9.21] Cherniavsky, O.F., On progressive buckling at thermal cycling, in Russian, 'Proc. All-Union Symp. on low-cycle fatigue at elevated temperatures', 3, Cheliabinsk, 1974.

[9.22] Gokhfeld, D.A., Load-carrying Capacity at Thermal Cycling, in Russian, Mashinostrojenie, 1970.

[9.23] Gokhfeld, D.A., Laptevsky, A.G., On producing Corrugations, in Russian, Author certificate Nr. 174 600, 1965, Buletin izobreteni i tovarnykh znakov, Nr 18, 1965.

[9.24] Gatewood, B.E., Thermal Stresses. McGraw-Hill, New York, 1957.

[9.25] Boley, B.A., Weiner, J.H., Theory of Thermal Stresses. Wiley, 1960.

[9.26] Birger, I.A., Circular Plates and Shells, in Russian, Oborongiz, Moscow, 1961.

[9.27] Feodosev, W.I., Thermal Strength of Connections in Liquid Fuel Rocket Engines, in Russian, Oborongiz, Moscow, 1963.

[9.28] Beszukhov, N.I., Bazhanov, W.L., Goldenblat, I.I., Nikolaenko, N.A., Sinukov, A.M., Analysis of Strength, Stability and Vibrations at Elevated Temperatures, in Russian, Mashinostrojenie, Moscow, 1965.

[9.29] Volmir, A.S., Stability of Elastic Systems, in Russian, Nauka, Moscow, 1967.

[9.30] Hoff, N., Buckling at high temperature. Journ. Roy. Aeronaut. Soc., V. 61, NS63, 1957.

[9.31] Griguluk, E.I., Some problems in stability of circular plates at nonuniform heating, in Russian, Inzhinernyj Sbornik, 6, Izd. AN SSSR, 1950.

[9.32] Van der Neut, A., Buckling caused by thermal stresses. 'High temperature effects in aircraft structures'. Pergamon Press, New York, 1958.

[9.33] Neale, Kenneth W., Schroeder, J., Plastic buckling of shells under cyclic loading. 'J. of the Eng. Mech. Div., Proc. ASCE', Febr. 1974.

[9.34] Horne, M.R., The progressive buckling of plates subjected to cycles of longitudinal strain. 'Quart. Frans. Inst. Naval Architects', 98, Nr I, 1956, 78-108.

[9.35] Gokhfeld, D.A., Kharitontshik, A.E., On progressive buckling at thermal cycling, in Russian, in 'Thermal stresses in structural members, 6, Naukova Dumka, Kiev, 1966.

[9.36] Rzhanitzin, A.R., Stability of rigid-plastic plates and shells, in Russian, in '4th All-Union Conf. on Stability and Structural Mechanics', 1972, 202-203.

9 Strain accumulation at thermal cycling

[9.37] Bolotin, W.W., Grigoluk, E.I., Stability of elastic and inelastic systems, in Russian, in 'Soviet Union Mechanics over 50 years', Nauka, 1972.
[9.38] Anderson, R.G., The deformation of an axially held strut during repeated thermal cycling at creep relaxation temperature. 'Creep Struct. 1970, Symp. Gothenburg', Springer, Berlin, 1970.
[9.39] Mulcahy, T.-M., Shoemaker, E.M., Column instabilities caused by cyclic loading. Int. J. Solids and Struct., 7, 1970.
[9.40] Fraeijs, de Veubeke, D., Buckling caused by creep. 'High temperature effects in aircraft structures'. Pergamon Press, New York, 1958.
[9.41] Hoff, N.J., A survey of the theories of creep buckling. Stanford University. Division of Aeronautical Engineering Research, Nr 80 (SUDAER N80), 1958.
[9.42] Rzhanitzin, A.R., Structural Analysis Taking Account of Plastic Properties, in Russian, Gosstrojizdat, Moscow, 1954.
[9.43] Hodge, Ph.G., Jr., Limit Analysis of Rotationally Symmetric Plates and Shells. Prentice-Hall, Englewood Cliffs, 1963.
[9.44] Maier, J.A., Shakedown matrix theory allowing for work-hardening and second-order geometric effects, in Foundation of Plasticity. (Warsaw, 1972), Noordhoff, Leiden, 1973.

10

Some contact problems in the shakedown theory

This branch of shakedown theory began to develop after the appearance of the paper [10.1] by Johnson. But the exclusively statical approach to the contact problems employed in the quoted paper and elsewhere made it difficult to predict the consequences of the violated shakedown conditions (alternating flow, progressive collapse) in the specific situations considered. The above circumstance together with the fact that no due considerations were given to the choice of collapse mechanisms resulted in further difficulties caused by comparisons of the analytical and test results, disregarding the proper development of the very theory.

It is attempted in this chapter to eliminate those inadequacies by employing the kinematical shakedown theorem.

10.1. Contact problems in the shakedown theory. Description of the problem. Statical method of analysis

In modern machines there are many parts working in contact. This contact is not infrequently of repeated character and accompanied by relative displacements of adjacent surfaces. Moving contact zones are encountered in ball and roller bearings, gear boxes, rolling mills, rail vehicles and so on. Experience shows that such repeated contacts can result in premature wear and unacceptable accumulation of strains leading to specific modes of failure.

A number of papers has been published on the subject. In particular, it was demonstrated experimentally in [10.2] that the running-in of two cylinders under conditions when the transmitted load exceeds a certain level, can end in excessive sub-surface deformations due to plastic shear. In repeated rolling of a ball or a roller on a semi-space we can often observe an indentation that becomes deeper and deeper as the

10 Some contact problems in the shakedown theory

material is bulging out to form an elevated rim [10.3, 10.4]. Strains can sometimes continue to accumulate over a great number of cycles despite gradual decrease in contact stresses due to growing impression.

Shearing in a surface layer can lead to its spalling. Such type of collapse was encountered in bearings [10.5] and then termed 'roller peeling', in toothed wheels [10.6], in rolling mills [10.7] etc.; only a limited number of cycles was found to be necessary to initiate such a failure, especially at thermal cycling [10.7].

Johnson [10.1] investigated the shakedown conditions for an elastic-perfectly plastic semi-space on which a rigid long cylinder was repeatedly rolling. The cylinder was pressed down by a force P uniformly distributed along its axis, Fig. 10.1. Contact was assumed to be frictionless, $T = 0$. Friction was accounted for in [10.8] in the case of a rolling contact between two cylinders; the traction force T was assumed to be proportional to the normal load P with a friction coefficient f. In [10.8] it was pointed out that the friction forces result in a decrease of the ultimate loads at which shakedown is still possible.

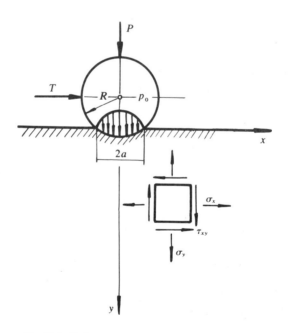

Fig. 10.1. Rolling of a rigid cylinder on a semi-space.

10.1. Description of the problem. Statical method

Both in [10.1] and [10.8] the shakedown analysis rests upon the Melan statical theorem. The contact zone and the contact stress distribution are assumed to be independent of the occurring plastic strains. However this assumption cannot be sufficiently justified in the general situation, since the contact strains can sometimes considerably affect the contact stresses. The range of validity of this assumption will be studied in the next section with the help of the kinematical approach.

Let us return to Fig. 10.1. On pressing down the cylinder of radius R with force P a narrow impression forms parallel to the cylinder axis. According to Hertz's theory [10.9], its width $2a$ at the elastic stage of the process is given by

$$a = \sqrt{\frac{4PR(1-\mu^2)}{E}}. \tag{10.1}$$

Contact pressure is distributed in accordance with the elliptical law,

$$p(x) = p_0\sqrt{1 - x^2/a^2}, \quad -a \leq x \leq a, \tag{10.2}$$

where $p_0 = 2P/(\pi a)$ is the maximum pressure in the middle of the depression.

The contact pressure exerts stresses in the semi-space; for a cylinder of substantial length they can be calculated under the conditions of plane strain. The stresses can be found from the formulae given in [10.10] for the frictional case, $f \neq 0$. We get

$$\begin{aligned}
\sigma_{x\tau}^{(e)} &= -\frac{2}{\pi}\int_{-a}^{a} p(\zeta)(x-\zeta)^2 \frac{y - f(x-\zeta)}{[y^2 + (x-\zeta)^2]^2}\,d\zeta, \\
\sigma_{y\tau}^{(e)} &= -\frac{2}{\pi}\int_{-a}^{a} p(\zeta)y^2 \frac{y - f(x-\zeta)}{[y^2 + (x-\zeta)^2]^2}\,d\zeta, \\
\tau_{xy\tau}^{(e)} &= -\frac{2}{\pi}\int_{-a}^{a} p(\zeta)y(x-\zeta) \frac{y - f(x-\zeta)}{y^2 + (x-\zeta)^2}\,d\zeta, \\
\sigma_{z\tau}^{(e)} &= \mu(\sigma_{x\tau}^{(e)} - \sigma_{y\tau}^{(e)}).
\end{aligned} \tag{10.3}$$

Related diagrams are plotted in Figs. 10.2 and 10.3 showing distributions of $\sigma_{x\tau}^{(e)}$, $\sigma_{y\tau}^{(e)}$ and $\tau_{xy\tau}^{(e)}$ at certain fixed depths of the semi-space as depending on the horizontal coordinate x. Their extrema are seen at some verticals as depending on the coordinate y. Solid lines correspond to the frictionless contact, $f = 0$, dashed lines to rough contact with the friction

10 Some contact problems in the shakedown theory

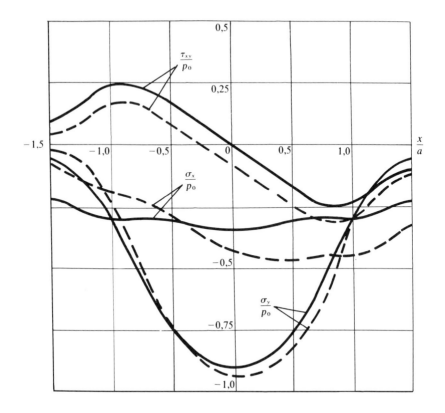

Fig. 10.2. Instantaneous distribution of "elastic" stresses at a depth of semi-space where shearing stresses reach their maxima ($f = 0$ – solid lines, $f = 0.2$ – dashed lines).

coefficient $f = 0.2$. The stresses in Fig. 10.2 are shown at such depths where the shearing stresses reach their absolute maxima: for $f = 0$ the depth is $y = 0.5a$, for $f = 0.2$ the depth is $y = 0.415a$. When the cylinder rolls in the x-direction, the curves in Fig. 10.2 show the history of stress changes at each point of the given depth. This is the case when the actual stress distribution is quasistationary relative to the coordinate system translating with the cylinder.

The contact loading is known to have a considerable hydrostatic component in the vicinity of the contact zone; high stress concentrations are also observed to occur. According to [10.10], in the absence of friction the maximum shear takes place below the centre of the contact zone, $x = 0$, at the depth $y = 0.78a$, Fig. 10.3, and it amounts to

10.1. Description of the problem. Statical method

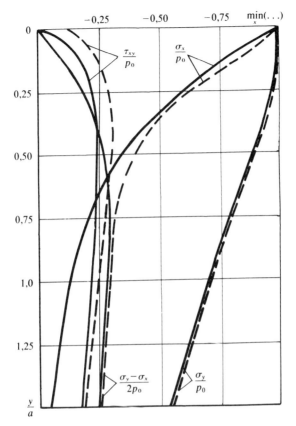

Fig. 10.3. Curves of extremum (absolute maximum) magnitudes of stress components as dependent on the distance of a layer from the boundary of semi-space (f = 0 – solid lines, f = 0.2 – dashed lines).

$$\tau_{max} = \left|\frac{\sigma_y - \sigma_x}{2}\right| = 0{,}3p_0. \tag{10.4}$$

Assuming the Tresca material (2.10) we find that the plastic flow will commence at the depth $y = 0{,}78a$ as soon as the force P is large enough; with increasing P the plastic region will shift upwards, towards the edge of the semi-space.

The intensity of the ultimate pressure at which the semi-space can still shake down to the repeated passages of a rolling cylinder was evaluated in [10.1]. The procedure was the following: First a field of

10 Some contact problems in the shakedown theory

time-independent self-equilibrating stresses ρ_x, ρ_y, ρ_z, ρ_{xy}, ρ_{yz}, ρ_{zx} was found which, on adding to the elastic stresses (10.3) in the case of $f = 0$, i.e. when friction is not taken into account, was contained inside the yield surface at all the instants of the cycle,

$$F(\sigma_1, \sigma_2, \sigma_3) \leq k^2. \qquad (10.5)$$

Here k is the plastic constant and $\sigma_1, \sigma_2, \sigma_3$ stand for the principal stresses. The residual stress field was postulated in the form

$$\rho_x = F_1(y), \quad \rho_z = F_2(y), \quad \rho_y = \rho_{xy} = \rho_{yz} = \rho_{zx} = 0, \qquad (10.6)$$

where $F_1(y)$, $F_2(y)$ are certain arbitrary functions. Obviously the residual stresses are independent of the coordinates x, z and the components ρ_{yz} and ρ_{xz} vanish. The identities $\rho_{xy} = 0$ and $\rho_y = 0$ follow from the plane problem equilibrium,

$$\frac{\partial \rho_x}{\partial x} + \frac{\partial \rho_{xy}}{\partial y} = 0,$$

$$\frac{\partial \rho_{xy}}{\partial x} + \frac{\partial \rho_y}{\partial y} = 0. \qquad (10.7)$$

The principal magnitudes of the total stresses $\sigma_{ij}^{(e)} + \rho_{ij}$ are

$$\sigma_1 = \frac{1}{2}(\sigma_x + \rho_x + \sigma_y) + \frac{1}{2}\sqrt{(\sigma_x + \rho_x - \sigma_y)^2 - 4\tau_{xy}^2},$$

$$\sigma_2 = \frac{1}{2}(\sigma_x + \rho_x + \sigma_y) - \frac{1}{2}\sqrt{(\sigma_x + \rho_x - \sigma_y)^2 + 4\tau_{xy}^2}, \qquad (10.8)$$

$$\sigma_3 = \mu(\sigma_x + \sigma_y) + \rho_z.$$

In the above notation the subscripts do not imply any natural ordering of the principal stresses.

However, by a suitable choice of the function $\rho_z = F_2(y)$ we can make the stress σ_3 an intermediate one, at least at the critical points and at the proper instants of times. Thus σ_3 is excluded from the Tresca yield condition (2.12) and inequality (10.5) takes the form

$$\frac{1}{4}(\sigma_x + \rho_x - \sigma_y)^2 + \tau_{xy}^2 \leq k^2, \qquad (10.9)$$

where $k = \tau_s$ is the yield stress in pure shear.

10.1. Description of the problem. Statical method

The residual stress $\rho_x = F_1(y)$ can be selected in such a way as to make, at suitable instants of time and at appropriate points of the semi-space, the first term in the left-hand side of (10.9) vanish. When assuming that the total stresses reach the yield surface at that instant whereas they are inside the elastic domain over the rest of the cycle, the shakedown pressure can be computed from

$$\max_{x,y}|\tau_{xy}| = k. \tag{10.10}$$

Friction can either be absent [10.1] or present [10.8] owing to the expressions (10.3) into which we can substitute either $f = 0$ or $f \neq 0$. In Table 10.1 there are shown, after [10.8], the shakedown pressures for various friction coefficients and the associated coordinates x, y of determining points on the curve $\tau_{xy} = \tau_{xy}(x, y)$; in Fig. 10.2 only those curves are shown that correspond to $f = 0$ and $f = 0{,}2$. The coordinate x plays in the present case of moving pressure the role of current time.

From Table 10.1 it follows that with increasing friction the depth y of the critical layer, where the shear stresses are at their largest, decreases from $0{,}5a$ for $f = 0$ to $0{,}415a$ for $f = 0{,}2$. The sought for ultimate pressure, (10.3) and (10.10), changes from $p_0 = 4k$ for $f = 0$ [10.1] to $p_0 = 3{,}21k$ for $f = 0{,}2$ [10.8].

In Johnson's paper a check was made on the bounds of total stresses at all instants of time by means of a graphical construction of stress profiles in the octahedral plane, Fig. 10.4. The Huber–Mises condition (2.9) is seen as a circle, the Tresca condition (2.12) as the circumscribed hexagon, the plastic shear being the same $k = \tau_s$. The residual stresses at the depth $y = 0{,}5a$ were $\rho_x = -0{,}134p_0$ and $\rho_z = -0{,}04p_0$. The points I_T, F, S_T on the stress trajectory correspond, respectively, to the semi-space points with the coordinate $x = \pm\infty$ (infinitely far from the contact zone), $x = \pm 0{,}25a$ and $x = 0$, i.e. right below the cylinder.

Table 10.1
Limiting pressures for various friction coefficients.

f	0	0,05	0,1	0,2
x/a	0,866	0,866	0,864	0,865
y/a	0,5	0,475	0,452	0,415
p_0/k	4,0	3,78	3,58	3,21

10 Some contact problems in the shakedown theory

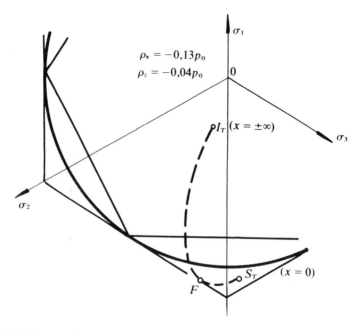

Fig. 10.4. Stress trajectory at the depth $y = 0{,}5a$ of the semi-space at rolling contact with cylinder.

When the roller moves, at all the points at depth $y = 0{,}5a$ the stresses are reached twice per cycle for each point of the stress profile except $x = 0$. The construction of the deviatoric stress profile presented in [10.1] fails to reflect the important circumstance that although the absolute values of all the stress components are the same in the two states, the signs of the shearing stresses are opposite. On accepting that the limit state (10.10) refers to the shearing stresses alone, we admit that the shakedown domain can be bounded due to the alternating flow. The results obtained in [10.1] were not discussed in this light. This fact seems to be characteristic for most of the analyses approached via the statical method.

In [10.8] no check was made on the admissibility of the stresses when the friction coefficient was different from zero.

In [10.1] it was also stated that the magnitude of the shakedown pressure $p_0 = 4k$ did not change when the semi-space obeyed the Huber–Mises yield criterion. The ratio of this value to the pressure p_s at which the elastic stresses reach the Huber–Mises yield surface for the first

10.2. Kinematical method of analysis

loading was found to be

$$\frac{p_0}{p_s} = \frac{4,0}{3,1} = 1,29.$$

When the non-linear relationship between the contact pressure and the applied load is adopted, we get

$$\frac{P_0}{P_s} = \left(\frac{p_0}{p_s}\right)^2 = 1,66.$$

Thus, in the case of linear contact the shakedown analysis implies that the load can be increased by 70 per cent as compared with the 'elastic' calculations, the safety margins being the same.

10.2. Contact problems in the shakedown theory. Kinematical method of analysis

The analyses of the previous section have not thrown adequate light on the consequences of violation of the shakedown conditions. There were neither considered any characteristic collapse mechanisms to ascertain whether the assumption of unaltered geometry in the contact zone is acceptable. No grounds are thus provided for designing experimental verification, nor selecting the relevant factors. In this respect the kinematical approach has definite advantages.[1]

Let us look again at the problem formulated in Sec. 10.1 and shown in Fig. 10.1, particularly considering the commencement of progressive collapse and alternating plastic flow. An admissible collapse mechanism is shown in Fig. 10.5a; it is assumed that a rigid surface layer of the semi-space, having the depth y_0, translates in the direction of the force T due to friction forces between the rolling cylinder and the upper surface of the layer.

Put analytically, we have

$$\Delta u_x = \Delta u_0' \quad \text{for } 0 \leq y \leq y_0,$$
$$\Delta u_x = 0 \quad \text{for } y_0 < y < \infty, \quad (10.11)$$
$$\Delta u_y = \Delta u_z = 0 \quad \text{for } 0 \leq y < \infty,$$

---

[1] The contact problems in this section were solved by A.R. Belyakov.

10 Some contact problems in the shakedown theory

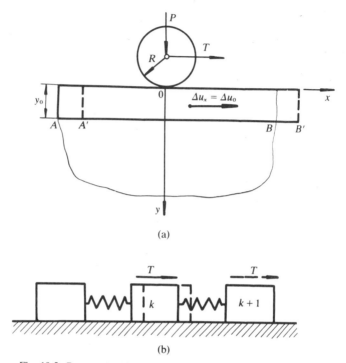

Fig. 10.5. Progressive slip mechanism and its mechanical model.

where $\Delta u_x, \Delta u_y, \Delta u_z$ are the displacement increments per cycle in the directions x, y, z, respectively, and $\Delta u_0'$ is the displacement increments discontinuity.

The assumed mechanism stems from the conjecture that an infinitely thin layer of concentrated plastic slip is present at the depth y_0, gradually spreading during each cycle (passage) along the x-axis due to the cylinder displacement. The process is modelled in Fig. 10.5b in which the surface layer is visualized as a series of dry friction elements linked by elastic springs. The horizontal traction force T is imagined to be applied element by element, thus moving in a step-wise fashion in the x-axis direction.

Let the force T be first applied to the element k. As soon as it overcomes the friction resistance, it moves to the right generating at the same time certain forces in the adjacent springs. Those forces are

10.2. Kinematical method of analysis

assumed to be so small as to leave the neighbouring elements at rest (the reasoning need not be extended, if we assume that a number of elements can simultaneously displace). Now, let us consider the transmission of the force T to the next element $k+1$. Its change is due to the fact that the element begins to translate at a slightly lower force as a result of an elastic link force present to the left. An analysis similar to that made in Sec. 9.2 shows that the process of transmitting the force T stabilizes rapidly, i.e. it becomes quasi-steady in the moving coordinate system, and at each stage the relevant element translates by the same amount. Thus, the whole system of elements and its links remains unaltered and translates in the direction of the force T. Clearly, the force T necessary to activate the assumed earthworm-like mechanism is essentially lower than that necessary for the entire system to be pushed as one piece, since an infinitely large force would be necessary to move a system consisting of unlimited number of elements.

Frictional rolling of a cylinder can thus be understood as generation of the 'slip front'; the necessary traction force is the main factor to be determined.

A modification of the above model makes it possible to describe the alternating plastic flow as well. This mode of failure sets a limit on the shakedown conditions in the absence of friction between the cylinder and the semi-space. A combination of alternating flow and strain accumulation can also be studied with the help of the modified pattern.

When the cylinder is long enough, plastic yielding takes place in the x, y-plane only, i.e. $\dot{\epsilon}''_z = \dot{\gamma}''_{xz} = \dot{\gamma}''_{yz} = 0$. For plane strain the Huber–Mises and the Tresca yield conditions coincide [10.11]. In this case they can be written, in terms of total stresses (including the residual ones) in the form similar to (10.9)

$$\left(\frac{\sigma_y - \sigma_x}{2}\right)^2 + \tau_{xy}^2 = k^2. \tag{10.12}$$

Hodographs of changes in the 'elastic' stresses per cycle at various depths are shown in Fig. 10.6 in the plane of the $(\sigma_y - \sigma_x)/2$, τ_{xy}-coordinates. Solid curves correspond to frictionless rolling, $f = 0$, and the dashed one to the friction coefficient $f = 0{,}2$. The hodographs for $f = 0$ are found to be symmetric with respect to the vertical axis; in the presence of friction they undergo both translation and rotation which is best seen for $y = 0{,}1a$. The principal stress directions do not remain fixed at any point of the semi-space since

10 Some contact problems in the shakedown theory

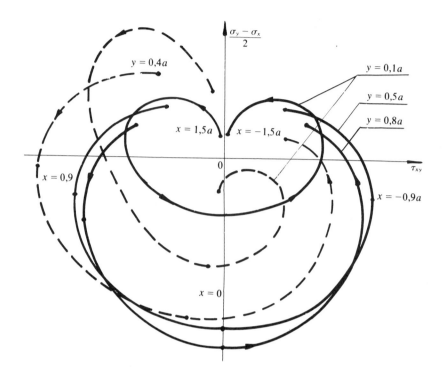

Fig. 10.6. Stress hodographs for rolling contact between cylinder and semi-space.

$$\operatorname{tg} 2\alpha_0 = \frac{2\tau_{xy}}{\sigma_y - \sigma_x} \tag{10.13}$$

is variable, Fig. 10.6.

The fictitious yield curves (see Chapter 2) for the yield condition (10.12) and the variable stresses (10.3) are shown in Fig. 10.7 for various depths y. The values of the average contact pressure p_0 were taken as the ultimate ones established via the statical approach in Sec. 10.1: $p_0 = 4k$ at $f = 0$ (solid lines in Fig. 10.7), $p_0 = 3{,}21k$ at $f = 0{,}2$ (dashed lines).

10.2. Kinematical method of analysis

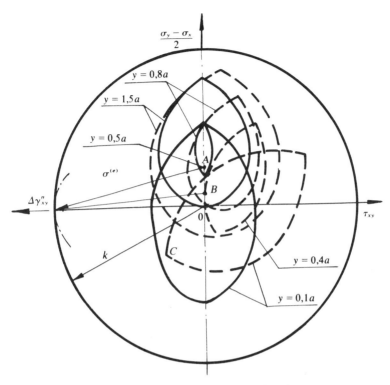

Fig. 10.7. Fictitious yield curves for points of semi-space.

In the first case the points at depth $y = 0,5a$ were found to be associated with the fictitious yield curves that, due to large amplitudes of shearing stresses, shrank to points. This means that in the problem dealt with by Johnson [10.1] the shakedown domain for the semi-space was bounded by the alternating flow. In the second case ($f = 0,2$, $p_0 = 3,21k$) no degeneration of the fictitious curve is observed and therefore the incremental collapse occurs.

According to the kinematical shakedown theorem (2.88) applied to the assumed collapse mechanism (10.11) and consisting in quasi-steady plastic slip in-between the surface layer of the depth y_0 and the rest of the semi-space, the condition for the commencement of progressive collapse takes the form

$$\Delta u_0' \tau_{xy*}^0 (p_0, y_0) = 0, \tag{10.14}$$

10 Some contact problems in the shakedown theory

where τ_{xy*}^0 are the shearing stresses which belong to the fictitious yield curve and are associated with the slip increment $\Delta\gamma_{xy}$, Fig. 10.7.

The upper bound on the ultimate, progressive collapse pressure p_0 can be improved by minimizing the solution of the equation (10.14) with respect to the depth y_0. The fictitious yield curve for those points of the semi-space that lie at the optimum depth is tangent to the vertical axis, $\tau_{xy*}^0 = 0$; the fictitious yield curves for other depths y_0 intersect this axis.

Due to the symmetry of the elastic stress hodographs (Fig. 10.6), in the frictionless case the fictitious yield curve can be tangent to the axis $\tau_{xy} = 0$ only when it simultaneously becomes degenerate. This happens, see Fig. 10.7, for $p_0 = 4k$ and $y_0 = 0{,}5a$, reflecting the fact that, in the absence of friction, the strains can accumulate only in connection with the alternating plastic flow and under some additional conditions as to the prevailing direction of deformations. In other words, the process of alternating flow at the depth $y_0 = 0{,}5a$ is an unstable one with respect to the strain accumulation.

If the plastic flow in the x, y-plane (at suitable depth y_0) occurs at each point for only one single location of the rolling cylinder, the progressive collapse condition (10.14) assumes the form (10.10) derived via the statical method. An analysis made by using the fictitious yield curves of Fig. 10.7 has shown that, in the presence of friction, repeated plastic flow for a new location of the cylinder is possible only at points of small depth (for instance, $y_0 < 0{.}1a$ for $f = 0{,}2$).

The ultimate pressures obeying (10.14) and the above condition are shown in Fig. 10.8 against the depth of the slip layer y_0. Line 1 ($f = 0$) has its minimum at $y_0 = 0{,}5a$, line 2 ($f = 0{,}2$) at $y_0 = 0{,}415a$.

The condition for degeneration of the fictitious yield curve corresponding to the beginning of alternating flow, states that the elastic shearing stress amplitude should amount to the double yield point in pure shear,

$$(\Delta\tau_{xy\tau}^{(e)})_{max} = 2k. \tag{10.15}$$

Departure from this condition takes place at $y_0 < 0{,}15a$ in the case $f = 0{,}2$: the elastic stress hodograph appears to be elongated in the vertical direction of the axis $(\sigma_y - \sigma_x)/2$ rather than in the horizontal direction.

In Fig. 10.8 there also can be seen the lines corresponding to alternating flow. In the frictional case the progressive slip deformations commence at lower pressures than the alternating flow. In the no-friction contact both shakedown conditions coincide, as pointed out earlier.

10.3. Strain accumulation in a semi-plane

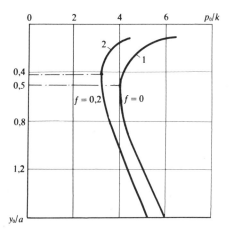

Fig. 10.8. Limiting conditions as regards progressive collapse and alternating flow.

10.3. Strain accumulation in a semi-plane on which a rigid disc is rolling

Consider a similar contact problem in plane stress, i.e. when a disc is rolling over a semi-plane. Friction will be disregarded (cf. Fig. 10.1, $T = 0$). As before, the vertical load per unit width measured in the z-direction, i.e. across the thickness is designated by P. The elastic stresses are determinable from (10.3) except for σ_z which equals to zero due to $\epsilon_z \neq 0$; changes in the contact due to inelastic strains will be ignored.

The material of the semi-plane is assumed to obey the Huber–Mises yield criterion,

$$\sigma_x^2 - \sigma_x\sigma_y + \sigma_y^2 + 3\tau_{xy}^2 = \sigma_s^2. \qquad (10.16)$$

Due to unconstrained lateral deformability, $\epsilon_z \neq 0$, the collapse mechanism differs from that envisaged for the semi-space. The following distribution of displacement increments is postulated;

$$\begin{aligned}
\Delta u_y &= \Delta u_0' = \text{const.} & \text{for } 0 \leq y \leq y_0, \\
\Delta u_y &= 0 & \text{for } y_0 < y < \infty, & \qquad (10.17) \\
\Delta u_x &= 0 & \text{for } 0 \leq y \leq \infty.
\end{aligned}$$

501

10 Some contact problems in the shakedown theory

This means that an upper layer of the semi-plane undergoes subsidence, see Fig. 10.9, possible due to a jump in the vertical displacements at the depth $y = y_0$ at the cost of local bulging of the semi-plane [10.11]. Thus, the sub-surface plastic strains are concentrated in the discontinuity plane. Although not being a unique one, this mechanism makes it possible to find an upper bound to the progressive collapse load P.

The basic relationships of the kinematical theorem reduce here to the form similar to (10.14),

$$\Delta u'_0 \sigma^0_{y*}(p_0, y_0) = 0, \qquad (10.18)$$

where σ^0_{y*} are the normal stresses belonging to the fictitious yield curve and associated with the direction of the displacement vector $\Delta u_y = \Delta u'_0$ at the depth $y = y_0$.

The fictitious yield curves for the boundary points ($y_0 = 0$) are shown in Fig. 10.10; at those points the elastic stresses $\sigma^{(e)}_{yr}$ reach their maxima while the shearing stresses disappear. Equation (10.18) is satisfied for the value of p_0 at which the fictitious yield curve becomes tangent to the axis $\sigma_y = 0$, see Fig. 10.10.

If the plastic flow takes place at each point of the semi-plane at only one location of the disc, then the fictitious yield curve implies that when

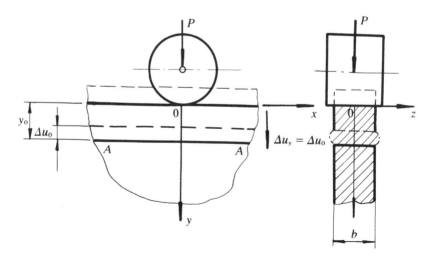

Fig. 10.9. *Progressive collapse mechanism of a semi-plane in rolling contact with a disc.*

10.3. Strain accumulation in a semi-plane

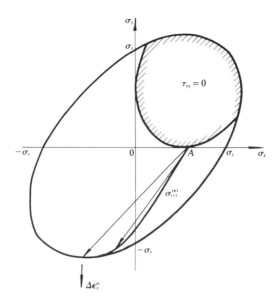

Fig. 10.10. Fictitious yield curves for the semi-plane.

point A corresponds to the point B of the actual yield surface (Fig. 10.10), for the Huber–Mises material the following stresses are relevant:

$$\sigma^0_{x*} = -\frac{\sigma_s}{\sqrt{3}} - (\sigma_x)_{x=0},$$

$$\sigma^0_{y*} = -\frac{2\sigma_s}{\sqrt{3}} - (\sigma_y)_{x=0}.$$
(10.19)

At the instant when the considered point is just below the centre of the disc, the stresses σ_y attain their absolute maxima. This occurs at $y = 0$ where $\sigma_y = -p_0$. Hence, on account of (10.18) and (10.19), we get $p_0 = (2/\sqrt{3})\sigma_s$. At that magnitude of contact pressure the progressive squashing commences of the upper layer with vanishing depth, $y_0 = 0$. The depth of the sinking layer increases with pressure.

Plastic deformation of the semi-plane at the first loading commences at $p_0 = \sigma_s$, the alternating flow enters the picture at $p_0 = 2\sigma_s$. Thus one-sign strains begin to accumulate before the alternating plasticity can set in; however it is the actual growth of the contact zone that retards the strain increments per passage.

10 Some contact problems in the shakedown theory

Experimental evidence shows that, depending on the conditions, one can observe both the slip of the surface layer (in plane strain) and the progressive sinking (in plane stress).

References

[10.1] Johnson, K.L., A shakedown limit in rolling contact. Proc. Fourth U.S. Congr. Appl. Mech., Berkeley, 1962.
[10.2] Crook, A.W., Simulated gear-tooth contacts: some experiments upon their lubrication and sub-surface deformations, Proc. Inst. Mech. Engrs, London, 1957, 171, 187.
[10.3] Moyar, Sinclair, Cumulative plastic strain in rolling contact. Trans. ASME, Ser. D, Nr 1, 1963.
[10.4] Orlov, A.W., Pinegin, S.W., Residual Stresses in Contact Loading, in Russian, Nauka, Moscow, 1971.
[10.5] Pinegin, S.W., Contact Strength and Rolling Resistance, in Russian, Mashinostrojenie, Moscow, 1969.
[10.6] Trubin, G.K., Contact Fatigue in Toothed Wheels, in Russian, Mashgiz, Moscow, 1962.
[10.7] Polukhin, W.P., Serviceability and Durability of Cold Rolling Molls, in Russian, Metalurgia, Moscow, 1976.
[10.8] Romalis, B.L., Contact shakedown in the presence of friction, in Russian, Mashinovedenje, Nr 1, 1973.
[10.9] Timoshenko, S.P., Theory of Elasticity. McGraw-Hill, New York.
[10.10] Saverin, M.M., Contact Strength of Materials, in Russian, Mashgiz, Moscow, 1946.
[10.11] Kachanov, L.M., Foundations of the Theory of Plasticity. North-Holland, Amsterdam, 1971.

11

Inelastic shakedown in steady stress cycles

When the service life of separate parts of heavy duty equipment is purposely limited, it appears reasonable to allow them to work beyond their shakedown domain and undergo cyclic plastic strains. In order to assess the life of a part, it is first necessary to predict the character of expected deformation, alternating plastic flow or the strain accumulation or the combination of both, since it predicts the type of failure mechanism to form (low-cycle or thermal fatigue, quasi-statical collapse, unserviceability due to unacceptable deflections). Thus a decisive factor must be singled out such as strain amplitudes of accumulated displacements. In principle, such problem can be solved by investigating the kinetics of the cyclic deformation in accordance with the prescribed loading history. However, the necessary computations appear to be too bulky, especially for a substantial number of cycles sufficient for making extrapolations and for models adequately reflecting the material properties, see e.g. [11.1]. Hence this approach is used only in exceptionally precise calculations, when a check on the already estimated parameters of the considered structure and loading is really necessary.

Experience shows that, at least as far as some simpler structural members are concerned, the stress cycle in the process of steady repeated loading does gradually stabilize and for materials with simple mechanical properties, the strain amplitudes and increments cease to change after a number of cycles. The asymptotic stabilization process was illustrated in Chapter 1 with the help of one- and two-parameter bar systems made of elastic-perfectly plastic material. It was further shown that in the simplest cases it was possible to obtain the asymptotic solution in a direct way by the additional loading method, without the necessity of studying the whole deformation history. In this manner we can, in particular, 'separate' the non-shakedown domains from the point of view of the type of expected cyclic deformation (cf. Secs. 5.6, 8.7). Since, as

11 Inelastic shakedown in steady stress cycles

the number of working cycles increases, the damage and deformations accumulated during the initial stages before the stabilization process, become less and less relevant, a simple analysis of the steady cycle would do to obtain information sufficient to assess the service life of the part considered. In an analogous fashion the time prior to failure can be evaluated on account of the steady-state creep stage only. This is why the direct analysis of steady cycles appears to be so useful in engineering applications. The following two statements of the problem are here possible:

a) Construction of complete shakedown diagrams of the type obtained in Sec. 8.7 (limiting steady cycles) with the determination of the boundaries corresponding to qualitatively different states of steady cycling such as shakedown, alternating flow, strain accumulation and the combination of the last two types of strain cycling.

b) Direct determination (i.e. without recourse to all of the loading history) of stresses and plastic strain rates in the stationary deformation cycle at the given parameters of external actions.

General proofs of the theorems on the existence of steady cycle stresses were given by Frederik and Armstrong [11.2] who based them on Drucker's postulate. An attempt to find inequalities of the energy type for the direct analysis of steady cycle was made by Mróz in [11.3]. The problem of existence of steady cycles was studied anew in the recent paper [11.4].

It is worth mentioning that, as regards the structures under the conditions of steady at repeated loading, the problem of direct analysis of stresses and strain rates over the steady cycle was formulated and solved by Shorr [11.5, 11.6] even before the existence theorem was proved. Later Ponter [11.7, 11.9, 11.1], Leckie [11.8] and others [11.10, 11.12] considered the problems of bounds on the cyclic loading parameters at which creep was not accompanied by short-term inelastic strains. They obtained certain relations useful for approximate evaluation of strains and energy dissipation in steady cycle. A theorem was proved (starting from a somewhat simpler power creep law) according to which the short-term plastic strains are absent from the steady cycle, provided the limiting cycle parameters established with no regard to creep, are diminished by using the multiplier $n/(n + 1)$, where n is the exponent in the creep rate expression. It was also shown that, at the values of loading parameters outside the appropriate domain, i.e. when both types of irreversible deformation can alternately occur over the cycle,[1] the intensity of strain accumulation increases

---
[1] Mutual influencing of those two types of deformation will not be considered here.

11.1. Existence and uniqueness of a steady stress cycle

rapidly. This rather obvious statement was confirmed experimentally in [11.8] where tests on aluminium frames are reported.

Certain cybernetical aspects of the theory of the steady stress cycle were discussed in [11.13].

In this chapter, after exposition of the theorems on the existence of steady cycles and on the uniqueness of associated stress states (after [11.2]), we shall formulate the general variational problem of direct analysis of the steady inelastic cycle characteristics (scleronomic deformation). Using suitable extremum principles we shall derive relationships necessary to fully construct the shakedown diagrams. An example will be considered in which the external actions remain quasi-steady in the moving coordinate system.

11.1. Existence and uniqueness of a steady stress cycle

Assume a body subjected to a cyclic mechanical load

$$P(\tau) = P(\tau + nT), \quad P(\tau) = [P_1(\tau), P_2(\tau), \ldots, P_k(\tau)] \qquad (11.1)$$

and cyclic temperature

$$t(\tau) = t(\tau + nT), \quad t(\tau) = t(x, y, z, \tau), \qquad (11.2)$$

where τ is the current time measured from the beginning of the cycle; T is the period; $n = 1, 2, \ldots$ stands for the number of full cycles; and x, y, z denote the spatial coordinates of a point in the body.

Let the material obey the Drucker postulate (2.7), namely

$$(\sigma_{ij} - \sigma_{ij*})\dot{\epsilon}_{ij}'' \geq 0, \quad (\sigma_{ij*} - \sigma_{ij})\dot{\epsilon}_{ij*}'' \geq 0, \qquad (11.3)$$

where σ_{ij} and σ_{ij*} are two admissible stress states, and $\dot{\epsilon}_{ij}''$, $\dot{\epsilon}_{ij*}''$ designate the plastic strain rates associated with the yield surface. From (11.3), we get

$$(\sigma_{ij} - \sigma_{ij*})(\dot{\epsilon}_{ij}'' - \dot{\epsilon}_{ij*}'') \geq 0. \qquad (11.4)$$

Drucker's postulate in the form (11.4) refers to scleronomic (perfectly plastic and strain-hardening) materials as well as to the rheonomic ones [11.14]. Hence the theorems to follow will be valid not only in the

11 Inelastic shakedown in steady stress cycles

situations of short-term plastic strains but also in the presence of creep effects.

The theorem on existence of steady cycles reads as follows:

At cyclic loading of a structure made of Drucker's material the stresses and the strain rates gradually stabilize to remain unaltered on passing to the next cycle.

To prove the above statement, let us consider a functional I depending on the difference between the actual stresses at two instants of time differing by a period T,

$$I = \int \frac{1}{2} [\sigma_{ij}(\tau + T) - \sigma_{ij}(\tau)][\epsilon'_{ij}(\tau + T) - \epsilon'_{ij}(\tau)] \, dv, \tag{11.5}$$

where ϵ'_{ij} are the strains corresponding to the stresses via Hooke's law in which, for simplicity, the tensor of elastic moduli is kept time-independent. The elastic energy potential I is positive when $\sigma_{ij}(\tau + T) \neq \sigma_{ij}(\tau)$ and vanishes only when at all points of the body the alteration of stresses over two consecutive cycles is fully identical.

The rate of energy (11.5) is equal to

$$\dot{I} = \int [\sigma_{ij}(\tau + T) - \sigma_{ij}(\tau)][\dot{\epsilon}'_{ij}(\tau + T) - \dot{\epsilon}'_{ij}(\tau)] \, dv. \tag{11.6}$$

At a certain instant at which the actual stresses are $\sigma_{ij}(\tau)$ let us decompose the total strain rates into the elastic, plastic and thermal parts,

$$\dot{\epsilon}_{ij} = \dot{\epsilon}'_{ij} + \dot{\epsilon}''_{ij} + \dot{\epsilon}'''_{ij}.$$

The expression (11.6) takes now the form

$$\dot{I} = \int [\sigma_{ij}(\tau + T) - \sigma_{ij}(\tau)][\dot{\epsilon}_{ij}(\tau + T) - \dot{\epsilon}_{ij}(\tau) - \dot{\epsilon}''_{ij}(\tau + T) +$$
$$+ \dot{\epsilon}'''_{ij}(\tau) - \dot{\epsilon}''_{ij}(\tau + T) + \dot{\epsilon}'''_{ij}(\tau)] \, dv. \tag{11.7}$$

The difference $\sigma_{ij}(\tau + T) - \sigma_{ij}(\tau)$ is clearly a self-equilibrating stress state since the body is loaded in exactly the same manner both at the instant τ and $\tau + T$. The total strain rates $\dot{\epsilon}_{ij}(\tau + T)$ and $\dot{\epsilon}_{ij}(\tau)$ are kinematically admissible. Thus, on account of the virtual work principle (2.15), we can write

11.1. Existence and uniqueness of a steady stress cycle

$$\int [\sigma_{ij}(\tau + T) - \sigma_{ij}(\tau)][\dot{\epsilon}_{ij}(\tau + T) - \dot{\epsilon}_{ij}(\tau)] \, dv = 0. \tag{11.8}$$

Since the thermal strain rates at the instant $\tau + T$ and τ are the same, see (11.2),

$$\dot{\epsilon}_{ij}'''(\tau + T) = \dot{\epsilon}_{ij}'''(\tau), \tag{11.9}$$

the energy rate (11.7) can be rewritten in view of (11.8) and (11.9) as

$$\dot{I} = -\int [\sigma_{ij}(\tau + T) - \sigma_{ij}(\tau)][\dot{\epsilon}_{ij}''(\tau + T) - \dot{\epsilon}_{ij}''(\tau)] \, dv. \tag{11.10}$$

Recalling (11.4), we conclude that $\dot{I}$ is non-positive. Thus the non-negative functional (11.5) is getting smaller as the time elapses and approaches zero as the difference between the stresses $\sigma_{ij}(\tau + T)$ and $\sigma_{ij}(\tau)$ decreases.

The above proof has obviously much in common with that of the known Melan theorem (see Sec. 2.2); the latter may formally be derived from the above by assuming that the differences in the functionals (11.5), (11.10) are written for an arbitrary number n of full cycles and that the steady state is of purely elastic character throughout the body, i.e.

$$\dot{\epsilon}_{ij}''(\tau + nT) \equiv 0.$$

Nevertheless, the Melan theorem has in fact a much broader meaning: It not only states the existence of elastic stabilization (shakedown) but also formulates the shakedown conditions in the form of a certain extremum problem whose solution bounds the limiting cycle parameters from below. The above given theorem on existence of inelastic shakedown is devoid of such useful properties.

The theorem on uniqueness of stresses in the steady cycle can be written as follows:

The stress distribution in the steady cycle does not depend upon the initial (prior to the first cycle) state of a structure and is unique in those regions in which the short-term plastic strain rates, or creep strains, are non-vanishing in steady cycle.

The proof is similar to that of the existence theorem. Let us assume that at a certain instant inside the steady cycle two different instantaneous stress distributions corresponding to the same magnitudes of external actions are possible and let us denote them by σ_{ij} and σ_{ij0}. It can be now demonstrated that the non-negative functional

11 Inelastic shakedown in steady stress cycles

$$I = \int \frac{1}{2}(\sigma_{ij} - \sigma_{ij0})(\epsilon'_{ij} - \epsilon'_{ij0}) \, dv$$

in which elastic strains are related to the stresses by Hooke's law has the time derivative

$$\dot{I} = -\int (\sigma_{ij} - \sigma_{ij0})(\dot{\epsilon}''_{ij} - \dot{\epsilon}''_{ij0}) \, dv$$

which, due to (11.4), is non-positive and hence the functional can only be decreasing. This decrease will not cease before the rates $\dot{\epsilon}''_{ij}$ and $\dot{\epsilon}''_{ij0}$ become equal to each other at those points of the body where they are non-vanishing over the cycle.

In both proofs the stresses σ_{ij} should be understood as those parts of the stress tensor that are relevant to the determination of plastic strain rates. This issue was already discussed in Chapter 4.

The existence theorem leaves open the problem of how fast the stabilization process is, i.e. how many cycles are necessary prior to the steady state. Nor does it indicate the amount of deformation accumulated over the process. The analysis, made in Chapter 1 for some simple bar systems yields no more than the qualitative description of the stabilization under various conditions. In general, the steady state is reached in an asymptotic manner. It is worthwhile to realize that in the structures made of strain-hardening materials no steady (non-zero) strain accumulation can take place if the material continues to strain-harden up to fracture. Similarly, for the cyclically strain-hardening materials the amplitude of inelastic strains will diminish unless the alternating plastic flow ceases or fracture occurs. Thus the elastic shakedown may appear to be the only cyclic steady state possible for a structure composed of the above discussed materials.

Starting from functionals of the type (11.5) and (11.10), an attempt was made in [11.2] to draw general conclusions on the character of the steady process. As a measure of average (over the body) rate of the stabilization a ratio of an increment ΔI per cycle period T to the energy level I was adopted. As seen from (11.5) and (11.10), this ratio depends, in turn, on the ratio of plastic strain increments per cycle $\Delta \epsilon''_{ij}$ to the elastic strains inside the cycle ϵ'_{ij}. Since ΔI, obtained by integrating (11.10), can be shown to constitute a certain part of the quantity I, we may expect that the stabilization process can be described by an exponential law. The numerical parameters of this law cannot be

11.2. Stress and strain rate analysis in steady cycles

established in general; they have to be found in each specific case anew. However, the stabilization process turns out to be, as a rule, rather rapid over the first cycles; after several or more cycles both the stresses and the strain rates in two consecutive cycles differ only negligibly. Thus the authors of [11.2] maintain that an analysis of the problem over the first couple of cycles suffices to predict the steady state in a satisfactory manner.

As late as in [11.15] it was shown, on an example of a bar system, that the stabilization process of exponential nature depends on parameters that vary with increasing number of cycles; therefore some predictions on the steady state, stemming from the analysis of the first cycles, can lead to substantial inaccuracies resulting in overestimating the load-carrying capacity of the structure in question. In these circumstances any extrapolations of the results concerning the kinetics of the first cycles must be made very carefully. Thus one should encourage efforts to develop methods of direct analysis of steady cycles.

11.2. Stress and strain rate analysis in steady cycles. Different formulations of the problem

In any real process the residual stresses and plastic strain rates should satisfy the following constraints:

a) Equilibrium conditions:

$$\rho^0_{ij,j} = 0 \quad \text{in the interior } V, \quad \rho^0_{ij} n_j = 0 \quad \text{on } S_p, \tag{11.11}$$

$$\dot{\rho}_{ij,j} = 0 \quad \text{in the interior } V, \quad \dot{\rho}_{ij} n_j = 0 \quad \text{on } S_p, \tag{11.12}$$

where ρ^0_{ij} are the residual stresses at the initial instant of the cycle, $\tau = 0$, simply called in what follows the initial ones; and $\dot{\rho}_{ij}$ stand for the residual stress rates.

b) Compatibility conditions in terms of the residual strain rates composed of plastic and elastic parts:

$$\dot{\epsilon}''_{ij} + A_{ijhk}\dot{\rho}_{hk} = \frac{1}{2}(\dot{u}_{i,j} + \dot{u}_{j,i}), \tag{11.13}$$

11 Inelastic shakedown in steady stress cycles

where $\dot{u}_i$ are the residual displacement rates.

c) Stress–strain (or stress–strain rate) relationships. In particular, for elastic-perfectly plastic material the known equations, see Chapter 2, can be written as

$$f_\alpha \left(\sigma_{ij}^{(e)} + \rho_{ij}^0 + \int_0^\tau \dot{\rho}_{ij}\, d\zeta \right) \leq 0, \quad (0 \leq \tau \leq T), \tag{11.14}$$

$$\dot{\epsilon}_{ij}'' = \sum_\alpha \lambda_\alpha \frac{\partial f_\alpha}{\partial \sigma_{ij}}, \quad \lambda_\alpha \geq 0, \quad (\alpha = 1, 2, \ldots, m), \tag{11.15}$$

$$\lambda_\alpha f_\alpha \left(\sigma_{ij}^{(e)} + \rho_{ij}^0 + \int_0^\tau \dot{\rho}_{ij}\, d\zeta \right) = 0, \quad \lambda_\alpha \dot{f}_\alpha \left(\sigma_{ij}^{(e)} + \rho_{ij}^0 + \int_0^\tau \dot{\rho}_{ij}\, d\zeta \right) = 0, \tag{11.16}$$

where $\sigma_{ij}^{(e)}$ denote the applied stresses calculated in accordance with the loading and temperature programs (11.1), (11.2) assuming that the material is purely elastic; $f(\sigma_{ij}) = 0$ is the yield condition; and λ_α is an indefinite non-negative multiplier. On account of (11.16), the value of λ_α is non-zero only when the stresses reach the yield surface and no unloading takes place.

The actual steady cycle should, in addition to the conditions (11.11)–(11.16), satisfy the equality

$$\rho_{ij}(T) = \rho_{ij}^0. \tag{11.17}$$

This means that the cycle must be closed, i.e. the residual stresses at its end, $\tau = T$, should attain the same values as at its beginning, $\tau = 0$. In order to determine the residual stresses and plastic strain rates in the steady cycle it is necessary to find a distribution of initial stresses that obeys the conditions (11.11)–(11.17). For perfectly rheonomic material, this initial stress field can be found by solving the system of differential and algebraic equations (11.11)–(11.13) supplemented by the algebraic relationships between the creep strain rates and the stresses. Shorr was the first to consider this formulation of the problem. He obtained a closed form solution to the system of equations in the cases of an axi-symmetric disc and thick-walled tube, both under cyclic loading and temperature. Further developments were made by Ponter, Leckie and others [11.7–11.12].

The determination of the steady cycle parameters under the conditions of short-term (scleronomic) plastic strains is more complex; one

11.2. Stress and strain rate analysis in steady cycles

has to solve the stress problem including not only equations but also inequalities of the type (11.14). As we know, the direct methods to solve the simultaneous systems of equations and inequalities have been developed to a lesser degree than the methods to solve extremum problems in which the maximum (minimum) of a cost function is to be found under certain constraints imposed on the variables in the form of equations and inequalities. Thus the problem of steady cycle should be formulated as the extremum one. The first attempt in this direction was made by Mróz [11.3]. However the possibilities to apply his relationships to the approximate evaluation of energy-type parameters of the steady cycle remain to be investigated.

To formulate the extremum problem, let us introduce the non-negative functional

$$I_1 = \int \frac{1}{2} A_{ijhk}[\rho_{ij}(T) - \rho_{ij}^0][\rho_{hk}(\tau) - \rho_{hk}^0] \, dv \geq 0 \qquad (11.18)$$

which vanishes only when condition (11.17) is satisfied, i.e. in the steady cycle. The functional can be rewritten in the form

$$I_1 = \int_0^T d\tau \int \left(A_{ijhk} \dot\rho_{ij} \int_0^\tau \dot\rho_{hk} \, d\zeta \right) dv, \qquad (11.19)$$

as can be readily seen by performing integration of the first part of (11.19).

The problem of determination of the steady cycle parameters can now be formulated as follows:

From among all the statically admissible (satisfying (11.11)) initial stress fields ρ_{ij}^0 there should be chosen one which makes the functional (11.19) attain its absolute minimum,

$$\min_{\rho_{ij}^0} I_1 = ? \qquad (11.20)$$

under the constraints (11.12)–(11.16). Indeed, this field and the associated (by means of (11.12)–(11.16)) plastic strain rates will develop in the actual steady cycle.

From the computational point of view the above formulation has much in common with that considered in Chapter 2 concerning shakedown in the presence of variable elastic moduli. Note that the variational formulation of the problem here considered includes the non-linear relations (11.16) and (11.19). This fact leads to certain inconveniences when numerical methods are applied. Moreover, it directly follows that the time-independent self-equilibrating stresses, generated for instance by thermal actions, have no effect whatever on the steady cycle characteristics.

11.3. Particular cases of the problem of steady stress cycle calculation. The Melan theorem. Strain accumulation conditions at advanced alternating plastic flow

It is easy to obtain the conditions for shakedown (i.e. the elastic stabilization) as a particular case of the general extremum problem of stress analysis in the steady cycle. In the latter we have

$$\dot{\epsilon}_{ij}'' \equiv 0, \quad \dot{\rho}_{ij} \equiv 0 \tag{11.21}$$

and the conditions (11.12), (11.13), (11.15), (11.16) are satisfied identically ($\lambda = 0$). The non-negative functional (11.19) attains its minimal value, equal to zero because of (11.21). Thus the cycle in which (11.21) are satisfied identically does actually occur, i.e. the structure will shake down, if the conditions (11.11), (11.14) are satisfied, the latter taking the form

$$f(\sigma_{ij}^{(e)} + \rho_{ij}^0) \le 0, \quad (0 \le \tau \le T). \tag{11.22}$$

The obtained result wholly coincides with the first statement of the Melan theorem with the help of which a set of external action parameters ensuring shakedown is determined.

It is also easy to specify one more particular formulation of the steady cycle problem in which there can be separated a domain of external actions parameters which ensures the absence of incremental collapse, i.e.

$$\Delta\epsilon_{ij}'' = \int_0^T \dot{\epsilon}_{ij}'' \, d\tau \equiv 0 \tag{11.23}$$

holds true at all the points of the body although the plastic strain rates can be non-vanishing:

Incremental collapse will not take place if an initial stress field ρ_{ij}^0 can be selected, satisfying the conditions (11.11)–(11.16) and (11.23).

11.3. The problem of steady stress cycle calculation

The above statement is a direct result of the general extremum problem for the steady cycle, since the functional (11.19) vanishes (i.e. condition (11.17) holds) as soon as the plastic strain increments per cycle become kinematically admissible; condition (11.23) corresponds to that requirement.

As was the case with the first statement of the Melan theorem, the conditions for the non-occurrence of incremental collapse determine a suitable set of external actions parameters. What matters in practical applications is the establishment of boundaries to this domain, i.e. the determination of the limiting parameters of the steady cycle. The maximum values of the external actions parameters can then be found at which conditions (11.11)–(11.16) and (11.23) are satisfied.

Example. We shall consider conditions for strain accumulation at advanced alternating flow. Let a two-parameter system shown in Fig. 11.1 ($F_1 = F_2 = 0.5 F_3 = F$, $l_1 = l_2 = \tfrac{1}{4} l_3 = l$) be subjected to a cyclic load $P(\tau)$, simultaneous to thermal cycling in bar 1:

$$0 \leq P(\tau) \leq P_*, \quad 0 \leq t_1(\tau) \leq t_*.$$

Both loading and temperature profiles in time are shown in Fig. 11.2.

The yield point σ_s and Young's modulus E, will be assumed constant for all the bars and independent of temperature.

We have to find maximum values of the force P_* and of the temperature t_* whose excess would result in progressive accumulation of strains. For definiteness, the temperature t_* will be treated as given which leaves us only with the task of finding the magnitude P_*.

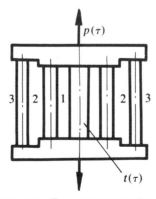

Fig. 11.1. Two-parameter system.

11 Inelastic shakedown in steady stress cycles

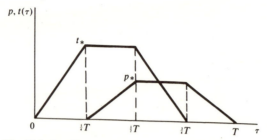

Fig. 11.2. Loading program for the two-parameter system.

To establish the limiting cycle parameter, as regards progressive collapse, we should employ now the following procedure:

Find the maximum of P_*,

$$\max_{\rho_{ij}^0} P_* = ?, \tag{11.24}$$

under the constraints (11.11)–(11.16) and (11.23) which assume now the form:

$$\sum_{i=1}^{3} \rho_i^0 F_i = 0, \quad (i = 1, 2, 3), \tag{11.25}$$

$$\sum_{i=1}^{3} \dot\rho_i F_i = 0, \tag{11.26}$$

$$\left(\frac{\dot\rho_i}{E} + \dot\epsilon_i''\right) l_i = \text{const.}, \tag{11.27}$$

$$-\sigma_s \leq \sigma_i^{(e)} + \rho_i^0 + \int_0^\tau \dot\rho_i \, d\zeta \leq \sigma_s, \tag{11.28}$$

$$\dot\epsilon_i'' = \lambda_i^+ - \lambda_i^-, \quad \lambda_i^+ \geq 0, \quad \lambda_i^- \geq 0, \tag{11.29}$$

$$\lambda_i^+\left(\sigma_i^{(e)} + \rho_i^0 + \int_0^\tau \dot\rho_i \, d\zeta - \sigma_s\right) = 0, \tag{11.30a}$$

$$\lambda_i^-\left(\sigma_i^{(e)} + \rho_i^0 + \int_0^\tau \dot\rho_i \, d\zeta + \sigma_s\right) = 0, \tag{11.30b}$$

$$\lambda_i^+(\dot\sigma_i^{(e)} + \dot\rho_i) = 0, \quad \lambda_i^-(\dot\sigma_i^{(e)} + \dot\rho_i) = 0, \tag{11.31}$$

$$\int_0^T \dot\epsilon_i'' \, d\tau = 0. \tag{11.32}$$

11.3. The problem of steady stress cycle calculation

The problem (11.24)–(11.32) is one of dynamic programming [11.16]. When the constraints (11.26)–(11.31) do not have to be satisfied at all instants of time but only at a finite number of values τ, we get a problem of quadratic programming, simply because the constraints (11.30)–(11.31) become quadratic whereas the others are linear.

Let us get an approximate solution by prescribing time intervals in which the plastic strain rates are non-vanishing and the directions in which they are generated. It is clear that an a priori adoption of those parameters confines the class of situations obeying the constraints (11.25)–(11.32) and hence the limiting cycle parameters will be underestimated.

Let, in particular,

$$\dot{\epsilon}_1'' = C_1, \quad C_1 < 0, \quad \dot{\epsilon}_2'' = \dot{\epsilon}_3'' = 0 \quad \text{for } a_1 \leq \frac{\tau}{T} \leq \frac{1}{4},$$

$$\dot{\epsilon}_1'' = -C_1, \quad \dot{\epsilon}_2'' = \dot{\epsilon}_3'' = 0 \quad \text{for } a_2 \leq \frac{\tau}{T} \leq \frac{3}{4},$$

(11.33)

and $\dot{\epsilon}_1'' = \dot{\epsilon}_2'' = \dot{\epsilon}_3'' = 0$ at all the remaining instants of time. On account of (11.30), (11.31) we get

$$\dot{\rho}_1 = -\dot{\sigma}_1^{(e)}, \quad \sigma_1^{(e)} + \rho_1^0 + \int_{a_1}^{\tau/T} \dot{\rho}_1 \, d\zeta = -\sigma_s \quad \text{for } a_1 \leq \frac{\tau}{T} \leq \frac{1}{4},$$

$$\dot{\rho}_1 = -\dot{\sigma}_1^{(e)}, \quad \sigma_1^{(e)} + \rho_1^0 + \int_{a_1}^{1/4} \dot{\rho}_1 \, d\zeta + \int_{a_2}^{\tau/T} \dot{\rho}_1 \, d\zeta = \sigma_s \quad \text{for } a_2 \leq \frac{\tau}{T} \leq \frac{3}{4}.$$

(11.34)

Equality (11.32) assumes due to (11.33), (11.34) the form

$$C_1\left(\frac{1}{4} - a_1\right) - C_1\left(\frac{3}{4} - a_2\right) = 0,$$

(11.35)

hence $a_2 = \frac{1}{2} + a_1$.

Having (11.34), it is easy to obtain the system of equations in the unknowns ρ_1^0 and a_1

$$\sigma_1^{(e)}(a_1) + \rho_1^0 = -\sigma_s,$$

$$\sigma_1^{(e)}(a_2) + \rho_1^0 + \int_{a_1}^{1/4} \dot{\sigma}_1^{(e)} \, d\zeta = \sigma_s.$$

(11.36)

11 Inelastic shakedown in steady stress cycles

Solving those equations with the elastic stresses

$$\sigma_1^{(e)} = -0{,}6\alpha Et(\tau) + 0{,}4P(\tau)/F,$$
$$\sigma_2^{(e)} = 0{,}4\alpha Et(\tau) + 0{,}4P(\tau)/F, \qquad (11.37)$$
$$\sigma_3^{(e)} = 0{,}1\alpha Et(\tau) + 0{,}1P(\tau)/F,$$

we arrive at

$$a_1 = \frac{2\sigma_s - 0{,}4P_*/F}{2{,}4\alpha Et_*}, \quad a_2 = \frac{1}{2} + \frac{2\sigma_s - 0{,}4P_*/F}{2{,}4\alpha Et_*}, \quad \rho_1^0 = \sigma_s - 0{,}4P_*/F. \qquad (11.38)$$

The equations (11.26), (11.27) and (11.34) provide at $a_1 \leq \tau/T \leq \frac{1}{4}$,

$$\dot{\rho}_1 = 2{,}4\alpha Et_*, \quad \dot{\rho}_2 = -1{,}6\alpha Et_*, \quad \dot{\rho}_3 = -0{,}4\alpha Et_*, \quad \dot{\epsilon}_1'' = -4\alpha t_*; \qquad (11.39)$$

and at $\frac{1}{2} + a_1 \leq \tau/T \leq \frac{3}{4}$,

$$\dot{\rho}_1 = -2{,}4\alpha Et_*, \quad \dot{\rho}_2 = 1{,}6\alpha Et_*, \quad \dot{\rho}_3 = 0{,}4\alpha Et_*, \quad \dot{\epsilon}_2'' = 4\alpha t_*. \qquad (11.40)$$

Let us now rewrite the conditions (11.28) with the help of the results (11.37)–(11.40) and with the stress ρ_3^0 expressed in terms of ρ_1^0 and ρ_2^0 as implied by (11.25). The results are shown in Table 11.1. Now, solving them with respect to ρ_2^0 and selecting the strongest (determining) ones by using the non-negativity of both P_* and t_*, we get:

$$\rho_2^0 + \sigma_s \geq 0, \qquad (11.41)$$

$$\rho_2^0 + \frac{7}{3}\sigma_s - 1{,}4P_*/3F \geq 0, \qquad (11.42)$$

$$\rho_2^0 + 3\sigma_s - 0{,}6P_*/F \geq 0, \qquad (11.43)$$

$$-\rho_2^0 - \frac{1}{3}\sigma_s - 0{,}4P_*/3F \geq 0, \qquad (11.44)$$

$$-\rho_2^0 + \sigma_s - 0{,}4P_*/F \geq 0. \qquad (11.45)$$

The above system of constraints constitutes, together with the cost function (11.24), a linear programming problem with one free variable.

Table 11.1

$0 \leq \tau/T \leq a_1$	$-\sigma_s \leq -2{,}4\alpha Et_* \dfrac{\tau}{T} - 0{,}4\dfrac{P_*}{F} + \sigma_s \leq \sigma_s$
	$-\sigma_s \leq 1{,}6\alpha Et_* \dfrac{\tau}{T} + \rho_2^0 \leq \sigma_s$
	$-\sigma_s \leq 0{,}4\alpha Et_* \dfrac{\tau}{T} + 0{,}2\dfrac{P_*}{F} - \dfrac{1}{2}\rho_2^0 - \dfrac{1}{2}\sigma_s \leq \sigma_s$
$a_1 \leq \tau/T \leq \dfrac{1}{4}$	$-\sigma_s \leq -\sigma_s \leq \sigma_s$
	$-\sigma_s \leq -\dfrac{0{,}8}{3}\dfrac{P_*}{F} + \rho_2^0 + \dfrac{4}{3}\sigma_s \leq \sigma_s$
	$-\sigma_s \leq +\dfrac{0{,}4}{3}\dfrac{P_*}{F} - \dfrac{1}{2}\rho_2^0 - \dfrac{1}{6}\sigma_s \leq \sigma_s$
$\dfrac{1}{4} \leq \tau/T \leq \dfrac{1}{2}$	$-\sigma_s \leq -0{,}4\dfrac{P_*}{F} + 1{,}6\dfrac{P_*}{F}\dfrac{\tau}{T} - \sigma_s \leq \sigma_s$
	$-\sigma_s \leq -\dfrac{2}{3}\dfrac{P_*}{F} + 1{,}6\dfrac{P_*}{F}\dfrac{\tau}{T} + \rho_2^0 + \dfrac{4}{3}\sigma_s \leq \sigma_s$
	$-\sigma_s \leq \dfrac{0{,}1}{3}\dfrac{P_*}{F} + 0{,}4\dfrac{P_*}{F}\dfrac{\tau}{T} - \dfrac{1}{2}\rho_2^0 - \dfrac{1}{6}\sigma_s \leq \sigma_s$
$\dfrac{1}{2} \leq \tau/T \leq \dfrac{1}{2} + a_1$	$-\sigma_s \leq -1{,}2\alpha Et_* + 2{,}4\alpha Et_* \dfrac{\tau}{T} + 0{,}4\dfrac{P_*}{F} - \sigma_s \leq \sigma_s$
	$-\sigma_s \leq 0{,}8\alpha Et_* - 1{,}6\alpha Et_* \dfrac{\tau}{T} + 0{,}4\dfrac{P_*}{F} + \rho_2^0 + \dfrac{4}{3}\sigma_s \leq \sigma_s$
	$-\sigma_s \leq 0{,}2\alpha Et_* - 0{,}4\alpha Et_* \dfrac{\tau}{T} + \dfrac{0{,}7}{3}\dfrac{P_*}{F} - \dfrac{1}{2}\rho_2^0 - \dfrac{1}{2}\sigma_s \leq \sigma_s$
$\dfrac{1}{2} + a_1 \leq \tau/T \leq \dfrac{3}{4}$	$-\sigma_s \leq \sigma_s \leq \sigma_s$
	$-\sigma_s \leq \rho_2^0 + 0{,}4\dfrac{P_*}{F} \leq \sigma_s$
	$-\sigma_s \leq 0{,}3\dfrac{P_*}{F} - \dfrac{1}{2}\rho_2^0 - \dfrac{1}{2}\sigma_s \leq \sigma_s$
$\dfrac{3}{4} \leq \tau/T \leq 1$	$-\sigma_s \leq 1{,}2\dfrac{P_*}{F} - 1{,}6\dfrac{P_*}{F}\dfrac{\tau}{T} + \sigma_s \leq \sigma_s$
	$-\sigma_s \leq 1{,}6\dfrac{P_*}{F} - 1{,}6\dfrac{P_*}{F}\dfrac{\tau}{T} + \rho_2^0 \leq \sigma_s$
	$-\sigma_s \leq 0{,}6\dfrac{P_*}{F} - 0{,}4\dfrac{P_*}{F}\dfrac{\tau}{T} - \dfrac{1}{2}\rho_2^0 - \dfrac{1}{2}\sigma_s \leq \sigma_s$

11 Inelastic shakedown in steady stress cycles

Solving it, we finally obtain

$$\frac{P_*}{F} = \frac{10}{3}\sigma_s, \quad \rho_2^0 = -\frac{7}{9}\sigma_s. \tag{11.46}$$

Those results are valid only when $0 \leq a_1 \leq \frac{1}{4}$, see (11.33), i.e. when, see (11.38),

$$0{,}4P_*/F + 0{,}6\alpha Et_* - 2\sigma_s \geq 0. \tag{11.47}$$

All the results are condensed in Fig. 11.3. The equality sign in (11.47), corresponding to the commencement of alternating flow, yields the straight line AB. The beginning of progressive collapse at the stage of an advanced alternating flow, described by (11.46), is depicted by the vertical line BD. The straight line BC is associated with the beginning of incremental collapse in the absence of alternating flow. For the appropriate analysis reference should be made to Chapters 3 and 4.

Thus, region I indicates shakedown, region II corresponds to alternating flow without accumulation of the one-sign strains, region III to the incremental collapse. It must be borne in mind that the solution supplies only a lower bound on the boundaries of region III. However the additional

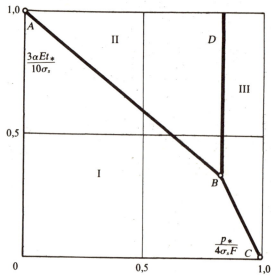

Fig. 11.3. Conditions for shakedown and progressive collapse of the two-parameter system.

11.4. Stress and strain rate analysis in steady cycles

loading method solutions (or direct investigations of the deformation) show that the above lower bound assessment coincides with the exact one.

In order to construct a complete shakedown diagram, similar to that for the bar of Sec. 5.6, a subregion in region III must be separated, corresponding to the commencement of alternating flow. In other words, the maximum values of the external actions parameters should be found at which

$$\text{sign } \dot{\epsilon}''_i = \text{const.} \qquad (11.48)$$

at all those instants of the cycle at which $\dot{\epsilon}''_i$ do not vanish. The present system of constraints on the extremum problem differs from that valid in the case of incremental collapse; condition (11.23) is replaced by (11.48) and the equality of the type (2.1).

However it must not be forgotten that under the adopted conditions, the structure behaves as a one-parameter system; the bars 2 and 3 constitute in fact a single element with a confined linear strain-hardening. In this case, as shown in Fig. 1.6, Chapter 1, the region of combined deformation is absent and line BD, Fig. 11.3, corresponds to an unstable state in which neither the amplitudes nor the increments of plastic strains per cycle are definite.

11.4. Stress and strain rate analysis in steady cycles. Another variational formulation and its consequences

1. Consider one more variant of the formulation of the steady cycle extremum problem for elastic-perfectly plastic bodies. Let a certain kinematically admissible cycle of plastic strain rates

$$\Delta\epsilon''_{ij0} = \frac{1}{2}(\Delta u_{i0,j} + \Delta u_{j0,i}) \qquad (11.49)$$

be associated with the residual stress rates $\dot{\rho}_{ij}$ derivable from

$$\dot{\epsilon}''_{ij0} + A_{ijhk}\dot{\rho}_{hk} = \frac{1}{2}(\dot{u}_{i\rho,j} + \dot{u}_{j\rho,i}). \qquad (11.50)$$

These stress rates are self-equilibrating, i.e.

$$\dot{\rho}_{ij,j} = 0, \quad \dot{\rho}_{ij}n_j = 0 \quad \text{on } S_p. \qquad (11.51)$$

11 Inelastic shakedown in steady stress cycles

Since (11.49) and (11.50) are simultaneously satisfied, the stress cycle is closed, i.e. the equality (11.17) holds good.

Further, let the initial self-equilibrating stresses ρ_{ij}^0,

$$\rho_{ij,j}^0 = 0, \quad \rho_{ij}^0 n_j = 0 \quad \text{on } S_p, \tag{11.52}$$

added to the elastic stresses and to the increments of residual stresses inside the cycle, constitute an admissible state at a generic point of the body and at any instant of time, that is

$$f\left(\sigma_{ij}^{(e)} + \rho_{ij}^0 + \int_0^\tau \dot{\rho}_{ij} \, d\zeta\right) \leq 0, \quad (0 \leq \tau \leq T). \tag{11.53}$$

The relationships (11.49)–(11.53) determine a set of stress and strain rate cycles to which an actual cycle has to belong. The latter should, in addition, obey the conditions imposed by the associated flow law.

Consider the functional

$$I_2 = \int_0^T d\tau \int \left[\sigma_{ij}^{(e)} + \rho_{ij}^0 + \int_0^\tau \dot{\rho}_{ij} \, d\zeta - \sigma_{ij0}\right] \dot{\epsilon}_{ij0}'' \, dv. \tag{11.54}$$

We have

$$I_2 = \int_0^T d\tau \int (\sigma_{ij}^{(e)} - \sigma_{ij0}) \dot{\epsilon}_{ij0}'' \, dv, \tag{11.54a}$$

as

$$\int_0^T d\tau \int \rho_{ij}^0 \dot{\epsilon}_{ij0}'' \, dV = \int \rho_{ij}^0 \Delta \epsilon_{ij0}'' \, dv = 0;$$

$$\int_0^T d\tau \int \left(\int_0^\tau \dot{\rho}_{ij} \, d\zeta\right) \dot{\epsilon}_{ij0}'' \, dv = -\int dv \int_0^T \left[A_{ijhk} \dot{\rho}_{ij} \int_0^\tau \dot{\rho}_{hk} \, d\zeta\right] d\tau = 0$$

in accordance with the virtual work principle and condition (11.50). Here σ_{ij0} is the stress state on the yield surface $f(\sigma_{ij0}) = 0$ associated with the strain rates $\dot{\epsilon}_{ij0}''$ by means of

$$\dot{\epsilon}_{ij0}'' = \sum_\alpha \lambda_\alpha \frac{\partial f_\alpha}{\partial \sigma_{ij0}}, \quad \lambda_\alpha \geq 0. \tag{11.55}$$

11.4. Stress and strain rate analysis in steady cycles

On account of (11.53), the integrand in the functional (11.54) can be rearranged to become

$$(\sigma_{ij}^{(a)} - \sigma_{ij0})\dot\epsilon_{ij0}'',$$

where $\sigma_{ij}^{(a)}$ is an admissible stress state.

The Drucker postulate (2.7) says that the above quantity is non-positive and vanishes only when $\sigma_{ij}^{(a)} = \sigma_{ij0}$. In consequence, for non-vanishing plastic strain rates $\dot\epsilon_{ij0}''$, the absolute maximum of the non-positive functional (11.54) can be reached only if the stresses

$$\sigma_{ij}^{(e)} + \rho_{ij}^0 + \int_0^\tau \dot\rho_{ij}\, d\zeta$$

belong to the yield surface and satisfy the equations (11.55).

Thus the first condition of the associated flow rule (11.16) can be replaced by the requirement that the functional (11.54) attains a maximum. The second condition (according to which the plastic strain rates must be equal to zero at the instant when unloading begins) is important in such cases only when the instantaneous values of strain rates have to be determined. Usually this requirement is not essential in applications. The problem of the steady cycle can now be reformulated as follows:

A strain rate field $\dot\epsilon_{ij0}''$ and initial stress field ρ_{ij}^0 should be selected that make the non-positive functional (11.54a) attain its absolute maximum and satisfy the conditions (11.49)–(11.53) and (11.55). The above established strain rates and residual stresses take place in the actual steady cycle.

Unlike in the variational formulation considered in the previous paragraph, the functional (11.54a) is a linear one related to the unknown rates $\dot\epsilon_{ij0}''$, if the yield conditions are piece-wise linear and the restrictions do not include the non-linear equalities (11.16). Thus after discretizing the problem, we can employ the procedure of linear programming.

2. Consider another formulation. Let us insist on finding the values of external actions parameters at which progressive accumulation of the one-sign strains is impossible (even in the conditions of alternating flow), i.e. on choosing a cycle which is a limiting one as regards incremental collapse. To do this, let us make two statements:

1) Progressive deformation will not occur if for all fields of plastic strain rates $\dot\epsilon_{ij0}''$ obeying the equations

$$\Delta\epsilon_{ij0}'' = \frac{1}{2}(\Delta u_{i0,j} + \Delta u_{j0,i}) \neq 0 \tag{11.56}$$

11 Inelastic shakedown in steady stress cycles

the functional (11.54a) remains negative while the conditions (11.50)–(11.53) and (11.55) are satisfied.

2) Progressive deformation will certainly take place under the prescribed program of external actions if at least one admissible cycle of plastic strain rates $\dot{\epsilon}''_{ij0}$ can be found securing the inequality

$$\int_0^T d\tau \int \left[\sigma_{ij}^{(e)} + \int_0^\tau \dot{\rho}_{ij}\, d\zeta - \sigma_{ij0} \right] \dot{\epsilon}''_{ij0}\, dv > 0 \tag{11.57}$$

together with (11.56), (11.55) and while the residual stress rates $\dot{\rho}_{ij}$ correspond to an actual steady cycle that produces no plastic strain increments. In other words, the rates $\dot{\rho}_{ij}$ in the expression (11.57) and the corresponding strain rates $\dot{\epsilon}''_{ij}$ (different from $\dot{\epsilon}''_{ij0}$) should obey (11.11) to (11.15) for

$$\Delta \epsilon''_{ij} = 0 \tag{11.58}$$

at all the points of the body and

$$\int_0^T d\tau \int [\sigma_{ij}^{(e)} - \sigma_{ij}] \dot{\epsilon}''_{ij}\, dv = 0. \tag{11.59}$$

The latter is equivalent to the equalities (11.16).

We should point out here the coincidence of the inequality (11.57) together with the conditions (11.55), (11.56), (11.11)–(11.15) with the requirements of the Koiter theorem, provided the residual stresses vary over the cycle, cf. Chapter 2. It is equally easy to disclose a direct relationship between condition (11.57) and the additional loading method considered in Sec. 5.5, as applied to the determination of incremental collapse conditions at advanced alternating flow, see Secs. 5.6 and 8.7.

From these two statements, we can deduce now the following formulation of the limiting cycle problem:

The minimum value of a parameter p of external actions should be found,

$$\min_{\dot{\epsilon}''_{ij0}} p = ?, \tag{11.60}$$

which satisfies also (11.55) to (11.59), (11.11) to (11.15); the remaining parameters being prescribed.

11.4. Stress and strain rate analysis in steady cycles

The constraints (11.58), (11.59), (11.11)–(11.15) serve here to determine the residual stress rates in the limiting cycle.

The problem of establishing the parameters of the limiting cycle (as regards incremental collapse) with the help of the above relationships can obviously be solved either for a given (deterministic) program of external actions or for actions varying arbitrarily within known limits.

The above two statements remain to be proved. The first one is almost evident. In the actual steady cycle the functional (11.54) becomes zero. In consequence, if at every non-zero plastic strain increments per cycle this functional is negative, it can only vanish for $\Delta \dot{\epsilon}''_{ij0} = 0$, i.e. the alternating flow or shakedown must take place in the actual steady cycle.

The second statement can conveniently be proved by disproving the contrary, as it was done in the first part of the Koiter theorem, see Sec. 2.3. If incremental collapse is impossible in the presence of the conditions (11.55)–(11.59) and (11.11)–(11.15), i.e. if either the alternating flow or elastic shakedown takes place in the actual steady cycle, then there must exist self-equilibrating stresses ρ^0_{ij} at which

$$\sigma^{(e)}_{ij} + \int_0^T \dot{\rho}_{ij} \, d\zeta + \rho^0_{ij} = \sigma^{(a)}_{ij}, \tag{11.61}$$

where $\sigma^{(a)}_{ij}$ denotes an admissible stress state. Now, employing the virtual work principle and the Drucker postulate, we can readily show that the integrand on the left-hand side of the inequality (11.57) is non-positive, i.e. the starting assumption on the impossibility of progressive collapse has resulted in a contradiction.

3) Let us now discuss another particular case of the variational (extremum) formulation. We shall insist on finding the boundaries of purely elastic behaviour (shakedown) of the structure considered. Suitable methods to establish the limiting cycle parameters can be worked out in a manner similar to that used in the case of progressive collapse.

In accordance with the general formulation of the steady cycle problem we aim at finding fields $\dot{\epsilon}^*_{ij0}$ and $\dot{\rho}^0_{ij}$ that obey (11.49)–(11.53) and make the functional (11.54) reach its absolute maximum.

Let us now extract from the general formulation of the conditions for steady cycles, the first statement of the Koiter theorem. Assume that a strain rate field $\dot{\epsilon}''_{ij0}$ has been found, satisfying condition (11.49), together with a corresponding, in the sense of (11.55), stress field σ_{ij0} such that the functional (11.54a) becomes positive,

11 Inelastic shakedown in steady stress cycles

$$\int_0^T d\tau \int (\sigma_{ij}^{(e)} - \sigma_{ij0}) \dot{\epsilon}_{ij0}'' \, dv > 0. \tag{11.62}$$

Since the system (10.50), (11.51) has always a solution, inequality (11.62) can hold only in the absence of any self-equilibrating stresses ρ_{ij}^0, at which the inequality (11.53) holds; otherwise the functional (11.54) would, as shown earlier, be non-positive. Thus, shakedown is impossible when inequality (11.62) is satisfied.

4. From the practical point of view it is of importance to investigate the effect of time-independent external actions on the distributions of residual stress rates and plastic strain rates in the conditions of alternative flow when the strain accumulation is absent.

The time-independent mechanical loads are known not to affect the conditions for the appearance of alternating flow, see Sec. 2.8. We shall show that an analogous statement is true at advanced alternating plasticity: The application of time-independent mechanical loads does not affect the distribution of either residual stress rates or plastic strain rates.

Consider two programs of cyclic external actions generating equal time-dependent elastic stresses and different time-independent ones; we simply assume that the latter are absent in the second process. Let us further assume that different steady cycles correspond to each loading program and that the strain increments per each cycle vanish,

$$\Delta \epsilon_{ij\alpha}'' \equiv 0, \quad \Delta \epsilon_{ij}'' \equiv 0. \tag{11.63}$$

Let us indicate from now on the first process by the subscript α.

In accordance with the Drucker postulate (11.4) and assuming that different flow regimes are generated in the considered steady cycles at the same instant of time, we can write the inequality

$$\left[\left(\sigma_{ij\tau}^{(e)} + \sigma_{ij\alpha}^{(e)} + \rho_{ij\alpha}^0 + \int_0^\tau \dot{\rho}_{ij\alpha} \, d\zeta \right) - \left(\sigma_{ij\tau}^{(e)} + \rho_{ij}^0 + \int_0^\tau \dot{\rho}_{ij} \, d\zeta \right) \right] (\dot{\epsilon}_{ij\alpha}'' - \dot{\epsilon}_{ij}'') \ge 0. \tag{11.64}$$

Integrating over the volume and the period of a cycle, we get

$$\int_0^T d\tau \int \left[\left(\sigma_{ij\tau}^{(e)} + \sigma_{ij\alpha}^{(e)} + \rho_{ij\alpha}^0 + \int_0^\tau \dot{\rho}_{ij\alpha} \, d\zeta \right) - \left(\sigma_{ij\tau}^{(e)} + \rho_{ij}^0 + \int_0^\tau \dot{\rho}_{ij} \, d\zeta \right) \right] (\dot{\epsilon}_{ij\alpha}'' - \dot{\epsilon}_{ij}'') \, dv \ge 0.$$

11.4. Stress and strain rate analysis in steady cycles

Using condition (11.63), we can rearrange the above inequality to obtain

$$\int_0^T d\tau \int \left[\int_0^\tau (\dot{\rho}_{ij\alpha} - \dot{\rho}_{ij}) \, d\zeta \right] (\dot{\epsilon}''_{ij\alpha} - \dot{\epsilon}''_{ij}) \, dv \geq 0. \tag{11.65}$$

An increment of the difference in stresses

$$\int_0^\tau (\dot{\rho}_{ij\alpha} - \dot{\rho}_{ij}) \, d\zeta$$

turns out to be self-equilibrating and the plastic strain rates are associated with appropriate residual stress rates by means of relationships of the type (11.50). Hence we can rearrange the left-hand side of inequality (11.65) with the help of the virtual displacements principle to obtain

$$-\int_0^T d\tau \int A_{ijhk} \left[\int_0^\tau (\dot{\rho}_{ij\alpha} - \dot{\rho}_{ij}) \, d\zeta \right] (\dot{\rho}_{hk\alpha} - \dot{\rho}_{hk}) \, dv =$$

$$= -\int \frac{1}{2} A_{ijhk} [(\rho_{ij\alpha}(T) - \rho^0_{ij\alpha}) - (\rho_{ij}(T) - \rho^0_{ij})] \times$$

$$\times [(\rho_{hk\alpha}(T) - \rho^0_{hk\alpha})(\rho_{hk}(T) - \rho^0_{hk})] \, dv. \tag{11.66}$$

Since we consider actual steady cycles in which $\rho_{ij\alpha}(T) \equiv \rho^0_{ij\alpha}$, $\rho_{ij}(T) \equiv \rho^0_{ij}$, the integral (11.66) vanishes. On account of (11.64), the integrand is non-negative and thus it has to vanish identically. This enables us to draw the conclusion that the flow régimes in the two steady cycles coincide at every instant of time and at all the points in the body where the plastic strain rates have non-zero values (to within an accuracy of stresses associated with yielding). From the above it immediately follows that the absolute values of plastic strains are equal in both steady cycles and so are their amplitudes per semi-cycle.

The insensitivity of alternating flow to the time-independent external actions simplifies not only the calculations of strain rates (amplitudes) necessary for the predictions of service life (when one-sign strains are known beforehand not to accumulate) but also the determination of incremental collapse conditions at advanced alternating flow. The latter problem has been reduced to the minimization of load parameters under the conditions (11.53) to (11.59), (11.11) to (11.15). It can be readily seen that these conditions may be rearranged to take the form in which it is unnecessary to predetermine stresses generated by

11 Inelastic shakedown in steady stress cycles

the constant external actions; an analogous procedure was applied in Chapter 2 to restate the Koiter theorem.

Employing the same notation for time-independent body forces and surface tractions, namely X_i^0, p_i^0, let us rearrange the left-hand side of the inequality (11.57) with the help of the virtual work equation. We obtain

$$\int X_i^0 \Delta u_{i0}\,dv + \int_{S_p} p_i^0 \Delta u_{i0}\,dS + \int_0^T d\tau \int \left(\sigma_{ij\tau}^{(e)} + \int_0^\tau \dot{\rho}_{ij}\,d\zeta - \sigma_{ij0} \right) \dot{\epsilon}_{ij0}^{\prime\prime}\,dv > 0. \tag{11.67}$$

Next, since constant load affects neither the distribution of plastic strain rates, and consequently also not the residual stress rates in the cycle in which $\Delta \epsilon_{ij}^{\prime\prime} = 0$, we can replace in the relationships (11.14) and (11.59) the total elastic stresses $\sigma_{ij}^{(e)}$ by those components $\sigma_{ij\tau}^{(e)}$ that are generated by variable external actions only. Thus the determination of the elastic stresses generated by constant external actions becomes unnecessary. This remark refers to the situation in which we seek the plastic strain amplitudes associated with the interval of load parameters in which the strain accumulation is absent.

11.5. Evaluation of steady cycle characteristics under quasi-stationary external actions

The analysis of structures under quasi-steady, with respect to a moving frame of reference, external actions considered in a moving coordinate system appears to be one of the most interesting problems in the shakedown theory because, as was already pointed out in Chapters 9 and 10 (contact problems), it is those actions that are critical for progressive collapse. In particular, situations can be indicated in which incremental collapse takes place as soon as the applied stresses exceed the single value of the yield point. Quasi-steady actions, both of thermal and mechanical nature, are often encountered in many applications. Endeavours to enhance the service parameters of machines and processes are sometimes leading to the acceptance of their performance well beyond the limits of elastic shakedown. In such cases to estimate the durability of a structure we have to determine the strain increments per cycle and, when alternating flow is present, the strain amplitudes as well.

As an example, let us consider a segment of a thin-walled tube far

11.5. Evaluation of steady cycle characteristics

away from each end of the whole tubular structure. The segment is subject to repeated passages of an axially symmetric temperature field whose profile resembles a thermal wave, see Fig. 11.4. Let us investigate the steady cycle in the light of the formulation given in the previous paragraph.

To establish the steady stress cycle, we have to start with finding distributions of both longitudinal and hoop plastic strain rates $\dot{\epsilon}''_x, \dot{\epsilon}''_\varphi$ as well as the initial stress fields ρ^0_x, ρ^0_φ which make the functional (11.54a) attain its maximum. It is a peculiarity of this problem that the instantaneous stress and strain rate fields are sufficient to completely describe the whole behaviour of the tube over the steady cycle. This is due to the fact that, as time elapses, both the stress and strain rate fields merely translate along the axis of the tube while remaining unchanged relative to the moving frame of reference. Thus the time integration of the functional (11.54a) is redundant; the integral assumes the form

$$I_2 = \int_{-\infty}^{\infty} d(\beta x) \int_{-h}^{h} [(\sigma^{(e)}_x - \sigma_x)\dot{\epsilon}''_x + (\sigma^{(e)}_\varphi - \sigma_\varphi)\dot{\epsilon}''_\varphi] \, dz. \tag{11.68}$$

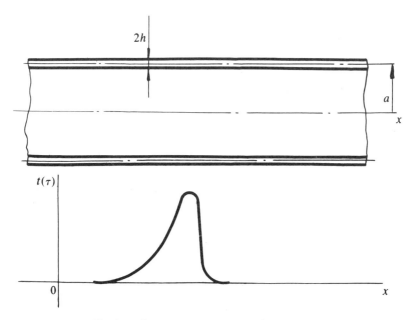

Fig. 11.4. Quasi-steady temperature field in a tube.

11 Inelastic shakedown in steady stress cycles

A dot signifies differentiation with respect to the coordinate βx which plays here the part of current time; recall that $\beta^4 = 3(1-\mu^2)/(4a^2h^2)$.

On account of the Kirchhoff–Love hypothesis, the constraints (11.49) take the form [11.17]

$$\Delta\epsilon''_\varphi = -\frac{\Delta w}{a}, \quad \Delta\epsilon''_x = \frac{d}{dx}(\Delta u) - z\frac{d^2}{dx^2}(\Delta w), \tag{11.69}$$

where $\Delta u, \Delta w$ are the increments of axial and hoop displacements per cycle.

Specifying the conditions for self-equilibrium of initial stresses (11.52) to the case of a hollow cylinder and expressing them in terms of internal forces, one gets [11.17]:

$$\frac{d^2 M^0_{x\rho}}{dx^2} + \frac{1}{a} N^0_{\varphi\rho} = 0, \quad N^0_{x\rho} = 0,$$

$$M^0_{x\rho} = \int_{-h}^{h} \rho^0_x z \, dz, \quad N^0_{\varphi\rho} = \int_{-h}^{h} \rho^0_\varphi \, dz, \quad N^0_{x\rho} = \int_{-h}^{h} \rho^0_x \, dz. \tag{11.70}$$

Similarly the constraints (11.50), (11.51) on the residual stress rates and plastic strain rates take the form

$$\dot\epsilon_{\varphi\rho} = -\frac{\dot w}{a}, \quad \dot\epsilon_{x\rho} = \frac{d\dot u}{dx} - z\frac{d^2\dot w}{dx^2},$$

$$\dot\epsilon_{x\rho} = \dot\epsilon''_x + \frac{1}{E}(\dot\rho_x - \mu\dot\rho_\varphi), \quad \dot\epsilon_{\varphi\rho} = \dot\epsilon''_\varphi + \frac{1}{E}(\dot\rho_\varphi - \mu\dot\rho_x),$$

$$\frac{d^2 \dot M_{x\rho}}{dx^2} + \frac{1}{a}\dot N_{\varphi\rho} = 0, \quad \dot N_{x\rho} = 0, \tag{11.71}$$

$$\dot M_{x\rho} = \int_{-h}^{h}\dot\rho_x z \, dz, \quad \dot N_{x\rho} = \int_{-h}^{h}\dot\rho_x \, dz, \quad \dot N_{\varphi\rho} = \int_{-h}^{h}\dot\rho_\varphi \, dz.$$

Again, current time is here represented by βx.

On adopting the Tresca yield condition (2.10) the constraints (11.53) for the tube take the form

$$\left| \sigma_x^{(e)} + \rho_x^0 + \int_{-\infty}^{\beta x} \dot\rho_x \, d\zeta \right| \leq \sigma_s,$$

$$\left| \sigma_\varphi^{(e)} + \rho_\varphi^0 + \int_{-\infty}^{\beta x} \dot\rho_\varphi \, d\zeta \right| \leq \sigma_s, \qquad (11.72)$$

$$\left| \sigma_x^{(e)} - \sigma_\varphi^{(e)} + \rho_x^0 - \rho_\varphi^0 + \int_{-\infty}^{\beta x} (\dot\rho_x - \dot\rho_\varphi) \, d\zeta \right| \leq \sigma_s.$$

Thus, the problem of describing the steady cycle consists in finding functions $\dot\epsilon_x''$, $\dot\epsilon_\varphi''$, ρ_x^0, ρ_φ^0 that satisfy the constraints (11.69) to (11.72), (11.55) and for which the non-positive functional (11.68) attains its maximum (zero) value.

Let us now introduce certain simplifications to the considered example. Since at quasi-steady, with respect to a moving frame of reference, external actions the history of stress and strain rate variations per cycle is the same for all the cross-sections, we can expect that the initial stresses ρ_x^0, ρ_φ^0 do not depend on the coordinate βx. Hence

$$M_{x\rho}^0 = \text{const.}, \quad N_{\varphi\rho}^0 = 0, \quad N_{x\rho}^0 = 0. \qquad (11.73)$$

Under the conditions (11.73) the differential equation (11.70) is satisfied identically and therefore it disappears from the system of constraints.

The obtained description of the steady state problem can be rendered in the language of the mathematical theory of optimum control. The functions $\dot\epsilon_x''$, $\dot\epsilon_\varphi''$, ρ_x^0, ρ_φ^0 play the part of controls whereas the functions $\dot\rho_x, \dot\rho_\varphi$ are the phase coordinates. When discretized, the problem becomes one in linear programming. As in the previous chapters on limit equilibrium and shakedown, discretization consists in replacing the differential equations of compatibility and equilibrium, appearing in the constraints (11.69) and (11.71), by linear algebraic equations, and the integrals by finite summations based on piece-wise linear approximation of the integrands. The constraints (11.72) are imposed only at a finite number of isolated points in the tube.

Considering the size of the set of constraints, the formulated problem does not differ from limit equilibrium problems for axi-symmetric shells. It can be easily solved using standard simplex method computer programs.

References

[11.1] Gokhfeld, D.A., Sadakov, O.S., A mathematical model of medium for analysing the inelastic deforming of structures subjected to repeated actions of load and tem-

11 Inelastic shakedown in steady stress cycles

perature. Third Intern. Conf. Struct. Mech. React. Technol., London, L 5/7, 1975.

[11.2] Frederick, C.O., Armstrong, P.J., Convergent internal stresses and steady cyclic states of stress. 'J. Strain Anal.', Nr 2, 1966.

[11.3] Mróz, Z., On the theory of steady plastic cycles in structures. 'Prepr. 1st Int. Conf. Struct. Mech. React. Technol.', Vol. 5, Part I, Berlin, 1971.

[11.4] Halphen, Bernard, Stress accommodation in elastic-perfectly plastic and viscoplastic structures. 'Mech. Res. Communs.', 2, Nr 5–6, 1975.

[11.5] Shorr, B.F., Cyclic creep of non-uniformly heated cylinders, in Russian, in 'Thermal stresses in structural members', 4, Naukova Dumka, Kiev, 1964.

[11.6] Shorr, B.F., Periodic processes and shakedown at creep, in Russian, in 'Strength and dynamics of aircraft engines', 4, Mashinostrojenie, Moscow, 1966.

[11.7] Ponter, A.R.S., On the relationship between plastic shakedown and the repeated loading of creeping structures. 'Trans. ASME, Mech.', Nr 2, 1971.

[11.8] Leckie, F.A., Limit and shakedown loads in the creep range. 'Therm. stress and thermal fatigue', London, 1971.

[11.9] Ponter, A.R.S., Deformation, displacement, and work bounds for structures in a state of creep and subject to variable loading. 'Trans. ASME, Journal of Applied Mech.', Nr 4, 1972.

[11.10] Martin, J.B., Williams, J.J., On the existence of an extremum principle for creeping structures subjected to cyclic loading. 'Int. J. Mech. Sci.', 13, Nr 3, 1971.

[11.11] Ponter, A.R.S., On the stress analysis of creeping structures subject to variable loading. 'Trans. ASME, Appl. Mech.', Nr 2, 1973.

[11.12] Marriot, D.L., Penny, R.K., Strain accumulation and rupture during creep under variable uniaxial tensile loading. 'J. Strain Anal.', 8, Nr 3, 1973.

[11.13] Kyaras, W., On cybernetic aspects of fracture mechanics, in Russian, in 'Strength of materials, Proc. of XXII Conf.' (Kaunas, 1972).

[11.14] Drucker, D.C., A definition of stable inelastic material. 'Trans. ASME, J. Appl. Mech'., Nr 81, 1959, 101.

[11.15] Ikrin, W.A., On plastic strain accumulation at cyclic actions, in Russian, Proc. of All-Union Symposium on low-cycle fatigue at elevated temperatures, 4, Cheliabinsk, 1974.

[11.16] Bellman, R., Dynamic Programming. Princeton Univ. Press, 1957.

[11.17] Timoshenko, S., Woinowsky-Kriger, S., Theory of Plates and Shells. McGraw-Hill, New York, 1959.

[11.18] Zukhovitsky, S.I., Avdeeva, L.I., Linear and Convex Programming, in Russian, Nauka, Moscow, 1967.

Index

Accumulated displacement 33, 35, 81
Accumulation of deformation (strain accumulation) 9, 16, 26, 36, 174
 in the absence of external loads 28, 407
Additional loading (overload) 288, 335, 342
 method 213, 214, 220 225, 240, 298, 339, 396
 volume (zone, region, domain) 203, 212, 256, 270, 285, 289, 293
Admissible control 134
Admissible cycle of plastic strain rates 56
Admissible stress state 46, 82
Alternating plasticity (plastic flow) 8, 10, 77, 155, 526
 ~ condition 77
 ~ domain 15
Amplitude of deformation 11
Arbitrary loading program 13, 215, 296
Associated flow rule 47, 67, 71, 381, 386
Asymptotic stabilization 26

Bar system 186, 204, 412, 466, 515
Bauschinger effect 31

Carring capacity 227, 264, 266
Coffin–Manson law 32, 401
Collapse
 Global ~ 229, 243, 250
 Incremental ~ (progressive deformation) 10, 17, 43, 78, 162, 190, 204, 216, 286, 520, 523
 ~ mechanism 199, 252, 304, 312, 324, 341, 382
 Partial ~ 231, 243, 251, 332
Combustion chamber 391
Compatibility equations (conditions) 4, 9, 18, 21, 45
Contact problems in shakedown theory 487
 for a half-plane 501
 for a semi-space 488
Continuity equations 1
Control
 ~ parameter 134, 137, 140
 ~ process 137, 141, 144
 ~ region 134
Cost function 136
Creep effect 395, 404
Creep limit 228
Crystallizer of a casting machine 425
Cutout coefficient 325
Cyclic loading 3
Cyclic stable materials 31
Cyclic (isotropic) strain-hardening (softening) 31, 32
Cylindrical shell 144, 417

Deformation preceding (prior) to the shakedown 12, 81
Degree of redundancy 2
Discretization of the continuum 164
Displacement discontinuities (jump) 75, 183, 197, 200
Displacement increment at shakedown process 84
Disk 227
 Plane 156, 175, 192, 237
 Turbine 227, 229, 235, 268
 of equal strength 236
Domain of admissible values of the constant generalized stress 96, 99
Dynamical actions 55, 61

Effect (influence) of creep on shakedown 36, 39, 245
Elastic-perfectly plastic material 1, 46, 48, 53

Index

Elastic strain energy 51
Elastic stresses 50, 57, 60, 61, 64, 108, 150
Equilibrium equations 3, 21, 44, 137, 158
Equilibrium line 6
Evolution of the stress-strain diagram 31

Fictitious interaction surface (curve) 96
   approximate method of ~ deformation 107, 109
   ~ for a bar 96, 99, 105, 110
   ~ for a conical shell 374, 376
   ~ for a cylindrical shell 347, 352
   ~ for a spherical shell 387
   ~ for a stretched rod 100
   ~ for shells 108
   ~ for plates 108, 114, 118, 120, 125
Fictitious yield surface 64, 66, 68, 69
   degeneration 77
Finite element method 164, 165
Free variable 146, 306
Frictional rolling of a cylinder 497
Fuel element 395

Generalized stresses in the shakedown problems 90, 96
Generalized variables 90, 374, 377, 381
   ~ in the incremental collapse problems 95
   ~ in the limit analysis 93

Hold-time 37

Incremental (progressive) buckling at thermal cycling 456
Inelastic deformation 1
Initial half-cycle 8
Initial, stress-free state 7
Instantaneous collapse 10, 78, 105, 213, 267, 322, 371, 396
Interaction surface 352, 374, 389
Internal forces diagram 7

Jordan elimination 147

Kinematically admissible
   ~ collapse mechanism 49, 72, 78
   ~ plastic strain increments 75, 383, 384
   ~ strain distribution 56, 60
   ~ strain rate 49

Kinematical methods 131, 183, 202, 238, 339, 366, 384, 495
Limit analysis 33, 49, 56, 138, 140, 229, 355, 357
Limit equilibrium state (load) 8, 19, 27, 76
Limit load 11, 14, 15, 49
Limit state 9, 10, 76
Limiting cycle 10, 14, 27, 28, 52, 63, 72, 83, 386
Load – carrying capacity 79, 372
Loading parameters 2, 149
Longitudinal stiffness 4
Long-term strength 228, 234, 245, 266
Low-cycle (plastic) fatigue (failure) 8

Masing principle 32
Mathematical programming 145, 185
   convex 145
   linear 145, 147, 185, 192, 193, 307, 345, 378
Moving
   ~ temperature fields 30
   ~ heat source 412
   mechanical load 366
Multiple stress concentrations 323

Nonisochronism (nonsimultaneity) 9, 13, 18, 27, 78
Nonisothermal shakedown problems 16, 53, 54, 60, 74
Non-shakedown (inadaptation) conditions 51, 58, 61, 71, 75

Objective function 145
One-parametric system (structure) 3, 18, 33
Optimal control 134, 137, 140, 363
   theory 134, 137
Optimal solution 147
Optimum trajectory 134, 140, 142

'Physical' shakedown 32, 33
Phase coordinates 134, 137, 142, 310
Plate
Pivotal element 147, 383
   ~ annular 137, 140, 142
   ~ circular 144, 283, 284, 292, 307
   ~ of arbitrary shape 313, 322
   ~ perforated 323
   ~ rectangular 313, 323

Index

~ sandwich 298
~ square 318
Plastic anisotropy 16
Plastic collapse 49, 371
Plastic deformation 5, 45
Plastic energy dissipation 49, 59
Plastic hinge 359
Plastic yielding 9
Pontragin maximum principle 134, 135, 309, 355
Principle of virtual work 49, 59, 71
Probability of incremental collapse 56
Process of stabilization 12, 26, 29, 505
Progressive warping 457, 472, 475
Properties of the fictitious yield surface 70

Quasi-static failure 14
Quasi-stationary (steady) temperature field 409, 417, 528

Radial-flow turbines 235, 261, 274
Ratcheting 408
Residual displacement (strain) 75, 398
Residual stress 35, 50, 84

Safe stress state 46
Safety factor 35, 82, 83, 229, 234, 266
Safety margin 35, 39, 236, 264, 268, 275
Self-equilibrium stresses 1
Shakedown 8, 396
  ~ conditions 50, 54, 58, 63, 65, 71, 185, 241
  ~ diagram 11, 100, 175, 242, 264, 275, 290, 334, 365, 398 (full diagram 396, 506)
  ~ domain 15, 39, 190, 222, 248, 260, 272, 296, 396
  ~ process 82, 83
  ~ theorems: statical 50, 54, 62, 127, 128
    kinematical 56, 58, 61, 74, 75, 183, 202
Shells 339
  Conical ~ 372, 434
  Cylindrical ~ 340, 355, 366, 396
  Sandwich ~ 391
  Spherical ~ 383

Simplex method 147, 152, 192, 345, 351, 378, 382
Slag container 202
Spherical vessel 213
Stabilized stress-strain diagram 32
Stabilized system 8, 30, 59
Starting solution 147, 383
Statical method 131, 133, 137, 339, 487
  Approximate ~ 166, 168, 386
Steady cycle 10, 17, 27, 57, 505
  Existence and uniqueness of a ~ 507
  Limiting ~ 506
  Variational formulation of the problems 513, 521
Strain hardening 31, 33, 35, 402, 404
Strain increments 224
Strain tensor 45
Stress-strain diagram (curve) 229, 246, 249
Stress tensor 43

Temperature-dependence of elastic moduli 36, 54, 61
Temperature-dependence of stress-strain diagram 36
Thermal actions (forces) 1, 36
Thermal deformation 12
Thermal ratcheting 412, 425
Thermal stresses 36, 211
Thermal wave 30
Thick-walled tube 169, 210
Thin-hollow sphere 219
Two-parameter-structure 20
  ~ loading 12

Ultimate load 138, 139
Uniqueness of stress state 79, 80

Variable action of torsional moment 100
Variable repeated loading 2

Working cycle 264
Working point 265

Yield condition 45, 48, 138, 144, 359
  Huber–Mises ~ 48, 69, 340
  Tresca–Saint Venant ~ 48, 66, 140, 345, 374
Yield line 313, 314, 322

535

Index

Yield load (force) 4, 25
Yield point 4
Yield stress 16

Yield surface 46, 47, 108
  Fictitious ~ 64, 193, 367, 385, 389
  ~ convexity 46

Colophon

typeface: times 10/12, 8/10
typesetter: European Printing Corporation
printer: Samsom–Sijthoff Grafische Bedrijven
binder: Abbringh